Return to S. MIN

W9-BMZ-327

LOW-POWER DIGITAL VLSI DESIGN

CIRCUITS AND SYSTEMS

ROCKWELL INTERNATIONAL, INC.
INFORMATION CENTER (NB)
4311 JAMBOREE ROAD (501-345)
NEWPORT BEACH, CA. 92660

LOW-POWER DIGITAL VLSI DESIGN

CIRCUITS AND SYSTEMS

by

Abdellatif Bellaouar
University of Waterloo

and

Mohamed I. Elmasry
University of Waterloo

KLUWER ACADEMIC PUBLISHERS
Boston / Dordrecht / London

Distributors for North America:
Kluwer Academic Publishers
101 Philip Drive
Assinippi Park
Norwell, Massachusetts 02061 USA

Distributors for all other countries:
Kluwer Academic Publishers Group
Distribution Centre
Post Office Box 322
3300 AH Dordrecht, THE NETHERLANDS

Consulting Editor: Jonathan Allen, Massachusetts Institute of Technology

Library of Congress Cataloging-in-Publication Data

A C.I.P. Catalogue record for this book is available
from the Library of Congress.

Copyright © 1995 by Kluwer Academic Publishers, Second Printing 1996

All rights reserved. No part of this publication may be reproduced, stored in a retrieval system or transmitted in any form or by any means, mechanical, photo-copying, recording, or otherwise, without the prior written permission of the publisher, Kluwer Academic Publishers, 101 Philip Drive, Assinippi Park, Norwell, Massachusetts 02061

Printed on acid-free paper.

Printed in the United States of America

CONTENTS

PREFACE

A major creative challenge facing today circuit and system VLSI designers is to design new generation products which consume minimum power. Power saving must be achieved without compromising high performance or minimum area. This has created a new design culture within the design community which we have just seen its preliminary results. The essence of this culture must be accessible to the new generation of designers.

The concern of power dissipation has been part of the design process since the early 1970s, but was less visible. High speed operation, and designing with minimum area, specially in memories, were the main design constraints. The state-of-the-art was driven towards lower delays and smaller chip area. Design tools were all geared towards achieving these two goals. Major milestones on chip integration and clock rates have been reported in technical conferences (e.g., IEEE International Solid-State Circuits Conference) and journals (e.g., IEEE Journal of Solid-State Circuits) from the late fifties till the early nineties. Power dissipation has taken a back seat as a figure of merit. However, as we approach the end of this century, power dissipation has become the main design concern in many applications. Two contributing factors were the area of portable electronics and the area of high-performance chips exceeding power dissipation limits.

This book addresses the design of low-power VLSI digital circuit and system design. The book starts with an introduction to the topic of low-power design. Followed with two supporting chapters on low-power process technology and device modeling. Circuit design for low-power is addressed in two chapters; one on CMOS and the other on BiCMOS. Low-power design applications are covered in subsequent chapters; one on low-power RAMs and the other on low-power subsystem designs. The subsystems include adders, multipliers, data path, regular structures and phase locked loops. The last chapter deals with overall low-power VLSI design methodology. The book addresses many design issues related to low-power; the concept of switching activity, the use of pass-transistor logic, designing using multi-and-low threshold voltage CMOS logic, the integration of on-chip voltage down converters, etc.

We hope that students and instructors find this book useful in their class-room instruction and also hope that it will be valuable to researchers working in this area.

Abdellatif Bellaouar
Mohamed I. Elmasry
Waterloo, Ontario
Canada

Acknowledgements

Firstly we would like to acknowledge the countless blessings of God Almighty throughout our lives. During the course of writing this book we have developed a greater appreciation for God's created biological processing circuits and systems in terms of low-power and low-energy design. Such systems provides a great aspiration to VLSI designers. The brain, with 30 Watts of active power and processing information at less than 0.01 pJ, is an excellent example of low-power processing/memory design. More research is needed to abstract low-power concepts from the brain and apply them to VLSI circuits and systems.

We would also like to thank our families whose support and endurance helped us to complete writing this book. A. Bellaouar, would like to acknowledge his wife. She was very patient and helpful when he spent over 16 hours/day to complete this manuscript.

We also extend our thanks to Mr. Carl Harris from Kluwer Academic Publishers for encouraging us to work on this new era of VLSI design.

We would like to thank our colleagues at the VLSI Research Group of the Department of Electrical and Computer Engineering at the University of Waterloo for their encouragement and support, in particular, Issam S. Abu-Khater. We are grateful to Joan Pache for carefully proof reading the book.

We appreciate the financial support to our research provided in part by NSERC, MICRONET, ITRC, CMC, BNR and NTE.

Finally, we appreciate the effort of those who assisted us in preparing the manuscript and the figures, in particular, Kamel Benaissa, Muhammed Elrabaa, Ahmed R. Fridi and Phil Regier. Also, we thank Dave Bartholomew from Graphic Services at the University of Waterloo for helping in the design of the book front cover.

To

My parents, my wife Ghania and my son Mouaadh Bellaouar

Elizabeth, Carmen, Samir, Nadia and Hassan Elmasry

1

LOW-POWER VLSI DESIGN: AN OVERVIEW

1.1 WHY LOW-POWER?

Historically, VLSI designers have used circuit speed as the "performance" metric. Large gains, in terms of performance and silicon area, have been made for digital processors, microprocessors, DSPs (Digital Signal Processors), ASICs (Application Specific ICs), etc. In general, "small area" and "high performance" are two conflicting constraints. The IC designers' activities have been involved in trading off these constraints. Power dissipation issue was not a design criterion but an afterthought. In fact, power considerations have been the ultimate design criteria in special portable applications such as wristwatches and pacemakers for a long time. The objective in these applications was minimum power for maximum battery life time.

Recently, power dissipation is becoming an important constraint in a design. Several reasons underlie the emerging of this issue. Among them we cite:

- Battery-powered systems such as laptop/notebook computers, electronic organizers, etc. The need for these systems arises from the need to extend battery life. Many portable electronics use the rechargeable Nickel Cadmium (NiCd) batteries. Although the battery industry has been making efforts to develop batteries with higher energy capacity than that of NiCd, a strident increase does not seem imminent. The expected improvement of the energy density is 40% by the turn of the century. With recent NiCd batteries, the energy density is around 20 Watt-hour/pound and the voltage is around 1.2 V. So, for example, for a notebook consuming a typical power of 10 Watts and using 1.5 pound of batteries, the time of operation between recharges is 3 hours. Even with the advanced battery technologies, such as Nickel-Metal Hydride (Ni-MH) which provide large energy density characteristics ($\sim$ 30 Watt-hour/pound), the life time of the battery is still low. Since battery technology has offered a limited improvement, low-power design techniques are essential for portable devices.

- Low-power design is not only needed for portable applications but also to reduce the power of high-performance systems. With large integration density and improved speed of operation, systems with high clock frequencies are emerging. These systems are using high-speed products such as microprocessors. The cost associated with packaging, cooling and fans required by these systems to remove the heat is increasing significantly. Table 1.1 shows the power consumption of various microprocessors that operate in the frequency range of 66-to-300 MHz. This table demonstrates that, at higher frequencies, the power dissipation is too excessive.

Table 1.1 Power dissipation of microprocessors.

Processor	Clock (MHz)	Technology (μm)	V_{DD} (V)	Power Peak (W)	Ref.
Intel Pentium	66	0.80	5.0	16	[1]
DEC Alpha 21064	200	0.75	3.3	30	[2]
DEC Alpha 21164	300	0.50	3.3	50	[3]
PowerPC 620	133	0.50	3.3	30	[4, 5]
MIPS R10000	200	0.50	3.3	30	[6]
UltraSparc	167	0.45	3.3	30	[4, 7]

- Another issue related to high power dissipation is reliability. With the generation of on-chip high temperature, failure mechanisms are provoked [8]. Among them, we cite silicon interconnect fatigue, package related failure, electrical parameter shift, electromigration, junction fatigue, etc..

- In addition, there is a trend to keep the computers from using more than 5% share of the total US power budget [9]. Note that 50% of office power is used by PCs. Since the processors' frequency is increasing, which results in increased power, then low-power design techniques are prerequisites.

The power dissipation issues and the devices' reliability problems, when they are scaled down to 0.5 μm and below, have driven the electronics industry to adopt a supply voltage lower than the old standard, 5 V. The new industry standard for IC operating voltage is 3.3 V (± 10%). The effect of lowering the voltage to much lower values can be impressive in terms of power saving. The power is not only reduced but also the weight and volume associated with batteries in battery-operated systems.

1.2 LOW-POWER APPLICATIONS

Low-power design is becoming a new era in VLSI technology, as it impacts many applications; such as:

- Battery-powered portable systems; for example notebooks, palmtops, CDs, language translators, etc. These systems represent an important growing market in the computer industry. High-performance capabilities, comparable to those of desktops, are demanded. Several low-power microprocessors have been designed for these computers. Table 1.2 shows some examples of these low-power processors. However, these circuits still consume significant power on the order of 1-to-3 Watts. These systems have their power

Table 1.2 Power dissipation of LP microprocessors.

Processor	Clock (MHz)	Technology (μm)	V_{DD} (V)	Power (W)	Ref.
PowerPC 603	80	0.5	3.3	2.2	[10]
IBM 486SLC2	66	0.8	3.3	1.8	[11]
MIPS R4200	80	0.64	3.3	1.8	[12]

  dissipation dominated by I/O devices such as hard disk drives and LCD displays. The total expected power dissipation of notebooks is 2 Watts with 4 pounds weight and daily recharge.

- Electronic pocket communication products such as; cordless and cellular telephones, PDAs (Personal Digital Assistants), pagers, etc. Table 1.3 shows a battery analysis for a handheld cellular system. Low-power is crucial for extending the battery life of these systems. Also, battery improvement is needed. The PDAs require a large amount of data processing with multimedia capabilities. The expected power of PDAs is around 0.5 Watt with 0.5 pound weight. Also the expected power for pagers is 10 mW with 0.125 pound weight.

Table 1.3 Battery analysis of a cellular system [13].

Item	Handheld Cellular
Example	Motorola Microtac
RF Power	600 mW
Battery cells	5-AA 750 mAH secondary NiCd
Battery life	75 minutes talk time 20 hours standby
Total power load	650 mA $\times$ 6 V = 3900 mW

- Sub-GHz processors for high-performance workstations and computers. 100 MHz systems and over are emerging, and 500 MHz and higher will be common by the end of the decade. Since the power consumed is increasing with the trend of frequency increase then processors with new architectures and circuits optimized for low-power are crucial.
- Other applications such as WLANs (Wireless Local Area Networks) and electronic goods (calculators, hearing aids, watches, etc.).

1.3 LOW-POWER DESIGN METHODOLOGY

In order to optimize the power dissipation of digital systems low-power methodology should be applied throughout the design process from system-level to process-level, while realizing that performance is still essential. During optimization, it is very important to know the power distribution within a processor. Thus, the parts or blocks consuming an important fraction of the power are properly optimized for power saving. Fig. 1.1 shows the different design levels of an integrated system. The process technology is under the control of the device/process designer. However, the other levels are controlled by the circuit designer.

1.3.1 Power Reduction Through Process Technology

One way to reduce the power dissipation is to reduce the power supply voltage. However the delay increases significantly, particularly when V_{DD} approaches

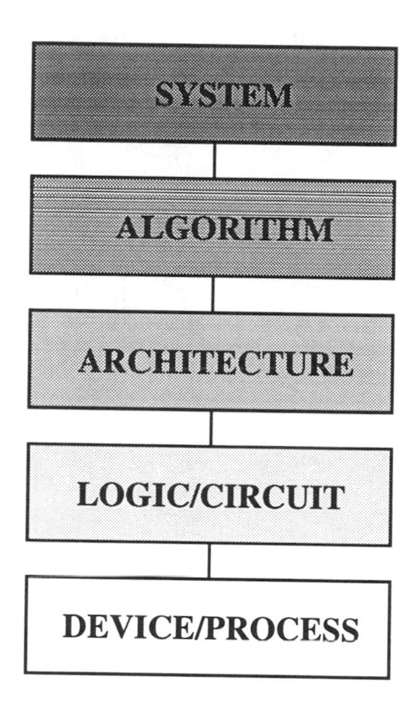

Figure 1.1 Power reduction design space.

the threshold voltage. To overcome this problem, the devices should be scaled properly. The advantages of scaling for low-power operation are the following:

- Improved devices' characteristics for low-voltage operation. This is due to the improvement of the current drive capabilities;
- Reduced capacitances through small geometries and junction capacitances;
- Improved interconnect technology;
- Availability of multiple and variable threshold devices. This results in good management of active and standby power trade-off; and
- Higher density of integration. It was shown that the integration of a whole system, into a single chip, provides orders of magnitude in power savings.

Table 1.4 shows the effect of scaling on microprocessor performance [14]. The power dissipation can be reduced by one order of magnitude at fixed frequency of operation.

Table 1.4 Scaling of the PowerPC 604 microprocessor.

L (μm)	0.50	0.35	0.25	0.15
L_{eff} (μm)	0.35	0.25	0.15	0.10
V_{DD} (V)	3.3	2.5	1.8	1.5
Area (mm^2)	8 × 10	5.6 × 7	4 × 5	2.5 × 3
Clock (MHz)	100	150	225	330
Power (W)	5.0	3.3	2.35	1.5
@ 100 MHz				
Area (mm^2)	6.4 × 8.4	4.5 × 6	3.2 × 4.2	2 × 2.5
Power (W)	5.0	2.2	1	0.45

1.3.2 Power Reduction Through Circuit/Logic design

To minimize the power at circuit/logic level, many techniques can be used such as:

- Use of more static style over dynamic style;
- Reduce the switching activity by logic optimization;
- Optimize clock and bus loading;
- Clever circuit techniques that minimize device count and internal swing;
- Custom design may improve the power, however, the design cost increases;
- Reduce V_{DD} in non-critical paths and proper transistor sizing;
- Use of multi-V_T logic circuits; and
- Re-encoding of sequential circuits.

1.3.3 Power Reduction Through Architectural Design

At the architecture level, several approaches can be applied to the design:

- Power management techniques where unused blocks are shutdown;
- Low-power architectures based on parallelism, pipelining, etc.;
- Memory partition with selectively enabled blocks;
- Reduction of the number of global busses; and
- Minimization of instruction set for simple decoding and execution.

1.3.4 Power Reduction Through Algorithm Selection

Among the techniques to minimize the power at the algorithmic level, we cite:

- Minimizing the number of operations and hence the number of hardware resources; and
- Data coding for minimum switching activity.

1.3.5 Power Reduction in System Integration

The system level is also important to the whole process of power optimization. Some techniques are:

- Utilize low system clocks. Higher frequencies are generated with on-chip phase locked loop; and
- High-level of integration. Integrate off-chip memories (ROM, RAM, etc.) and other ICs such as digital and analog peripherals.

1.4 THIS BOOK

This book is an early contribution to the field of low-power digital VLSI circuit and system design. It targets two types of audiences; the senior undergraduate and postgraduate university students and the VLSI circuit and system designer working in industry. In this book we have tried to cover the basics, from the process technologies and device modeling to the architecture level, of VLSI systems. The fundamentals of power dissipation in CMOS circuits are presented to provide the readers with sufficient background to be familiar with the low-power design world. Several practical circuit examples and low-power techniques, mainly in CMOS technology, are discussed. Also low-voltage issues for digital CMOS and BiCMOS circuits are emphasized. This book also provides an extensive study of advanced CMOS subsystem design. Various power minimization techniques, at the circuit, logic, architecture and algorithm levels, are presented. Finally, the book includes a rich list of references, treating advanced topics, at the end of each chapter. This allows the readers to study, in depth, any topics they find interesting.

This book is organized into eigth chapters. The first chapter is an introduction to low-power design. The other chapters are presented in the following sections.

1.4.1 Low-Voltage Process Technology

Chapter 2 deals with CMOS bulk, bipolar, BiCMOS and CMOS Silicon On Insulator (SOI) process technologies. Several CMOS technologies (N-well and twin-tub) and low-voltage CMOS enhancement are reviewed. Bipolar technology with emphasis on advanced structures is considered. The topic of the isolation techniques used for both bipolar and CMOS is addressed. Three BiCMOS technologies, with different performance/cost, are presented. Complementary BiCMOS structure, where a vertical isolated PNP transistor is merged with an NPN transistor in a CMOS process. The design rules of a 0.8 μm BiCMOS process is supplied. Finally, SOI technology is reviewed for low-voltage and low-power applications.

1.4.2 Low-Voltage Device Modeling

Chapter 3 addresses the topic of device modeling. This topic is of interest to those readers who need to analyze, design and/or simulate circuits. It introduces commonly used models of both MOS and bipolar devices. In this chapter we consider simple analytical models which can be used for circuit analysis and design of deep-submicrometer MOSFETs at low-voltage. Also, a simple model to compute the leakage current, hence the static power dissipation, of MOS-

FETs is discussed. The SPICE[1] device models of an 0.8 μm CMOS/BiCMOS process are also presented. This should help the reader to appreciate the meaning of the model parameters as well as to analyze the power and delay of the low-voltage circuits presented throughout the book. Supply voltage scaling, due to reliability and power dissipation issues, is presented.

1.4.3 Low-Voltage Low-Power VLSI CMOS Circuit Design

Chapter 4 focuses on CMOS logic circuit design. The sources of power dissipation in these circuits are reviewed. Simple models for delay and power dissipation estimation are presented. The concept of switching activity is introduced and examples are given. The power dissipation due to spurious transitions is described. Several CMOS design styles, such as pseudo-NMOS, dynamic and NO RAce (NORA) logics, are studied. Guidelines for low-power physical design are presented. Other circuit variations of the static complementary CMOS, which are suitable for low-power applications, are discussed. This includes the pass-transistor logic family such as Complementary Pass-transistor Logic (CPL), Dual Pass-transistor Logic (DPL), and Swing Restored Pass-transistor Logic (SRPL). Also an overview of clocking strategy in VLSI systems is covered. Included in this chapter is one important area which is the I/O circuits. The power dissipation of the I/O circuits is also analyzed. Finally, techniques to reduce static and dynamic power components for CMOS design are also reviewed. This chapter is intended to provide the readers sufficient background in low-power circuit design.

1.4.4 Low-Voltage VLSI BiCMOS Circuit Design

A variety of BiCMOS logic circuits suitable for 3.3 and sub-3.3 V are presented in Chapter 5. The chapter starts with the introduction of the conventional BiCMOS (totem-pole) gate which was used in 5 V applications. The degradation of this gate, with supply voltage scaling, is demonstrated. The BiNMOS family suitable for low-voltage applications (3.3 - 2 V range) is introduced. It is shown that it provides better performance and delay-power product than CMOS, at these voltages, even at low fan-out. Other logic families, for low power supply voltage operation, are also discussed. Finally, this chapter presents several low-voltage applications of BiCMOS.

[1] SPICE is the most commonly used circuit simulator.

1.4.5 Low-Power CMOS Random Access Memory Circuits

The objective of Chapter 6 is two-fold. It is intended to present circuit techniques for active and standby power reduction in static and dynamic RAMs, and to apply the concepts behind these techniques for other applications because RAMs have seen a remarkable and rapid progress in power reduction. These techniques are applied to the architectural and circuit levels. Several advanced circuit structures and memory organizations are described. Circuits, operating at a power supply as low as 1 V, are also discussed. The Voltage Down Converters (VDCs) used as DC-DC converters are also treated. Their low-power aspects are investigated.

1.4.6 VLSI CMOS SubSystem Design

Chapter 7 presents a subsystem view of CMOS design. A variety of building blocks of VLSI systems such as adders, multipliers, ALUs, data path, ROMs, PLAs, etc. are discussed. Several options of each subsystem are presented with power dissipation emphasis. The use of PLL in high-speed CMOS systems for deskewing the internal clock is also examined. Low-power issues of CMOS subsystems are also included.

1.4.7 Low-Power VLSI Design Methodology

In Chapter 8 advanced techniques to reduce the dynamic power component at several levels of design are presented. Lowering the power supply voltage while maintaining the performance is one technique for power reduction addressed extensively in this chapter. It is shown that low-power techniques at the high-level (algorithmic and architectural) of the design lead to a power saving of several orders of magnitude. Several examples are included to give the reader a clear picture of low-power design aspects. In addition, the power estimation techniques, at the circuit, logical, architectural and behavioral levels, are overviewed. The goal of power estimation is to optimize power, meet requirements and know the power distribution through the chip.

REFERENCES

[1] Special Report, "The New Contenders," IEEE Spectrum, pp. 20-25, December 1993.

[2] D. W. Dobberpuhl et al., "A 200-MHz 64-b Dual-Issue CMOS Microprocessor", IEEE J. Solid-State Circuits, vol. 27, no. 11, pp. 1555-1567, November 1992.

[3] W. J. Bowhill et al., "A 300MHz 64b Quad-Issue CMOS RISC Microprocessor," IEEE International Solid-State Circuits Conf., Tech. Dig., pp. 182-183, February 1995.

[4] Technology 1995: Solid State, IEEE Spectrum, pp. 35-39, January 1995.

[5] D. Bearden, et al., "A 133 MHz 64b Four-Issue CMOS Microprocessor," IEEE International Solid-State Circuits Conf., Tech. Dig., pp. 174-175, February 1995.

[6] MIPS Press release, 1994.

[7] A. Charnas, et al., "A 64b Microprocessor with Multimedia Support," IEEE International Solid-State Circuits Conf., Tech. Dig., pp. 178-179, February 1995.

[8] C. Small, "Shrinking Devices Put the Squeeze on System Packaging," EDN, vol. 39, no. 4, pp. 41-46, February 1994.

[9] P. Verhofstadt, "Keynote Address," IEEE Symposium on Low Power Electronics, Tech. Dig., October 1994.

[10] G. Gerosa, et al., "A 2.2 W 80 MHz Superscalar RISC Microprocessor," IEEE Journal of Solid-State Circuits, vol. 29, no. 12, pp. 1440-1454, December 1994.

[11] R. Bechade, et al., "A 32b 66MHz Microprocessor," IEEE International Solid-State Circuits Conference, Tech. Dig., pp. 208-209, February 1994.

[12] N. K. Yeung, Y-H. Sutu, T. Y-F. Su, E. T. Pak, C-C Chao, S. Akki, D. D. Yau, and R. Lodenquai, "The Design of a 55SPECint92 RISC Processor under 2W," IEEE International Solid-State Circuits Conference, Tech. Dig., pp. 206-207, February 1994.

[13] S. Lipoff and A. D. Little, "Evaluation of New Battery Technology in Selected Applications," IEEE Workshop on Low-power Electronics, Phoenix, AZ, August 1993.

[14] J. M. C. Stork, "Technology Leverage for Ultra-Low Power Information Systems," IEEE Symposium on Low Power Electronics, Tech. Dig., pp. 52-55, October 1994.

2

LOW-VOLTAGE PROCESS TECHNOLOGY

This chapter serves as an introduction to IC fabrication of CMOS bulk, bipolar BiCMOS and CMOS SOI devices including sub-micron devices for low-voltage applications. Section 2.1 is a review of CMOS process technologies. Examples for an N-well CMOS process and a twin-tub CMOS process are considered. Section 2.2 deals with bipolar technology with emphasis on advanced bipolar structures. The topic of the isolation techniques used for both bipolar and CMOS is addressed in Section 2.3. In Section 2.4 we discuss the similarities between advanced CMOS and advanced bipolar transistor structures to demonstrate how both technologies are indeed converging. The BiCMOS technologies are introduced in Section 2.5. with emphasis on CMOS-based processes. Three BiCMOS technologies, with different performance/cost, are presented. Section 2.6. introduces a complementary BiCMOS structure, where a vertical isolated PNP transistor is merged with an NPN transistor in a CMOS process. In Section 2.7, a table with the design rules of a generic 0.8 μm BiCMOS process is supplied. Finally, in Section 2.8, SOI technology is reviewed for low-voltage applications.

2.1 CMOS PROCESS TECHNOLOGY

The idea of CMOS was first proposed by Wanlass and Sah [1]. In the 1980's, it was widely acknowledged that CMOS is the technology for VLSI because of its unique advantages, such as low power, high noise margin, wider temperature and voltage operation range, overall circuit simplification and layout ease. The development of VLSI in the 80's has driven the integration density to millions of transistors on a single chip.

In this section we review two CMOS bulk technologies: N-well and twin-tub processes. Other processes such as retrograde-well technology is not discussed.

2.1.1 N-well CMOS Process

In the N-well CMOS process, the P-channel transistor is formed in the N-well itself and the N-channel in the P-substrate. Fig. 2.1 illustrates cross-sectional views and process steps of a typical N-well process.

The process starts by growing an oxide on the wafer. The oxide is then patterned to open N-well windows. Phosphorus atoms are implanted into the silicon followed by a high-temperature annealing to diffuse the well [Fig. 2.1(a)]. The LOCOS (LOCal Oxidation of Silicon)[1] technique is used to isolate the different active areas. After removing the nitride used in the LOCOS process, a photoresist layer is deposited and is then patterned by a P-well mask (new mask). This is followed by low energy ion implantation of boron (B I/I) to adjust the threshold voltage of the N-channel transistor [Fig. 2.1(b)]. A second ion implantation can be applied to eliminate punchthrough in the short channel device. Similarly, the threshold voltage of the P-channel transistor is adjusted [Fig. 2.1(c)]. A thin gate oxide is then grown and a layer of polysilicon is deposited and doped with phosphorus. The polysilicon is patterned to form the gates of all the transistors and interconnection layer [Fig. 2.1(d)]. The source and drain regions are then implanted by using a photoresist mask. Boron is used for the P^+ regions of the P-channel transistors and arsenic for N-channel transistors [Fig. 2.1(e)]. The N^+ and P^+ regions are also used N- and P- wells contacts, respectively. The photoresist is removed and a thick oxide is deposited by Chemical Vapor Deposition (CVD) as an isolation layer between the polysilicon layer and the subsequent metal layer. Contact holes are opened in the oxide layer and metal (usually aluminum) is deposited on the whole wafer. At this stage, the metal is patterned and annealed at relatively low-temperature (450 C) [Fig. 2.1(f)]. One or two other metal layers are usually added. At the end, the wafer is passivated and windows are patterned over the metal bonding pads to provide electrical contacts with pins.

---

[1] For more details on the LOCOS isolation see Section 2.3.1.

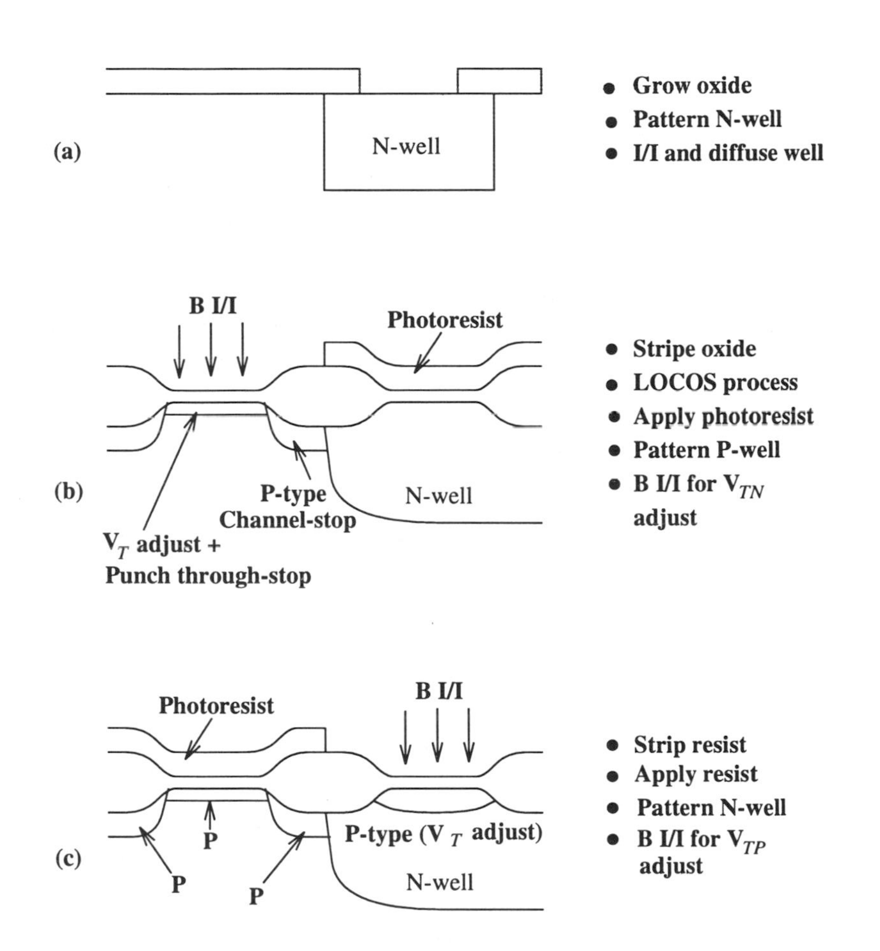

Figure 2.1 Cross-sectional views of an N-well process.

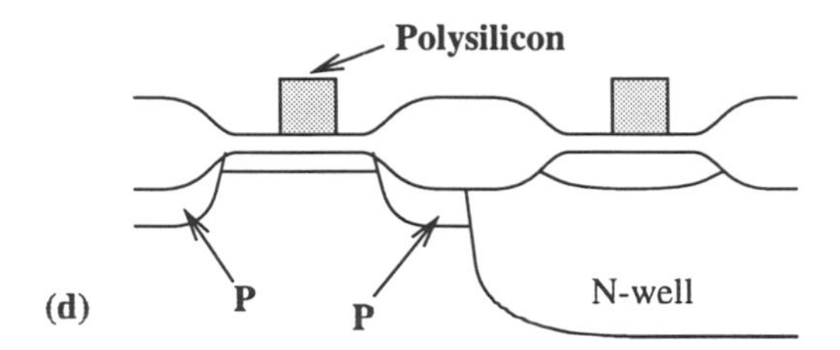

- Strip resist/oxide
- Grow gate oxide
- Deposit polysilicon
- Apply photoresist and pattern
- Strip resist

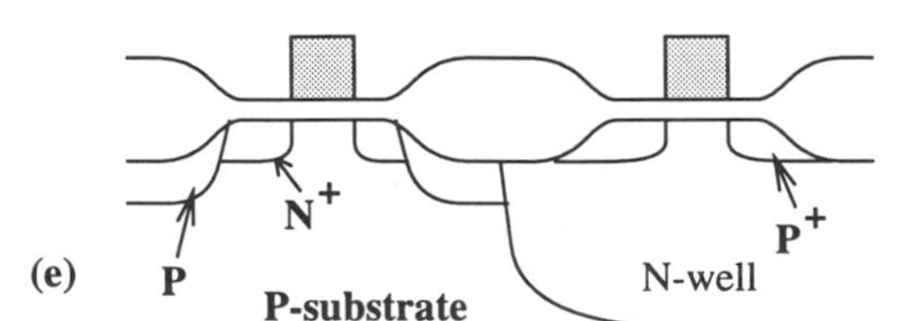

- Apply photoresist
- Pattern S/D regions for P-channel
- B I/I for P^+S/D
- Strip photoresist
- Repeat for N^+S/D
- Strip photoresist

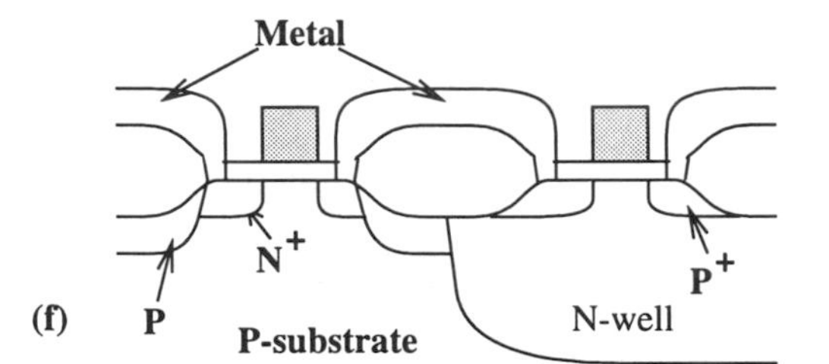

- Grow oxide
- Etch contact holes
- Deposit metal
- Pattern metal
- Metal anneal

Figure 2.1 *(continued)*

2.1.2 Twin-Tub CMOS Process

An alternative approach for CMOS devices fabrication is to use two separate wells (tubs) for N- and P-channel transistors in a lightly doped N- or P-type substrate. This "twin-tub" CMOS technology uses a single mask that allows it to form two independently doped and self-aligned tubs [2] ; hence both CMOS devices types are optimized independently. This flexibility in selecting the substrate type with no change in the process flow is the major advantage of twin-tub CMOS. This technology is also more attractive when the devices are scaled down to submicron dimensions.

Fig. 2.2 shows the major steps involved in a typical twin-tub process. The starting material is a lightly doped P-epitaxial material over a P^+ substrate to reduce latch-up. In addition to the conventional N-tub process, another N-type (arsenic) shallow implant is used to increase the surface doping of the N-tub to prevent punchthrough (for short channel devices). It is also used to form the channel-stoppers[2] for the P-channel transistors [Fig. 2.2(a)]. The photoresist is stripped and a selective oxidation of the N-tub is performed. The nitride/pad oxide layers are removed to implant boron, which is driven in to form the P-tub. This is followed by a second boron ion implantation for the channel-stoppers for the N-channel device [Fig. 2.2(b)]. The N-tub oxide is then stripped. So far only one mask (N-tub mask, MASK#1) is required for self-aligned wells and channel-stopper processes. Both tubs are driven in. LOCOS isolation is developed to isolate between the devices using MASK#2, which defines the active areas. After the LOCOS process, boron is implanted through the pad oxide (used in the LOCOS) to reduce the threshold voltage of the P-channel transistor using MASK#3. This process results in a *buried-channel* PMOS transistor. The pad oxide is then removed. The remaining steps are similar to those used in the N-well process where MASK#4 is needed to pattern the polysilicon [Fig. 2.2(c)]. MASK#5 and MASK#6 are required to form the N^+ and P^+ sources/drains (S/D), respectively. MASK#7 for contact openings, and MASK#8 for patterning the metal [Fig. 2.2(d)].

The fabrication of submicron MOS transistors requires additional process steps to avoid hot carrier effects. Fig. 2.3 illustrates a CMOS twin-tub structure with Lightly Doped Drain (LDD). Both NMOS and PMOS devices have *lightly doped* extensions to the source and drain regions. The electric field near the drain is reduced due to its light doping. This prevents the generation of hot carriers. The major process steps to fabricate the LDD structure are shown in Fig. 2.4.

2.1.3 Low-Voltage CMOS Technology

Scaled CMOS has been recognized as the technology suitable for low-power battery operated systems demanding high-speed operations. Conventional scaled CMOS technology undergoes a drastic reduction in speed when the power supply is reduced to 1 V and sub-1 V. If the threshold voltage is scaled aggressively, the subthreshold leakage current increases drastically, which causes limitations for battery applications. Hence, high-performance low-power scaled CMOS technology is needed for ultra-low voltage operation. One key in achieving low-power CMOS devices is the reduction of the junction capacitances as well as

[2] For more details on the channel-stoppers refer to Section 2.3.

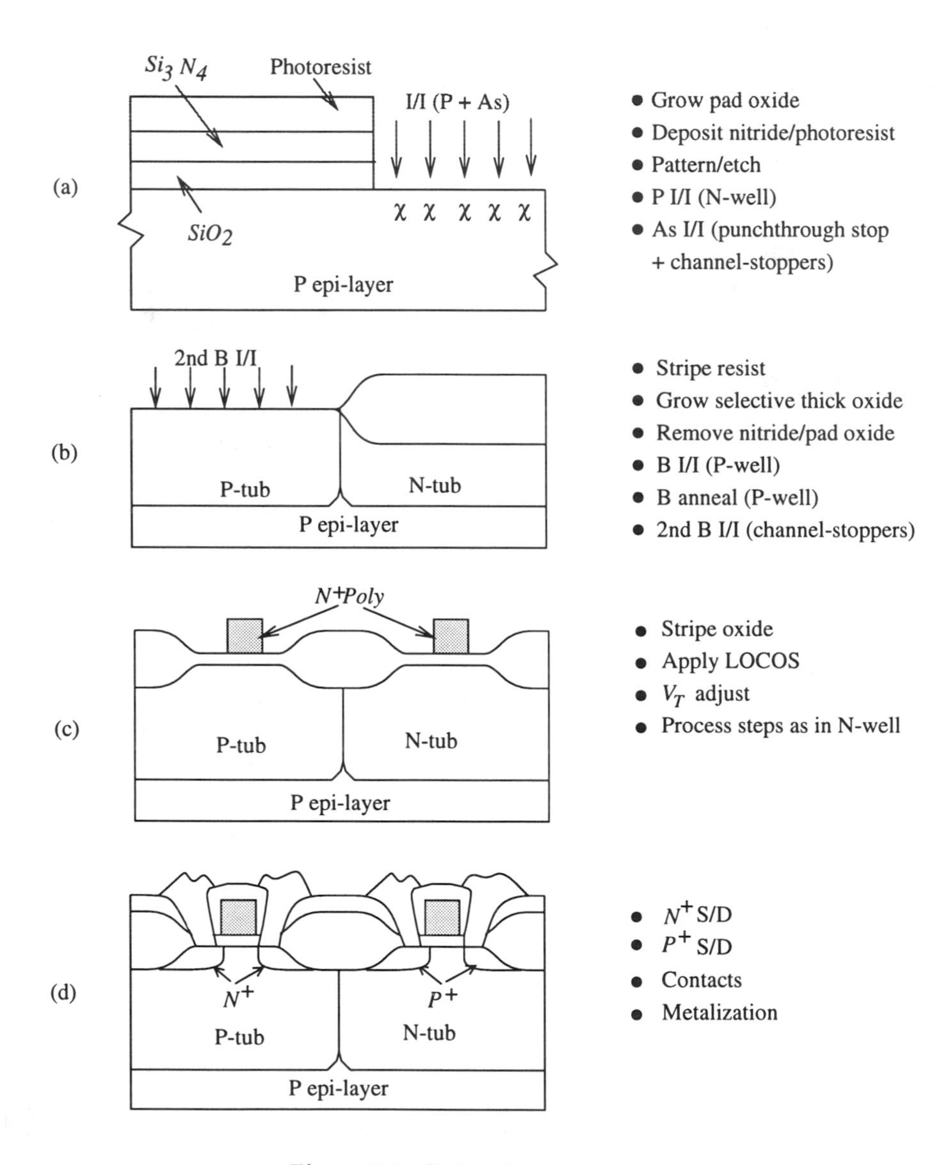

Figure 2.2 Twin-tub process sequence.

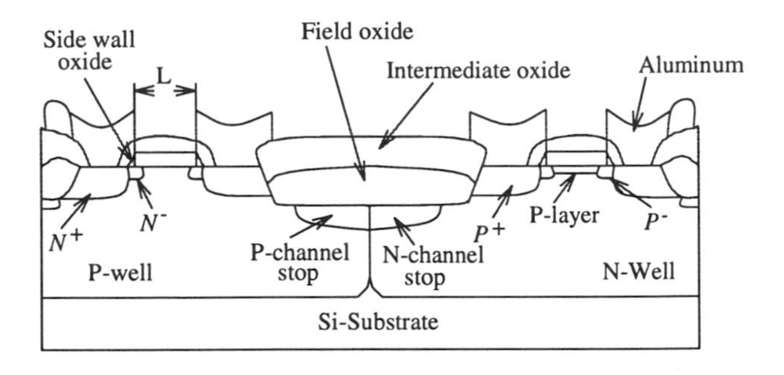

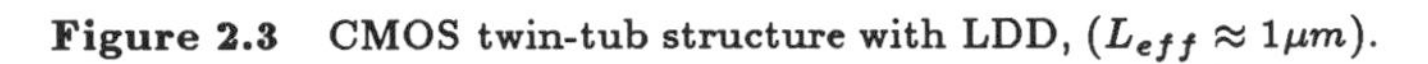

Figure 2.3 CMOS twin-tub structure with LDD, ($L_{eff} \approx 1\mu m$).

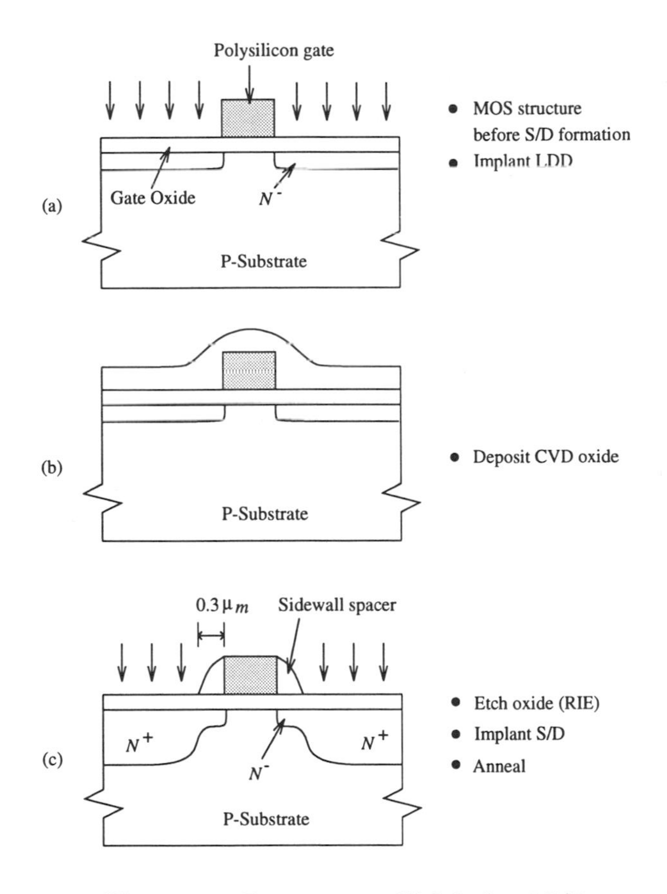

Figure 2.4 Process steps of lightly doped S/D.

other parasitic capacitances. Also, the subthreshold current should be reduced when low threshold voltage ($V_T \leq 0.3V$) is used.

Extensions and variations of standard CMOS process have been proposed to enhance the performance of devices at low-voltage [3, 4]. These devices have good short channel behavior, low junction capacitance and reduced parasitic resistance. The power supply choice depends on performance/reliability/power trade-offs. Reduced power supply is needed for low-power applications, but a deep-submicron CMOS device with ultrathin gate oxide and low threshold voltage should be used to improve performance. Table 2.1 shows the speed achieved at low-voltages using deep-submicron processes.

Table 2.1 Performance comparison at low-voltage.

Name [Ref.]	CMOS Process	Voltage (V)	Delay (ps)
IBM [3]	0.10 μm	1.5	22.0
AT&T [4]	0.10 μm	2.0	12.7
NEC [5]	0.15 μm	1.9	33.0
Fujitsu [6]	0.10 μm	2.0	21.0
Toshiba [7]	0.15 μm	1.0	50.0
Toshiba [8]	0.35 μm	1.5	52.0

An example of improved performance CMOS technology suitable for low-voltage is the one proposed by Toshiba [8] called CMOS Shallow Junction Well FET (SJET). Fig. 2.5 shows the cross-sectional view of the CMOS-SJET process. The N-well and P-well depths are very shallow and comparable to the maximum depletion layer width in the channel. With this CMOS-SJET structure the depletion layer of the NMOS device, for example, is extended compared to the original one and reaches the depletion layer of the P-well and the N-type substrate. As a result, the total depletion layer width is increased and low depletion capacitance, C_D, is obtained. This leads to the reduction of the subthreshold slope (see Section 3.3.2). Thus, the threshold voltage can be reduced at low power supply voltage compared to the conventional CMOS process. Furthermore the wells are designed to reduce junction capacitance of the S/D regions by 40 to 55 % compared to the conventional one. The structure of Fig. 2.5 also uses dual polysilicon gate N^+ and P^+, to optimize the threshold voltages of the MOS devices. Also W-polycide gates are used to reduce the poly sheet resistance. The delay of the CMOS-SJET inverter is 2.5 times better than that of conventional CMOS using the same gate size (0.5 μm technology) at 1.5 V power supply. The power-delay product of a CMOS-SJET gate at

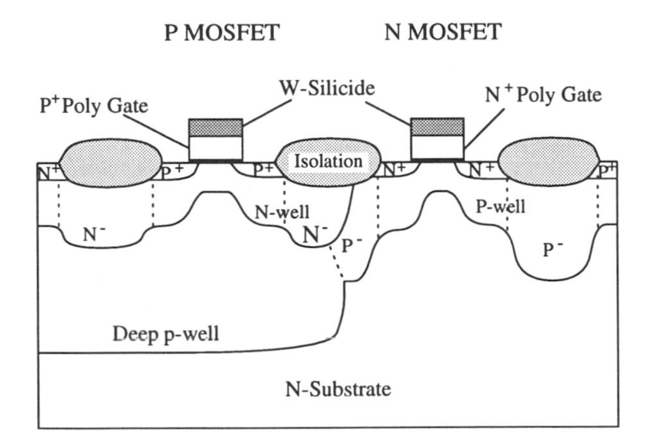

Figure 2.5 CMOS shallow junction well structure [8].

1.5 V using 0.35 μm technology is 1.3 fJ which is 1/3 times improvement of that for conventional CMOS devices. However, the main drawback with the CMOS-SJET is the large body effect due to its retrograde doping profile.

2.2 BIPOLAR PROCESS TECHNOLOGY

The technology of epitaxial growth gave rise to the economical manufacturing of monolithic bipolar ICs as it allows a high-quality thin film of semiconductor to be grown on the top of a substrate. Junction-isolation and epitaxy techniques triggered the progress of bipolar technology. Although, most of the focus has been on the development of CMOS for the last ten years, yet, we find that bipolar technology has achieved significant progress as well. Impressive high-speed results were demonstrated at the 1985 ISSCC (International Solid-State Circuits Conference) and thereafter. ECL (Emitter Coupled Logic) gate delay of 15 ps have been reported [9]. It was shown that advanced silicon bipolar technologies, although quite complex, could be integrated at the LSI level and operate at frequencies above those of CMOS circuits. Since then, the interest in advanced bipolar processes has increased. The key features for such technologies are: i) self-aligned base, ii) advanced isolation techniques such as deep-trench, and iii) polysilicon emitter contact.

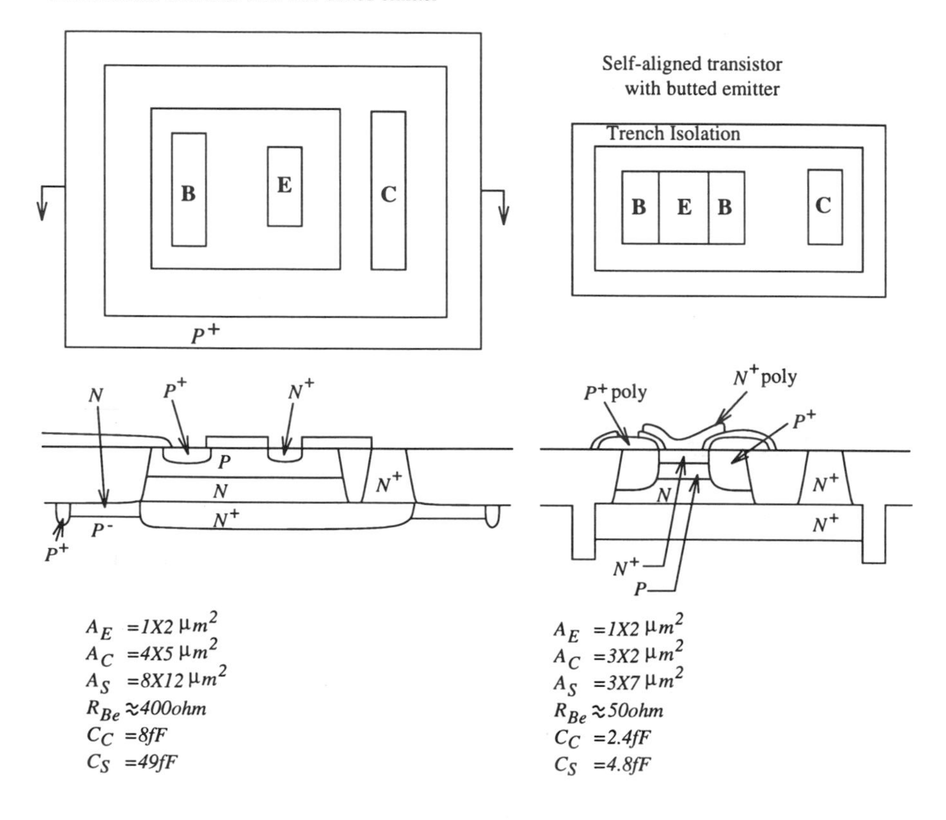

Figure 2.6 Comparison of a conventional and trench-isolated self-aligned bipolar devices with 1 μm feature size [10].

Fig. 2.6 depicts the differences between the structures of a conventional bipolar transistor[3] and an advanced one [10]. The parasitical capacitances and resistances have improved by reducing the areas of the emitter, base and collector. This has been achieved by using self-alignment and trench isolations.

Further reduction in parasitics can be achieved by using the so-called *SICOS SIde wall base electrode COntact Structure* process [Fig. 2.7] [11]. The extrinsic base regions required for base contacts in the conventional bipolar structures

[3] In this context, the term "conventional bipolar transistors" refers to those bipolar transistors which have a metal emitter contact with oxide-isolation.

Table 2.3 *(continued)*

10.	Metal 2 (M2)	
	10.1 minimum width	2λ
	10.2 minimum spacing	3λ
	10.3 maximum current density	$1\ mA/\mu m$
11.	Via (VIA)	
	11.1 minimum size	$3\lambda \times 3\lambda$
	11.2 minimum spacing	3λ
	11.3 minimum M1 or M2 overlap of VIA	1λ
	11.4 minimum VIA to CO spacing	2λ
	11.5 minimum PO to VIA spacing	2λ
	11.6 minimum PO overlap of VIA	2λ

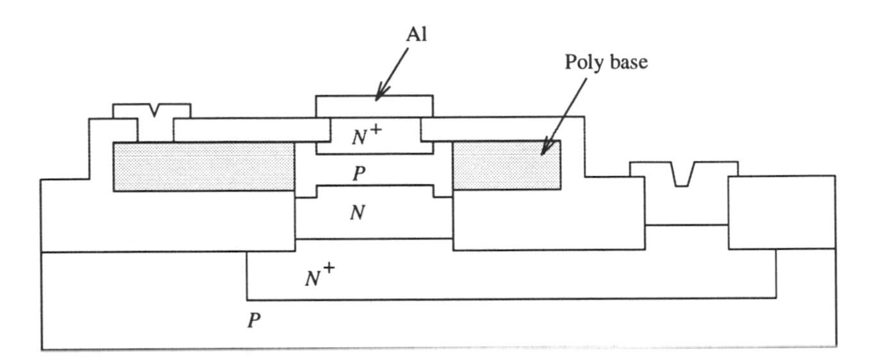

Figure 2.7 Cross-sectional view of the SICOS bipolar device structure [11].

have been replaced by the side wall base electrodes. This allows the base area to be almost as large as the emitter. The SICOS structure is suitable for VLSI applications because of its density and low parasitics.

One of the features of advanced bipolar transistors is the replacement of aluminum by polysilicon for the contact of the emitter. This step has led to noticeable improvement in the current gain of bipolar transistors. For further reading on polysilicon emitter BJTs refer to [10, 12, 13].

In this section, we introduce a typical Double-Polysilicon Self-Aligned (DPSA) process technology as an example of the advanced bipolar technologies[4].

Any bipolar process typically starts with creating the buried layers and the epitaxial layer. Fig. 2.8 illustrates the major steps of the epitaxial growth with an N^+ buried layer (BL). This buried layer is introduced to reduce the collector resistance of a bipolar transistor. While the epitaxial layer offers the high-quality silicon host for the bipolar transistor. The steps involved in Fig. 2.8 are the following. First, an oxide layer is grown on the substrate and is then patterned using the buried layer mask. The photoresist on the oxide serves as a mask against etching and ion implantation. After etching the oxide, the exposed regions of the silicon surface are implanted by arsenic or antimony to form the N^+ buried layers. The photoresist is then removed and an annealing step is carried out. All oxide is then stripped. An N-epitaxial layer is grown

[4] A review of conventional bipolar technology using the junction isolation techniques can be found in [14].

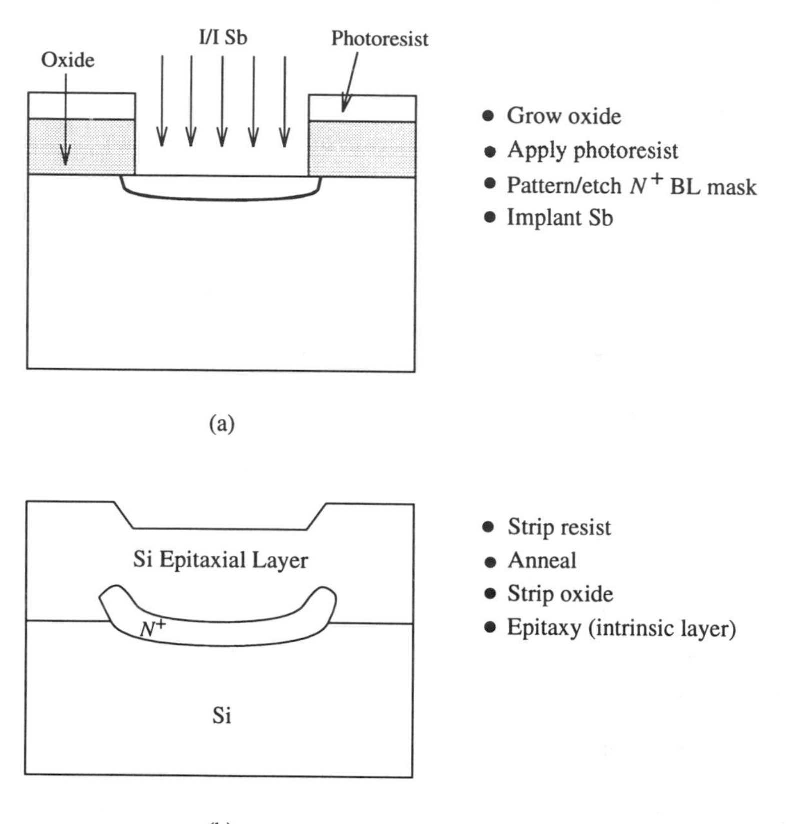

Figure 2.8 Buried and epitaxial layers growth.

on the substrate as shown in Fig. 2.8(b). The thickness of this epitaxial layer can be as low as 0.8 μm for advanced digital bipolar technology. The problems limiting the scaling down of the thickness of epitaxial layer are the autodoping and out-diffusion of the buried layer.

Fig. 2.9 illustrates the sequence of a DPSA process assuming a starting structure with N^+ buried layer, N-epitaxial layer and isolation oxide as shown in Fig. 2.9(a). First, photoresist is deposited and patterned to define the collector contact region (deep N^+ collector sink). This region is then implanted with phosphorus to increase its doping level. The photoresist is stripped and

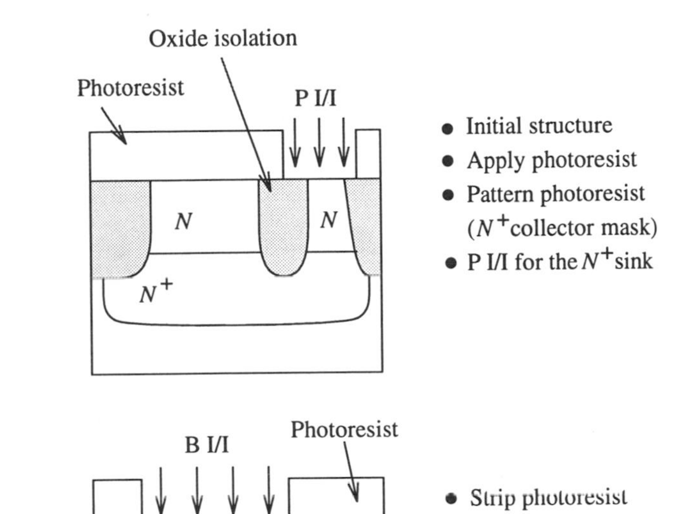

B I/I
Photoresist

N
N+
N^+

- Strip photoresist
- Grow oxide
- Apply photoresist
- Pattern resist (base mask)
- B I/I (P-base)

CVD Oxide
P^+ Poly
(c)

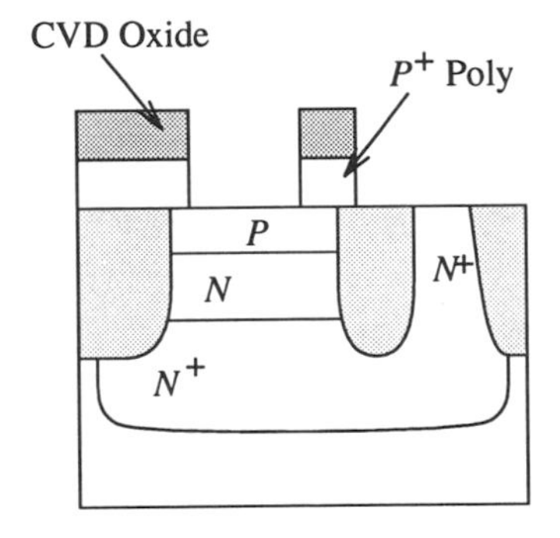

- Strip photoresist/oxide
- Deposit P^+polySi/oxide
- Pattern/etch oxide/polySi

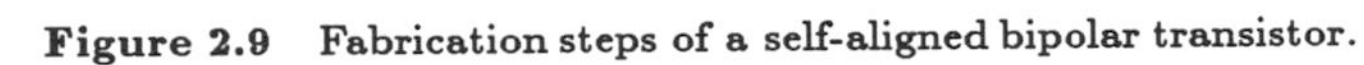
Figure 2.9 Fabrication steps of a self-aligned bipolar transistor.

(d)

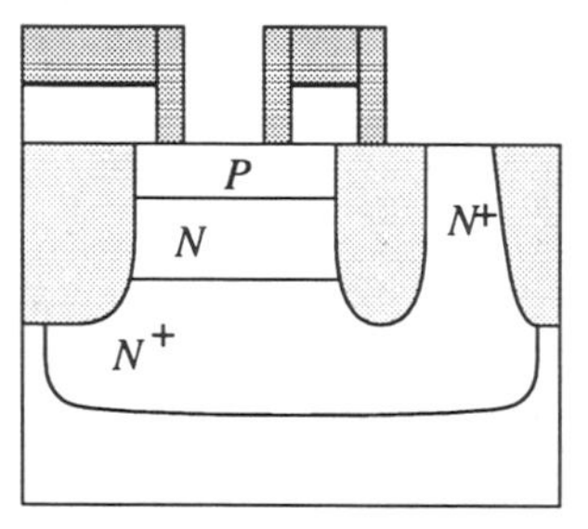

- Deposit CVD oxide
- RIE etch of oxide

(e)

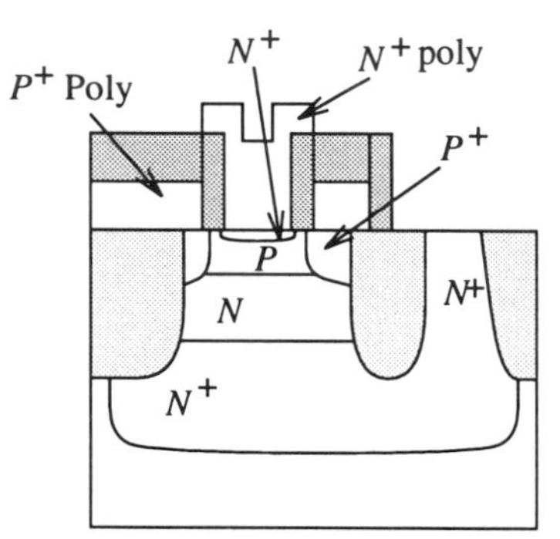

- Deposit the second level of polysilicon
- P I/I (N^+ poly)
- Anneal
- Pattern/etch N^+ polySi

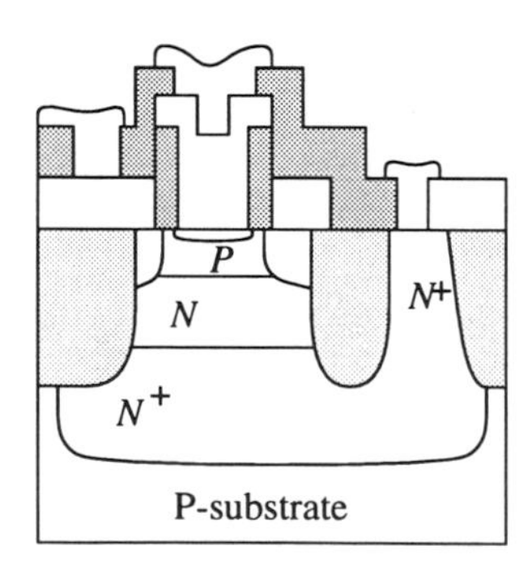

- Deposit oxide
- Open contact holes
- Deposit metal
- Pattern/etch metal

Figure 2.9 *(continued)*

a P-type base is implanted through a pre-implantation oxide as shown in Fig. 2.9(b). The resist and the oxide are then removed. A combination of P^+ polysilicon and oxide layers are deposited over the wafer. These layers are then etched as shown in Fig. 2.9(c). A CVD oxide is deposited over the wafer. The oxide is then dry etched using reactive ion etching (RIE). The P^+ polysilicon is walled with the oxide (called sidewall spacer) [Fig. 2.9(d)]. The second level of polysilicon is deposited and implanted with phosphorus that will ultimately form the diffused emitter junction. At this stage, the wafer is annealed to drive the dopants from the P^+ and N^+ polysilicon layers. Fig. 2.9(e) illustrates the structure after patterning the N^+ polysilicon. The P^+ diffusion under the polysilicon forms the extrinsic base. The contact openings to the P^+ and N^+ polysilicon, and collector are etched. This is followed by the metallization step. At the end, the metal is patterned as shown in Fig. 2.9(f).

The advantage of bipolar devices is their high-speed performance. However, there are not suitable for battery backup systems because they consume high DC current. Many logic circuit techniques have been proposed for low-power and low-voltage operation, particularly for telecommunications applications [15, 16].

2.3 ISOLATION IN CMOS AND BIPOLAR TECHNOLOGIES

2.3.1 CMOS Device Isolation Techniques

Isolation in an integrated circuit means to electrically isolate similar or different transistors. In a CMOS chip, where more than one million transistors can be integrated, 1 μA/transistor of leakage current due to a bad isolation can lead to a few watts of DC power consumption. Moreover this leakage current provokes susceptibility to the latch-up as will be discussed in Section 3.1.6.

Isolation in CMOS is required to separate the devices electrically by eliminating the inversion layers, which might be induced by the interconnection layer between the transistors. The principle of isolation in CMOS is based on a field oxide formation between two active areas [Fig. 2.10]. The width of the isolation region should be minimized to attain dense layout and particularly for VLSI circuits.

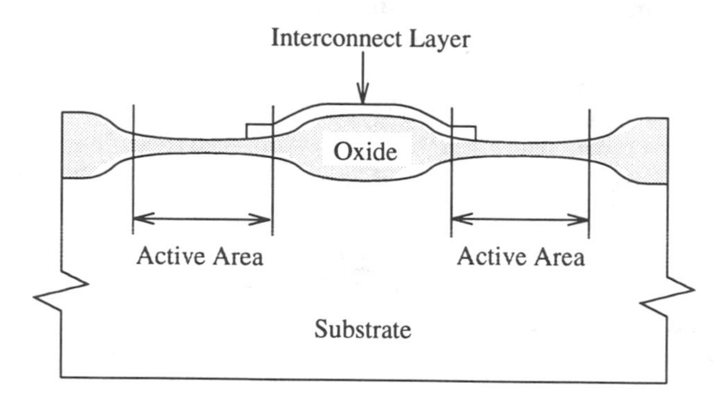

Figure 2.10 Field oxide isolation in MOS integrated circuits.

Several isolation techniques have been proposed and used. The most popular are LOCOS (LOCal Oxidation of Silicon) [17], trench isolation [18, 19, 20, 21], and selective epitaxy [22]. Selective epitaxy is not studied in this chapter.

2.3.1.1 *Local Oxidation of Silicon (LOCOS)*

LOCOS is a relatively simple process for the isolation of active devices in CMOS technology. It is realized by forming a thick field oxide (FOX) between the active areas. FOX is very thick ($0.4 - 0.6\ \mu m$), hence the corresponding field threshold voltage is high. The condition for preventing an inversion layer under FOX and between two active regions is that this field threshold voltage should be higher than the highest power supply voltage used on chip. The field threshold voltage can be further increased by raising the doping level under the FOX. This can be achieved by selectively implanting the regions over which the FOX is subsequently grown. These regions are commonly known as *channel-stoppers.*

The steps of the LOCOS process are illustrated in Fig. 2.11. A pad oxide of 40 nm is grown and is followed by chemical vapor deposition of a 100 nm thick nitride layer, which masks the active region. The pad oxide is called stress-relief-oxide (SRO) because it protects the silicon from stress caused by the nitride during subsequent high temperature processes. Silicon nitride is used as a mask to protect the active region from oxidation. A layer of photoresist is applied to the wafer and then patterned using the mask of the active areas. The nitride/oxide layers are etched [Fig. 2.11(a)]. A P-type dopant is

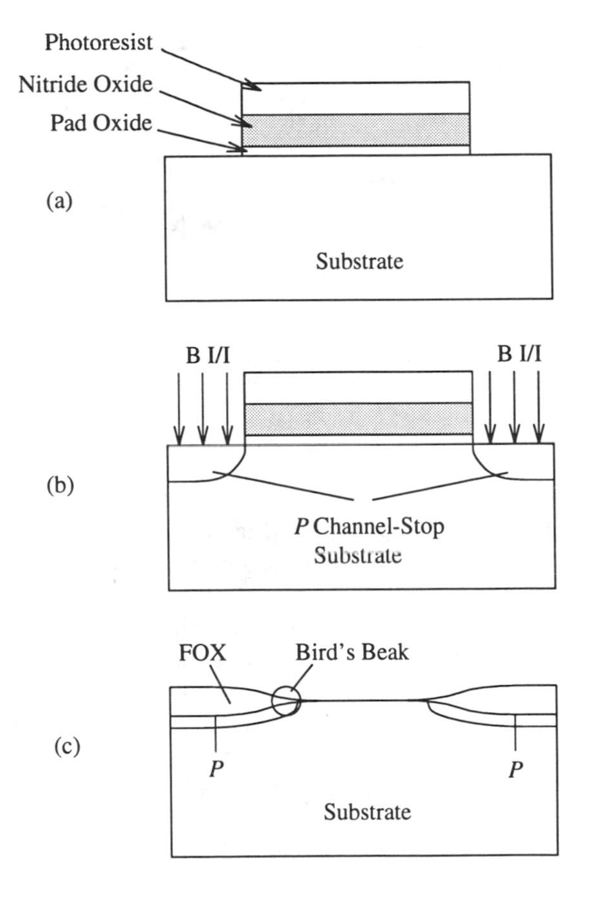

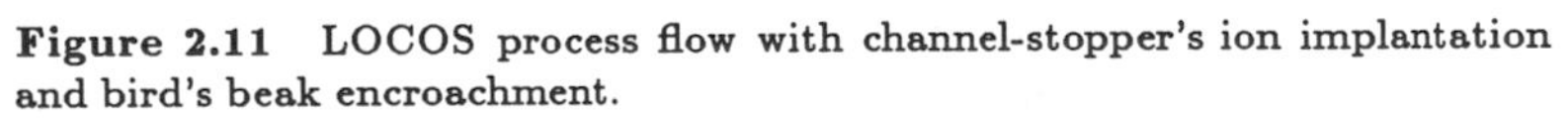
Figure 2.11 LOCOS process flow with channel-stopper's ion implantation and bird's beak encroachment.

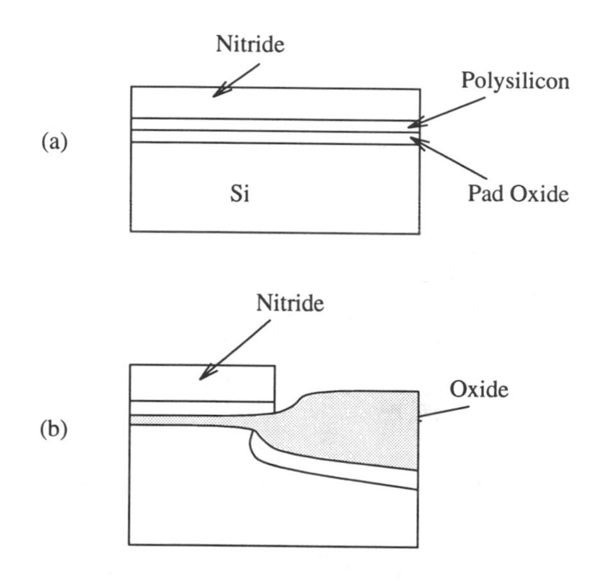

Figure 2.12 Poly buffered LOCOS process.

implanted to form the channel-stoppers [Fig. 2.11(b)]. The photoresist, which is used for protection against ion implantation, is stripped and a thick thermal oxide is grown; i.e. FOX. Only local oxidation is realized because the nitride masks the regions beneath it. At the end, the nitride/oxide are removed [Fig. 2.11(c)]. During this LOCOS process, 56% of the FOX thickness is under the silicon surface because the oxidation consumes some of the silicon. This process is called semi-recessed LOCOS isolation. One problem associated with this process is the lateral extension of the field oxide under the nitride during the oxidation, forming what is called bird's beak encroachment [Fig. 2.11(c)]. A typical value of this encroachment is 0.5 μm/side. This encroachment limits the scaling of the active areas and the channel width of the MOS device. Moreover, this bird's beak introduces imprecise channel widths.

The *Poly Buffer LOCOS* process was developed to reduce the bird's beak encroachment [23]. In this modified LOCOS process, the nitride mask thickness has been increased to 240 nm and a polysilicon stress relief buffer layer of 50 nm has been added between the nitride and a 10 nm pad oxide [Fig. 2.12(a)]. This arrangement prevents deep lateral extension of the field oxide under the nitride layer [Fig. 2.12(b)]. A 0.8 μm field oxide thickness results in 0.15 μm/side of encroachment and 2.2 μm minimum isolation pitch. Other techniques to solve the problem of the bird's beak encroachment can be found in [24, 25, 26].

2.3.1.2 Trench Isolation

Trench Isolation is another alternative to LOCOS isolation process. This technology has been accepted relatively quickly by the industry [27]. It addresses the isolation problem between opposite type devices (like N-channel and P-channel MOSFETs in CMOS technology). The advantages of the trench isolation are: i) no bird's beak encroachment, ii) latch-up free structure, and iii) planar surface.

Fig 2.13 illustrates the steps of the trench isolation process. First, the pad oxide, the nitride and the thick oxide layers are patterned using the mask of the active areas. The thick oxide serves as a mask in the trench processing [Fig. 2.13(a)]. A deep trench is formed by dry etching (RIE). This is followed by a boron implant to create the P^+ channel-stoppers at the bottom of the trench. The top thick oxide is removed, and the trench sidewalls are oxidized [Fig. 2.13(b)]. The polysilicon is deposited over the whole wafer, filling the trenches. The polysilicon is used as the trench dielectric because it uniformly fills the trenches better than other dielectrics. The surface polysilicon is then etched to yield the structure shown in Fig. 2.13(c). The wafer is oxidized using the nitride as a mask. The nitride is finally removed as illustrated in Fig. 2.13(d). At this stage, conventional processing can be used to integrate the CMOS devices.

Although trench isolation permits reduction of the separation between the active regions; it has several drawbacks: i) it is a costly process because of the large number of processing steps, and ii) it can not be used as an isolation region for the inactive parts of the chip. In this case, LOCOS is usually used. The description of other trench isolation processes can be found in [28].

2.3.2 Bipolar Device Isolation Techniques

The first technique used for bipolar isolation was based on collector/substrate junction isolation [Fig. 2.14]. The N-wells (N collectors) of the adjacent transistors were separated by P^+ isles, which are deeply diffused to reach the P-type substrate. By tying these isles and the substrate to the most negative voltage, the junctions between them and the N-type collectors are reverse biased. Thus,

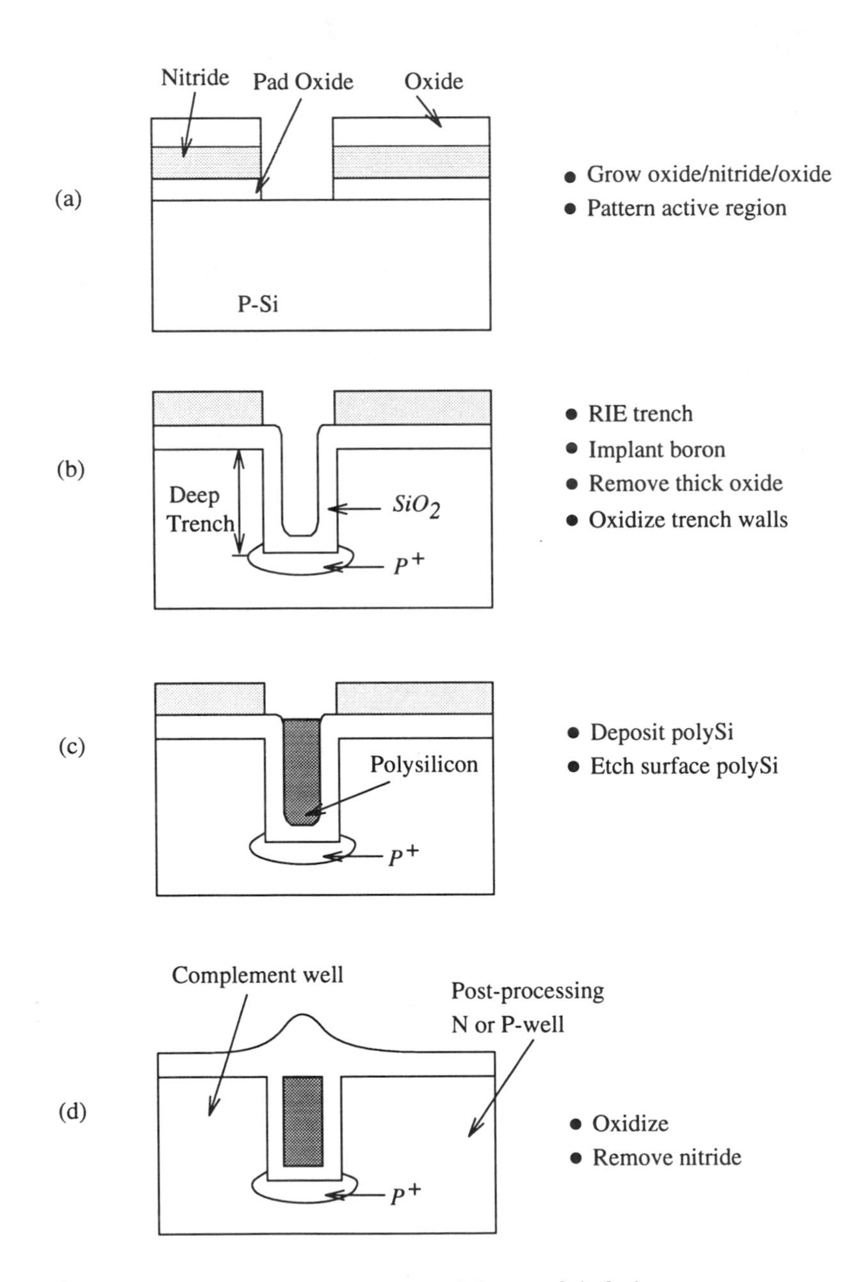

Figure 2.13 The major process steps of the trench isolation.

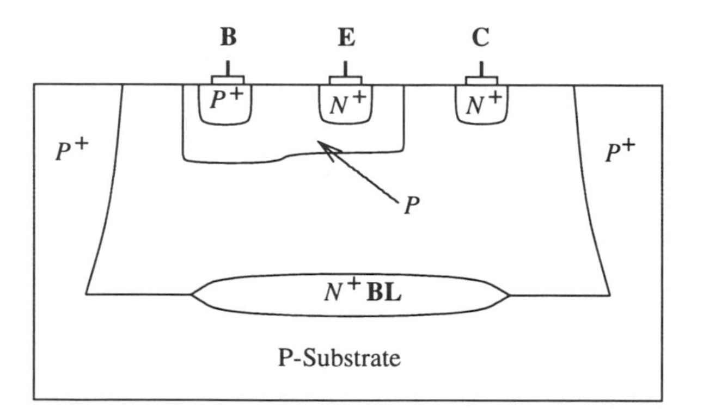

Figure 2.14 Cross-section of a junction isolated NPN bipolar transistor (with a buried layer).

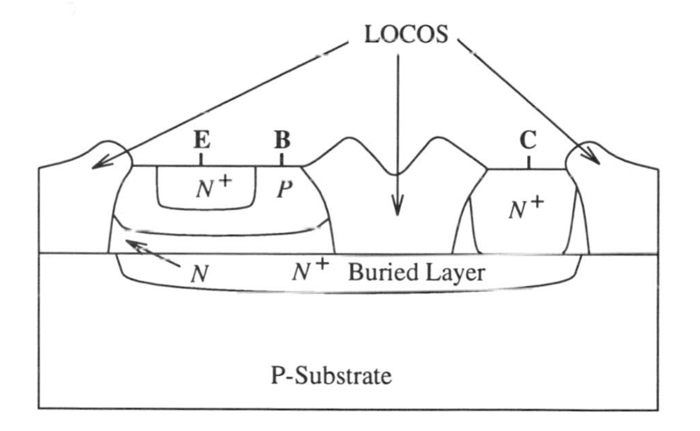

Figure 2.15 Cross-sectional view of an NPN bipolar transistor with LOCOS isolation.

all the components in different N-wells (N collectors) are isolated. The area consumed by the isolation isles is large relative to the transistor area.

The packing density of the bipolar technology can be improved by replacing the junction isolation with LOCOS isolation. An additional advantage of LOCOS isolation is the reduction of the parasitic collector-substrate capacitance. Fig 2.15 illustrates the cross-sectional view of an NPN bipolar transistor with LOCOS isolation. The area occupied by the oxide isolation is proportional to the epitaxial layer thickness. As the epitaxial thickness is being reduced for higher device performance the oxide isolation area becomes smaller, which means that LOCOS may become a practical isolation technique for advanced bipolar and BiCMOS technologies.

Fig. 2.16 illustrates the process steps for oxide isolation in a bipolar process. After epitaxy growth, a thin layer of SiO_2 is grown and a layer of Si_3N_4 is deposited. A photoresist layer is applied and patterned with an isolation mask [Fig. 2.16(a)]. Then the nitride/pad oxide layers and approximately half of the epitaxial layer are dry etched. Boron implant is performed to form the channel-stopper [Fig. 2.16(b)]. The photoresist is then removed and the wafer is oxidized to grow the thick isolation oxide. This oxide is called *recessed oxide.* The Si_3N_4 and the pad oxide are stripped at this stage. The resulting structure is almost planar. In this structure the bird's beak is formed as in the MOS case [Fig. 2.16(c)].

In the early 1980's, new isolation techniques such as grooves and trenches [29, 30, 31] were demonstrated. These techniques reduced the collector-substrate capacitance and increased the packing density. Hence they improve circuit speeds. The fabrication process is the same as the one described in CMOS trench isolation.

2.4 CMOS AND BIPOLAR PROCESSES CONVERGENCE

An interesting exchange of process technology know-how between the CMOS and the bipolar domains has taken place over the years. We have seen that epitaxial and buried layers have been used for CMOS to mute the latch-up. At the same time LOCOS, which was originally developed for CMOS, has been used for isolating bipolar transistors. The use of polysilicon for creating self-aligned MOS transistors was later adapted for self-aligned poly emitter bipolar transistors. Another example of the convergence between bipolar and CMOS is the use of oxide spacers in CMOS for formation of LDD regions, while, it has been used in bipolar to reduce the separation between the base contact and the emitter. The convergence of both technologies made the attractive idea of merging bipolar and CMOS seem more rational and feasible than ever.

Many of the steps of the advanced CMOS and bipolar processes are similar, hence, they can be shared for the fabrication of MOS and bipolar transistors

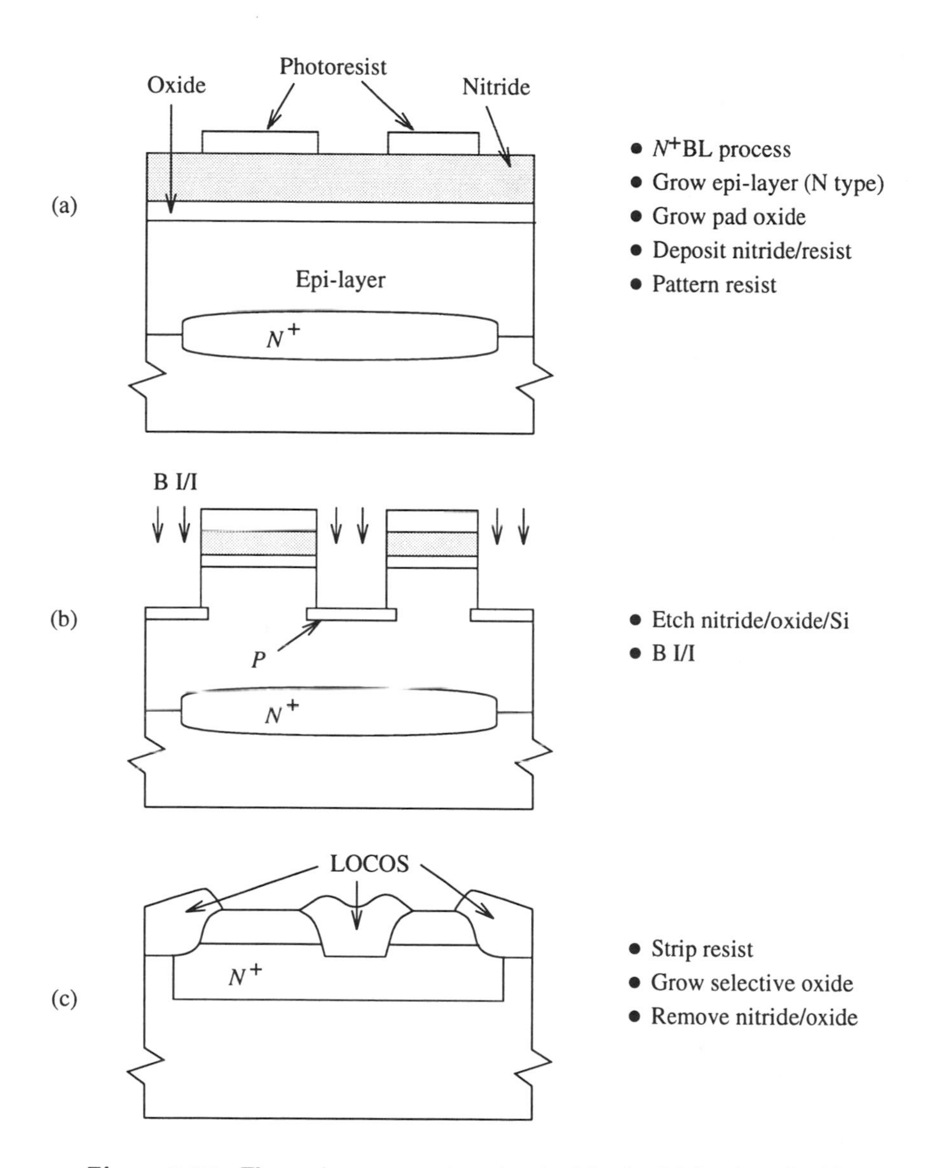

Figure 2.16 The major process steps involved in the fabrication of oxide-isolation isles in a bipolar process.

when they are integrated in a BiCMOS process. Some examples of these steps are:

1. The N-well, which can be used as the body of the PMOS transistor and as the N-collector of the NPN transistor;
2. The N^+ buried layer of the NPN can be used to form a retrograde well for the PMOS to reduce the latch-up susceptibility;
3. The polysilicon can be used for the CMOS gates and for the emitter contacts;
4. The shallow P-type implantation can be shared by the PMOS S/D and the self-aligned extrinsic base of the NPN transistor;
5. The shallow N-type implantation can be shared by the NMOS S/D and the emitter of the NPN transistor; and
6. The final annealing steps match.

However, as more steps are being shared by the different devices, the device characteristics have to be compromised. There is a tradeoff between the process complexity and device quality.

2.5 BICMOS TECHNOLOGY

Although the idea of merging bipolar and CMOS on the same chip originated 20 years ago [32], it was not feasible from a practical point of view because of the lack of adequate process technology. With the technological progress achieved in recent years, this idea has been revived. There are many techniques to merge bipolar and CMOS devices as reported in the literature [33, 34, 35, 36, 37, 38]. There are two ways of classifying BiCMOS processes. One way is to classify them according to the baseline process. A CMOS-based BiCMOS process is a CMOS baseline process, to which a bipolar transistor is added. Similarly, a bipolar-based BiCMOS process is a bipolar baseline process, to which CMOS transistors are added. In both cases, the added device would have to be compromised, which means that its characteristics can not be optimized. Alternatively, BiCMOS processes can be classified according to their cost/performance. In this regard, three categories can be identified:

1. Low-cost;
2. Medium-performance; and
3. High-performance (high-speed).

In this section, we present three examples of BiCMOS processes. The first one represents a low-cost process. It needs only one mask to incorporate the bipolar device in a CMOS-based process. The second example shows a medium-performance BiCMOS process, which requires 3 extra masks to a CMOS process. The third example illustrates a high-performance process in which polysilicon emitter and self-aligned structures are used.

2.5.1 Example 1: Low-Cost BiCMOS Process

In a low-cost BiCMOS process, a bipolar transistor is added to a CMOS process with minimum additional process steps. A typical N-well CMOS/bipolar process sequence is listed in Fig. 2.17(a). The N-well of the PMOS is used for the collector of the vertical NPN. The base is implanted in a separate step using an additional mask. The P^+ S/D and the extrinsic base share the same implantation step. The emitter and the N^+ S/D of the NMOS are also implanted in the same step. Fig. 2.17(b) illustrates the cross-section of an N-well BiCMOS structure. The process complexity is comparable to that of the CMOS. However, there are many trade offs in designing the emitter, base, and collector of the NPN. If the CMOS process is optimized, some of the bipolar device parameters, such as the breakdown voltage and the gain, may be satisfactory, but many others are degraded. For example, due to the absence of the buried layer and the deep N^+ collector in the NPN, the collector resistance is high. Hence, the cut-off frequency is low, the current drive is poor, and the collector-emitter saturation voltage is high.

2.5.2 Example 2: Medium-Performance BiCMOS Process

Fig. 2.18 shows a cross-sectional view of a BiCMOS structure, which can be realized by adding an NPN to a baseline twin-tub CMOS process. This structure has an N^+ buried layer and a deep N^+ collector sink which enhance the collector conductivity. The N^+ buried layer, under the PMOS, with the uniform N-well form a desired retrograde N-well. Similarly, the P^+ buried layer creates a retrograde P-well for the NMOS transistor. It also acts as an isolation

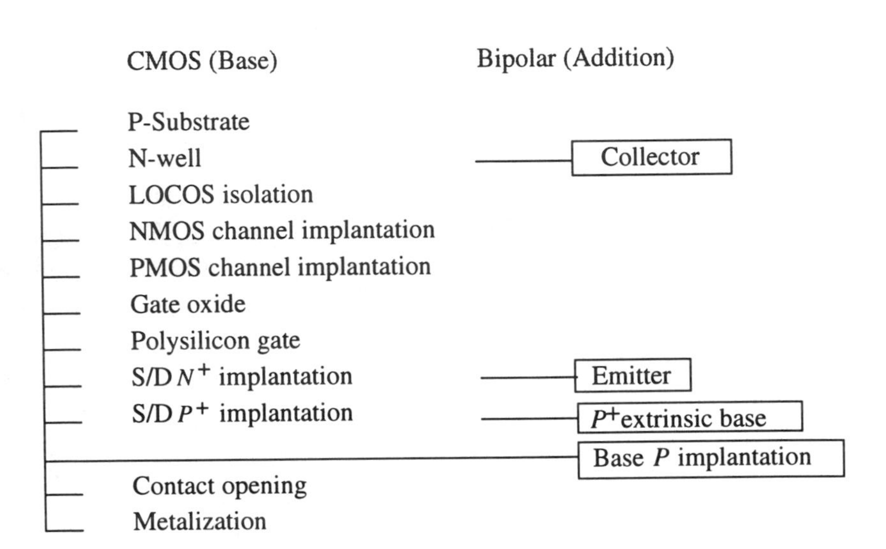

(a)

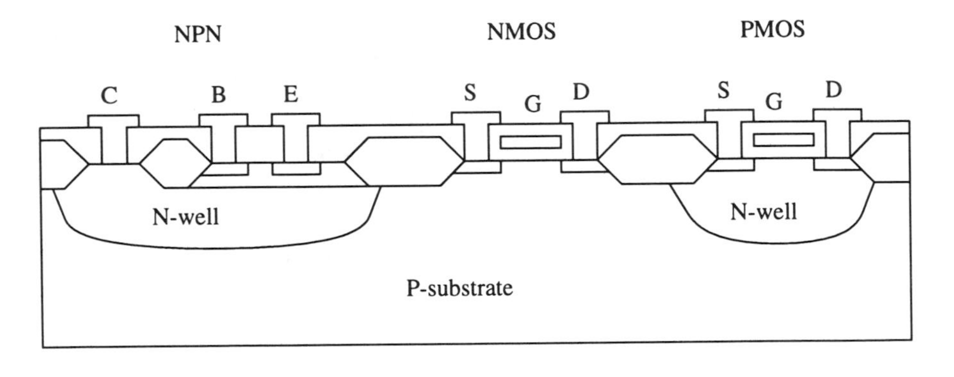

(b)

Figure 2.17 CMOS based BiCMOS process: (a) Process flow; (b) Device cross-sectional view.

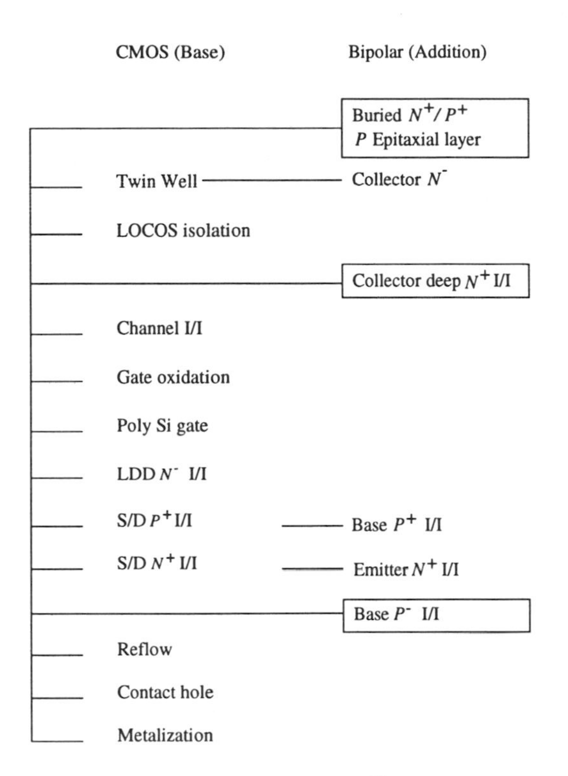

(a)

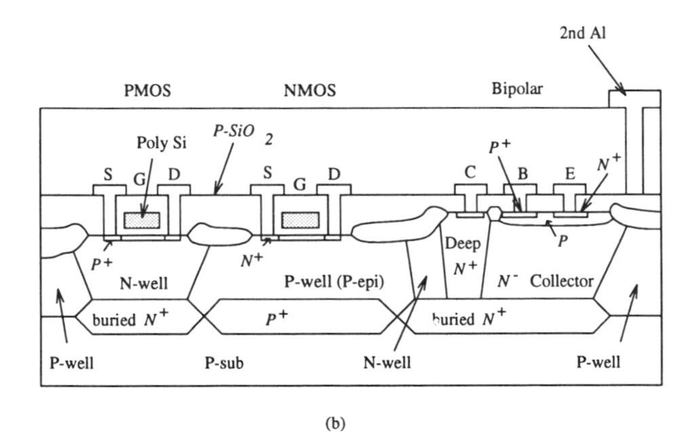

(b)

Figure 2.18 Medium-performance BiCMOS process (a) Process flow; (b) Device cross-sectional view.

region between the N^+ buried layers. A thin epitaxial layer ($1\ \mu m\ -\ 2\ \mu m$) is used to increase the cutoff frequency of the NPN transistor and to reduce the required width of the isolation isles between the bipolar transistors. The N collector is formed at the same time with N-well of the PMOS transistor. After the formation of LOCOS a deep N^+ sink is implanted and driven in. The P^+ extrinsic base is implanted at the same time with P^+ S/D regions of the PMOS transistor. The N^+ emitter and the N^+ S/D share the same implantation step. In this process an aluminum emitter contact is used. Therefore, the size of the emitter is larger compared to the case where a self-aligned polysilicon emitter contact is used. This process uses only 3 extra masks to form the bipolar transistor. The first mask is needed for N^+ buried layer. The second mask is used to implant the N^+ deep collector, and the third one for the base implantation.

The BiCMOS process described above can be optimized to be used for high performance circuits. The collector resistance is low in comparison to the low-cost process (example 1). For a $0.8\ \mu m$ process, the cut-off frequency (f_t) of a bipolar can be as high as $5\ GHz$.

2.5.3 Example 3: High-Performance BiCMOS Process

A high-performance BiCMOS process can be achieved by replacing the N^+ S/D implant, used to form the emitter in example (2), by a doped polysilicon emitter. One extra mask is required to open the emitter window of the bipolar transistor. The ion implantation of the poly emitter and MOS gates is developed simultaneously. As shown in Fig. 2.19, four additional mask levels (N^+ buried layer, N^+ deep collector, P-base, and emitter window) are required to obtain an advanced BiCMOS.

After the formation of the N^+/P^+ buried layers, the conventional twin-tub process is carried out. LOCOS is developed to isolate the devices. The deep collector N^+ is implanted and driven in, and the P-base is then patterned and implanted. The threshold voltages of the MOS transistors are adjusted by additional ion implantations. After the gate oxide growth, a thin polysilicon is deposited as shown in Fig. 2.20(a). The emitter window is then patterned and a second polysilicon layer is deposited [Fig. 2.20(b)]. The polysilicon is then doped by implantation and patterned to define the CMOS gates and polysilicon emitter [Fig. 2.20(c)]. Next, implants are selectively carried out to form the LDD regions for CMOS. Before implanting the N^+/P^+ S/D regions, a sidewall

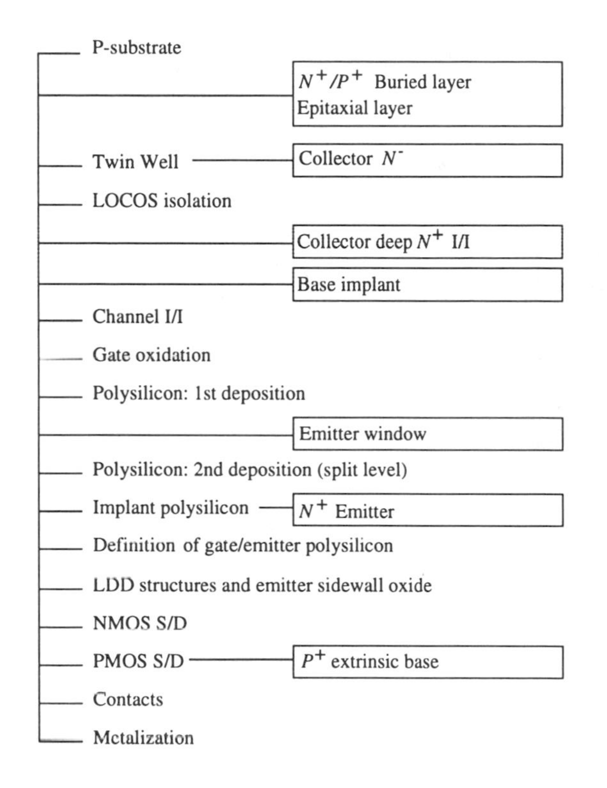

(a)

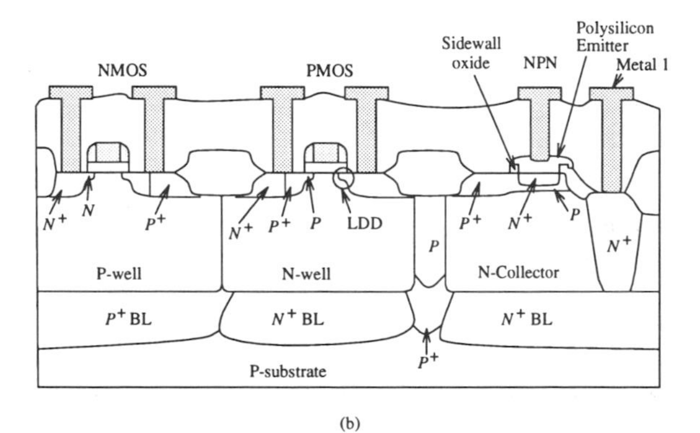

(b)

Figure 2.19 High-performance BiCMOS process: (a) Process flow; (b) Device cross-sectional view.

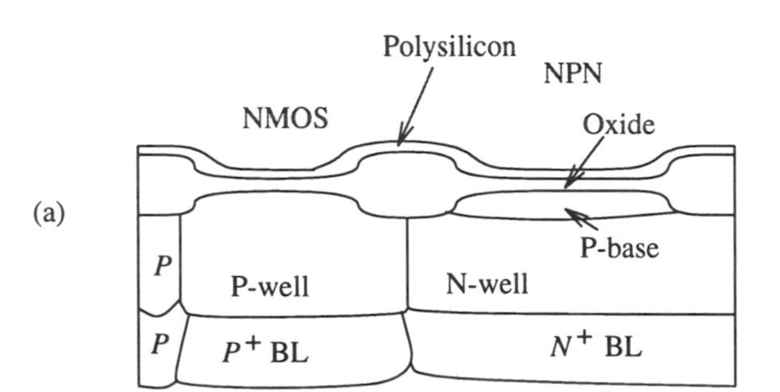

- Initial structure
- Grow thin gate oxide
- Deposit thin LPCVD poly (200 nm)

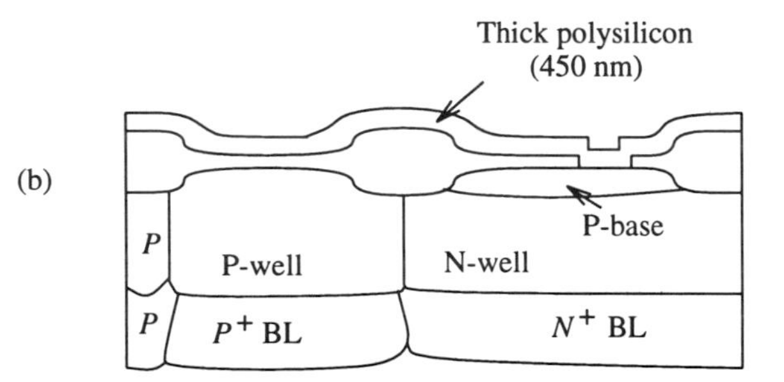

- Apply photoresist
- Pattern emitter
- Etch poly/oxide
- Strip resist
- Deposit LPCVD poly (250 nm) 2nd part of split poly

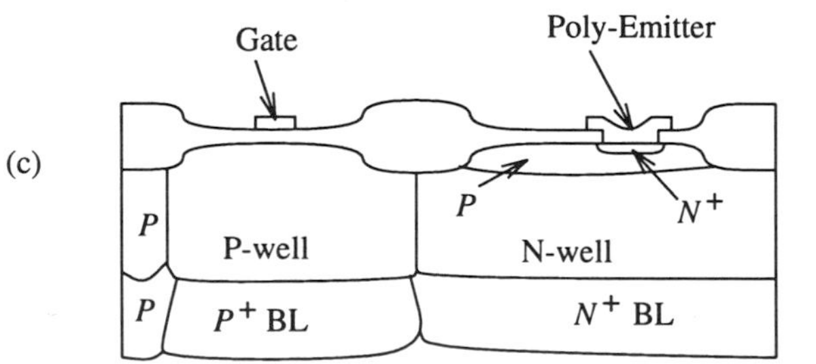

- Implant As/P
- Apply photoresist
- Pattern poly
- Dry etch poly
- Strip resist
- Anneal

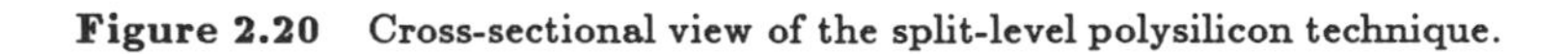

Figure 2.20 Cross-sectional view of the split-level polysilicon technique.

oxide is formed near the emitter and gate edges. Fig. 2.19(b) shows the final cross-section of this BiCMOS process.

The BJTs realized in the presented high-performance BiCMOS process have low collector resistance (because of the buried layer and deep sink), high current gain (because of the poly emitter contact) and low parasitic capacitances (because of the self-alignment). With this BiCMOS process f_t's greater than 5 GHz can be achieved.

BiCMOS technology has a relatively high cost and complexity, because it requires a total of 15 masks for sub-micron process. Several solutions have been proposed to reduce the number of process steps to lower process complexity and cost. Recently one idea [40] has resulted in low-cost 0.35 μm BiCMOS technology which needs only 11 masks by using W-plug trench collector sink. This technology is suitable for 3.3 V power supply voltage and promising for low-power mixed-signal applications.

Recently BiCMOS technologies with high NPN f_t's transistor, from 10-to-30 GHz, have been reported [39, 40, 41]. The applications of these technologies are, for example, for low-voltage (3 V and sub-3 V) and high-speed logic circuits. Another application of BiCMOS is mixed analog/digital ICs ranging from telecommunication circuits and high-speed networks to wireless systems. Among these applications, BiCMOS can be used for low-power high-frequency portable systems. Bipolar devices can be used for high-frequency and high-speed parts with low-power innovative circuits, and CMOS can be used for low-speed ultra-low-power parts.

2.6 COMPLEMENTARY BICMOS TECHNOLOGY

In a Complementary BiCMOS (CBiCMOS) process both vertical NPN and PNP transistors are merged with CMOS on the same chip. Recent investigations indicate that CBiCMOS allows for improving the performance of BiCMOS gates at low supply voltages [42, 43, 44]. Moreover for wireless applications, where high-speed and low-power characteristics are required, CBiCMOS technology is one of the solution. The added PNP device to conventional BiCMOS can be used to efficiently design low-voltage circuits. Further discussion on CBiCMOS circuits can be found in Section 5.3.2. Although, to date, the NPN has shown superior performance to that of PNP, future trend indicates that PNP performance is approaching that of NPN. Some of the problems associated with the PNP transistor are its high collector resistance, low current gain, and high base transit time.

It has been recently reported that CBiCMOS processes can offer NPNs with f_t's of 8-20 GHz and PNPs with 2-7 GHz f_t [45, 46, 47, 48, 49, 50]. Fig. 2.21 shows a cross-sectional view and process flow of a CBiCMOS [46]. The N^+ buried layer of the NPN transistor creates a retrograde well for the PMOS transistor. The P^+ buried layer is only used for isolation isles between NPN transistors. After the epitaxial layer growth, twin-well and LOCOS processes are performed. The P-well of the NMOS device is used as the collector of PNP transistor. A second high energy (600 keV) boron ion implantation is carried out to form the retrograde well (2nd P-well) for the NMOS and the P^+ buried layer for PNP device. The S/D implants of MOS transistors are used simultaneously for the extrinsic bases of the NPN and the PNP transistors. The emitters of the NPN and the PNP are formed by the self-aligned contact doping technique to simplify the process flow. Finally, the metal is deposited and patterned.

Complementary BiCMOS offers a technology with versatile devices. It adds flexibility for mixed bipolar/MOS circuit design. The CBiCMOS technology promises further improvements to BiCMOS circuits performance.

2.7 BICMOS DESIGN RULES

In this section, a set of lambda-based design rules of a typical BiCMOS process[5] (for 0.8 μm, $\lambda = 0.4\ \mu m$) is presented. The corresponding device parameters are presented in Chapter 3.

the minimum length of the MOS gate is 2λ and the minimum length and width of the bipolar emitter contact is 2λ and 4λ respectively. Table 2.2 describes the basic masks used in the layout design of BiCMOS devices. The rest of the masks are generated automatically.

Table 2.3 lists the design rules for the (design) masks only of a typical BiCMOS technology in terms of the parameter λ. The corresponding graphical representation of design rules is illustrated in Plate I. Plate II shows the layouts of minimum size PMOS, NMOS and bipolar transistors in a 0.8 μm BiCMOS technology.

[5] The given design rules are typical of a generic 0.8 μm high-performance BiCMOS process.

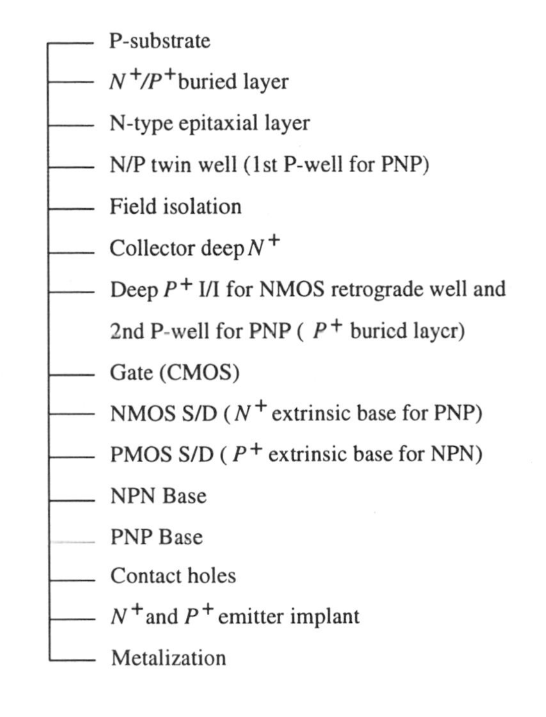

(a)

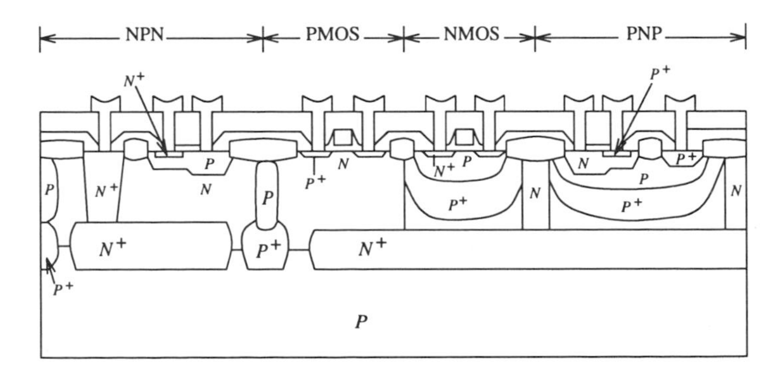

(b)

Figure 2.21 (a) Fabrication process flow; (b) Cross-sectional view of CBiCMOS [46].

Table 2.2 Basic BiCMOS Design Masks.

N-well (NW)	The NW mask is used to define the N substrate (bulk) of the PMOS and the N-collector of the NPN transistor.
N^+ deep collector (CN)	The CN mask defines the area which is exposed for the N^+ sink implantation.
P base (CP)	The CP mask defines the region which is to receive an P-implant to create the base diffusion.
Polysilicon (PO)	The PO mask defines the gate and the emitter electrodes, and the polysilicon interconnect layer.
Emitter window (EW)	The EW mask defines the opening for the emitter window.
N^+ and P^+ (DN and DP)	The DN (DP) mask defines the N^+ (P^+) source and drain regions of the N-channel (P-channel) device within the P-well (N-well), and the body contact regions in the N-well (P-well) respectively.
Contact (CO)	The CO mask defines the contact openings.
Metal 1 (M1)	The M1 mask defines the metal 1 interconnects.
Via (VIA)	The VIA mask defines the openings of the via that connects metal 1 to metal 2.
Metal 2 (M2)	The M2 mask defines the metal 2 interconnects.

Table 2.3 Design rules for a typical BiCMOS Process.

		Dimensions
1.	N-well (NW)	
	1.1 minimum width	12λ
	1.2 minimum spacing	12λ
2.	N^+ -diffusion (DN)	
	2.1 minimum width	3λ
	2.2 minimum spacing	3λ
	2.3 minimum NW overlap of DN	0λ
	2.4 minimum NW to external DN spacing	6λ
3.	P^+ -diffusion (DP)	
	3.1 minimum width	3λ
	3.2 minimum spacing	3λ
	3.3 minimum NW overlap of DP	4λ
	3.4 minimum NW to external DP spacing	4λ
	3.5 minimum space to DN (same potential)	0λ
	3.6 minimum space to DN (different potential)	3λ
4.	N-collector plug (CN)	
	4.1 minimum width	4λ
	4.2 minimum spacing	12λ
	4.3 minimum space to NW	10λ
	4.4 minimum NW overlap of CN	3λ
	4.5 minimum space to DN	6λ
	4.6 minimum space to DP	5λ
5.	P-base diffusion (CP)	
	5.1 minimum width	4λ
	5.2 minimum spacing	4λ
	5.3 minimum NW overlap of CP	3λ
	5.4 minimum space to CN	5λ
	5.5 minimum space to DN	3λ
	5.6 minimum space to DP	3λ

Table 2.3 *(continued)*

6.	Polysilicon (PO)	
	6.1 minimum width	2λ
	6.2 minimum spacing	3λ
	6.3 minimum space to DP or DN	2λ
	6.4 gate overhang of DP or DN	2λ
	6.5 minimum space to CN or CP	1λ
7.	Emitter window (EW)	
	7.1 minimum width	2λ
	7.2 minimum length	4λ
	7.3 minimum spacing	3λ
	7.4 minimum CP overlap of EW	2λ
	7.5 minimum poly overlap of EW	2λ
8.	Contact (CO)	
	8.1 minimum size (single)	$2\lambda \times 2\lambda$
	8.2 minimum size (double)	$2\lambda \times 4\lambda$
	8.3 minimum spacing	3λ
	8.4 minimum DN or DP overlap of CO	1λ
	8.5 minimum space to gate	2λ
	8.6 minimum PO overlap of CO	1λ
	8.7 minimum CN or CP overlap of CO	1λ
	8.8 minimum PO to CO spacing in P base	2λ
	8.9 minimum poly emitter CO to CP spacing	2λ
9.	Metal 1 (M1)	
	9.1 minimum width	2λ
	9.2 minimum spacing	3λ
	9.3 minimum M1 overlap of CO	1λ
	9.4 maximum current density	$1\ mA/\mu m$

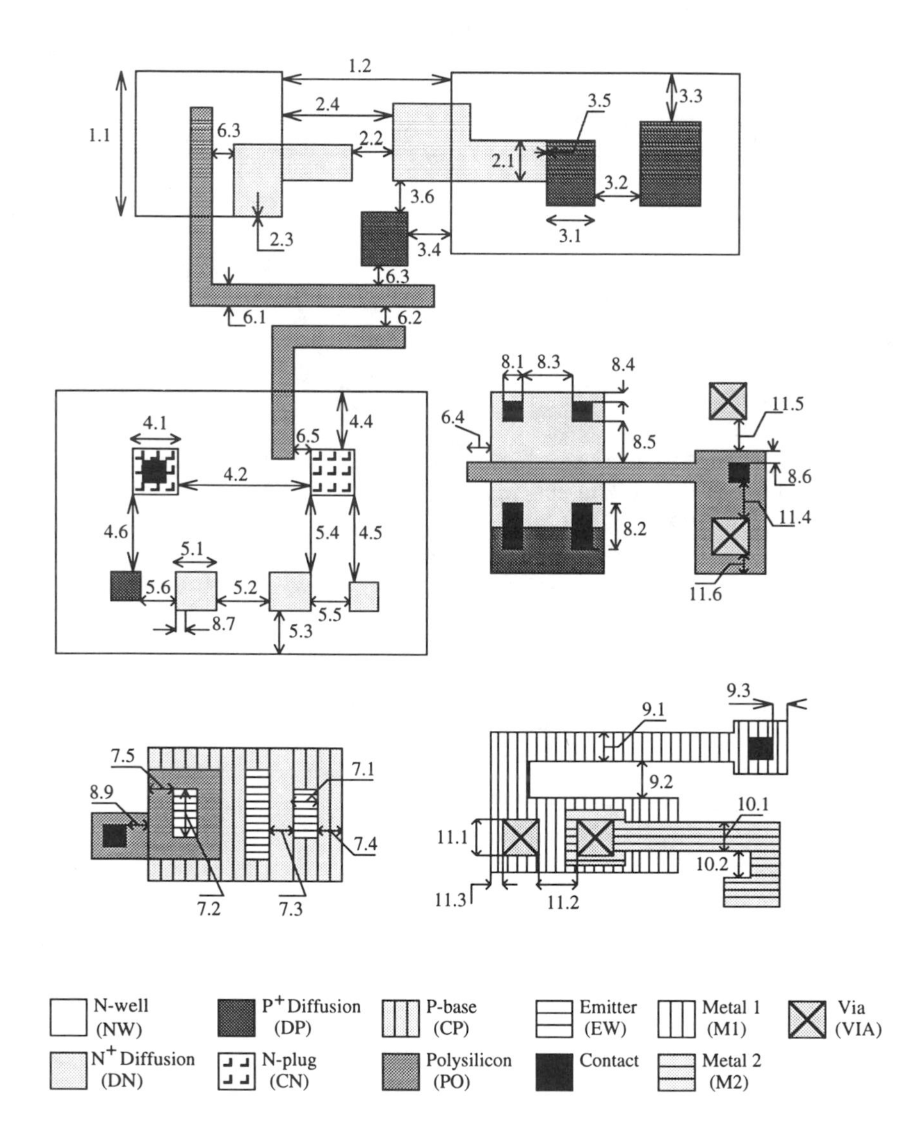

Plate I: Design Rules of Table 2.3.

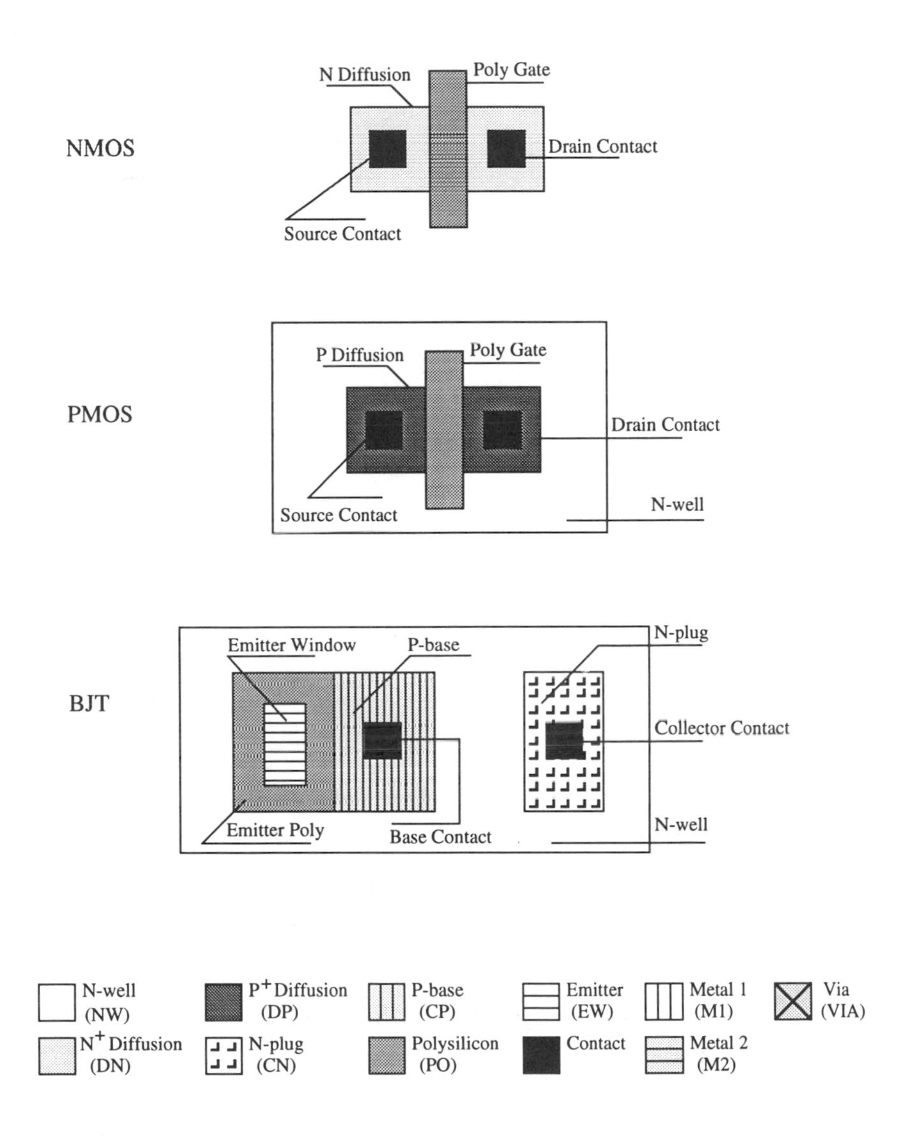

Plate II: Layouts of minimum size PMOS, NMOS and bipolar transistors.

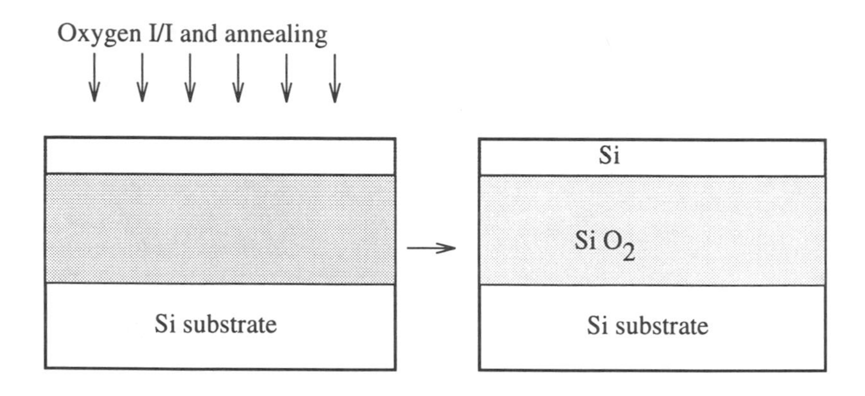

Figure 2.22 SIMOX material formation.

2.8 SILICON ON INSULATOR

Silicon On Insulator (SOI) has recently received renewed interest for low-voltage and low-power applications. This is due to the reduction of the cost and improvement of its performance at lower voltage. The emergence of thin-film SOI CMOS processes have demonstrated excellent characteristics for deep-submicron ULSI applications.

Many techniques existent to grow silicon on insulator [51]. The most mature technique is the epitaxial growth of Silicon On Sapphire (SOS). Many LSI/VLSI circuits have been fabricated using SOS technology. SOI can also be produced by using what is called SIMOX (Separation by IMplanted OXygen) [52] technology. It is fabricated simply by the formation of buried oxide (SiO_2) by implantation of oxygen underneath the surface of the silicon as illustrated in Fig. 2.22. Dose and energy of oxygen ions are as high as $2 \times 10^{18} cm^{-2}$ and 200 KeV respectively. A subsequent thermal annealing at high temperature is performed to improve the quality of the silicon overlayer. The buried oxide can be several hundreds of nm thick and the thin silicon layer can have several tens of nm thickness. Compared to SOS, SOI SIMOX materials have better defect density and thin silicon layer control. The dislocation density can be lower than $10^4 cm^{-2}$. One important phenomenon which exists in CMOS SOI devices is the kink effect. It consists of a "kink" which appears in the output characteristics of an SOI MOSFET, as illustrated in Fig. 2.23. It is due mainly to the floating substrate of an NMOS device. An explanation of this phenomena can be found in [51].

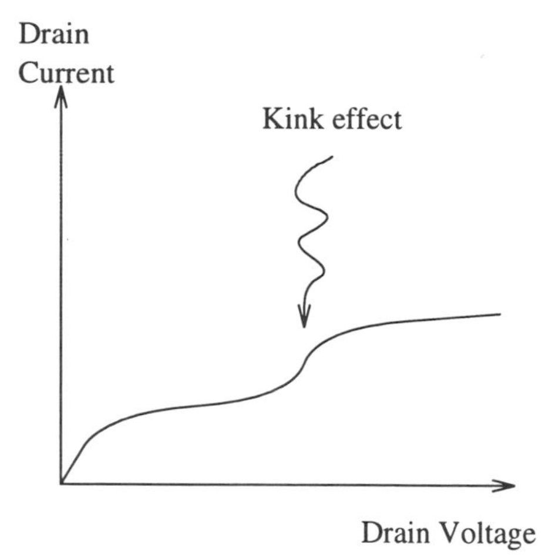

Figure 2.23 Kink effect in the output characteristic of an SOI MOS device.

The SOI SIMOX is now mature material and represents a potential technology for low-power applications. Several LSI/VLSI circuits have been fabricated in SOI/SIMOX, particularly for low-power application. Such circuits include PLL (Phase Locked Loop) for wireless terminals applications [54], and 1.2-GHz frequency divider under 1-V power supply [55]. The SOI technology was applied also to design a fully pipelined 512-Kb SRAM [53]. This SRAM worked successfully down to 0.7 V with an access time less than 5 ns.

Fig. 2.24 shows a thin film SOI/SIMOX CMOS process cross-section. The process starts by the formation of buried oxide in silicon wafer as explained above in [Fig. 2.24(a)]. Then, an oxide is grown on the surface silicon and a nitride layer is deposited. Silicon nitride is used as a mask to protect the active region from oxidation. The nitride/oxide layers are patterned and a LOCOS isolation is applied [Fig. 2.24(b)]. At the end, the nitride/oxide layers are removed. This is followed by P I/I to adjust the threshold voltage of the N-channel transistor. Similarly, the threshold voltage of the P-channel transistor is adjusted by I/I. A thin gate oxide is then grown and a layer of polysilicon is deposited and doped with phosphorus. Then the P^+ source and drain regions of the PMOS are patterned and implanted with boron [Fig. 2.24(c)]. Similarly, the N^+ S/D regions of the NMOS are patterned and implanted with phosphorus. A thick oxide is then deposited as an isolation layer between the polysilicon and the subsequent metal layer. The oxide is etched at contact locations. Next, the

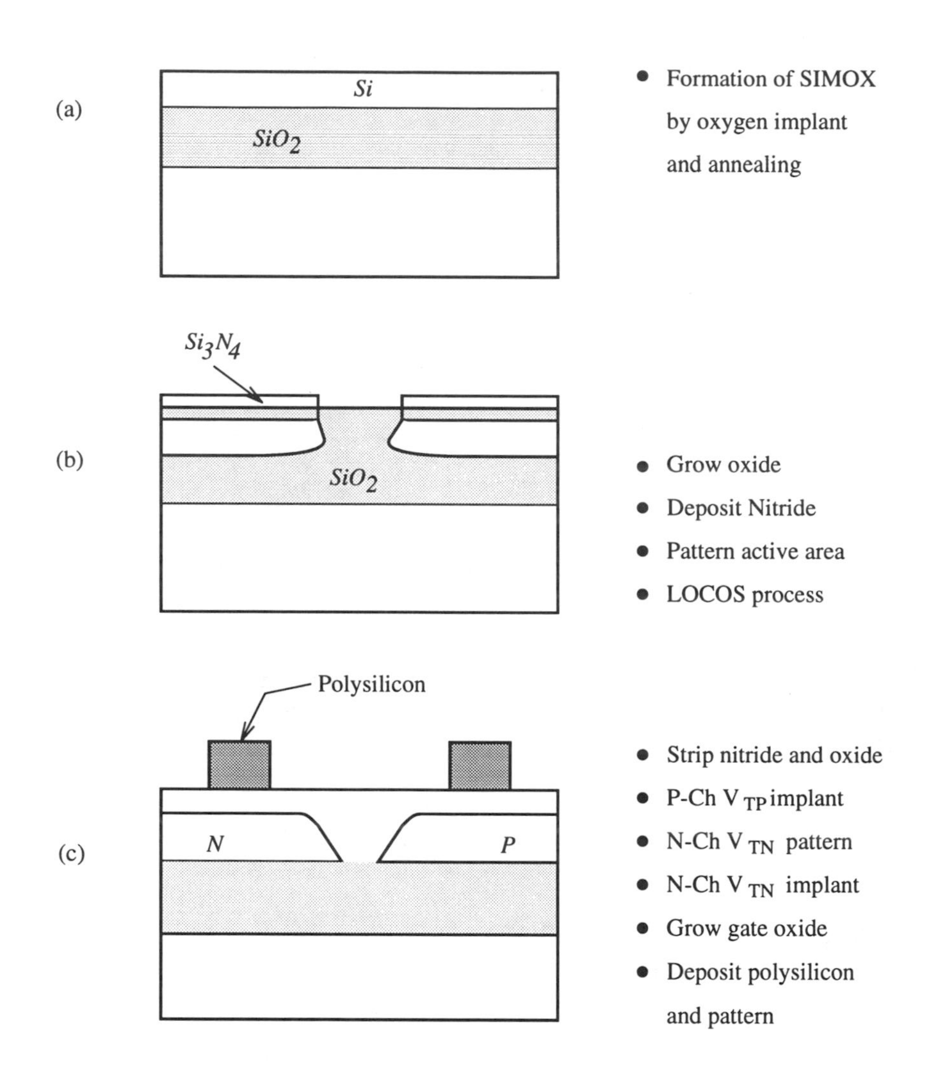

Figure 2.24 Main process steps of CMOS thin film SOI/SIMOX devices.

metal layer (aluminum) is deposited over the whole surface. Finally, the metal is etched and annealed.

This simple process description shows that the SOI process is much simpler than bulk CMOS. For instance, the wells are no longer needed, and the punchthrough implants are also unnecessary if thin-film SOI is used. Fig. 2.25 shows a

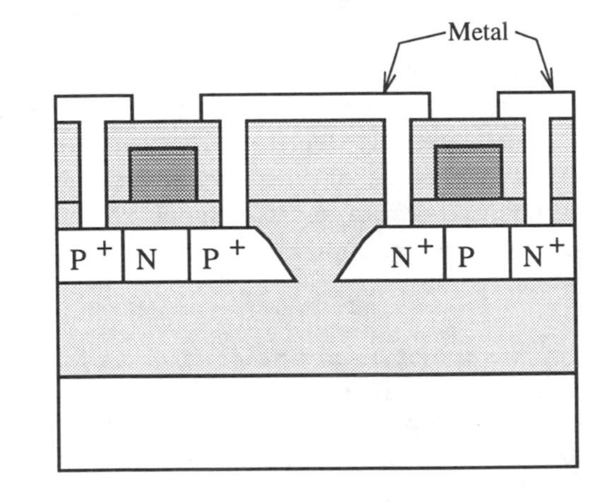

Figure 2.24 *(continued)*

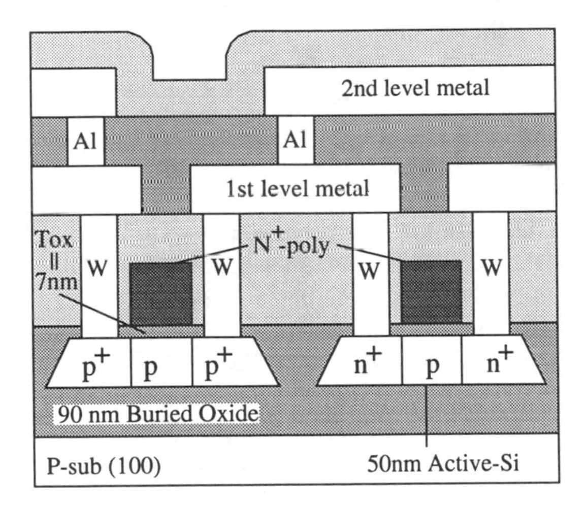

Figure 2.25 Cross section of an ultrathin-film 0.25 μm CMOS/SIMOX structure.

cross-section of deep-submicron (0.25 μm) CMOS/SIMOX devices [56] for low-voltage applications. This process uses advanced fabrication techniques such as a new LOCOS technique, gate spacer, and CVD W plugs, etc.

Due to the dielectric isolation, the MOS devices have several advantages over bulk CMOS such as : absence of latch-up, high packing density and lower parasitic capacitances. SOI reduces the circuit capacitance by 30% [57]. It has been discovered that if the silicon (containing the devices) is made sufficiently thin ($< 100nm$), the MOSFET's devices are fully depleted [51] even when $V_{GS} = 0$. Fully depleted thin film SOI MOS devices offer attractive characteristics for CMOS applications such as immunity from short channel effect, absence of kink effect, superior subthreshold leakage and high drain saturation current (due to low channel doping) [58, 59, 60].

Unfortunately, the technology has minor disadvantages such as floating body effects which result in i) floating body induced threshold voltage lowering and ii) low drain-to-source breakdown voltage. For 1 V power supply this is not a problem. However for 3 V operation this could be an important limitation. Also, the threshold voltage is very sensitive to the thickness uniformity of the superficial silicon. In addition, the low thermal conductivity of the oxide underneath the thin film silicon layer is a severe problem when the SOI circuit is operating at high-frequency. Therefore technological improvements are still needed to solve these limitations.

2.9 CHAPTER SUMMARY

In this chapter, we have studied the process technologies of CMOS and bipolar devices. We have shown that the advanced CMOS and bipolar processes are converging, and many process techniques can be shared for the fabrication of both devices. The different options for merging bipolar and CMOS devices are then discussed. Three examples for BiCMOS processes with different complexities are presented. The complementary BiCMOS process is also considered. A table of design rules for a state-of-the-art BiCMOS technology is given for layout exercises. Several advanced technologies such as CMOS SOI/SIMOX and CMOS-SJET are reviewed for low-voltage operation.

REFERENCES

[1] A F.M. Wanlass, and C.T. Sah, "Nanowatt Logic using Filed-Effect MOS Triodes," International Solid-State Circuits Conference Tech. Dig., pp.32-33, 1963.

[2] L.C. Parrillo, R.S. Payne, R.E. Davis, G.W. Reutlinger, and R.L. Field, "Twin-Tub CMOS: A Technology for VLSI Circuits," International Electron Devices Meeting Tech. Dig., pp. 752-755, December 1980.

[3] Y. Taur et al., "High-Performance 0.1 μm CMOS Devices with 1.5 V Power Supply," International Electron Devices Meeting Tech. Dig., pp. 127-130, December 1993.

[4] K. F. Lee et al., "Room Temperature 0.1 μm CMOS Technology with 11.8 ps Gate Delay", International Electron Devices Meeting Tech. Dig., pp. 131-134, December 1993.

[5] K. Takeuchi et al., "0.15 μm CMOS with High Reliability and Performance", International Electron Devices Meeting Tech. Dig., pp. 883-886, December 1993.

[6] T. Yamazaki, K. Goto, T. Fukano, Y. Nara, T. Sugii, and T. Ito, "21 ps Switching 0.1 μm-CMOS at Room Temperature using High Performance Co Salicide Process," International Electron Devices Meeting Tech. Dig., pp. 906-908, December 1993.

[7] H. Oyamatsu, K. Kinugawa, and M. Kakumu, "Design Methodology of Deep Submicron CMOS Devices for 1 V Operation," Symposium on VLSI Technology Tech. Dig., pp. 89-90, 1993.

[8] H. Yoshimura, F. Matsuoka, and M. Kakumu, "New CMOS Shallow Junction Well FET Structure (CMOS-SJET) for Low Power-Supply Voltage," International Electron Devices Meeting Tech. Dig., pp. 909-912, December 1992.

[9] T. Uchino, T. Shiba, T. Kikuchi, Y. Tamaki, A. Watanabe, Y. Kiyota, and M. Honda, "15-ps ECL/74-GHz f_t Bipolar Technology," International Electron Devices Meeting Tech. Dig., pp. 67-70, December 1993.

[10] T.H. Ning, and D.D. Tang, "Bipolar Trends," Proc. IEEE, vol. 74, no. 12, pp. 1669-1677, December 1986.

[11] T. Nakamura, T. Miyazaki, S. Takahashi, T. Kure, T. Okabe, and M. Nagata, "Self-Aligned Bipolar Transistor with Polysilicon Sidewall Base Electrode for High Packing Density and High Speed," IEEE Journal of Solid-State Circuits, vol. 17, no. 2, pp. 226-230, April 1982.

[12] T.H. Ning, and R. D. Isaac, "Effect of Emitter Contact on Current Gain of Silicon Bipolar Devices," IEEE Electron Device Letters, ED-27, pp. 2051-2055, November 1980.

[13] A.K. Kapoor and D.J. Roulston, "Polysilicon Emitter Bipolar Transistors," IEEE Press Book, 1989.

[14] M.I. Elmasry, "Digital Bipolar Integrated Circuits," John Wiley & Sons, New York, 1983.

[15] B. Razavi, Y. Ota and R. G. Swartz, "Design Techniques for Low-Voltage High-Speed Digital Bipolar Circuits," IEEE J. Solid-State Circuits, vol. 29, no. 3, pp. 332-339, March 1994.

[16] W. Wilhelm and P. Weger, "Low-Power Bipolar Logic," International Solid State Circuits Conf. Tech. Dig., pp. 94-95, February 1994.

[17] E. Kooi, J.G. Van Lierop, and J.A. Appels, "Formation of Silicon Nitride at a Si-SiO_2 Interface during Local Oxidation of Silicon and During Heat Treatment of Oxidized Silicon in NH_3 Gaz," J. Electrochem. Soc., vol. 123, p. 1117, 1976.

[18] R.D. Rung, H. Momose, and Y. Nagakubo, "Deep-Trench Isolated CMOS Devices," International Electron Devices Meeting Tech. Dig., pp. 6-9, December 1982.

[19] T. Yamaguchi, S. Morimoto, G. Kawamoto, H.K. Park, and G.C. Eiden, "High-Speed Latch-up Free 0.5 μm-Channel CMOS using Self-Aligned Ti-Si and Deep-Trench Isolation Technologies," International Electron Devices Meeting Tech. Dig., pp. 522-525, December 1983.

[20] R.D. Rung, "Trench Isolation Prospects for Application in CMOS VLSI," International Electron Devices Meeting Tech. Dig., pp. 574-577, December 1984.

[21] H. Mikoshiba, T. Homma, and K. Hamano, "A New Trench Isolation Technology as a Replacement for LOCOS," International Electron Devices Meeting Tech. Dig., pp. 578-581, December 1984.

[22] P. Singer, "Selective Epitaxial Growth Finds New Applications," Semiconductor International, p. 15, January 1988.

[23] R.A. Chapman, et al., "An 0.8 *mum* CMOS Technology for High-Performance Logic Applications," International Electron Devices Meeting Tech. Dig., pp. 362-365, December 1987.

[24] K.Y. Chiu, R. Fang, J. Lin, and J.L. Moll, "The SWAMI- A Defect Free and Near-Zero Bird's Beak Local Oxidation Technology for VLSI," Symp. on VLSI Technology Tech. Dig., pp. 28-29, 1982.

[25] K.Y. Chiu, J.L. Moll, and J. Manoliu, "A Bird's Beak Free Local Oxidation Technology Feasible for VLSI Circuits Fabrication," IEEE Trans. on Electron Devices, vol. ED-29, pp. 536-540, 1982.

[26] J. Hui, P. Vande Voorde and J. Moll, "Scaling Limitations of Submicron Local Oxidation Technology," International Electron Devices Meeting Tech. Dig., pp. 392-395, December 1985.

[27] H.B. Pogge, "Trench Isolation Technology," Bipolar Circuits and Technology Meeting Tech. Dig., pp. 18-25, September 1990.

[28] Y. Niitsu, "Latch-up Free CMOS Structure using Shallow Trench Isolation," International Electron Devices Meeting Tech. Dig., pp. 509-512, December 1985.

[29] H. Yamamoto, O. Mizuno, T. Kubota, M. Nakamae, H. Shiraki, and Y. Ikushima, "High-Speed Performance of a Basic ECL Gate with 1.25 Micron Design Rule," Symp. on VLSI Technology Tech. Dig., pp. 38-39, 1981.

[30] Y. Tamaki, T. Shiba, N. Honma, S. Mizuo, and A. Hayazaka, "New U-Groove Isolation Technology for High-Speed Bipolar Memory," Symp. VLSI Technology Tech. Dig., pp. 24-25, 1983.

[31] D.D. Tang, P.M. Solomon, T.H. Ning, R.D. Isaac, and R.E. Burger, "1.25 *mum* Deep-Groove-Isolated Self-Aligned Bipolar Circuits," IEEE Journal of Solid-State Circuits, vol. SC-17, pp. 925-931, 1982.

[32] H.C. Lin, J.C. Ho, R.R. Iyer, and K. Kwong, "CMOS-Bipolar Transistor Structure," IEEE Trans. Electron Devices, vol. ED-26, no. 11, pp. 945-951, November 1969.

[33] T. Ikeda, A. Watanabe, Y. Nishio, I. Masuda, N. Tamba, M. Okada, and K. Ogiue, "High-Speed BiCMOS Technology with a Buried Twin Well Structure," IEEE Trans. on Electron Devices, vol. ED-34, no. 6, pp. 1304-1309, June 1987.

[34] H. Momose, K.M. Cham, C.I. Drowley, H.R. Grinolds, and H.S. Fu, "0.5 Micron BiCMOS Technology," International Electron Devices Meeting Tech. Dig., pp. 838-840, December 1987.

[35] A.R. Alvarez, J. Teplik, D.W. Schucker, T. Hulseweh, H.B. Liang, M. Dydyk,and I. Rahim, "Second Generation BiCMOS Gate Array Technology," Bipolar Circuits and Technology Meeting Tech. Dig., pp. 113-117, 1987.

[36] B. Bastani, C. Lage, L. Wong, J. Small, R. Lahri, L. Bouknight, T. Bowman, J. Manoliu, and T. Tuntasood, "Advanced 1 Micron BiCMOS Technology for High Speed 256k SRAM's," Symp. on VLSI Technology Tech. Dig., pp. 41-42, 1987.

[37] T. Yamaguchi and T.H. Yuzuriha, "Process Integration and Device Performance of a Submicron BiCMOS with 16-GHz f_t Double Poly-Bipolar Devices," IEEE Trans. on Electron Devices, vol. 36, no. 5, pp. 890-896, May 1989.

[38] C. K. Lau, C-H Lin and D.L. Packwood, "Sub-micron BiCMOS Process Design for Manufacturing," Bipolar/BiCMOS Circuits and Technology Meeting Tech. Dig., pp. 76-83, 1992.

[39] C. H. Wang and J. Van Der Velden, "A Single-Poly BiCMOS Technology with a 30 GHz Bipolar f_t," Bipolar/BiCMOS Circuits and Technology Meeting Tech. Dig., pp. 234-237, October 1994.

[40] H. Yoshida, H. Suziki, Y. Kinoshita, K. Imai, T. Akimoto, K. Tokashiki, and T. Yamazaki, "Process Integration Technology for Low Process Complexity BiCMOS using Trench Collector Sink," Bipolar/BiCMOS Circuits and Technology Meeting Tech. Dig., pp. 230-233, October 1994.

[41] J. M. Sung et al., "BEST2- A High Performance Super-Aligned 3V/5V BiCMOS Technology, with Extremely Low Parasitics for Low-Power Mixed-Signal Applications," IEEE Custom Integrated Circuits Conf. Tech. Dig., pp. 15-18, May 1994.

[42] H.J. Shin, "Performance Comparison of Driver Configurations and Full-Swing Techniques for BiCMOS Logic Circuits," IEEE Journal of Solid-State Circuits, vol. 25, no.3, pp. 863-865, June 1990.

[43] S.H.K. Embabi, A. Bellaouar, M.I. Elmasry, and R.A. Hadaway, "New Full-Voltage-Swing BiCMOS Buffers," IEEE Journal of Solid-State Circuits, vol. SC-26, pp. 150-153, February 1991

[44] M. Hiraki, K. Yano, M. Minami, K. Sato, N. Matsuzaki, A. Watanabe, T. Nishida, K. Sasaki, and K. Seki, "A 1.5-V Full-Swing BiCMOS Logic Circuit," IEEE Journal of Solid-State Circuits, vol. 27, no. 11, pp. 1568-1574, November 1992.

[45] Y. Kobayashi, C. Yamaguchi, Y. Amemiya, and T. Sakai, "High Performance LSI Process Technology: SST CBiCMOS," International Electron Devices Meeting Tech. Dig., pp. 760-763, December 1988.

[46] K. Higashitani, H. Honda, K. Ueda, M. Hatanaka, and S. Nagao, "A Novel CBi-CMOS Technology by DIIP Process," Symp. on VLSI Technology Tech. Dig., pp. 77-78, 1990.

[47] T. Maeda, K. Ishimaru, and H. Momose, "Lower Submicron FCBiMOS (Fully Complementary BiMOS) Process with RTP and MeV Implanted 5GHz Vertical PNP Transistor," Symp. on VLSI Technology Tech. Dig., pp.79-80, 1990.

[48] W.R. Burger, C. Lage, B. Landau, M. DeLong, and J. Small, "An Advanced 0.8 Micron Complementary BiCMOS Technology for Ultra-High Speed Circuit Performance," Bipolar Circuits and Technology Meeting Tech. Dig., pp. 78-81, December 1990.

[49] S.W. Sun, et al., "A Fully Complementary BiCMOS Technology for Sub-Half-Micrometer Microprocessor Applications," IEEE Trans. Electron Devices, vol. 39, no. 12, pp. 2733-2739, December 1992.

[50] T. Ikeda, T. Nakashima, S. Kubo, H. Jouba, and M. Yamawaki, "A High Performance CBiCMOS with Novel Self-Aligned Vertical PNP," Bipolar/BiCMOS Circuits and Technology Meeting Tech. Dig., pp. 238-240, October 1994.

[51] J. P. Colinge, "SOI Technology: Materials to VLSI," Kluwer Academic Publishers, 1991.

[52] K. Izumi, M. Doken, and H. Ariyoshi, "CMOS Device Fabricated on Buried SiO_2 layers Formed by Oxygen Implanted into Silicon," Electron. Lett., vol. 14, pp. 593-594, 1978.

[53] G.G. Shahidi, T.H. Ning, R.H. Dennard and B. Davari, "SOI for Low-Voltage and High-Speed CMOS," International Conf. SSDM, Japan, pp. 265-267, 1994.

[54] Y. Kado, T. Ohno, M. Harada, K. Deguchi, and T. Tsuchiya, "Enhanced Performance of Multi-GHz PLL LSIs using Sub-1/4-micron Gate Ultrathin

Film CMOS/SIMOX Technology with Synchrotron X-ray Lithography", IEDM Tech. Digest, pp. 243-246, December 1993.

[55] M. Fujishima, K. Asada, Y. Omura and K. Izumi, "Low-Power 1/2 Frequency Dividers using 0.1-μm CMOS Circuits Built with Ultrathin SIMOX Substrates," IEEE Journal of Solid-State Circuits, vol. 28, no. 4, pp. 510-512, April 1993.

[56] T. Ohno, Y. Kado, M. Harada, and T. Tsuchiya, "A High-Performance Ultra-Thin Quarter-Micron CMOS/SIMOX Technology," IEEE Symposium on VLSI Technology Tech. Dig., pp. 25-26, 1993.

[57] Y. Yamaguchi, A. Ishibashi, M. Shimizu, T. Nishimura, K. Tsukamoto, K. Horie, and Y. Akasaka, "A High-Speed 0.6-μm 16K CMOS Gate Array on a Thin SIMOX Film," IEEE Trans. Electron Devices, vol. 40, no. 1, pp. 179-186, January 1993.

[58] J. P. Colinge, "Subthreshold Slope of Thin Film SOI MOSFET's," IEEE Trans. Electron Device Letters, pp.274-276, September 1988.

[59] J. C. Sturm, K. Tokunaga, and J. P. Colinge, "Increased Drain Saturation Current in Ultrathin SOI MOS Transistors," IEEE Electron Device Letters, vol. 9, no. 9, pp. 460-?, September 1988.

[60] Y. Omura, S. Nakashima, K. Izumi, and T. Ishii, "0.1-μm Gate Ultrathin Film CMOS Devices using SIMOX Substrate with 80-nm Thick Buried Oxide Layer," IEDM Tech. Dig., pp. 675-678, December 1991.

3

LOW-VOLTAGE DEVICE MODELING

The objective of this chapter is two-fold. It is intended to review the basics of the MOS transistor, which is a prerequisite for Chapters 4. to 7., and to introduce commonly used models of both MOS and bipolar devices [Sections 3.1, 3.2, and 3.5]. In this chapter we consider simple analytical models which can be used for circuit analysis and design of deep-submicrometer MOSFET's at low-voltage. Also, a simple model to compute the leakage current of MOSFET's is presented [Section 3.3]. The more sophisticated SPICE device models are also presented to allow the reader to appreciate the meaning of the model parameters as well as the capabilities and limitations of these models. The SPICE parameters for the 0.8 μm CMOS/BiCMOS process presented in Chapter 2 are included in this chapter for readers who are interested in designing and simulating low-voltage CMOS circuits as well as BiCMOS circuits. In Section 3.4, supply voltage scaling due to reliability and power dissipation issues is presented.

3.1 MOSFET STRUCTURE AND OPERATION

Fig. 3.1[1] shows cross-sections and views of an N-channel MOS transistor. By applying a positive voltage on the gate V_{GS}, a depletion layer is induced in the channel. Further increase in V_{GS} results in a surface inversion layer. The

[1] The effective channel width W_{eff} is equal to $W_{mask} - 2W_d - x_w$ and the effective channel length is equal to $L_{mask} - 2L_d - x_l$. W_{mask} and L_{mask} are the drawn channel width and length. $2W_d$ ($2L_d$) is the total reduction in the channel width (length) due to the lateral diffusion. x_w and x_l represent the effect of the processes of masking and etching on the channel width and length respectively.

surface charge of the semiconductor (Q_S $coul/cm^2$) is equal in magnitude to the charge of the gate electrode (Q_G $coul/cm^2$). Thus, we have

$$Q_S = -Q_G = -(V_{GS} - V_{FB} - \phi_s)C_{ox} \tag{3.1}$$

where V_{GS} is the gate-source voltage and ϕ_s is the semiconductor surface potential. C_{ox} is the gate oxide capacitance per unit area and is given by

$$C_{ox} = \frac{\epsilon_o}{t_{ox}} \tag{3.2}$$

where ϵ_o is the oxide permittivity and t_{ox} is the gate oxide thickness. The flatband voltage V_{FB} is given by

$$V_{FB} = \phi_{ms} - \frac{Q_0}{C_{ox}} \tag{3.3}$$

Q_0 is the total of all charges in the oxide and near the interface oxide/silicon. This charge is positive. The work function difference between the gate electrode and the semiconductor ϕ_{ms} depends on the type of the electrode and the doping concentration of the semiconductor. For an aluminum electrode, we have

$$\phi_{ms} = -0.61 + \phi_f \tag{3.4}$$

For N^+ polysilicon electrode, we have

$$\phi_{ms} = -0.55 + \phi_f \tag{3.5}$$

The *fermi potential* ϕ_f in Equations (4.4) and (4.5) is given by

$$\phi_{fp} = -V_t \ln(\frac{N_a}{n_i}) \qquad for\ P-type\ Si \tag{3.6}$$

$$\phi_{fn} = +V_t \ln(\frac{N_d}{n_i}) \qquad for\ N-type\ Si \tag{3.7}$$

where $V_t = KT/q$. The charge Q_S is the sum of the charge in the depletion layer Q_B and the inversion layer Q_I. Therefore;

$$V_{GS} = V_{FB} + \phi_s - \frac{Q_B + Q_I}{C_{ox}} \tag{3.8}$$

The bulk depletion charge (per unit area) consists of ionized acceptors (P-type substrate) or donors (N-type substrate). The depletion charge of a P-type bulk, with zero bias bulk-source voltage ($V_{BB} = 0$), is given by

$$Q_{B0} = -qN_aW_D \tag{3.9}$$

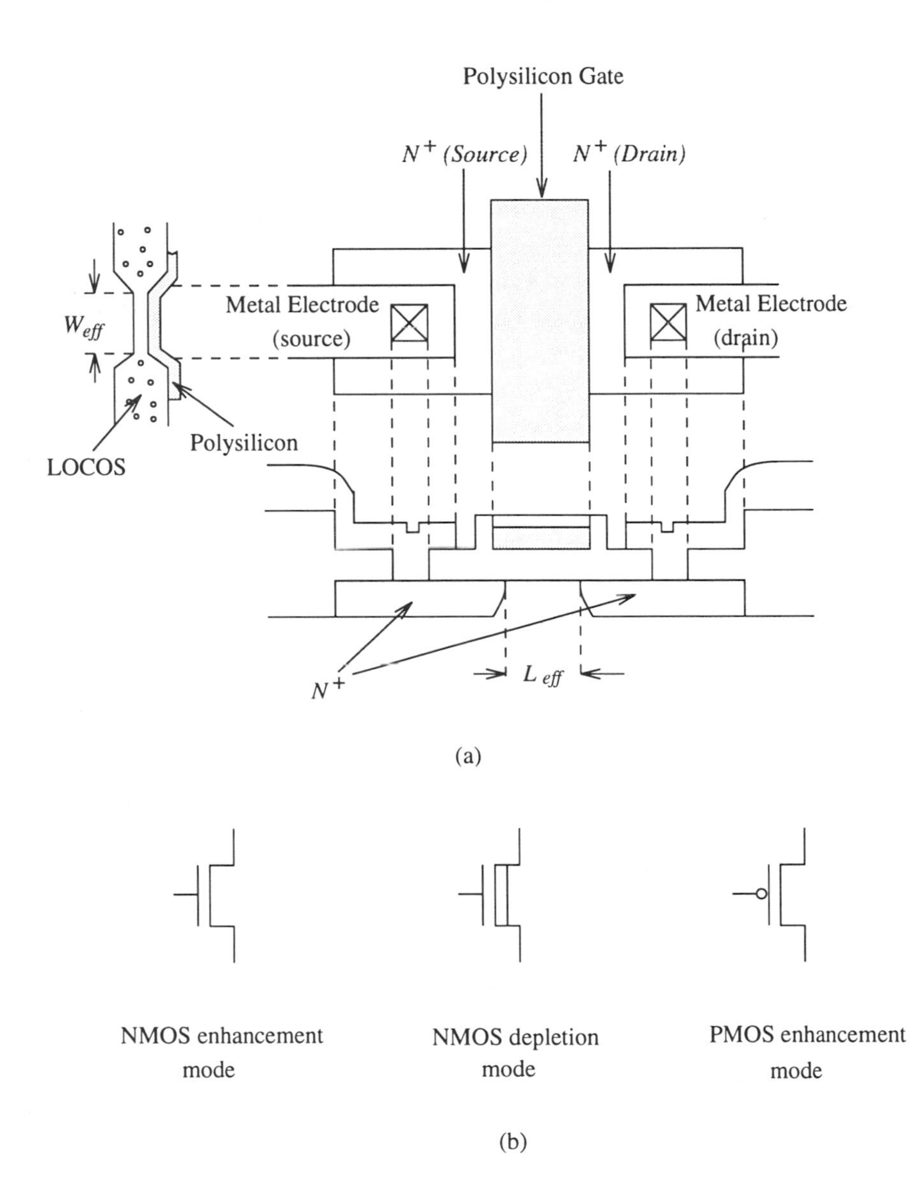

Figure 3.1 (a) The layout and cross-sectional views of an NMOS transistor; (b) Symbols of different types of MOS transistors.

where the q is the electron charge and N_a is the donor concentration. The width of the depletion layer in the bulk (W_D) is given by

$$W_D = \left\{2\frac{\epsilon_s}{qN_a}\phi_s\right\}^{1/2} \tag{3.10}$$

The turn-on (or threshold) voltage of an NMOS transistor is defined as the gate-source voltage at which the surface potential ϕ_s is equal to $2|\phi_f|$. This condition also defines what is known as the strong inversion[2]. At the onset of strong inversion we can assume that $Q_S \approx Q_B$. Using Equation (3.8), we can write the following expression of the threshold voltage

$$V_{TO} = V_{FB} + \phi_s - \frac{Q_{BO}}{C_{ox}} \tag{3.11}$$

Q_{BO} is equal to $-qN_aW_{Dm}$, where $W_{Dm} = W_D(\phi_s = 2|\phi_f|)$[3]. Thus, the threshold voltage can be rewritten as

$$V_{TO} = V_{FB} + 2|\phi_f| + \frac{\sqrt{2q\epsilon_s N_a(2|\phi_f|)}}{C_{ox}} \tag{3.12}$$

If the bulk-source is reverse biased ($|V_{BB}| > 0$), the threshold voltage becomes

$$V_T = V_{FB} + 2|\phi_f| + \frac{\sqrt{2q\epsilon_s N_a(|V_{BB}| + 2|\phi_f|)}}{C_{ox}} \tag{3.13}$$

This equation can be rewritten as

$$V_T = V_{TO} + \gamma(\sqrt{|V_{BB}| + 2|\phi_f|} - \sqrt{2|\phi_f|}) \tag{3.14}$$

where the body effect coefficient γ is given by

$$\gamma = \frac{\sqrt{2q\epsilon_s N_a}}{C_{ox}} \tag{3.15}$$

Consider an N-channel MOS transistor with $N_a = 2 \times 10^{16} cm^{-3}$, $t_{ox} = 10\ nm$, $V_{FB} = -1.04\ V$, then $2|\phi_f| = 0.735\ V$. Equation (3.12) gives $V_{TO} \approx -0.1\ V$.

[2] The surface is considered to be strongly inverted when the concentration of the inversion charge *at the surface* is equal or greater than the concentration of the ionized acceptors. Based on that definition it can be shown that at the onset of strong inversion $\phi_s = 2|\phi_f|$.

[3] The increase of the depletion region width W_D in the bulk is negligible beyond the onset of strong inversion. Hence, we can assume that the depletion region width reaches its maximum (W_{Dm}) at the onset of strong inversion. Therefore, under strong inversion conditions, $W_D = W_{Dm} = \sqrt{2\epsilon_s(2|\phi_f|)/qN_a}$.

This value is negative and is not suitable for digital circuits where a positive V_{TO} is required for switching. To get a reasonable V_{TO}, the device surface is implanted with boron. The implanted dose D_I causes V_{TO} to increase by the amount qD_I/C_{ox}. The threshold voltage is hence given by

$$V_{TO} = V_{FB} + 2|\phi_f| + \gamma\sqrt{2|\phi_f|} + q\frac{D_I}{C_{ox}} \tag{3.16}$$

Consider now the previous example, with $D_I = 1.725 \times 10^{12} cm^{-2}$ and $\gamma = 0.238\ V^{1/2}$ we find that V_T is equal to $0.7\ V$ when $|V_{BB}| = 0\ V$ and is equal to $0.98\ V$ when $|V_{BB}| = 3.3\ V$.

The symbols of the NMOS and PMOS transistors are shown in Fig. 3.1(c). Typical values of the V_T are $-2.5\ V$ to $-4\ V$ for depletion-mode NMOS devices. For low-voltage CMOS they are $0.3\ V$ to $0.8\ V$ for enhancement-mode NMOS devices, $-0.3\ V$ to $-0.8\ V$ for enhancement-mode PMOS devices.

When $V_{GS} < V_{TO}$, the transistor is in the *cutoff region*, since no inversion layer exists, as shown in Fig. 3.2(a). The drain current is, therefore, approximately zero. When $V_{GS} > V_{TO}$, the channel is formed and a drain current flows from the drain to the source [Fig. 3.2(b)]. The transistor is in the *linear region* (also called *ohmic region*) when V_{GD} (*i.e.* $V_{GS} - V_{DS}$) $\geq V_T$. When $V_{GS} > V_T$ and $V_{DS} > V_{GS} - V_T$ (i.e. $V_{GD} < V_T$) the channel is pinched off as illustrated in Fig. 3.2(c) and the device enters the *saturation region*. The drain-source voltage which causes the channel to pinchoff at the drain edge is commonly known as the saturation drain-source voltage V_{DSsat} and is equal to $V_{GS} - V_T$.

The voltage drop between the pinchoff point and the source is V_{DSsat}. Any V_{DS} higher than V_{DSsat} will appear between the pinchoff point and the drain. If we assume that the distance between the pinchoff point and the drain is extremely small compared with the overall length, then for $V_{DS} > V_{DSsat}$ the drain current is constant. The carriers which reach the pinchoff point are swept across to the drain by the potential $(V_{DS} - V_{DSsat})$ between the drain and the end of the channel.

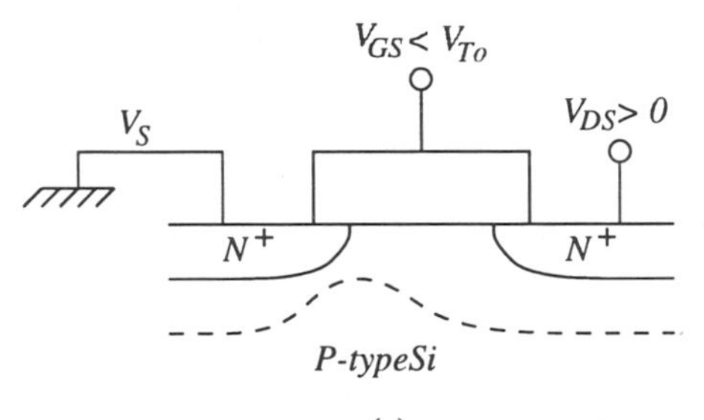

(a)

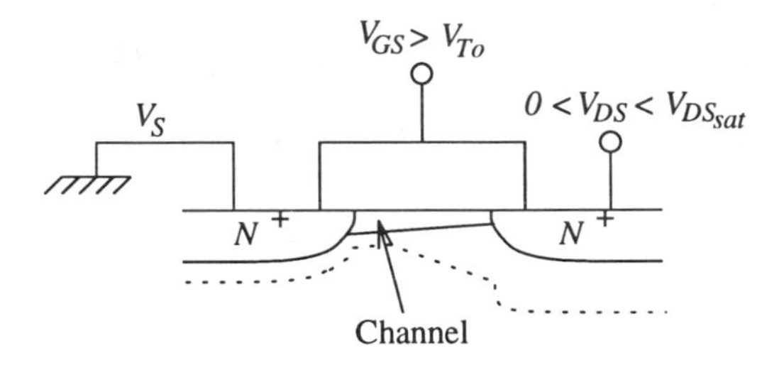

(b)

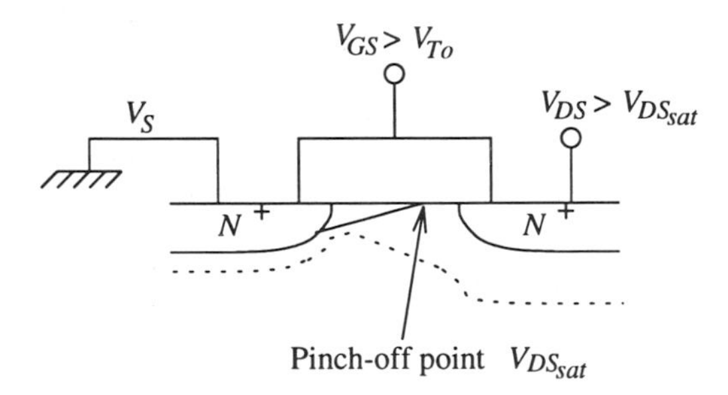

(c)

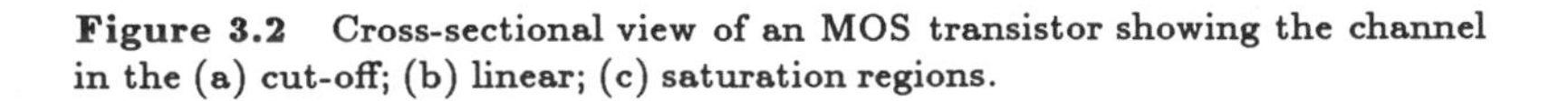
Figure 3.2 Cross-sectional view of an MOS transistor showing the channel in the (a) cut-off; (b) linear; (c) saturation regions.

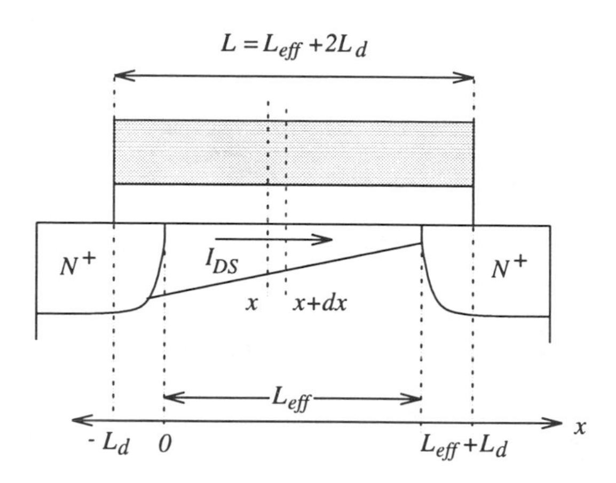

Figure 3.3 Cross-sectional view of an MOS transistor with the channel in the linear region.

3.2 SPICE MODELS OF THE MOS TRANSISTOR

3.2.1 The Simple MOS DC Model

Let us now analyze the simple DC model describing the I-V characteristics of an MOS transistor.

From Fig. 3.3 it can be shown that the element dx has a resistance

$$dR = -\frac{dx}{\mu W_{eff} Q_I(x)} \tag{3.17}$$

We assume that the mobility (μ) of the electrons in the channel of an NMOS device is constant. A current I_{DS} crossing the incremental resistance dR causes a voltage drop of

$$dV = I_{DS} dR \tag{3.18}$$

Substituting from Equation (3.17) in Equation (3.18) and integrating from the source to the drain, we obtain

$$I_{DS} \int_0^{L_{eff}} dx = -\mu W_{eff} \int_0^{V_{DS}} Q_I(x) dV \tag{3.19}$$

To solve this integration, we need to express the electron inversion charge density $Q_I(x)$ in terms of V. From Equation (3.8), we have

$$Q_I(x) = -\left[V_{GS} - V_{FB} + \frac{Q_{B0}}{C_{ox}} - \phi_s(x)\right] C_{ox} \qquad (3.20)$$

The surface potential ϕ_s at any point x along the channel is equal to $2|\phi_f| + V(x)$. By substituting for $V_{FB} - Q_{B0}/C_{ox} + 2|\phi_f|$ by V_{T0} [Equation (3.11)] in Equation (3.20) we get

$$Q_I(x) = -[V_{GS} - V_{T0} - V(x)]C_{ox} \qquad (3.21)$$

The surface potential at the drain is larger than that at the source by V_{DS}. Therefore, the magnitude of Q_I decreases with the distance across the channel. This is why the inversion layer is triangular as illustrated in Fig. 3.3. Assuming that Q_{B0} is constant across the channel and substituting for Q_I from Equation (3.21) into Equation (3.19), we obtain

$$I_{DS} = \mu \frac{W_{eff}}{L_{eff}} C_{ox} \int_0^{V_{DS}} [V_{GS} - V_{T0} - V(x)]dV \qquad (3.22)$$

or

$$I_{DS} = \mu \frac{W_{eff}}{L_{eff}} C_{ox}[(V_{GS} - V_{T0})V_{DS} - 1/2V_{DS}^2] \qquad (3.23)$$

or

$$I_{DS} = k_p \frac{W_{eff}}{L_{eff}}[(V_{GS} - V_{T0})V_{DS} - 1/2V_{DS}^2] \qquad (3.24)$$

where k_p is a process-dependent parameter defined as $k_p = \mu C_{ox}$. Equation (3.24) is valid only for $V_{DS} \leq V_{DSsat}$ (ohmic region). When V_{DS} exceeds V_{DSsat} the drain-source current saturates. The saturation current can be found by substituting for V_{DS} by V_{DSsat} in Equation (3.24) and is hence given by

$$I_{DS} = \frac{k_p}{2} \frac{W_{eff}}{L_{eff}}(V_{GS} - V_{T0})^2 \qquad (3.25)$$

The characteristics of an MOS transistor based on Equations (3.24) and (3.25) are shown in Fig. 3.4. The current equations (3.24) and (3.25) have to be modified if the bulk-source voltage is greater than zero by replacing V_{T0} by V_T [see Equation (3.14)]. Note that when V_{DS} is small (say 50 mV), Equation (3.24) can be approximated by

$$I_{DS} = k_p \frac{W_{eff}}{L_{eff}}(V_{GS} - V_{T0})V_{DS} \qquad (3.26)$$

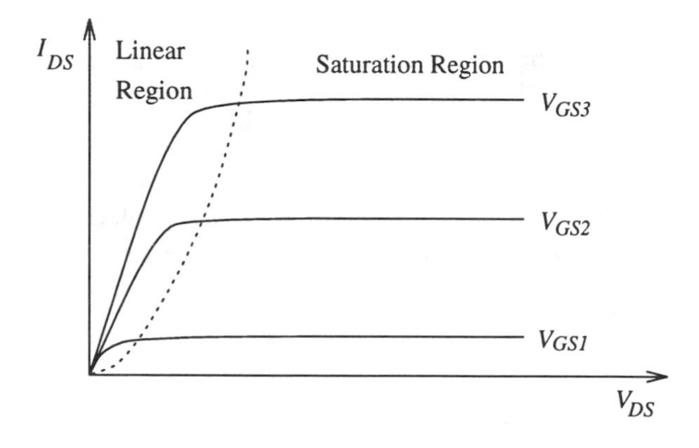

(a)

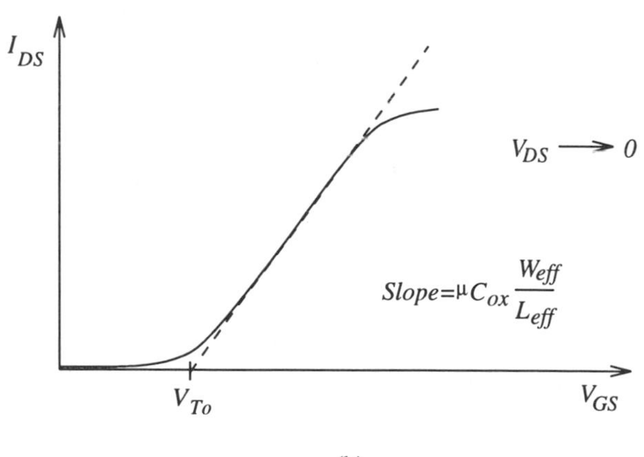

(b)

Figure 3.4 (a) Output characteristic of an MOS transistor (I_{DS} vs V_{DS}); (b) The transfer characteristics of an MOS transistor (I_{DS} vs V_{GS}).

This equation expresses a linear relationship between I_{DS} and V_{GS}. Using linear extrapolation, V_{T0} and $k_p \frac{W_{eff}}{L_{eff}}$ can be determined as shown in Fig. 3.4(b).

The measured I-V characteristics show that the drain current, in the saturation region, is a weak function of V_{DS}. This is due to the channel length modulation phenomenon which can be explained as follows. Let us define

$$L'_{eff} = L_{eff} - \Delta L \tag{3.27}$$

where ΔL is width of the depletion layer between the pinchoff point and the drain as shown in Fig. 3.5. The voltage across this depletion layer is $V_{DS} - V_{DSsat}$, therefore ΔL can be written as

$$\Delta L = \left\{ 2 \frac{\epsilon_s}{qN_a} (V_{DS} - V_{DSsat}) \right\}^{1/2} \tag{3.28}$$

The corrected saturation current becomes

$$I_{DS} = \frac{k_p}{2} \frac{W_{eff}}{L'_{eff}} (V_{GS} - V_T)^2 = \frac{k_p}{2} \frac{W_{eff}}{L_{eff}} (V_{GS} - V_T)^2 \frac{1}{\left[1 - \frac{\Delta L}{L_{eff}}\right]} \tag{3.29}$$

If we assume that $\frac{\Delta L}{L_{eff}} \ll 1$, then we can rewrite the current as

$$I_{DS} = \frac{k_p}{2} \frac{W_{eff}}{L_{eff}} (V_{GS} - V_T)^2 \left(1 + \frac{\Delta L}{L_{eff}}\right) \tag{3.30}$$

The ratio $\frac{\Delta L}{L_{eff}}$ can be related to V_{DS} by the following empirical relation

$$\frac{\Delta L}{L_{eff}} = \lambda V_{DS} \tag{3.31}$$

The channel modulation factor λ is very small. A typical value of λ is 0.01 V^{-1}.

The drain current model described, so far, is known as the LEVEL 1 (MOS1) model in SPICE[4]. This model is also called the Shichman-Hodges model. However, this model is still very simple[5] to account for state-of-the-art CMOS devices and might lead to a 100% error in the current particularly for low-voltage deep-submicrometer CMOS devices. However, k_p (or μ) can be used as a fitting parameter to reduce this error. This model is most suitable for preliminary analysis.

[4] SPICE2G6 or 3B1 or 3C1.

[5] This model was used in the 70's.

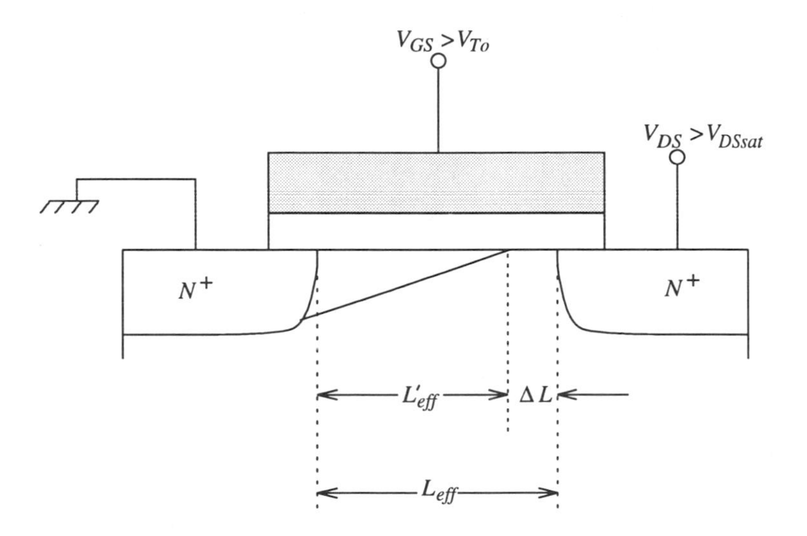

Figure 3.5 Channel length modulation phenomenon in an MOS transistor.

3.2.2 Semi-Empirical Short-Channel Model (LEVEL 3)

The MOS3 model (or MOS LEVEL 3) has been developed for short- and narrow- channel MOS ($L \leq 2\mu m$, $W \leq 2\mu m$) [1]. The MOS3 model has the following features (compared to MOS1):

- A model for mobility degradation with the vertical and the horizontal electric fields;
- A model for the threshold voltage of short- and narrow- channel devices (the (Drain Induced Barrier Lowering (DIBL) effect is accounted for);
- An improved model for the channel length modulation phenomenon;
- Weak inversion conduction (subthreshold conduction).

The threshold voltage expression is given by [1]

$$V_T = V_{FB} + 2|\phi_F| - \sigma V_{DS} + \gamma F_S \sqrt{2|\phi_F| + |V_{BB}|} + F_N(2|\phi_F| + |V_{BB}|) \quad (3.32)$$

γ in this expression is given by Equation (3.15). This expression includes:

- The static feedback effect coefficient σ (Due to DIBL effect) [2]

$$\sigma = \eta \frac{8.15 \times 10^{-22}}{C_{ox} L_{eff}^3} \tag{3.33}$$

where η is an empirical coefficient;

- The correction factor for short-channel effect is based on a modified trapezoidal approach for calculating the charge Q_B [Fig. 3.6]. The correction factor can be obtained from [3]

$$F_S = 1 - \frac{x_j}{L_{eff}} \left[\frac{L_d + W_c}{x_j} \left\{ 1 - \frac{W_D}{x_j + W_D} \right\}^{1/2} - \frac{L_d}{x_j} \right] \tag{3.34}$$

where W_c, the depletion layer width of a cylindrical junction and is given by

$$W_c = 0.0831353 + 0.8013929 \frac{W_D}{x_j} - 0.0111077 (\frac{W_D}{x_j})^2 \tag{3.35}$$

- The correction factor for narrow-channel MOS is given by

$$F_N = \pi \epsilon_s \frac{\delta}{4 C_{ox} W_{eff}} \tag{3.36}$$

3.2.2.1 Mobility degradation:

The mobility degradation due to the vertical electric field is modeled by the following simple equation [4]

$$\mu_s = \frac{\mu_0}{[1 + \theta (V_{GS} - V_T)]} \tag{3.37}$$

where θ is an empirical constant which depends on the oxide thickness. A typical value of θ is 0.05. To account for the effect of lateral average electric field, the effective mobility is related to the drain-source voltage and the channel length by [4]

$$\mu_{eff} = \frac{\mu_s}{1 + \mu_s \frac{V_{DS}}{v_{max} L_{eff}}} \tag{3.38}$$

In this expression, when the device operates in the saturation, V_{DS} is replaced by V_{DSsat}.

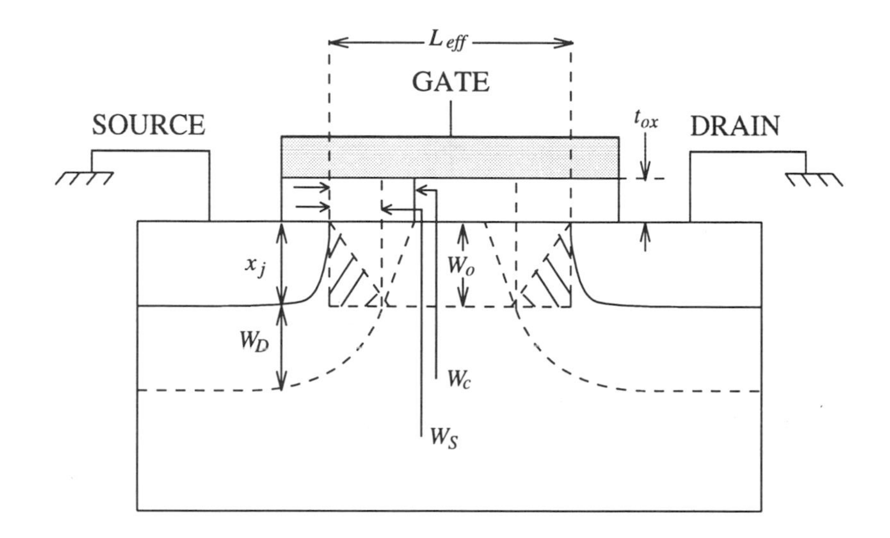

Figure 3.6 The model for the charge distribution in an MOS transistor showing that the bulk charge can be approximated by a trapezoid.

3.2.2.2 Channel length modulation

When $V_{DS} \geq V_{DSsat}$, the channel length is modulated by an amount ΔL. This channel length reduction is formulated in MOS3 by Baum's model [5]. In this model the voltage across the depletion surface of length ΔL is modeled by $\kappa(V_{DS} - V_{DSsat})$. κ is a fitting parameter.

3.2.2.3 Drain current

In the LEVEL 1 model of SPICE, the drain current in the weak inversion region was assumed zero. The modeling of the subthreshold current in LEVEL 3 is based on the analysis by Swanson and Meindl [6]. The drain current in weak inversion, which is basically a diffusion current, is given by

$$I_{DS} = I_{on} e^{[(V_{GS} - V_{on})/nV_t]} \tag{3.39}$$

where

$$V_{on} = V_T + nV_t \tag{3.40}$$

and

$$n = 1 + \frac{qN_{FS} + C_d}{C_{ox}} \tag{3.41}$$

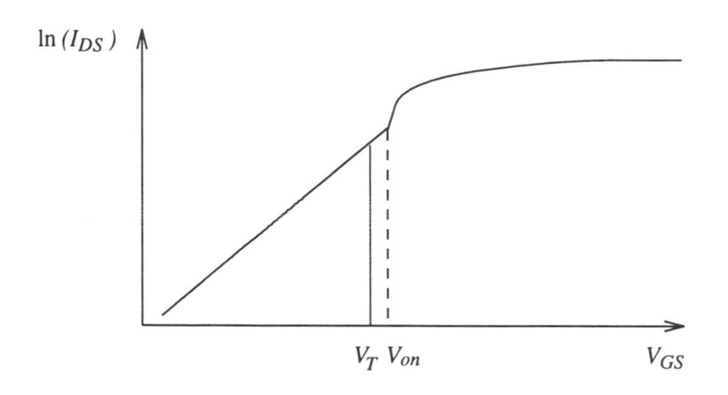

Figure 3.7 Subthreshold and drift current of an MOS transistor vs the gate-source voltage.

where

$$C_d = \frac{dQ_B}{dV_{BB}} \tag{3.42}$$

and N_{FS} is a curve fitting parameter. V_{on} marks the point between the weak and strong inversion modes. Typical values of n range from 1.0 to 2.5. I_{on} is related to the current of Equation (3.39) by taking $V_{GS} = V_{on}$.

Fig. 3.7 illustrates the transfer characteristics of the weak inversion and drift model. The voltage V_{on} insures the continuity of the current, but it is clear from the figure that at $V_{GS} = V_{on}$ a discontinuity exists in the derivative. Therefore, the MOS3 model is not precise in simulating the intermediate region where the diffusion and drift currents are comparable.

In the strong inversion, the drain current can be expressed as

$$I_{DS} = \mu_{eff} C_{ox} \frac{W_{eff}}{L_{eff}} \int_0^{V_{DS}} [V_{GS} - V_T(x)] dV \tag{3.43}$$

The threshold voltage along the channel is given by

$$V_T(x) = V_T + \gamma F_S(\sqrt{|2\phi_f| + |V_{BB}| + V(x)} - \sqrt{|2\phi_f| + |V_{BB}|}) + F_N V(x) \tag{3.44}$$

Using Taylor series expansion, we have

$$V_T(x) = V_T + (1 + F_B) V(x) \tag{3.45}$$

where

$$F_B = \frac{\gamma F_S}{4}(2|\phi_f| + |V_{BB}|)^{-1/2} + F_N \tag{3.46}$$

By substituting for V_T from Equation (3.45) in Equation (3.43), and integrating we obtain the following expression for the drain current

$$I_{DS} = \mu_{eff} C_{ox} W_{eff} L_{eff} \left[V_{GS} - V_T - \frac{1 + F_B}{2} V_{DS} \right] V_{DS} \tag{3.47}$$

The saturation voltage, which takes into account the carrier velocity saturation effect, is given by

$$V_{DSsat} = V_{sat} + V_c - \sqrt{V_{sat}^2 + V_c^2} \tag{3.48}$$

where

$$V_{sat} = (V_{GS} - V_T)/(1 + F_B) \tag{3.49}$$

$$V_c = \nu_{max} L_{eff} / \mu_s \tag{3.50}$$

Table 3.1 shows the CMOS device and HSPICE parameters correspondence. Typical values for parameters of LEVEL 3 are shown in Table 3.2 for MOS devices of the 0.8 μm BiCMOS process described in Chapter 2.

The LEVEL 3 model approximates the device physics and relies on the proper choice of the empirical parameters to accurately reproduce the device characteristics.

3.2.3 BSIM Model (LEVEL 4)

BSIM (Berkeley Short-Channel IGFET Model) is a simple and accurate short channel MOS transistor model [7]. It is implemented in SPICE as LEVEL 4. The model was tested for effective channel length down to 1 μm. This model includes:

- Vertical field dependence of carrier mobility;
- Carrier velocity saturation;
- Drain-induced barrier lowering effect;
- Non-uniform doping in the channel surface and sub-surface regions effect;

Table 3.1 CMOS device parameter and HSPICE correspondence.

Parameter	SPICE Keyword	Description
-	LEVEL	Model level
V_{T0}	VTO	Zero-bias threshold voltage
t_{ox}	TOX	Gate oxide thickness
N_a	NSUB	Substrate doping
N_{FS}	NFS	Surface fast state density
μ_0	UO	Surface mobility
v_{max}	VMAX	Maximum drift velocity of carriers
η	ETA	Static feedback on threshold voltage
κ	KAPPA	Saturation field factor
θ	THETA	Mobility degradation factor
δ	DELTA	Width effect on threshold voltage
x_j	XJ	Junction depth
C_j	CJ	Zero-bias bulk junction capacitance
J_s	JS	Bulk junction saturation current
J_{sw}	JSW	Sidewall bulk junction saturation current
M_j	MJ	Bulk junction grading coefficient
ϕ_j	PB	Junction potential
C_{jsw}	CJSW	Zero-bias side wall capacitance
M_{jsw}	MJSW	Sidewall capacitance grading coef.
C_{GD0}	CGDO	Gate-drain overlap capacitance
C_{GS0}	CGSO	Gate-source overlap capacitance
C_{GB0}	CGBO	Gate-bulk overlap capacitance
-	RD	Drain ohmic resistance
-	RS	Source ohmic resistance
L_d	LD	Lateral diffusion from drain or source
W_d	WD	Lateral diffusion along the width
x_l	XL	Masking and etching effects on W
x_w	XW	Masking and etching effects on L
-	ACM	Area calculation method
-	LDIF	Lateral diffusion beyond the gate

Table 3.2 HSPICE MOSFET model parameters (LEVEL=3) (0.8 μm BiCMOS process).

SPICE Keyword	N_Channel	P_Channel	Units
LEVEL	3	3	-
VTO	0.8	-0.9	V
TOX	17.5×10^{-9}	17.5×10^{-9}	m
NSUB	3.23×10^{16}	3.37×10^{16}	cm^{-3}
NFS	820×10^{9}	764×10^{9}	$cm^{-2}.V^{-1}$
UO	503	165	$cm^2/(V.s)$
VMAX	150×10^{3}	190×10^{3}	m/s
ETA	45×10^{-3}	121×10^{-3}	-
KAPPA	6.7×10^{-3}	1.45	$1/V$
THETA	63.4×10^{-3}	135×10^{-3}	$1/V$
DELTA	0.728	0.336	-
XJ	275×10^{-9}	230×10^{-9}	m
CJ	250×10^{-6}	450×10^{-6}	F/m^2
JS	5×10^{-4}	5×10^{-4}	A/m^2
JSW	5.5×10^{-9}	5.5×10^{-9}	A/m
MJ	0.50	0.50	-
PB	0.92	0.92	V
CJSW	205×10^{-12}	212×10^{-12}	F/m
MJSW	0.30	0.30	-
CGDO	274×10^{-12}	215×10^{-12}	F/m
CGSO	274×10^{-12}	215×10^{-12}	F/m
CGBO	571×10^{-12}	571×10^{-12}	F/m
RD	596	1189	Ω
RS	596	1189	Ω
LD	59.5×10^{-9}	0.	m
WD	0.	0.	m
XL	0.	0.	m
XW	0.	0.	m
ACM	2	2	-
LDIF	940×10^{-9}	1×10^{-6}	m

- Depletion charge sharing by the drain and source;
- Channel-length modulation;
- Dependence of some electrical parameters on drain and substrate biases;
- Better modeling of weak-, medium-, and strong- inversion regions and elimination of the discontinuity problem in the drain-current; and
- Geometric dependencies;

3.2.3.1 Threshold voltage:

The threshold voltage is given by

$$V_T = V_{FB} + \phi_s + K_1\sqrt{\phi_s + |V_{BB}|} - K_2(\phi_s + |V_{BB}|) - \eta V_{DS} \quad (3.51)$$

The two parameters, K_1 and K_2, model the effect of non-uniform doping of the substrate on the threshold voltage. Typical values for K_1 and K_2 are 1 $V^{1/2}$ and 0.12 respectively. The factor η models the DIBL effect and accounts for the channel-length modulation effect. It is a function of V_{DS} and V_{BB}.

3.2.3.2 Drain current:

When $V_{DS} \leq V_{DSsat}$ we have

$$I_{DS} = \frac{\mu_0}{1 + U_0(V_{GS} - V_T)} \frac{\frac{C_{ox}W_{eff}}{L_{eff}}}{(1 + \frac{U_1}{L_{eff}}V_{DS})} \left((V_{GS} - V_T)V_{DS} - \frac{a}{2}V_{DS}^2\right) \quad (3.52)$$

where

$$a = 1 + g\frac{K_1}{2}(\phi_s + |V_{BB}|)^{-1/2} \quad (3.53)$$

and

$$g = 1 - \frac{1}{1.744 + 0.836(\phi_s + |V_{BB}|)} \quad (3.54)$$

The parameters $U_0 = U_0(V_B)$, $U_1 = U_1(V_B)$ and $\mu_0 = \mu_0(V_{DS}, V_B)$ are bias sensitive. For $V_{DS} > V_{DSsat}$, the drain current is given by

$$I_{DS} = \frac{\mu_0}{1 + U_0(V_{GS} - V_T)} C_{ox} \frac{W_{eff}}{L_{eff}} \frac{(V_{GS} - V_T)^2}{2aK'} \quad (3.55)$$

where

$$K' = \frac{1 + v_c + \sqrt{1 + 2v_c}}{2} \tag{3.56}$$

and

$$v_c = \frac{U_1}{L_{eff}} \frac{(V_{GS} - V_T)}{a} \tag{3.57}$$

The drain-source saturation voltage is given by

$$V_{DSsat} = \frac{V_{GS} - V_T}{a\sqrt{K'}} \tag{3.58}$$

"a" is called body-effect coefficient.

3.2.3.3 Subthreshold current:

In BSIM, the total drain current is modeled as the linear sum of a strong-inversion component and a weak-inversion component I_w. I_w is expressed as

$$I_w = \frac{I_{exp}\, I_{lim}}{I_{exp} + I_{lim}} \tag{3.59}$$

where

$$I_{exp} = \mu_0 C_{ox} \frac{W_{eff}}{L_{eff}} V_t^2 e^{1.8} e^{(V_{GS}-V_T)/nV_t} \left(1 - e^{-V_{DS}/V_t}\right) \tag{3.60}$$

and

$$I_{lim} = \mu_0 \frac{C_{ox}}{2} \frac{W_{eff}}{L_{eff}} (3V_t)^2 \tag{3.61}$$

The factor $e^{1.8}$ is empirical to achieve the best fit. The subthreshold parameter n is a function of V_{DS} and V_B.

3.2.3.4 Sensitivity Factors of Model Parameters:

BSIM uses the following formula to account for the sensitivity of each parameter to the width and length of the channel

$$P = P_0 + \frac{LP_0}{L_{eff}} + \frac{WP_0}{W_{eff}} \tag{3.62}$$

where P_0 is an arbitrary parameter, LP_0 and WP_0 are the L and W sensitivity factors of P_0.

Another deep-submicrometer MOSFET's model called BSIM3 [8] has been developed for circuit simulation. It uses an improved threshold voltage, drain current and channel-length modulation models. The model is also simple and has a small number of parameters (≈ 25).

3.2.4 MOS Capacitances

In transient simulation, MOS capacitances are very important for CMOS and BiCMOS circuits analysis. The MOS capacitances can be divided into two types of lumped capacitors:

- the depletion capacitors of the bulk-drain and bulk-source pn junctions (C_{BD} and C_{BS}) [Fig. 3.8].
- the capacitors associated with the gate (C_{GS}, C_{GD}, C_{GB}, C_{GSov}, C_{GDov} and C_{GBov}) [see Fig. 3.8, except for C_{GBov}].

3.2.4.1 Junction Depletion Capacitances

The bulk-source and the bulk-drain junctions have a bottom area A_S and A_D respectively and a sidewall with a perimeter P_S and P_D respectively. Each of the bottom area and the sidewall contributes to the total depletion capacitance. The bottom area capacitance is measured per unit area, while the sidewall capacitance is measured per unit perimeter. Both of these components are voltage dependent. As these junctions are normally reverse biased, we will consider the case when the bulk-source and bulk-drain voltages (V_{BS} and V_{BD}) are less than or equal to $0.5\phi_j$ (ϕ_j is the junction built-in potential).

The total bulk-source and bulk-drain capacitances can be expressed by the following relations [1]

$$C_{BD} = \frac{C_j A_D}{(1 - \frac{V_{BD}}{\phi_j})^{M_j}} + \frac{C_{jsw} P_D}{(1 - \frac{V_{BD}}{\phi_j})^{M_{jsw}}} \tag{3.63}$$

$$C_{BS} = \frac{C_j A_S}{(1 - \frac{V_{BS}}{\phi_j})^{M_j}} + \frac{C_{jsw} P_S}{(1 - \frac{V_{BS}}{\phi_j})^{M_{jsw}}} \tag{3.64}$$

The exponential factors M_j and M_{jsw} are in the order of 0.3-0.5. C_j is the zero-bias capacitance of the bottom junction per unit area and C_{jsw} is the zero-bias capacitance per unit perimeter.

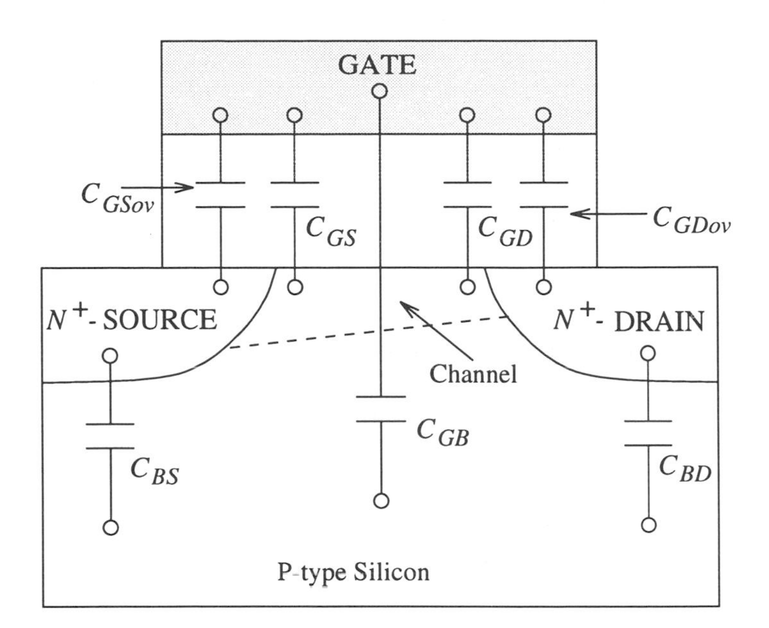

Figure 3.8 MOSFET capacitance model.

3.2.4.2 Gate Capacitances

The gate capacitances can be divided into two categories:

- *The fixed overlap capacitances*: gate-drain (C_{GDov}), gate-source (C_{GSov}), and gate-bulk (C_{GBov}) overlap capacitances. Both C_{GSov} and C_{GDov} exist due to the lateral diffusion of the source and drain under the gate. They are usually given per unit width as C_{GS0} and C_{GD0}. The total gate-source and gate-drain overlap capacitance is given by:

$$C_{GSov} = C_{GS0}\, W_{eff} \tag{3.65}$$

$$C_{GDov} = C_{GD0}\, W_{eff} \tag{3.66}$$

where C_{GS0} and C_{GD0} are equal to $C_{ox} L_d$. The capacitor C_{GBov} is due to the overlap of the gate oxide and the bulk along the channel length at both ends of the active area of the transistor. This capacitance is typically normalized to the effective channel length, the total C_{GBov} is hence given by

$$C_{GBov} = C_{GB0}\, L_{eff} \tag{3.67}$$

where C_{GB0} is equal to $C_{ox}W_d$.

- *The nonlinear capacitance due to the charge of the bulk or the channel.* This capacitance is actually distributed but can be modeled by lumped capacitances. In the case when the channel does not exist the capacitance can be expressed as

$$C_{GB} = C_{ox}W_{eff}L_{eff} \tag{3.68}$$

When the device is in the linear region the channel is extending uniformly from the source to the drain. The channel shields the bulk and the capacitance exists only between the gate and the channel. The gate-bulk capacitance goes to zero. The gate-channel capacitance can be expressed in terms of two equal lumped capacitances, a gate-source and a gate-drain capacitance, which are denoted C_{GS} and C_{GD} and are given by

$$C_{GS} = C_{GD} = \frac{1}{2}C_{ox}W_{eff}L_{eff} \tag{3.69}$$

Finally, when the device enters saturation, the channel at the drain pinches off and hence the gate-drain capacitance component becomes zero while the gate-source capacitance can be expressed by

$$C_{GS} = \frac{2}{3}C_{ox}W_{eff}L_{eff} \tag{3.70}$$

Fig. 3.9 depicts the change of the capacitance components as a function of the gate-source voltage (assuming that the source-bulk voltage is zero). The total gate-source capacitance is given by the summation of the C_{GSov} and C_{GS}, and similarly, the total gate-drain capacitance is given by the summation of C_{GDov} and C_{GD}.

The above described capacitance model can be used for circuit analysis and circuit design. SPICE uses a charge-control model, which was developed by Ward and Dutton [9]. This model is based on the actual distribution of charge in the MOS structure and its conservation.

3.3 CMOS LOW-VOLTAGE ANALYTICAL MODEL

The MOS models discussed previously have been developed for circuit simulators. These models (e.g. BSIM) involve large numbers of parameters whose values must be derived from device measurements. With these models it is difficult to develop an intuitive understanding of the device behavior. Therefore,

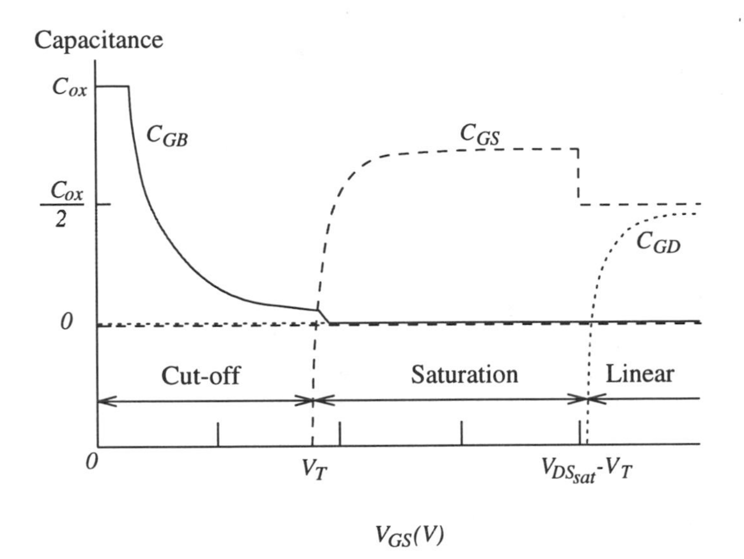

Figure 3.9 MOSFET capacitances as function of gate to source voltage.

an analytical drain current model valid for submicrometer MOSFETs operating at low-voltage is needed for hand calculation and first order circuit analysis, with reasonable accuracy.

3.3.1 Threshold Voltage Definitions

The threshold voltage, V_T, has some definitions which are important for the estimation of the static power dissipation. The first definition is the extrapolated threshold voltage from the characteristic $I_{DS} - V_{GS}$ [see Section 3.2.1]. Another one is the constant-current (i.e., @10 nA per width unit) threshold voltage. These voltages do not have the same value [10, 11]. The extrapolated V_T has approximately 0.2 V more than the constant-current one [11]. The extrapolated threshold voltage should be scaled down proportionally to the supply voltage. This is because the drive (saturation) current depends on $(V_{DD} - V_T(extrapolated))$.

3.3.2 Subthreshold Current

When the threshold voltage is scaled for low power supply voltage operation, subthreshold current increases significantly. This current is a limiting factor for battery operated circuits. As shown in Fig. 3.10, the drain current in the subthreshold region can be modeled by

$$I_{DSsub} = \frac{W_{eff}}{W_o} I_o 10^{(V_{GS}-V_T)/S} \tag{3.71}$$

where V_T here is the constant-current threshold voltage. I_o and W_o are the drain current and the gate width to define V_T. S is the subthreshold swing parameter, which is the gate voltage swing required to reduce the drain current by one decade. The current I_o is related to V_{DS} by

$$I_o = I_o'(1 - e^{V_{DS}/V_t}) \tag{3.72}$$

The subthreshold swing is given by [12]

$$S \approx 2.3V_t \left(1 + \frac{C_d}{C_{ox}}\right) \quad V/decade \tag{3.73}$$

where C_d is the depletion-layer capacitance of the source/drain junctions. Thus, S has a theoretical minimum limit which is 60 $mV/decade$.

The leakage current, due to the subthreshold conduction, is computed from I_{DSsub} when $V_{GS} = 0$. Then

$$I_{leak} = \frac{W_{eff}}{W_o} I_o 10^{-V_T/S} \tag{3.74}$$

Using the examples of Fig. 3.10, typical values for constant-current and extrapolated threshold voltages are 0.3 V and 0.5 V respectively. The parameter S is equal to 75 $mV/decade$ and the leakage current is equal to 1 $pA/\mu m$.

When estimating the static power dissipation, the worst-case leakage current has to be evaluated. In this case, the worst case threshold voltage, V_{Twc} has to be used where

$$V_{Twc} = V_T - \Delta V_T \tag{3.75}$$

ΔV_T is the variation of the threshold voltage due to the process parameters fluctuation such as the oxide thickness, doping profile, junction depth, gate and width lengths, etc. ΔV_T can be as high as 50 mV on the same wafer and 150 mV for different wafers. This results in almost two decades of leakage

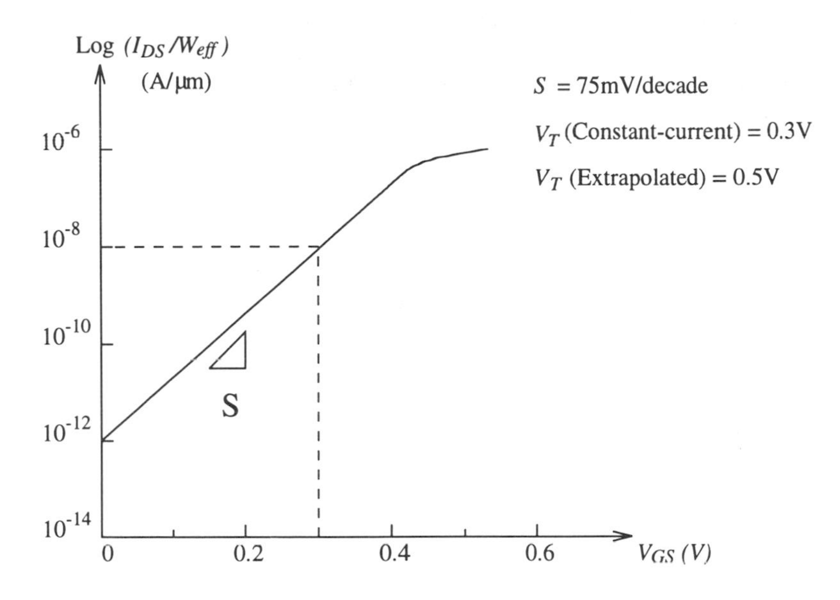

Figure 3.10 Subthreshold current of NMOS at room temperature.

current increase. Also the temperature effect has to be considered when leakage current is computed. The temperature affects both V_T and S. A typical value of the temperature coefficient of the threshold voltage is 1.6 mV decrease per degree Celsius. The subthreshold swing, S increases by 0.25 $mV/(decade.C)$ [See Equation 3.73]. For example, if the temperature increases from 25 C to 75 C, the threshold voltage decreases by 80 mV and the leakage current equals 30 $pA/\mu m$ (initial extrapolated $V_T = 0.5\ V$). This value is 30 times higher than that at 25 C. Both the temperature and process effects can result in a drastic increase of the worst-case static power dissipation. Note that this variation of V_T greatly affects the delay of CMOS circuits at low supply voltage, since the drive current is proportional to $(V_{DD} - V_T)$.

3.3.3 Low-Voltage Drain Current

A part of this model is based on the one proposed by [13]. For long-channel devices, the carrier drift velocity v is related to the horizontal electric field E by a simple linear relation ($v = \mu E$) where the carrier mobility is constant. For short-channel devices, the mobility is no longer a constant and is a function of the vertical electric field in the inversion layer. At this point we prefer to use the symbol μ_v for the mobility to denote its dependence on the vertical electric field. Also, the velocity (v) is no longer proportional to E but is given by the following two-region piecewise empirical model [14]

$$v = \begin{cases} \mu_v E/(1 + E/E_c) & if \;\; E \leq E_c \\ v_{sat} & if \;\; E > E_c \end{cases} \tag{3.76}$$

where

$$E_c = \frac{2v_{sat}}{\mu_v} \tag{3.77}$$

where the saturation velocity v_{sat} is equal to $8 \times 10^6 \; cm/s$ for electrons (NMOS device) and $6.5 \times 10^6 \; cm/s$ for holes (PMOS device).

The drain current in triode region ($V_{DS} \leq V_{DSsat}$) is given by [13]

$$I_{DS} = \mu_v C_{ox} \frac{W_{eff}}{L_{eff}} \frac{1}{(1 + \frac{V_{DS}}{E_c L_{eff}})} (V_{GS} - V_T - \frac{V_{DS}}{2}) V_{DS} \tag{3.78}$$

The saturation current can be expressed by

$$I_{DSsat} = v_{sat} C_{ox} W_{eff} (V_{GS} - V_T - V_{DSsat}) \tag{3.79}$$

By equating (3.78) and (3.79) we can derive the following expression for V_{DSsat}

$$V_{DSsat} = (1 - K)(V_{GS} - V_T) \tag{3.80}$$

where

$$K = \frac{1}{1 + \frac{2v_{sat}}{\mu_v} \frac{L_{eff}}{(V_{GS} - V_T)}} \tag{3.81}$$

The drain current in the saturation can be rewritten as

$$I_{DSsat} = K v_{sat} C_{ox} W_{eff} (V_{GS} - V_T) \tag{3.82}$$

Note that V_T, in the current equation, is the extrapolated threshold voltage. The mobility μ_v for electrons can be expressed [15]

$$\mu_n = 240 \sqrt{0.06 t_{ox} / (V_{GS} + V_T)} \qquad for \;\; N^+ \;\; poly - gate \tag{3.83}$$

and for holes

$$\mu_p = \begin{cases} 65 \left[0.06 t_{ox} / (V_{GS} - V_T)\right]^{1/3} & for \;\; P^+ \;\; poly - gate \\ 65 \left[0.06 t_{ox} / (V_{GS} - V_T - 1)\right]^{1/3} & for \;\; N^+ \;\; poly - gate \end{cases} \tag{3.84}$$

where t_{ox} is in $\AA$ and the mobility in $cm^2/(Vs)$. This analytical model can be used for gate length down to deep-submicron range.

3.4 CMOS POWER SUPPLY VOLTAGE SCALING

Scaling device feature size has been used to increase packing density and speed. MOSFET scaling can follow three theories:

1. Constant Electric Field (CE) scaling [16].
2. Constant Voltage (CV) scaling [17].
3. Quasi-Constant Voltage (QCV) scaling [17].

Table 3.3 Scaling laws of the MOS device.

Parameter	Expression	Const. Field	Const. Voltage	Quasi-Const. Voltage
Dimensions	W, L, x_j	k_h	k_h	k_h
Gate oxide	t_{ox}	k_h	$k_h^{0.5}$	k_h
Doping	N	k_h	k_h	k_h
Voltage	V_T, V_{DD}	k_h	1	$k_h^{0.5}$
Capacitance	$C_G = AC_{ox}$	k_h	$k_h^{1.5}$	k_h
Current	I_{DS}	$k_h^{0.5}$	$k_h^{-0.5}$	$k_h^{-0.25}$
Gate Delay	$t_d \propto C_G V_{DD}/I_{DS}$	$k_h^{1.5}$	k_h^2	$k_h^{1.75}$
Dynamic Power	$P_d = I_{DS}V_{DD}$	$k_h^{1.5}$	$k_h^{-0.5}$	$k_h^{0.25}$
Dynamic Energy	$E_d = P_d t_d$	k_h^3	$k_h^{1.5}$	k_h^2

In the CE scheme all horizontal and vertical dimensions and voltages scale linearly with the same factor. In the CV scheme, the dimensions are scaled, while the voltages are kept constant. This scenario has been the most commonly used. While the constant electric field scaling is natural from the device physics point of view, the constant voltage scaling is more practical from the systems standpoint. Changing the supply voltage every technology generation (when the feature sizes are scaled) is too expensive because multiple power supply generators will be required for each PC board. However, as the channel length scales below about 0.6 μm the 5 V supply voltage must be reduced for reliability reasons (e.g. hot carrier effects, breakdown, etc). The quasi-constant voltage scaling is an intermediary scheme between the CE and CV views. The scaling factors of the horizontal dimensions and the voltage are denoted by k_h and k_u, respectively. Table 3.3 summarizes the scaling of the important device parameters according to the three theories as a function of the horizontal scaling factor (k_h). Note that in the QCV scheme, the dimensions scale more aggressively than the voltage ($k_u = {k_h}^{0.5}$).

For the drain current, the following average value is used

$$I_{DS} \propto W/LC_{ox}(V_{GS} - V_T)^{1.5} \tag{3.85}$$

This expression is not far from the one proposed by [15]. Table 3.3 shows the effect of device scaling on the delay, power and energy. It is assumed that a gate drives other gates, where the load is mainly the gate capacitance. The threshold voltage is scaled proportional to V_{DD} scaling. The gate delays improve with scaling for all the scenarios, but with a better rate in the CV scheme. However, the dynamic power, at maximal frequency, of the gate increases by a factor $k_h^{0.5}$ in the case of CV. For the CE scheme, the power is reduced by a high factor equal to $k_h^{0.5}$. Also in this Table, the dynamic energy dissipated by a gate is reported. This is independent of frequency. For all schemes, it has improved significantly, particularly for the CE case.

Scaling the supply voltage is an efficient way to reduce the power consumption. However, to get a better performance at low-voltage the device sizes and the threshold voltage have to be properly scaled. For a fixed sub-micron technology, the supply voltage can not be reduced aggressively, otherwise the speed is degraded. However, for each fixed technology generation, there is a lower limit power supply voltage V_{DDmin} [18]. For V_{DD}'s higher than this minimum limit the speed does not improve significantly. Typical values for V_{DDmin} are, 3.3 V and 2.5 V for L_{eff} of 0.5 μm and 0.3 μm, respectively. On the other hand, the higher limit of V_{DD} is driven by the reliability and the power dissipation limitation. The value of this V_{DD} is proportional to the square root of design rules ($\sqrt{L}$) [19]. For 0.6 μm and 0.3 μm design rules with LDD structure, these high limits are 4.5 V and 3.3 V, respectively.

3.5 MODELING OF THE BIPOLAR TRANSISTOR

3.5.1 BJT Structure and Operation

Fig. 3.11 shows a cross-sectional view of a NPN bipolar junction transistor with geometrical layout and the corresponding symbols for NPN and PNP. To understand the basic operation of the bipolar transistor, one dimensional representation of the active region can be used. Fig. 3.12(a) illustrates a typical profile of the one-dimensional section of the active region [Fig. 3.12(b)]. The N^+PN^- sandwich forms the heart of BJT.

Consider an NPN transistor with $V_{BE} > 0.5V$ and $V_{BC} < 0V$ (forward-active mode). The corresponding energy band diagram is shown in Fig. 3.12(c). When the N^+P (emitter-base) junction is forward-biased, electrons are injected from the emitter into the base (current I_{nE}). A small fraction of these electrons recombine in the neutral base (I_{rB})[6]. The rest of the electrons, of which the current I_{nC} is constituted, diffuse through the base towards the reverse-biased base-collector junction where they are swept by the electric field into the base-collector depletion layer. On the other hand, some of the holes in the base are injected into the N^+ emitter region resulting in a current I_{pE}. This component is small compared to I_{nE} because the holes' concentration in the base is much smaller than the electron concentration in the emitter. The emitter-base depletion layer can be a site for the recombination between the injected electrons and holes resulting in a current I_{rscl}. Moreover, some holes are swept into the base due to the generation in the base-collector depletion region, but this component is very small ($\approx 10^{-17} A/\mu m^2$). The terminal currents can be written as follows

$$I_C = I_{nC} \tag{3.86}$$

$$I_B = I_{pE} + I_{rscl} + I_{rB} \tag{3.87}$$

$$I_E = I_{nE} + I_{rscl} + I_{pE} \tag{3.88}$$

Note that it has been assumed that the base and collector currents are flowing in the device, while the emitter current is flowing out of it [Fig. 3.12]. The emitter injection efficiency, which is defined as the ratio of the electron's current injected into the base to the total emitter current, is given by

$$\gamma_E = \frac{I_{nE}}{I_{nE} + I_{rscl} + I_{pE}} \tag{3.89}$$

[6] In order to reduce I_{rB}, the width of the neutral base should be narrow (i.e. $W_B \ll L_p$, where L_p is the diffusion length of the electrons in the base).

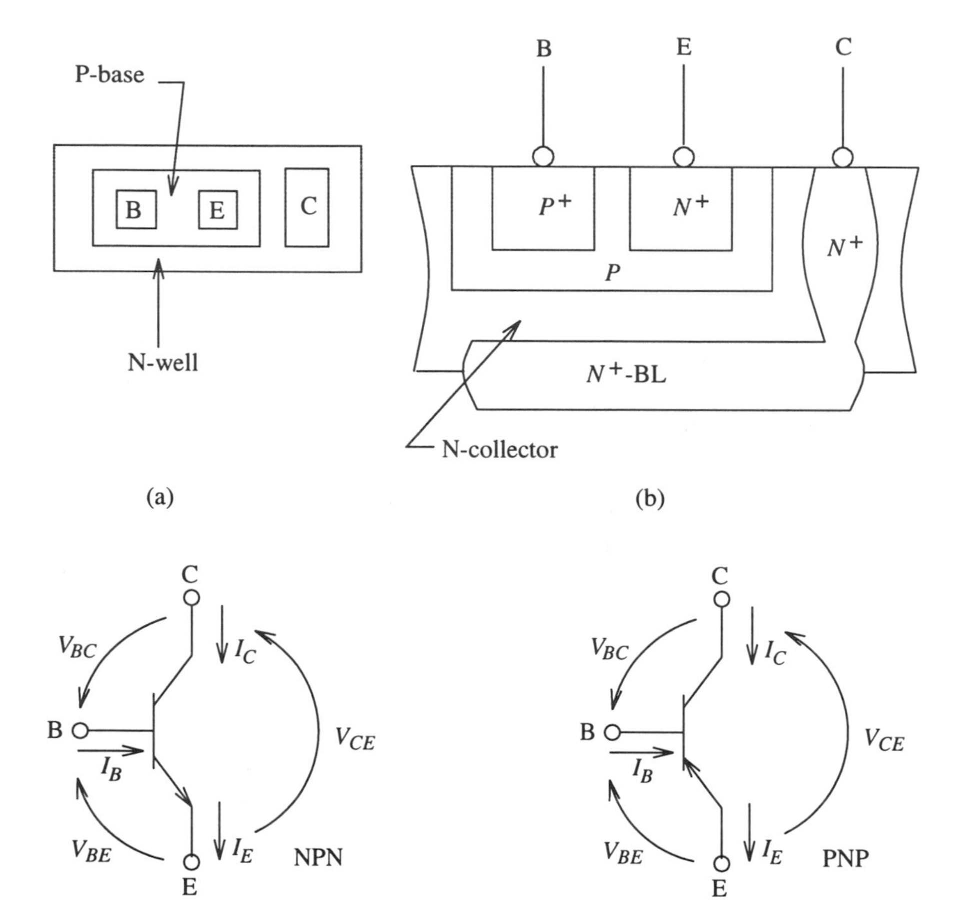

Figure 3.11 (a) The layout; (b) the cross-sectional view; (c) the symbol of an NPN BJT; (d) the symbol of an PNP BJT.

This ratio γ_E has to be near unity; that is, the emitter current should mostly be due to electrons for an NPN transistor. The ratio

$$\beta = \frac{I_C}{I_B} \tag{3.90}$$

is defined as the DC current gain.

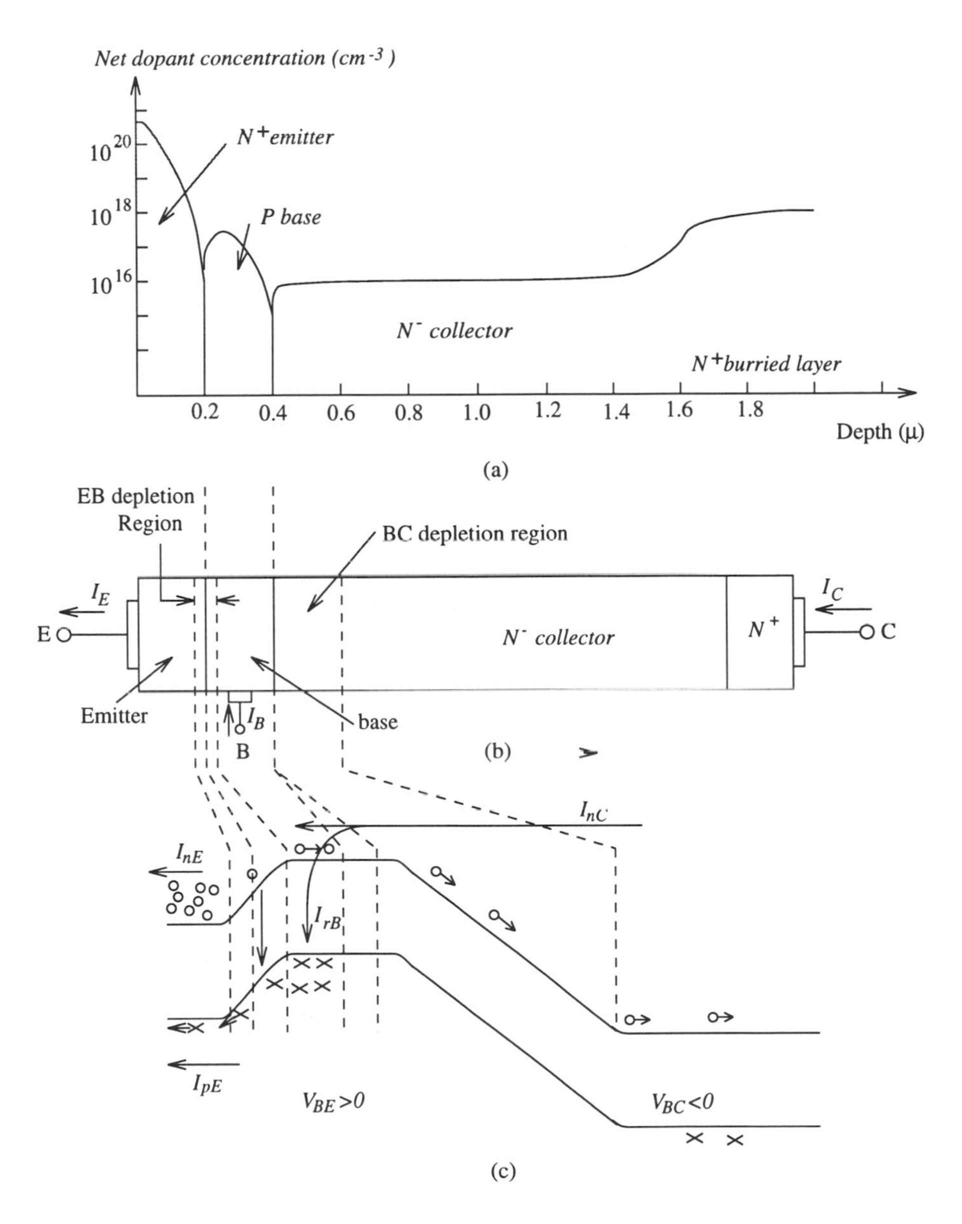

Figure 3.12 (a) Doping profile of a BJT; (b) one dimensional cross-sectional view of a BJT in the active region; (c) energy band diagram of a BJT in the active region.

When the emitter-base junction is reverse-biased and the collector-base junction is forward-biased, the transistor is in the inverse region where the emitter and collector may be exchanged. When both junctions are reverse-biased the transistor is in the *cutoff* region. But when they are forward-biased, the device is said to be in the saturation region. In this situation, both junctions are injecting into the base, the small electric fields in the two depletion regions sweep the carriers into the emitter and collector regions. Both junctions collect as well as emit.

3.5.2 Ebers-Moll Model

In this section, we present the Ebers-Moll (EM) model, which is a simple DC model of the bipolar transistor. The Ebers-Moll model can be used for hand calculations and first order circuit analysis. The derivation of the model equations, in this section, is based on the analysis by Roulston [20]. In Section 3.5.1, we have discussed the device operation in the forward active region only. For a general analysis, we assume that the base-emitter and the base-collector junctions are forward biased. In the following discussion we will neglect the currents due to recombination in the space charge layers and in the base. This implies that $I_{nC} = I_{nE}$ [7], hence, Equation (3.88) reduces to

$$I_E = I_{nC} + I_{pE} \tag{3.91}$$

The current due the holes injected from the base into the emitter is given by [20]

$$I_{pE} = \frac{q\ A_E\ D_{pE}\ p_{nE0}}{W_E} \left[e^{V_{BE}/V_t} - 1\right] \tag{3.92}$$

where p_{nE0} is the equilibrium hole concentration in the emitter and W_E is the neutral emitter width. The current I_{nC} is dominated by the diffusion current in the base and is proportional to the gradient of the minority carriers (electrons) in the neutral base. Because the neutral base width (W_B) is very thin, this gradient is approximately a constant. Therefore, we can write I_{nC} as [20]

$$I_{nC} = q\ A_E\ D_{nB} \left[\frac{n_B(0) - n_B(W_B)}{W_B}\right] \tag{3.93}$$

where $n_B(0)$ and $n_B(W_B)$ are the electron concentrations at the edges of the emitter-base and collector-base depletion regions respectively [see Fig. 3.13]. Note that the slope of the electrons in the base is given by the term between the brackets as demonstrated by Fig. 3.13.

[7] By applying KCL (i.e. $I_C + I_B - I_E = 0$), we can easily see that I_{rB} is the difference between I_{nE} and I_{nC}. If the recombination in the base is neglected ($I_{rB} = 0$), we can assume that $I_{nE} \approx I_{nC}$.

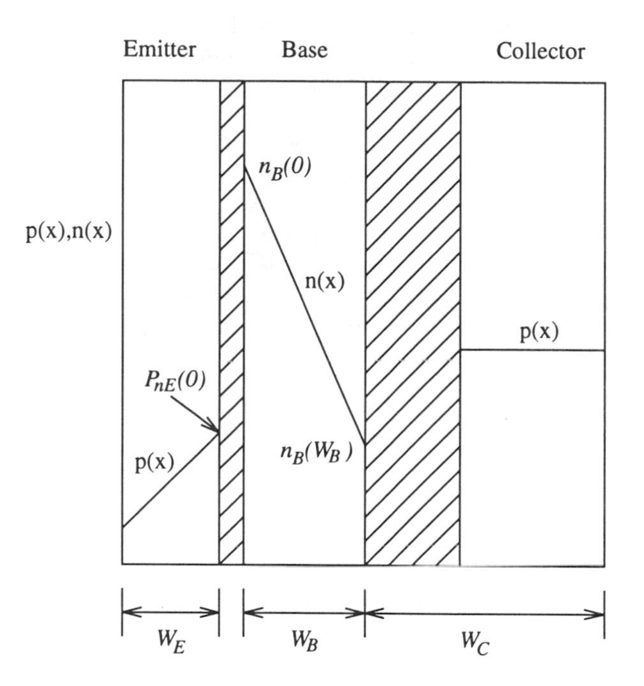

Figure 3.13 Minority carrier distribution in a BJT with both junctions forward biased.

Using the junction law, the electron concentrations $n_B(0)$ and $n_B(W_B)$, can be expressed in terms of V_{BE} and V_{BC} respectively. The current I_{nC} can hence be given by [20]

$$I_{nC} = \frac{q\ A_E\ D_{nB} n_i^2}{W_B\ N_B} \left[e^{V_{BE}/V_t} - e^{V_{BC}/V_t} \right] \tag{3.94}$$

where N_B is the base impurity concentration.

The collector current is given by

$$I_C = I_{nC} - I_{pC} \tag{3.95}$$

The current I_{pC} is due to the holes injected from the base to the collector[8]. The base-collector junction is basically a P^+NN^+ structure as shown in Fig.

[8]Note that I_{pC} was not included in Equation (3.86) because in deriving Equation (3.86) we have assumed that the collector-base junction was reverse biased.

3.12(a). An expression for I_{nC} can be derived from the analysis of a P^+NN^+ diode. The reader is adviced to consult with reference [20] for the details of this analysis. The current I_{pC} is given by

$$I_{pC} = \frac{q\, A_E\, p_{nCO}\, W_C}{\tau_{epi}} \left[e^{V_{BC}/V_t} - 1\right] \tag{3.96}$$

where p_{nCO} is the equilibrium hole concentration in the collector, W_C is the epitaxial thickness under the base and τ_{epi} is the hole lifetime in the epitaxial layer. By substituting from Equations (3.92) and (3.94) in Equation (3.91) and from Equations (3.94) and (3.96) in Equation (3.95) we get the following equations for I_E and I_C

$$I_E = I_f - \alpha_r I_r \tag{3.97}$$

$$I_C = -I_r + \alpha_f I_f \tag{3.98}$$

where

$$I_f = qA_E \left(\frac{D_{pE}\, p_{nEO}}{W_E} + \frac{D_{nB}\, n_{BO}}{W_B}\right) (e^{V_{BE}/V_t} - 1) \tag{3.99}$$

$$I_r = qA_E \left(\frac{D_{nB}\, n_{BO}}{W_B} + \frac{p_{nCO}\, W_C}{\tau_{epi}}\right) (e^{V_{BC}/V_t} - 1) \tag{3.100}$$

$$\alpha_f I_f = \alpha_r I_r = \frac{q\, A_E\, D_{nB}\, n_{BO}}{W_B} \tag{3.101}$$

Equations (3.97) and (3.98) are called the Ebers-Moll equations. Fig. 3.14 shows the equivalent circuit of the BJT based on the Ebers-Moll equations.

The Ebers-Moll model described above is general and can be used for any region of operation by substituting for V_{BE} and V_{BC} by the appropriate values. In the forward active region, assuming that $V_{BE} = 0.8\ V$ and $V_{BC} < 0.3\ V$ the emitter and collector current of Equations (3.97) and (3.98) reduce to

$$I_E = I_f \approx I_{Es}\, e^{V_{BE}/V_t} \tag{3.102}$$

and

$$I_C = \alpha_f I_f \approx \alpha_f I_{Es}\, e^{V_{BE}/V_t} = \alpha_f I_E \tag{3.103}$$

where the reverse saturation current of the base-emitter junction I_{Es} can be derived from Equation (3.99) and is given by

$$I_{Es} = qA_E \left(\frac{D_{pE}\, p_{nEO}}{W_E} + \frac{D_{nB}\, n_{BO}}{W_B}\right) \tag{3.104}$$

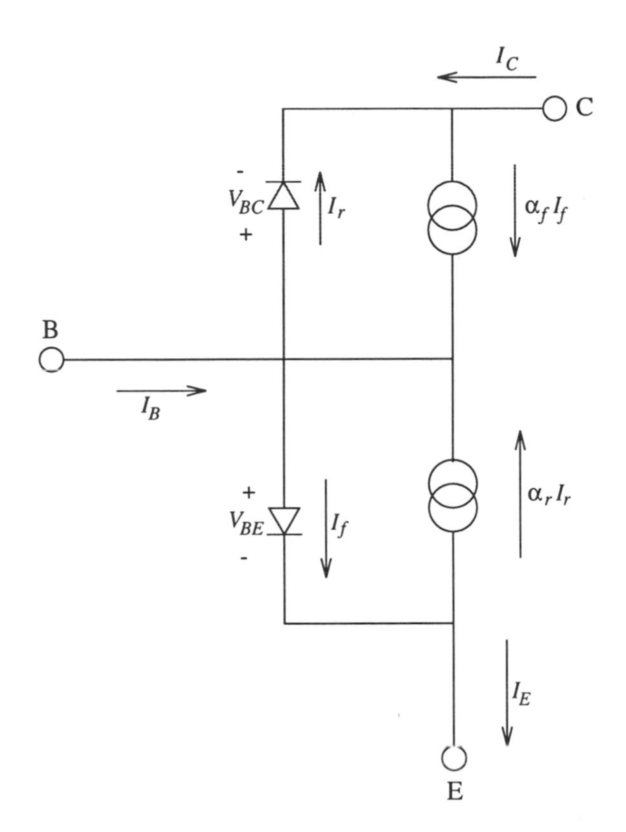

Figure 3.14 Equivalent DC circuit of the BJT based on the Ebers-Moll model.

It can easily be shown that the base current can be expressed as

$$I_B = \frac{1 - \alpha_f}{\alpha_f} I_C \tag{3.105}$$

Equations (3.102), (3.103) and (3.105) are the well-known current equations of a forward biased bipolar transistor. Note that Equation (3.105) yields the famous relation between α_f and the DC forward current gain β [$\beta = \alpha_f/(1 - \alpha_f)$].

The simple Ebers-Moll model lacks accuracy for the following three reasons

1. It does not account for the parasitic resistors of the emitter, base and collector.

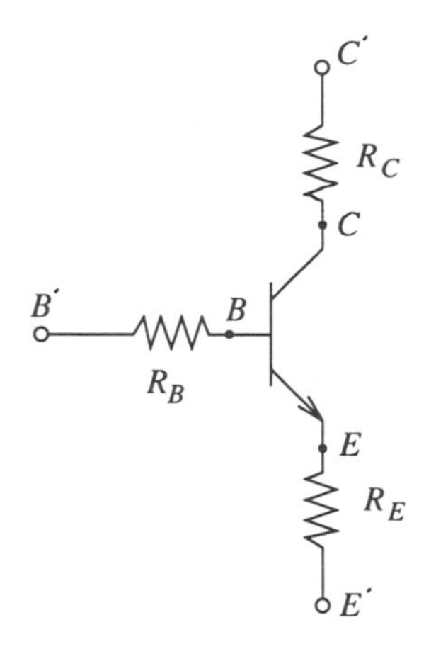

Figure 3.15 The extrinsic resistances of a bipolar transistor.

2. It does not account for the Early effect, which causes the collector current to increase as the collector-emitter voltage increases.
3. It does not account for the effect of the high collector currents on the current gain.

Next, we will discuss the modeling of each phenomena separately.

3.5.2.1 The Parasitical Resistors of a Bipolar Transistor

Fig. 3.15 shows the modification of the EM model by the addition of the base resistance R_B, the collector resistance R_C and the emitter resistance R_E. These extrinsic components represent the transistor's parasitic resistances from their active region to their base, collector and emitter terminals, respectively.

The effect of the parasitic resistances is important because the voltage drop across them contribute to the external base-emitter and collector-emitter voltages $V_{B'E'}$ and $V_{C'E'}$ respectively, as shown by the following two equations

$$V_{B'E'} = V_{BE} + R_B I_B + R_E I_E \qquad (3.106)$$

$$V_{C'E'} = V_{CE} + R_C I_C + R_E I_E \qquad (3.107)$$

The drop across the parasitic resistors has to be accounted for to get more accurate results from the EM model. Neglecting these drops may even lead to erroneous results. For example, if the external collector-emitter voltage is found to be equal to 2 V one may deduce that the BJT operates in the active region. However, if $R_C = 1.8K$ and $R_E = 0.1K$ and $I_C \approx I_E = 1\ mA$, then the intrinsic collector-emitter voltage (V_{CE}) is 0.1 V. This implies that the bipolar transistor is actually saturated. This phenomenon is known as *Quasi-Saturation.*

3.5.2.2 The Early Effect

The Early effect refers to the base width modulation due to the change of the collector base reverse voltage (in the forward active region). As the collector-base reverse voltage increases, the base-collector depletion layer widens. The resulting reduction in the neutral base width causes the current gain to increase which, in turn, leads to an increase in the collector current [see Fig. 3.16]. This effect can be modeled by introducing the Early voltage (V_{Af}) in the expression of the collector current as follows

$$I_C = I_s e^{V_{BE}/V_t}\left(1 + \frac{V_{CE}}{V_{Af}}\right) \tag{3.108}$$

The inverse of the forward Early voltage $1/V_{Af}$ is analogous to the coefficient λ in an MOS transistor. A typical value of V_{Af} is 50 V. The AC output resistance of the BJT in the forward active region is related to the Early voltage and is given by

$$r_o = \frac{V_{Af}}{I_C} \tag{3.109}$$

The Early effect in the inverse active region can be modeled by using the reverse Early voltage (V_{Ar}) which characterizes the slope of the collector current in that region (inverse active region).

3.5.2.3 High Current Effects

The current gain and the cut-off frequency are degraded due to high collector current. Fig. 3.17 shows the effect of the collector current on the gain. This degradation can be referred to the high level injection in the base (Webster effect) and/or the base pushout (Kirk effect). For a detailed discussion on these phenomenon, the reader is advised to consult reference [20]. In the case, where the injection level in the base is high (Webster effect) the collector

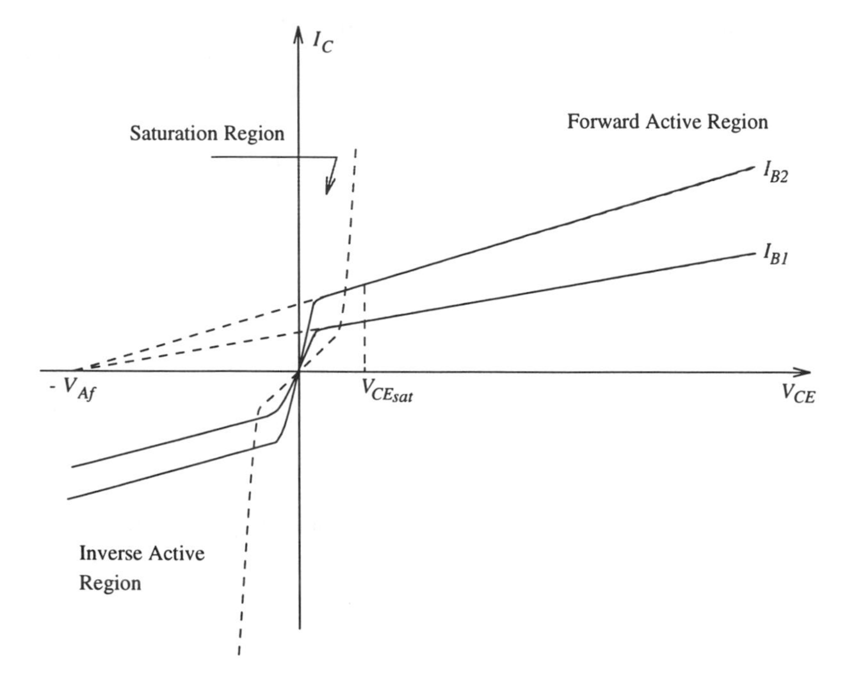

Figure 3.16 The I-V characteristics of a BJT.

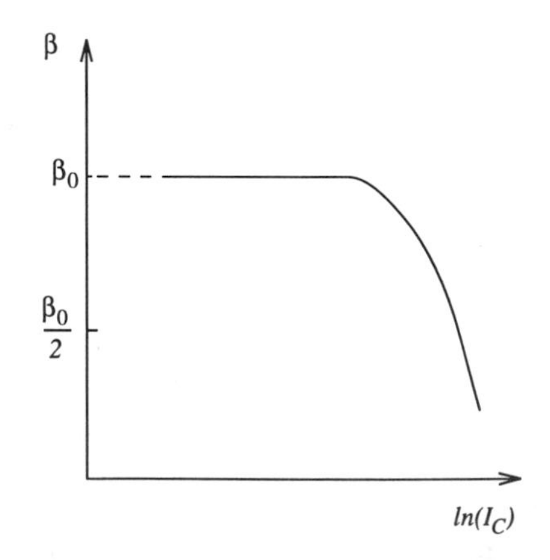

Figure 3.17 Current gain vs collector current at medium and high currents.

current can be expressed as [20]

$$I_C = \sqrt{I_s I_{Kf}}\, e^{V_{BE}/2V_t} \tag{3.110}$$

where the forward knee current I_{Kf}[9] is defined as the collector current at which its slope in the Gummel plot changes from 1 to 1/2 [see Fig. 3.18]. This current marks the onset of high level injection. The degradation of the current gain, when $I_C > I_{Kf}$, can be described by the following relation [20]

$$\beta = \frac{I_C}{I_B} = \beta_0 \frac{I_{Kf}}{I_C} \tag{3.111}$$

where β_0 is the value of the gain when $I_C < I_{Kf}$. The modeling of the Kirk effect is very complex. However, a simple model for the current gain, which can be used in first order circuit analysis, is given below [21]

$$\beta = \frac{\beta_0}{1 + \frac{I_C}{I_{Kf}}} \tag{3.112}$$

The accuracy of the simple EM model can be enhanced by accounting for the parasitic resistors, the Early effect and high current effect which can be modeled by simple analytical expressions as shown above.

3.5.3 Bipolar Models in SPICE

Two BJT models are implemented in SPICE. The Ebers-Moll model and a more sophisticated one, which is based on the Gummel-Poon (GP) model [22]. The second model includes the following second order effects:

- Very low current effect on the gain.
- Base width modulation effect.
- High-level injection effects (the Kirk effect is not included)
- Base resistance variation with current.

The GP model is based on one-dimensional analysis. It is valid for all regions of operation: cutoff, forward-active, inverse-active, and saturation. The GP-based bipolar model is illustrated by the equivalent circuit shown in Fig. 3.19.

[9] A typical value of I_{Kf} per unit area is 1 $mA/\mu m^2$.

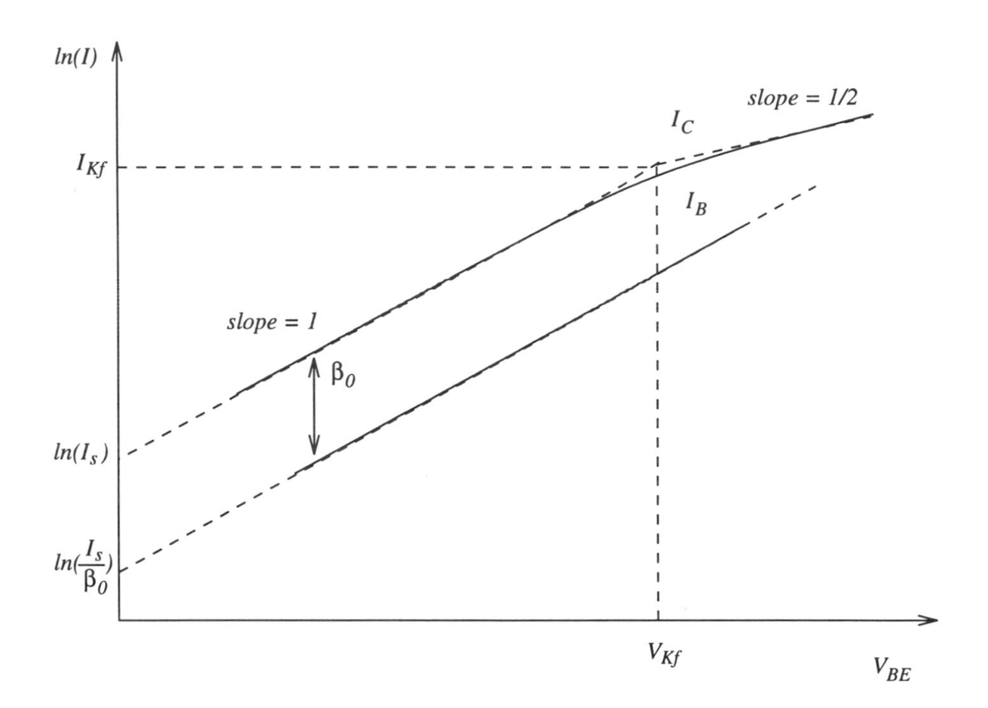

Figure 3.18 Variation of $\ln(I_C)$, $\ln(I_B)$ as function of V_{BE}.

The two back-to-back diodes on the right represent the intrinsic base-emitter and base-collector junctions and their currents are given by [23]

$$I_{CC} = \frac{I_s}{q_b}(e^{V_{BE}/n_f V_t} - 1) \tag{3.113}$$

$$I_{EC} = \frac{I_s}{q_b}(e^{V_{BC}/n_r V_t} - 1) \tag{3.114}$$

where I_s is given by [23]

$$I_s = \frac{q\ D_{nB}\ n_i^2\ A_E}{\int_0^{W_B} N_B(x)dx} \tag{3.115}$$

The forward and reverse current emission coefficient (n_f *and* n_r), which are introduced in Equations (3.113) and (3.114), are used to model the low currents. The parameter q_b (base charge factor) accounts for the high current and base

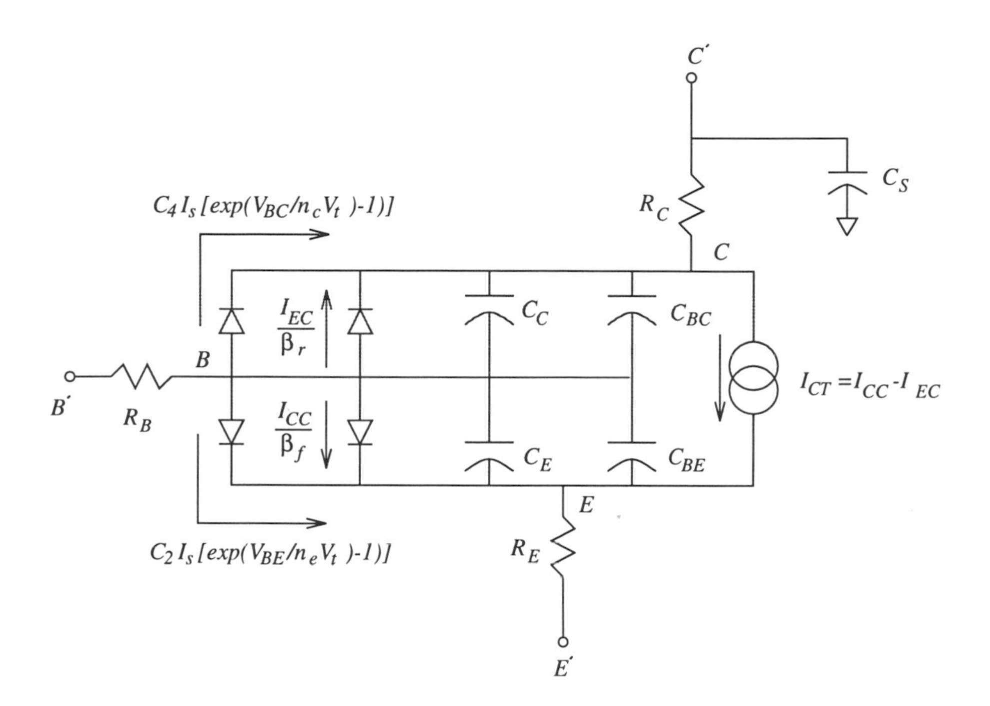

Figure 3.19 The GP-based model of a bipolar transistor.

width modulation effects. It is given by [23]

$$q_b = \frac{q_1}{2}[1 + \sqrt{1 + 4q_2}] \tag{3.116}$$

q_1 models the effects of base width modulation and can be expressed as

$$q_1 = \frac{1}{(1 - \frac{V_{BC}}{V_{Af}} - \frac{V_{BE}}{V_{Ar}})} \tag{3.117}$$

q_2 models the high-level injection and is expressed as

$$q_2 = \frac{I_s}{I_{Kf}}(e^{V_{BE}/n_f V_t} - 1) + \frac{I_s}{I_{Kr}}(e^{V_{BC}/n_r V_t} - 1) \tag{3.118}$$

The general expression of q_b [Equation (3.116)] can be simplified for low-level and high-level injection conditions.

$$q_b = \begin{cases} q_1 & if \quad q_2 \ll q_1^2/4 \quad (low-level-injection) \\ \sqrt{q_2} & if \quad q_2 \gg q_1^2/4 \quad (high-level-injection) \end{cases} \tag{3.119}$$

The two back-to-back diodes on the left [Fig. 3.19] account for the currents caused by the recombination of carriers in the emitter-base and the collector-base space-charge layers and other recombinations. These currents can be modeled by [23]

$$C_2 I_s (e^{V_{BB}/n_e V_t} - 1) \tag{3.120}$$

$$C_4 I_s (e^{V_{BB}/n_c V_t} - 1) \tag{3.121}$$

where C_2, C_4, n_e and n_c have been introduced to fit the measured currents.

Further improvements to this model are possible by the inclusion of three parasitic resistances (R_C, R_B, R_E); three junction capacitances (C_E, C_C, C_S); and two diffusion capacitances (C_{dE}, C_{dC}) as shown in Fig. 3.19.

The model of the base resistance takes into account the effect of the current (current crowding) through the following expression [24]

$$R_B(z) = R_{Bm} + 3(R_B - R_{Bm}) \frac{\tan(z) - z}{z \tan(z)^2} \tag{3.122}$$

where the variable z is given by

$$z = \frac{-1 + \sqrt{1 + 144\, I_B/(\pi^2 I_{rB})}}{\frac{24}{\pi^2}\sqrt{(I_B/I_{rB})}} \tag{3.123}$$

R_B represents the low-current maximum resistance and R_{Bm} high-current minimum resistance.

The junction depletion capacitance is a function of the junction voltage (V). This function can be approximated by the following two expressions

$$C_{j\,dep} = C_j (1 - \frac{V}{\phi_j})^{-M_j} \qquad if\ V < FC\phi_j \tag{3.124}$$

$$C_{j\,dep} = C_j \frac{1 - FC(1 + M_j) + M_j \frac{V}{\phi_j}}{(1 - FC)^{1+M_j}} \qquad if\ V > FC\phi_j \tag{3.125}$$

The empirical factor FC has a value between 0 and 1. Its default value in SPICE is 0.5. Note that Equations (3.124) and (3.125) apply for a reverse and forward biased junction respectively.

The diffusion capacitances model the charge associated with injected carriers. For example, the electrons injected in the base have a corresponding storage charge

$$Q_{DE} = \tau_f I_{CC} \tag{3.126}$$

The forward transit time τ_f is current-dependent and is given by an empirical expression[24]

$$\tau_f = \tau_{f0}\left(1 + XTF\left(\frac{I_{CC}}{I_{CC} + ITF}\right)^2 e^{V_{BC}/1.44VTF}\right) \quad (3.127)$$

Where VTF is a fitting parameter to model the change of τ_f as a function of V_{BC} (or V_{CE}), ITF models the change due to I_C and XTF controls the increase of τ_f. I_{CC} is the collector current in the absence of the high-current effects which corresponds to that of Ebers-Moll model.

The diffusion capacitance (associated with the injected electrons from the emitter into the base, when the base-emitter junction is forward biased) is given by

$$C_{DE} = \frac{\partial Q_{DE}}{\partial V_{BE}} \quad (3.128)$$

Similarly, the base-collector junction has a diffusion capacitance, which is given by

$$C_{DC} = \frac{\partial Q_{DC}}{\partial V_{BC}} \quad (3.129)$$

where

$$Q_{DC} = \tau_r I_{EC} \quad (3.130)$$

Although the SPICE models account for most of the first and second order effects, they are not highly accurate. This originates from some weaknesses in the theory on which the models are based. As the device features are scaled down the currently available models become less accurate. The physics and the theory of the scaled devices is more complex. Hence, accurate modeling becomes very difficult. One way around that problem is to chose the model parameters such that simulated device characteristics agree with measurements. In practice, the models' parameters are extracted automatically using parameter analyzers with software tools to obtain the best fit. As a result, the values of the extracted parameters may not correspond to their actual values. For example, it is common to find a discrepancy of 20% between the measured current gain of a bipolar transistor and that listed in the SPICE file. Another approach, which is equivalent to tweaking the parameters, is to use empirical models (e.g. BSIM model), in which the empirical (fitting) parameters can be optimized to get the best fit between simulation and measurements.

Typical GP parameters , for the 0.8 μm BiCMOS presented in Chapter 2., are shown in Table 3.4 and 3.5.

Table 3.4 Bipolar device parameter and HSPICE correspondence.

Parameter	SPICE Keyword	Description
I_s	IS	Saturation current
β_f	BF	Ideal maximum forward gain
β_r	BR	Ideal maximum reverse gain
n_f	NF	Forward current-emission coefficient
n_r	NR	Reverse current-emission coefficient
V_{Af}	VAF	Forward early voltage
V_{Ar}	VAR	Reverse early voltage
I_{Kf}	IKF	Forward-knee current
I_{Kr}	IKR	Reverse-knee current
I_{se}	ISE	Base-emitter leakage saturation current
I_{sc}	ISC	Base-collector leakage saturation current
n_e	NE	Base-emitter leakage emission coefficient
n_c	NC	Base-collector leakage emission coefficient
R_E	RE	Emitter resistance
R_C	RC	Collector resistance
R_B	RB	Base resistance at zero current
I_{rB}	IRB	Base current where $R_B = R_B(0)/2$
R_{Bm}	RBM	Minimum high-current base resistance
C_E	CJE	Base-emitter zero-bias depletion cap.
ϕ_{je}	VJE	Base-emitter built-in potential
M_{je}	MJE	Base-emitter junction grading factor
C_C	CJC	Base-collector zero-bias depletion cap.
ϕ_{jc}	VJC	Base-collector built-in potential
M_{jc}	MJC	Base-collector junction grading factor
C_S	CJS	Collector-substrate zero-bias cap.
ϕ_{js}	VJS	Collector-substrate built-in potential
M_{js}	MJS	Collector-substrate junction grading factor
-	XCJC	Internal base fraction of base-collector cap.
FC	FC	Coefficient for forward-bias depletion cap.

Table 3.4 *(continued)*

τ_f	TF	Forward transit time
XTF	XTF	TF bias-dependent coefficient
VTF	VTF	TF base-collector voltage dependence coef.
ITF	ITF	TF high current parameter
τ_r	TR	Reverse transit time
-	XTB	Forward and reverse *beta* temperature exponent
-	XTI	Saturation current temperature exponent
-	EG	Energy gap
-	KF	Flicker noise coefficient
-	AF	Flicker noise exponent

Table 3.5 HSPICE BJT model parameters (0.8 μm BiCMOS process).

SPICE Keyword	Value	Units
IS	2×10^{-18}	A
BF	100	-
BR	1	-
NF	1	-
NR	1	-
VAF	50	V
VAR	5	V
IKF	5×10^{-3}	A
IKR	0.	A
ISE	0.	A
ISC	0.	A
NE	1.5	-
NC	2.0	-

Table 3.5 *(continued)*

RE	30	Ω
RC	87	Ω
RB	650	Ω
IRB	0.	A
RBM	650	Ω
CJE	1.51×10^{-14}	F
VJE	0.87	V
MJE	0.265	-
CJC	1.75×10^{-14}	F
VJC	0.70	V
MJC	0.37	-
CJS	4.03×10^{-14}	F
VJS	0.64	-
MJS	0.285	-
XCJC	0.28	-
FC	0.5	-
TF	12.5×10^{-12}	s
XTF	916.2	-
VTF	1.6	-
ITF	8.7×10^{-2}	s
TR	4×10^{-9}	s
XTB	1.4	-
XTI	3.5	-
EG	1.11	eV
KF	2.9×10^{-9}	-
AF	2.0	-

3.5.4 Chapter Summary

In this Chapter, we have reviewed the fundamentals of the MOS and bipolar devices. The most common device models used in SPICE have been presented. The key device parameters of each model have been defined and explained, so that the reader is familiar with the details of these models and can appreciate the importance of the different model parameters. The reader is given a list of model parameters, for a typical 0.8 μm BiCMOS process, that can be used for circuit simulations. These model can be used even at low-voltage operation. Moreover, a simple analytical model valid for submicrometer MOSFET's has been discussed.

REFERENCES

[1] A. Vladimirescu, and S. Liu, "The simulation of MOS Integrated Circuits using SPICE2," Memo. No. UCB/ERL M80/7, Univ. California, Berkeley, October 1980.

[2] H. Masuda, M. Nakai and M. Kubo, "Characteristics and Limitations of Scaled Down MOSFET's Due to Two Dimensional Field Effect," IEEE Trans. on Electron Devices, Vol. ED-26, pp. 980-986, 1979.

[3] R.L.M. Dang, "A Simple Current Model for Short-Channel IGFET and Its Application to Circuit Simulation," IEEE Journal of Solid-State Circuits, vol. SC-14, pp. 358-367, 1979.

[4] G. Merkel, J. Borel and N.Z. Cupcea, "An Accurate Large Signal MOS Transistor Model for Use in Computer-Aided Design," IEEE Trans. on Electron Devices, vol. ED-19, 1972.

[5] G. Baum and H. Beneking, "Drift Velocity Saturation in MOS Transistors," IEEE Trans. on Electron Devices, vol. ED-17, pp. 481-482, 1970.

[6] R.M. Swanson and J.D. Meindl, "Ion-Implanted Complementary MOS Transistors in Low-Voltage Circuits," IEEE Journal of Solid-State Circuits, vol. SC-7, pp. 146-153, 1972.

[7] B.J. Sheu, D.L. Scharfetter, P.-K. Ko, and M.C. Jeng, "BSIM: Berkeley Short-Channel IGFET Model for MOS Transistors," IEEE Journal of Solid-State Circuits, vol. SC-22, pp. 558-566, 1987.

[8] J. H. Huang, Z. H. Liu, M. C. Jeng, P. K. Ko, and C. Hu, "A Robust physical and Predictive Model for Deep-Submicrometer MOS Circuit Simulation," IEEE Custom Integrated Circuits Conf., Tech. Dig., pp. 14.2.1-14.2.4, May 1993.

[9] D.E. Ward and R.W. Dutton, "A Charge-Oriented Model for MOS Transistors Capacitances," IEEE Journal of Solid-State Circuits, vol. SC-13, pp. 703-707, 1978.

[10] Y. P. Tsividis, "Operation and Modeling of the MOS Transistor," Mc Graw-Hill, 1988.

[11] T. Sakata et al., "Subthreshold-Current Reduction Circuits for Multi-Gigabit DRAM's," IEEE Journal of Solid-State Circuits, vol. 29, no. 7, pp. 761-769, July 1994.

[12] S.M. Sze, "Physics of Semiconductor Devices," John Wiley & Sons, 1981.

[13] C.G. Sodini, P.-K. Ko, and J.L. Moll, "The effect of High Fields on MOS Device and Circuit Performance," IEEE Trans. on Electron Devices, Vol. ED-31, No. 10, pp. 1386-1393, October 1984.

[14] B. Hoefflinger, H. Sibbert, and G. Zimmer, "Model and Performance of Hot-Electron MOS Transistor for VLSI," IEEE Trans. on Electron Devices, Vol. ED-26, pp. 513, 1979.

[15] C. hu, "Low-Voltage CMOS Device Scaling," IEEE International Solid-State Circuits Conf., Tech. Dig., pp. 86-87, 1994.

[16] R.H. Dennard, et al., "Design of Ion Implanted MOSFETs with Very Small Physical Dimensions," IEEE Journal of Solid-State Circuits, vol. SC-9, pp. 256-266, October 1974.

[17] P.K. Chatterjee, et al., "The Impact of Scaling Laws on the Choice of N-Channel or P-Channel for MOS VLSI," IEEE Electron Device Letters, Vol. EDL-1, pp. 220-223, October 1980.

[18] M. Kakumu, "Process and device Technologies of CMOS Devices for Low-Voltage Operation," IEICE Trans. Electron., vol. E76-C, no. 5, pp. 672-680, May 1993.

[19] M. Kakumu, M. Kinugawa, and K. Hashimoto, "Choice of Power-Supply Voltage for Half-Micrometer and Lower Submicrometer CMOS Devices," IEEE Trans. Electron devices, vol. 37, no. 5, pp. 1334-1342, May 1990.

[20] D.J. Roulston, "Bipolar Semiconductor Devices," McGraw-Hill Publishing Company, 1990.

[21] K. Nakazato, et al., "Characteristics and Scaling Properties of n-p-n Transistors with a Sidewall Base Contact Structure," IEEE Trans. on Electron Devices, vol. ED-32, no 2, pp. 328-332, February 1985.

[22] H.K. Gummel and H.C. Poon, "An Integral Charge Control Model of Bipolar Transistors," Bell Syst. Tech. J., vol. 49, 1970.

[23] I. Getreu, "Modeling the Bipolar Transistor," Tektronix, Inc., 1976.

[24] P. Antognetti and G. Massobrio, "Semiconductor Device Modeling with SPICE," McGraw Hill, 1988.

4

LOW-VOLTAGE LOW-POWER VLSI CMOS CIRCUIT DESIGN

In this chapter we introduce the CMOS logic gate with the development of simple models for delay and power dissipation estimation. These analysis permit us to understand the mechanisms that control the performance, particularly the power dissipation, of a logic circuit. Several CMOS design styles, such as pseudo-NMOS, dynamic logic and NORA, are presented. Other circuit variations of the static complementary CMOS, which are suitable for low-power applications, are discussed. These include the pass-transistor logic families such as Complementary Pass-transistor Logic (CPL), Dual Pass-transistor Logic (DPL), and Swing Restored Pass-transistor Logic (SRPL). Also an overview of clocking strategy in VLSI systems is covered. Included in this chapter is one important area which is the I/O circuits. The power dissipation of the I/O circuits is also analyzed. Finally, low-power techniques for CMOS design are also reviewed at the transistor-level. We will cover the low-power issues at subsystem/system/architecture levels in Chapter 6, 7 and 8 in more detail. Several books treat in detail other CMOS circuit design aspects [1, 2, 3]. The reader can refer to them.

Many issues existing in todays advanced CMOS circuit structures are considered; such as:

- Power dissipation components of a CMOS gate and their importance;
- Concept of switching activity;
- Power dissipation in I/O circuits;
- Single-phase clocking strategy;
- Clock skew issue;

- Clock distribution in VLSI systems;
- Ground bouncing; and
- Low-power circuit techniques and design guidelines.

4.1 CMOS INVERTER: DC CHARACTERISTICS

Fig. 4.1 shows the basic complementary MOS inverter. Before deriving the DC-transfer characteristics of this inverter (the output voltage versus the input voltage), lets understand the operation of this circuit.

- When the input is HIGH, which means at V_{DD}, we have

$$V_{GSn} = V_{in} = V_{DD} \tag{4.1}$$

$$V_{GSp} = V_{in} - V_{DD} = 0 \tag{4.2}$$

  In this case, $V_{GSn} > V_{Tn}$ and $|V_{GSp}| < |V_{Tp}|$. The PMOS is OFF and the NMOS is ON. The NMOS transistor N provides a current path to ground. The final stable value of the output voltage V_o is

$$V_o = 0 \tag{4.3}$$

  At the steady state, the DC current from V_{DD} to the ground is controlled by the subthreshold current of the PMOS P, since this device is OFF and the NMOS N has a V_{DS} equals to zero. We assume that the junctions leakage is negligible. If V_{Tp}[1] is low enough (lower for example than -0.5 V), the subthreshold current is negligible ($< 1\ pA/\mu m$ width). If V_{Tp} (negative) is high, the subthreshold is not negligible and can be as high as $1\ \mu A/\mu m$ for $V_{Tp} = -0.05$ V [see Section 3.3.2]. In this case the output is not exactly at zero and can have a value of tens of mV. In this section we assume that the subthreshold current is not important. Low-V_T CMOS circuits are treated in Section 4.10.

- Similarly, when V_{in} is low ($0V$) $V_{GSn} < V_{Tn}$ and $|V_{GSp}| > |V_{Tp}|$. The PMOS transistor is ON and the NMOS transistor is OFF. The output ...iven by

$$V_o = V_{DD} \tag{4.4}$$

  ...ume that the leakage current is negligible.

...hreshold voltage.

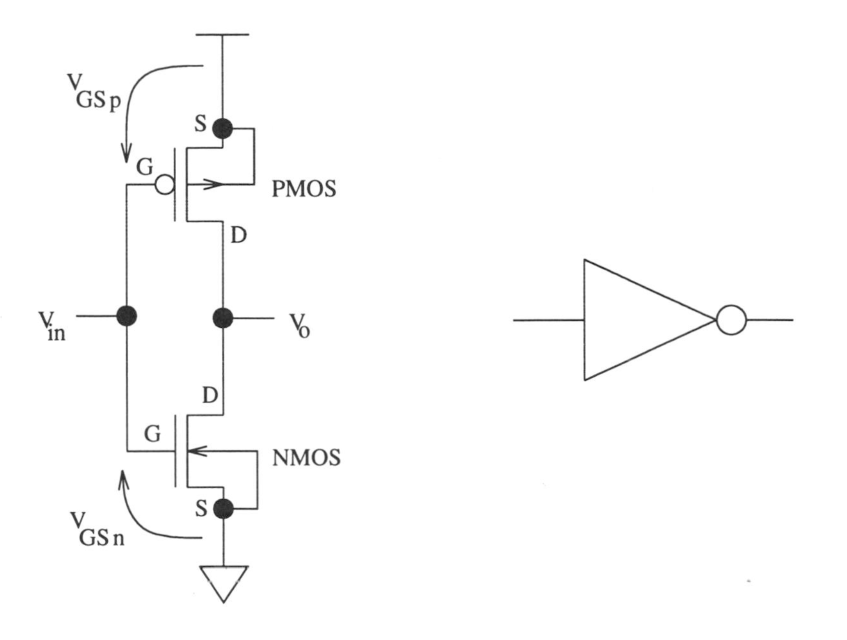

Figure 4.1 A CMOS Inverter

The logic levels of the CMOS inverter are close to V_{DD} and ground and the logic swing is equal to V_{DD}. This is a main feature of CMOS gates.

4.1.1 Transfer Characteristics

In this section we discuss the DC characteristics of the CMOS inverter of Fig. 4.1. Fig. 4.2 shows the DC transfer characteristic with the different regions of operation. For simplicity we use, for the MOS devices, the simple current models presented in Section 3.2.1. The circuit operation can be divided into five regions:

> **Region (A) :** $0 \leq V_{in} < V_{Tn}$
> The NMOS transistor is operating in the subthreshold region and the current is assumed zero. Hence the PMOS current is also zero. The PMOS transistor is in the linear region. Thus, $V_o = V_{DD}$.

Region (B) : $V_{Tn} < V_{in} < V_{inv}$
V_{inv} is defined as the input voltage at which the gain of the inverter is maximum and is also defined as the gate threshold voltage. In this region, the NMOS transistor is operating in the saturation region and the PMOS is in the linear region. Since the current in both devices is the same (in absolute value), we have

$$I_{DSp} = -I_{DSn} \tag{4.5}$$

The PMOS current is given by

$$I_{DSp} = -\beta_p \left[(V_{in} - V_{DD} - V_{Tn})(V_o - V_{DD}) - 1/2(V_o - V_{DD})^2\right] \tag{4.6}$$

Where

$$\beta_p = k_p \frac{W_{eff}}{L_{eff}} \tag{4.7}$$

$$V_{GSp} = V_{in} - V_{DD} \tag{4.8}$$

and

$$V_{DSp} = V_o - V_{DD} \tag{4.9}$$

The saturation current of the NMOS is given by

$$I_{DSn} = \beta_n \frac{(V_{in} - V_{Tn})^2}{2} \tag{4.10}$$

where

$$\beta_n = k_n \frac{W_{eff}}{L_{eff}} \tag{4.11}$$

and

$$V_{GSp} = V_{in} \tag{4.12}$$

Using equations (4.5), (4.6) and (4.10), the output voltage is given by

$$V_o = (V_{in} - V_{Tp}) + \sqrt{(V_{in} - V_{Tp})^2 - 2(V_{in} - \frac{V_{DD}}{2} - V_{Tp})V_{DD} - \frac{\beta_n}{\beta_p}(V_{in} - V_{Tn})^2} \tag{4.13}$$

This equation of V_o versus V_{in} is plotted in Fig. 4.2 region (B).

Region (C) : $V_{in} = V_{inv}$
Both the NMOS and PMOS transistors are in the saturation region. In this case, the PMOS current can be given by

$$I_{DSp} = -\beta_p \frac{(V_{in} - V_{Tp})^2}{2} \tag{4.14}$$

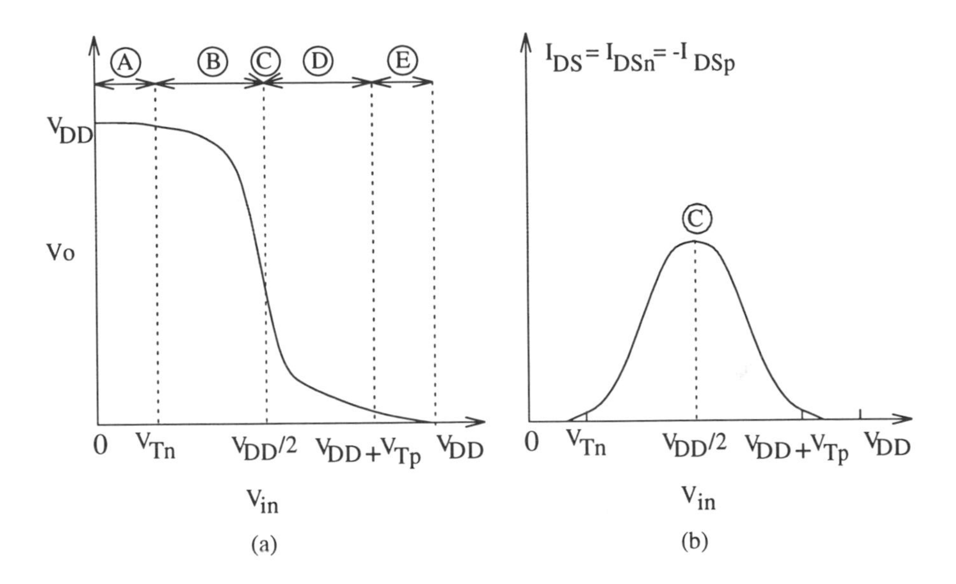

Figure 4.2 DC transfer characteristics of the CMOS inverter: (a) voltage; (b) current

The NMOS saturation current is given in Equation (4.10). By equalizing the absolute value of the two drain currents we have

$$V_{inv} = \frac{V_{DD} + V_{Tp} + V_{Tn}\sqrt{\beta}}{1 + \sqrt{\beta}} \tag{4.15}$$

where

$$\beta = \frac{\beta_n}{\beta_p} \tag{4.16}$$

This equation is very useful from a design point of view. Note, from this equation, that the logic threshold voltage of this gate is set by the designer; since the parameters β_n and β_p are dependent on W_{eff} and L_{eff}. Moreover, the region (C) is defined for only one point of V_{in}. For symmetrical NMOS and PMOS devices we have

$$V_{Tn} = V_{Tp} \tag{4.17}$$

If the designer set

$$\beta_n = \beta p \tag{4.18}$$

In a CMOS process,

$$\frac{k_n}{k_p} = \frac{\mu_n}{\mu_p} \approx 2-3 \tag{4.19}$$

This ratio is a typical example. The designer should set the size ratio as

$$\left\{\frac{W_eff}{L_{eff}}\right\}_p = 2.5\left\{\frac{W_{eff}}{L_{eff}}\right\}_n \tag{4.20}$$

We obtain

$$V_{in} = V_{inv} = \frac{V_{DD}}{2} \tag{4.21}$$

An inverter with this V_{inv} is sometimes called a symmetrical gate. The output voltage in this case is not necessary equal to $V_{DD}/2$ and is given by the following inequality

$$V_{in} - V_{Tn} < V_o < V_{in} + V_{Tp} \tag{4.22}$$

In reality, V_o is set by the slight dependence of I_{DS} versus V_{DS}.

Region (D) : $V_{inv} < V_{in} < V_{DD} + V_{Tp}$
In this region the NMOS is in the linear region while the PMOS is in the saturation region. Similar analysis used in region (B) can be applied. The output voltage is given by

$$V_o = (V_{in} - V_{Tn}) - \sqrt{(V_{in} - V_{Tn})^2 - \frac{\beta_p}{\beta_n}(V_{in} - V_{DD} - V_{Tp})^2} \tag{4.23}$$

Region (E) : $V_{DD} + V_{Tp} < V_{in} \leq V_{DD}$
In this region the NMOS transistor is ON, and in the linear region, and the PMOS is operating in the subthreshold region. If we assume that this current is too small then

$$V_o = 0 \tag{4.24}$$

The current flowing from V_{DD} to ground, versus the input voltage, is plotted in Fig. 4.2(b). It reaches its maximum when both the MOS transistors are in saturation. It is important to note that for $V_{in} = V_{inv}$ the DC power dissipation would be maximal.

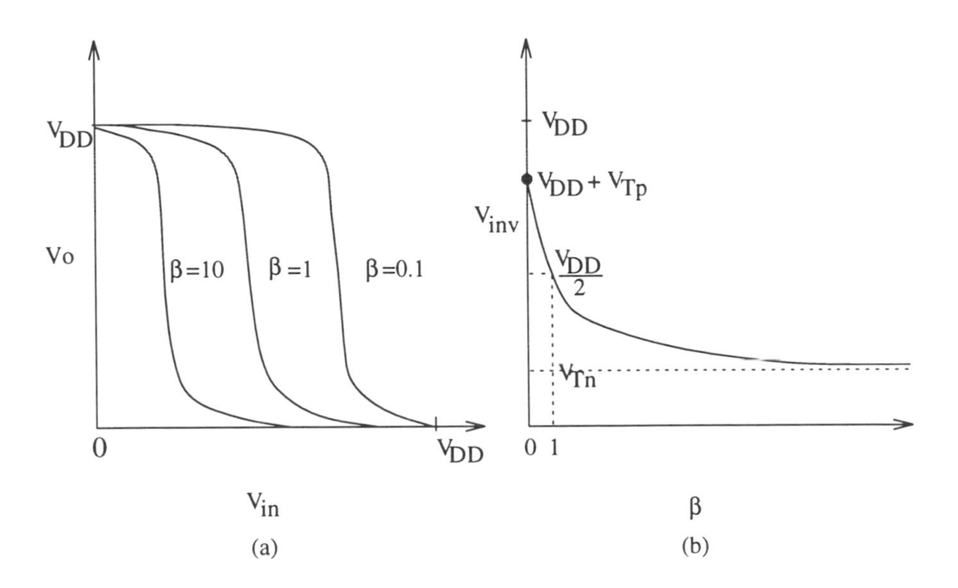

Figure 4.3 Effect of the ratio β on the (a) DC transfer characteristic; (b) threshold voltage of the CMOS inverter

4.1.2 Effect of β

As we discussed before, the ratio β controls the threshold voltage of the CMOS inverter. This parameter is set by the circuit designer through the transistor sizes. Other parameters such as the mobility and the threshold voltage of devices are set during the fabrication and the circuit designer can not change them. Fig. 4.3 illustrates the dependence of DC transfer characteristics and the threshold voltage of the CMOS inverter on the ratio β. Increasing β decreases the voltage V_{inv}. V_{inv} has a practical maximum less than $V_{DD} + V_{Tp}$ and practical minimum greater than V_{Tn}. Practical values mean that β can not have zero or infinite. In general, the circuit designer tries to set $\beta = 1$ for symmetrical operation unless the gate is used to switch an input swing different than a CMOS swing (from ground to V_{DD}).

4.1.3 Noise Margins

Noise margin is an important parameter in logic design. It is defined as the allowable noise voltage on the input so that the output is not affected. In other

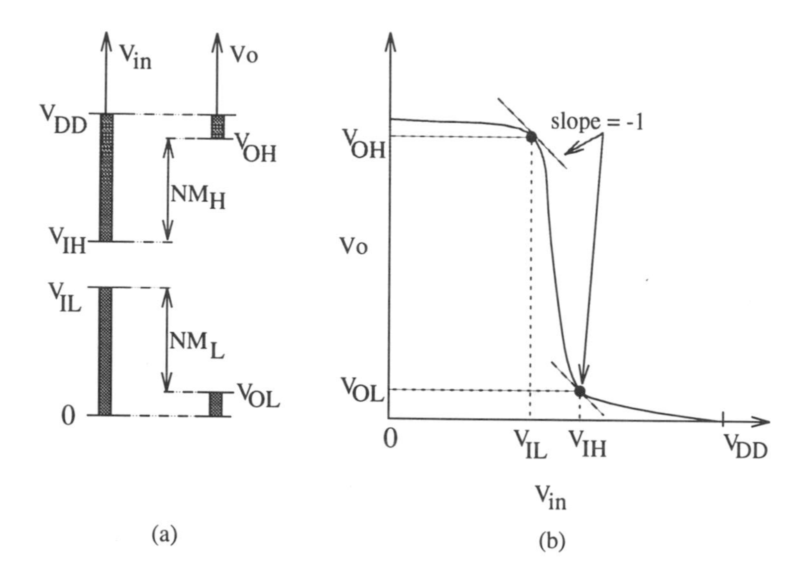

Figure 4.4 Definition of the noise margin voltages

words, we would define the valid logic levels such that they are restored when they propagate through a digital circuit. The logic levels can be extracted from the DC characteristic. As illustrated in Fig. 4.4 we define the logic levels at the input by

- Logic 0 : for $0 \leq V_{in} \leq V_{IL}$
- Logic 1 : for $V_{IH} \leq V_{in} \leq V_{DD}$

and at the output by

- Logic 0 : for $0 \leq V_o \leq V_{OL}$
- Logic 1 : for $V_{OH} \leq V_o \leq V_{DD}$

The LOW noise margin is defined by

$$NM_L = |V_{IL} - V_{OL}| \tag{4.25}$$

and the HIGH noise margin is defined by

$$NM_H = |V_{OH} - V_{IH}| \tag{4.26}$$

The V_{IL} and the V_{IH} levels can be defined as the points where the slope of the DC transfer characteristics is -1, i.e.,

$$\frac{dV_o}{dV_{in}} = -1 \tag{4.27}$$

These values can be deduced using equations (4.13) and (4.23). To have good noise margins, it is desirable to have V_{IL} and V_{IH} each near the other, around the point $V_{DD}/2$.

For CMOS circuits, the HIGH output voltage level V_{OH}, can be defined by letting $V_{OH} = V_{DD}$ and $V_{OL} = 0$. The CMOS logic inverter has fairly ideal transfer function and it tends to have very good noise margins. In some applications, either NM_H or NM_L is compromised to have good speed of operation.

4.1.4 Minimum Power Supply

To obtain the maximum power saving in CMOS logic circuits, the power supply voltage should be reduced. So, what is the lowest practical supply voltage at which CMOS will operate? In 1972, Swanson and Meindl [4] demonstrated that the minimum supply voltage is given by

$$V_{DDmin} = 8kT/q \tag{4.28}$$

At room temperature this value is equal to 0.2 V. This demonstrates that CMOS is a good candidate for ultra-low-power applications.

4.1.5 Example of Noise Margins

For an inverter with $W_p = 2W_n = 4\ \mu m$ (in 0.8 μm CMOS technology), and using a threshold voltage $V_T = V_{Tn}=|V_{Tp}|=0.5$ V, we have the following values for NM_L and NM_H. At 3.3 V power supply voltage, $N_{ML} = 1.15$ V and $NM_H = 1.45$ V. However at 1.5 V, $NM_L = 0.60$ V and $NM_H = 0.65$ V. So the noise level should be kept low, particularly at low power supply voltage.

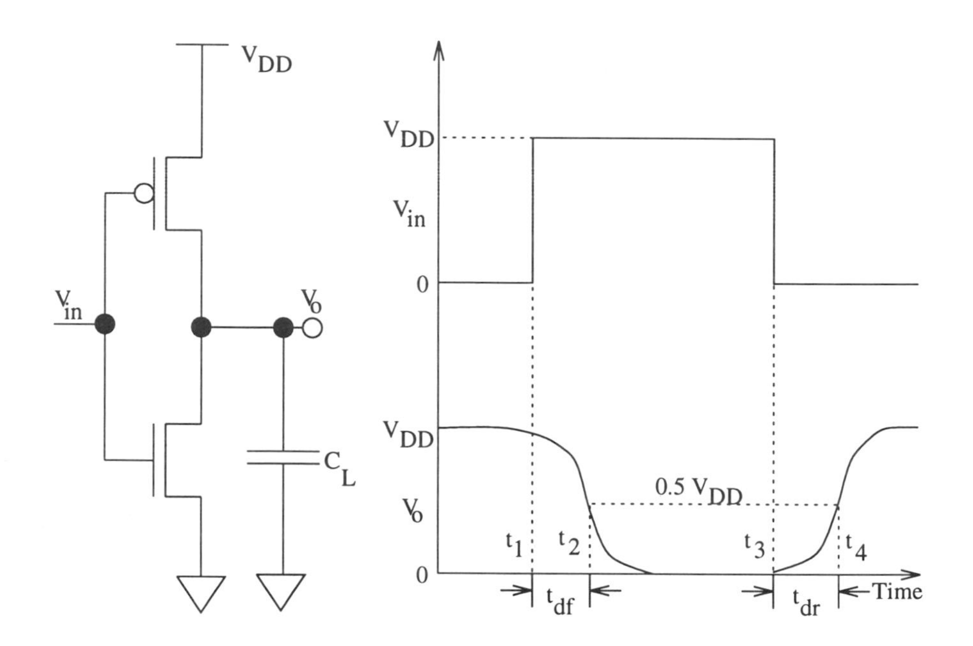

Figure 4.5 CMOS inverter and switching characteristic

4.2 CMOS INVERTER: SWITCHING CHARACTERISTICS

In this section, we present the transient behavior of the CMOS inverter. A very simple analytic model for delay is developed. The objective of this analysis is to understand the parameters that affect the speed of the gate. We assume that the input has a step waveform. The delay t_d, is the time difference between the mid point of the input swing and the mid point of the swing of the output signal. Referring to Fig. 4.5,

- t_{dr} is the 50% delay when the output is rising; and
- t_{df} is the 50% delay when the output is falling.

The power dissipation issue during the switching is considered in Section 4.3.

4.2.1 Analytic Delay Models

The load capacitance shown in Fig. 4.5 at the output of the CMOS inverter represents the total of the input capacitance of driven gates, the parasitic capacitance at the output of the gate itself and the wiring capacitance. In Section 4.4, we discuss the estimation of this load capacitance. For simplicity we assume for 50% delay, that the MOS current is averaged, and is equal to the saturation current. The equation of the saturation used in this section is the one given by Equation (3.82) Section 3.3.3. This saturation current is well modeled for short-channel devices.

4.2.1.1 Fall Delay

When the input goes from low (ground) to high (V_{DD}), initially the output is at V_{DD}, the pull-down NMOS of Fig. 4.5 is in the saturation region. We assume that when the output falls to $V_{DD}/2$, the NMOS drain current is approximated by the saturation current I_{DSsatn}. Referring to the equivalent circuit of Fig. 4.6(a), the delay is computed from the following differential equation

$$C_L \frac{dV_o}{dt} + I_{DSsatn} = 0 \tag{4.29}$$

where

$$I_{DSsatn} = K_n \nu_{sat} C_{ox} W_{effn}(V_{GSn} - V_{Tn}) \tag{4.30}$$

We assume that the factor K_n does not change. By integrating Equation (4.29) from $t = t_1$, corresponding to $V_o = V_{DD}$, to $t = t_2$, corresponding to $V_o = V_{DD}/2$, and substitution of (4.30) into (4.29) we obtain

$$t_{df} = \frac{0.5 C_L V_{DD}}{K_n \nu_{sat} C_{ox} W_{effn}(V_{DD} - V_{Tn})} \tag{4.31}$$

Note from this equation that the delay is inversely proportional to the width of the MOS transistor. So by sizing the gate we can reduce the delay of the gate alone.

4.2.1.2 Rise Delay

When the input goes from high (V_{DD}) to low (ground), initially the output is at zero. The pull-up PMOS transistor operates in the saturation region. Similarly using the equivalent circuit of Fig. 4.6(b), the rise delay is given by

$$t_{dr} = \frac{0.5 C_L V_{DD}}{K_p \nu_{sat} C_{ox} W_{effp}(V_{DD} - |V_{Tp}|)} \tag{4.32}$$

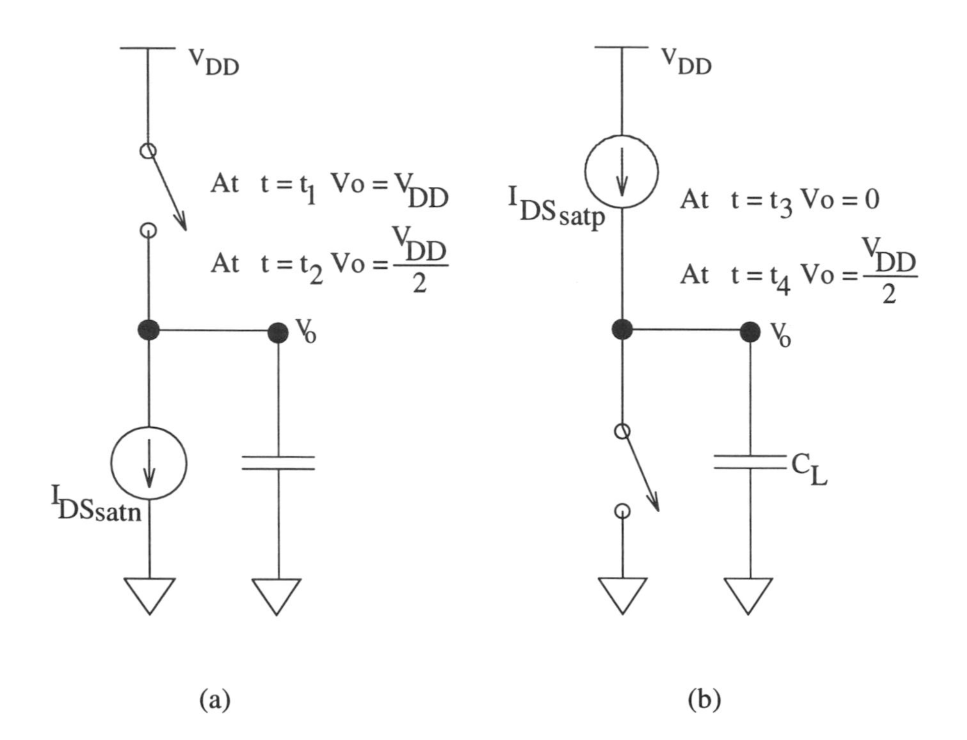

Figure 4.6 Equivalent circuit for delay calculation: (a) fall; (b) rise

From the above equation we can deduce that the rise delay is greater than the fall delay for equally sized MOS transistors. So W_{effp} should be sized such that the two saturation currents are almost equal in order to get symmetrical rise and fall delays.

4.2.1.3 Delay Time

By definition, the delay time (sometimes called propagation delay) is given by

$$t_d = \frac{1}{2}(t_{df} + t_{dr}) \tag{4.33}$$

Hence, for $V_{Tn} = -V_{Tp} = V_T$ the delay is given by

$$t_d = \left[\frac{1}{K_n W_{effn}} + \frac{1}{K_p W_{effp}}\right] \frac{0.5 C_L V_{DD}}{\nu_{sat} C_{ox}(V_{DD} - V_T)} \tag{4.34}$$

Or the equation can be written as

$$t_d = constant.\frac{C_L V_{DD}}{(V_{DD} - V_T)} \tag{4.35}$$

The constant is slightly affected by V_{DD} through the parameter K. This equation shows a simple analytic expression for the delay time. We can observe that the delay is linearly proportional to the total load capacitance. Secondly, the delay increases when the power supply is scaled down. When V_{DD} approaches the threshold voltage of the device, the delay increases drastically. If the threshold voltage is scaled down with the supply voltage and the oxide thickness is scaled down too, then the delay can improve with V_{DD} scaling. From the CMOS circuit designer point of view, the only parameters that can be controlled to optimize the speed of CMOS gates are:

- The width of the MOS transistor;
- The load capacitances (input of the next stages, wiring, etc.); and
- The supply voltage V_{DD}.

Fig. 4.7(a) shows the simulated effect of the power supply voltage on the delay of an inverter with $fanout = 3$, using the device parameters given in Chapter 3. We buffer the input voltage with one inverter stage to obtain accurate results. The delay is almost stable at high V_{DD}, however when V_{DD} approaches the threshold voltage of the NMOS and PMOS devices, it increases drastically as expected by Equation (4.35). Therefore, the threshold voltage should be reduced to overcome this problem. In Fig. 4.7(b), the delay of the inverter is plotted versus the ratio V_T/V_{DD} at $V_{DD} = 2.5$ V. For $V_T/V_{DD} > 0.5$, the delay increases rapidly. In order to maintain improvement in circuit performance at reduced power supply voltage, V_T/V_{DD} must be ≤ 0.2.

4.2.2 Delay Characterization with SPICE

A data sheet for the delay of a cell (i.e., CMOS inverter) can be easily prepared using SPICE. For example the load capacitance or the fanout of a CMOS inverter is swept during the simulation, and the relation of the type $t_d = a + b.C_o$(or $fanout$) can be obtained. Fig. 4.8 shows the delay vs. the external load capacitance C_o. Other parameters can be extracted also.

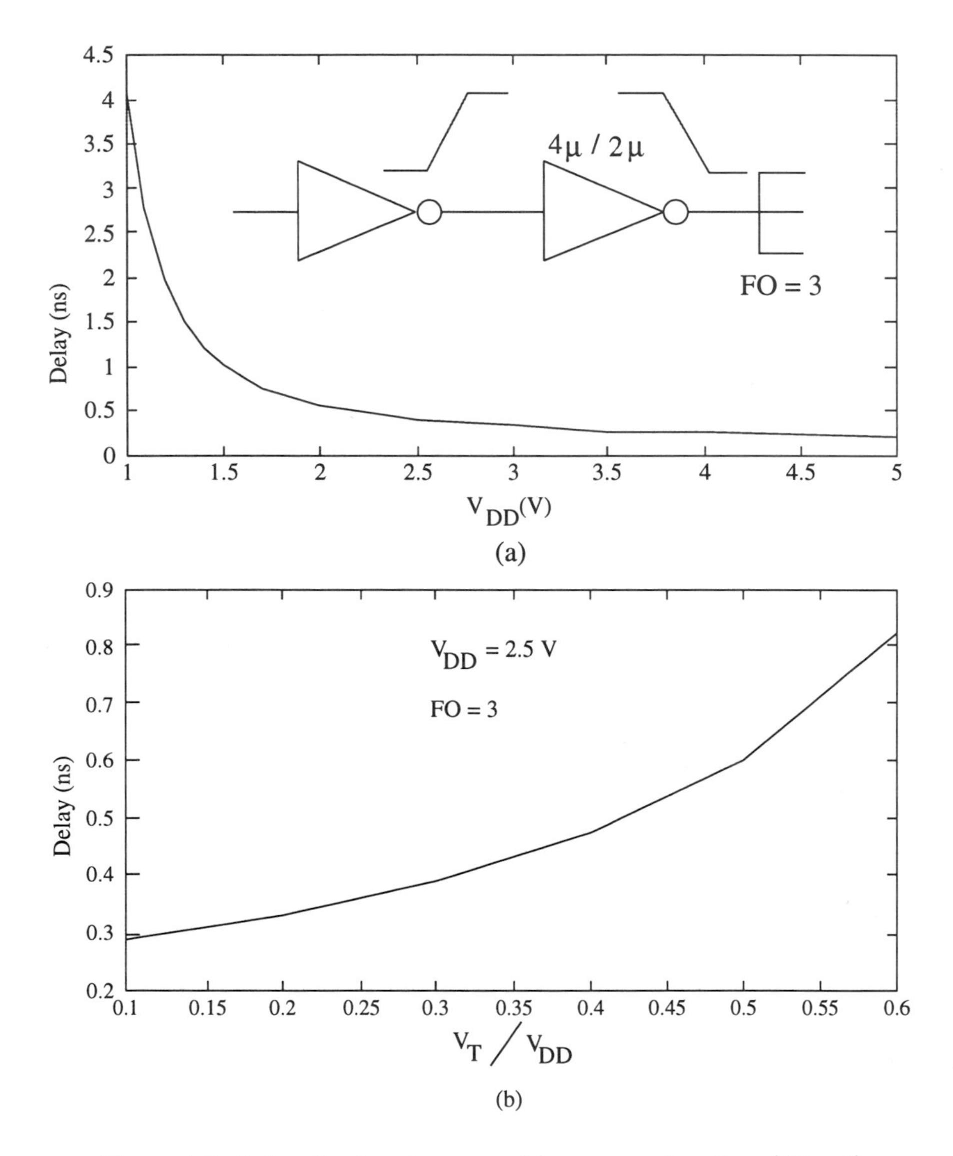

Figure 4.7 Delay of an inverter versus: (a) power supply voltage (FO = 3); (b) ratio of threshold voltage to power supply voltage

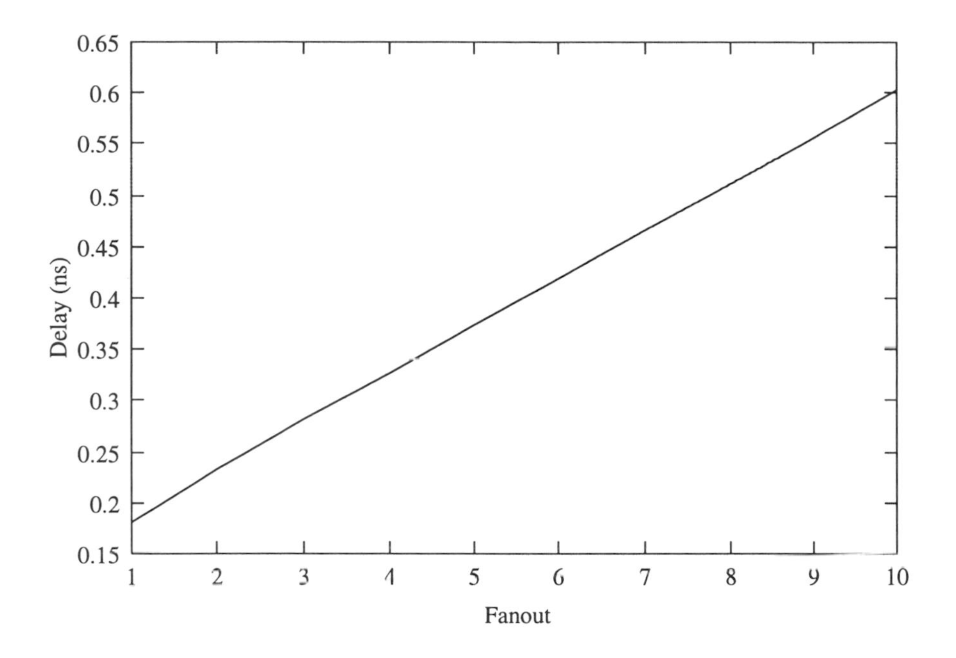

Figure 4.8 Inverter delay versus fanout @$V_{DD} = 3.3$ V

4.3 POWER DISSIPATION

To minimize the power consumption of a CMOS circuit, the various power components and their effect must be identified. There are two types of power dissipation. One is the maximum power dissipation which is related to the peak of the instantaneous current and the other is the average power dissipation. The peak current has an effect on the supply voltage noise due to the power line resistance. It can cause heating of the device, thus resulting in performance degradation. From the battery lifetime point of view, the average power dissipation is more important.

There are three power dissipation components within the CMOS inverter. These are:

1. Static power caused by the leakage current I_{leak} and other static current I_{st} due to the value of the input voltage;
2. Dynamic power caused by the total output capacitance C_L; and

3. Dynamic power caused by the short-circuit current I_{sc} during the switching transient.

Sometimes component (2) and (3) are merged as total dynamic power.

4.3.1 Static Power

This component is split sometimes into two other components. The sources of static power dissipation, in a complementary CMOS inverter, are leakage currents (P_{s1}) and current drawn from the supply due to the input voltage (P_{s2}). Hence the total static power is given by

$$P_s = P_{s1} + P_{s2} \tag{4.36}$$

Leakage current consists of MOS junction leakage currents. Fig. 4.9 shows the parasitic diodes in a CMOS inverter. The body ties in this structure, such as the parasitic diodes, are not conducting (i.e, reverse biased and/or at zero voltage). The current in a diode is given by

$$I_d = I_s(\exp \frac{qV_d}{nkT} - 1) \tag{4.37}$$

where n is the emission coefficient of the diode (sometimes equal to 1) and V_d is the applied voltage to the diode. Note that the current parameter I_s increases with temperature. The total power dissipation due to these leakage currents is given by

$$P_{s1} = \sum_i I_{di} V_{DD} \tag{4.38}$$

A typical value of this leakage current I_d is 1 $fA/$ device junction. This value is too small to have any effect on the static power, because if we have one million devices, the total contribution to the power would be $\sim 0.01\ \mu W$. This first component of the static power is neglected, in the analysis, through all the chapters of this book except Chapter 6 in the case of memory design.

We consider now the second component of the static power which is a function of the input voltage V_{in}. Assume that the input of the pull-down NMOS, of the inverter, is at a voltage $0 \leq V_{in} < V_T$. In this case the current is given by the subthreshold expression (Fig. 4.10)

$$I_{DS} = I_0 \frac{W_{eff}}{W_0} 10^{\frac{(V_{in} - V_T)}{S}} \tag{4.39}$$

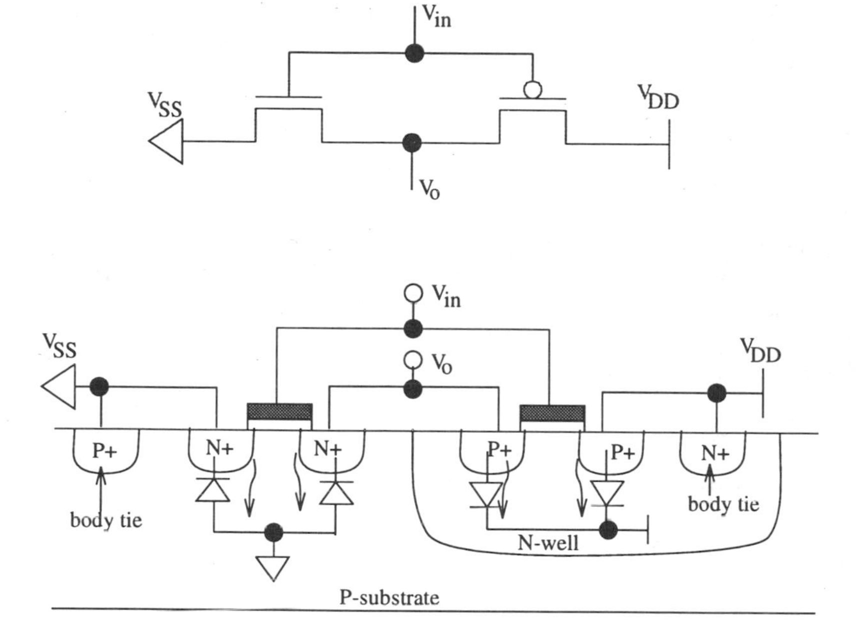

Figure 4.9 Leakage current in a CMOS inverter

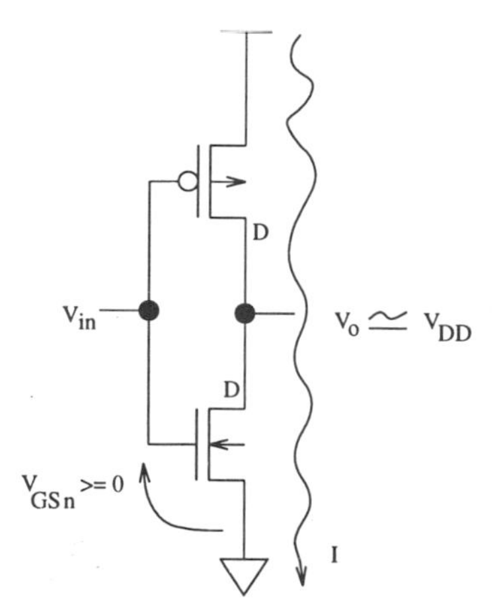

Figure 4.10 Static current in a CMOS inverter when the input is higher than 0

where V_T is the constant-current threshold voltage. For $V_{in} > V_T$ the current is given by expressions discussed in Chapter 3. The corresponding static power dissipation is given by

$$P_{s2} = I_{DSmean} V_{DD} \tag{4.40}$$

The mean value of the current is for both the PMOS and NMOS transistors. For example if $V_{in} = 0$, $V_T = 0.15$ V, $W_{eff} = 10\ \mu m$ and $S = 75\ mV/decade$, this current is 1 nA. For 1 million devices integrated, the total static power would be important (1 mA of current). Note that this current increases drastically with the increase of temperature [see Section 3.3.2]. This value, in standby mode, is not permitted for battery-operated applications. CMOS circuits have been known to consume energy only during switching. But this is not true now, since low-V_T CMOS is used for low-voltage operation. Some CMOS circuits, which exhibit a high DC current, are discussed in Section 4.6.

4.3.2 Dynamic Power of the Output Load

In this section we estimate the power dissipation due to the total output load capacitance C_L. This power is due to the currents needed to charge and discharge C_L as shown in Fig. 4.11 and 4.12. We assume a step input so neither the PMOS and NMOS are on simultaneously. The average dynamic power P_d required to charge and discharge a capacitance C_L at a switching frequency $f = 1/T$ (Fig. 4.12) is given by

$$P_d = \frac{1}{T}\int_0^T i_o(t) v_o(t) dt \tag{4.41}$$

The output current is given during charging phase by

$$i_o = i_p = C_L \frac{dv_o}{dt} \tag{4.42}$$

and during the discharge phase by

$$i_o = i_n = -C_L \frac{dv_o}{dt} \tag{4.43}$$

Then Equation (4.41) becomes

$$P_d = \frac{1}{T}\left[\int_0^{V_{DD}} C_L v_o dv_o - \int_{V_{DD}}^{0} C_L v_o dv_o\right] \tag{4.44}$$

Finally the dynamic power dissipation is

$$P_d = \frac{C_L V_{DD}^2}{T} = C_L V_{DD}^2 f \tag{4.45}$$

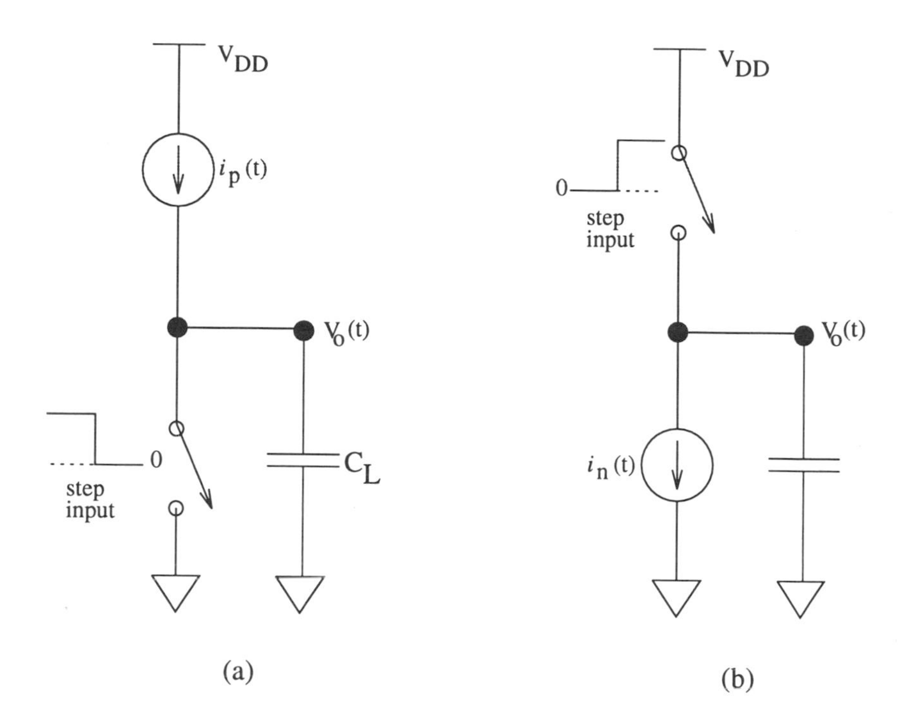

Figure 4.11 Equivalent circuit for power calculation

This equation shows that the power dissipation is proportional to the operating frequency. Moreover, the reduction of the power supply drastically reduces the power dissipation. Ideally, 3.3 V supply voltage reduces the power dissipation by 56% compared to that of 5 V. Moreover, at 1 V the power is reduced by 96% compared to 5 V. The expression of dynamic power in Equation (4.45) is valid only for an inverter. However, for a complex gate the concept of switching activity is introduced [see Section 4.5.3].

During the first output transition (charging) from $0 \rightarrow V_{DD}$, the energy drawn from the power supply is $E_d = C_L V_{DD}^2$. For this transition, the energy stored in the load capacitor is

$$E_{cap} = \int_0^{V_{DD}} C_L v_o dt = \frac{1}{2} C_L V_{DD}^2 \tag{4.46}$$

This means that during the output transition $0 \rightarrow V_{DD}$, half of the energy drawn from the supply is stored in the capacitor and the other half is consumed

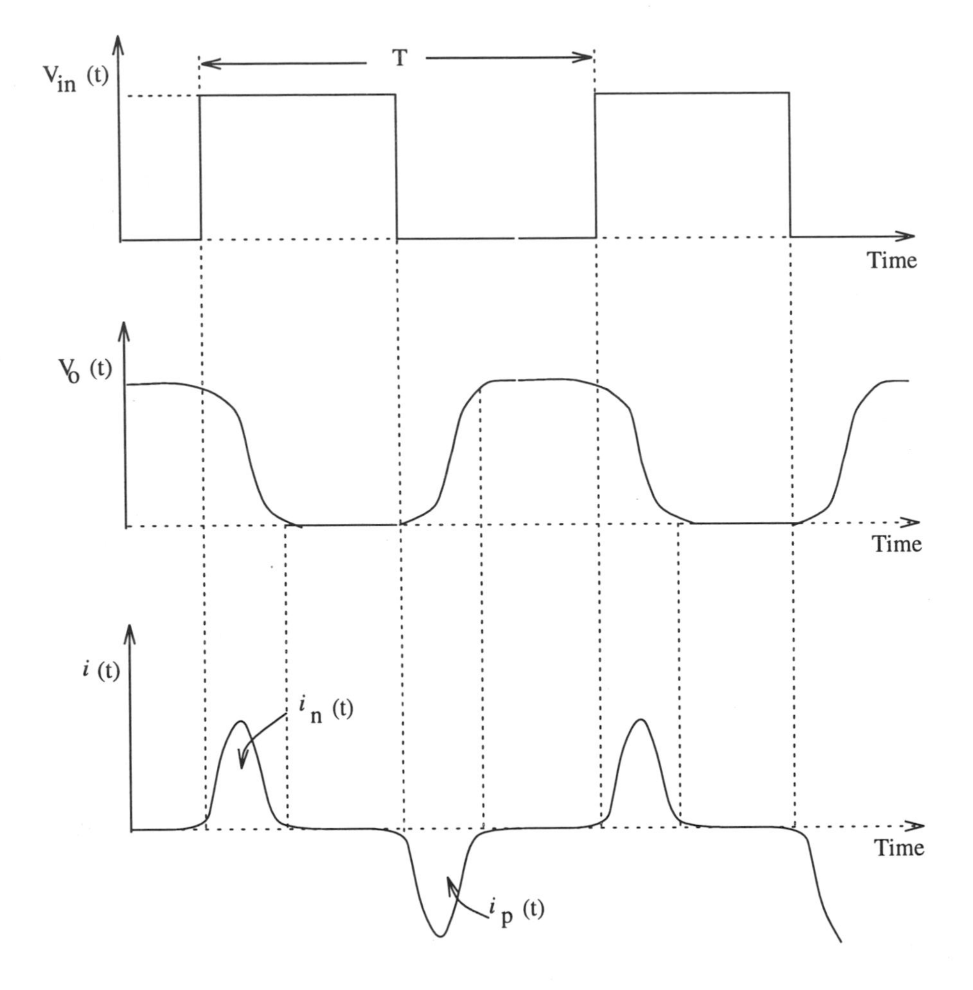

Figure 4.12 Voltage and current waveforms

by the pull-up PMOS transistor. For the output transition $V_{DD} \rightarrow 0$, the energy $(1/2\ C_L V_{DD}^2)$ stored in the capacitor is consumed by the pull-down NMOS transistor and no current is drawn from the supply.

4.3.2.1 Energy vs. Power

It is important to distinguish between *energy* and *power*. If for example, for a CMOS gate we reduce its clock rate (f), its power consumption will be reduced by the same proportion. However, its energy will still be the same. Assume that the gate is powered with a battery to perform computations. The time required to complete the computation, with low clock rate, will be increased. Therefore, after the computation the battery will be just as dead as if the computation had been performed at high clock rate. So *low-energy* design is more important than *low-power* design. The factor of merit in this case can be defined as the *product of energy times the delay*. The conventional term, low-power, is used through out this book to mean that we design for low-energy.

4.3.3 Short-Circuit Power Dissipation

Even if there were no load capacitance on the output of the inverter and the parasitics are negligible, the gate would still dissipate switching energy. If the input changes slowly, both the NMOS and PMOS transistors are ON, an excess power is dissipated due to the short-circuit current. Fig. 4.13 shows the short-circuit currents as the inverter switches as function of the fall time of the input. We are assuming that the rise time of the input is equal to the fall time.

$$P_{sc} = I_{mean} V_{DD} \tag{4.47}$$

To estimate I_{mean} we use the simple model of the short-circuit current of Fig. 4.14 [5]. Also we assume that the inverter has symmetrical devices, which means that $\beta_n = \beta_p = \beta$ and $V_{Tn} = -V_{Tp} = V_T$. We also assume that the rise time is equal to the fall time of the input signal ($\tau_r = \tau_f = \tau$). The mean short-circuit current in the unloaded inverter is

$$I_{mean} = 2 \times \frac{1}{T}\left[\int_{t_1}^{t_2} i(t)dt + \int_{t_2}^{t_3} i(t)dt\right] \tag{4.48}$$

Due to symmetry we have

$$I_{mean} = \frac{4}{T}\left[\int_{t_1}^{t_2} i(t)dt\right] \tag{4.49}$$

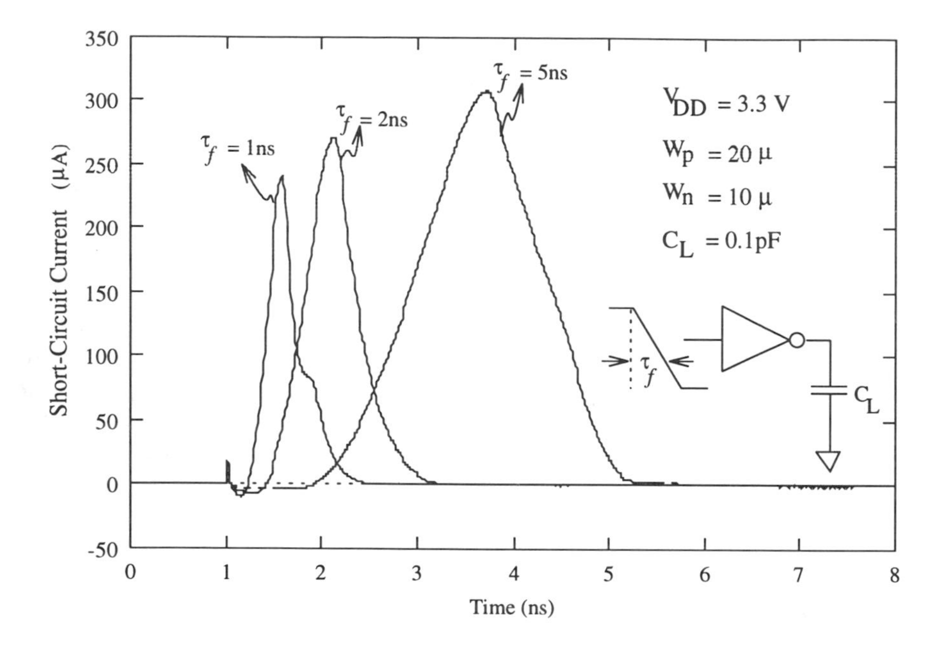

Figure 4.13 Short-circuit current function of the input slope

The NMOS transistor is operating in saturation, hence the above equation becomes

$$I_{mean} = \frac{4}{T}\left[\int_{t_1}^{t_2} \frac{\beta}{2}(V_{in}(t) - V_T)^2 dt\right] \tag{4.50}$$

The input voltage is given by

$$V_{in}(t) = \frac{V_{DD}}{\tau}t \tag{4.51}$$

It can be derived from Fig. 4.14 that

$$t_1 = \frac{V_T}{V_{DD}}\tau \quad and \quad t_2 = \frac{\tau}{2} \tag{4.52}$$

Then the integral leads to

$$P_{sc} = \frac{\beta}{12}(V_{DD} - 2V_T)^3 \tau f \tag{4.53}$$

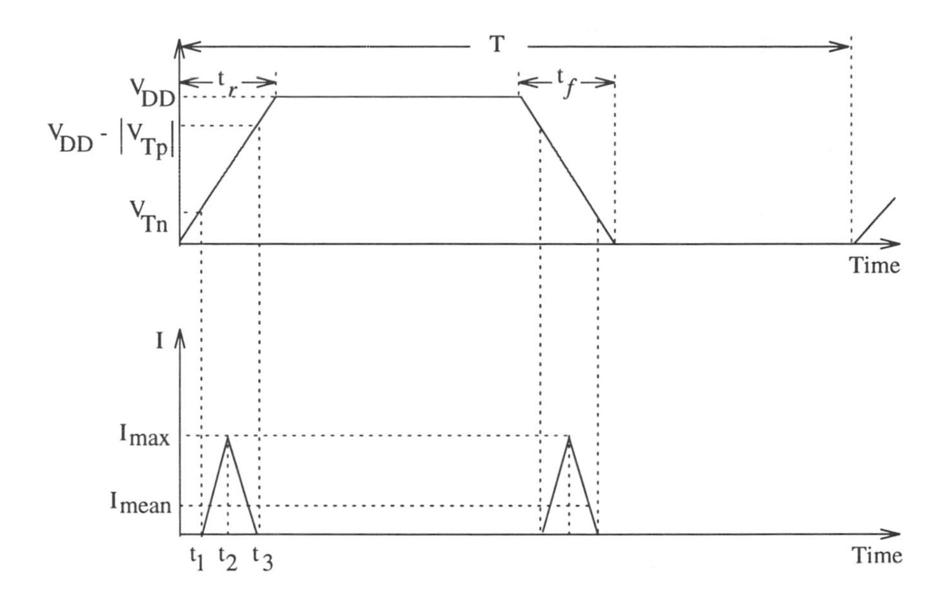

Figure 4.14 Input voltage and short-circuit current model

This equation shows that the short-circuit power dissipation is also proportional to the frequency. The only parameters that can be controlled by the circuit designer at given frequency and power supply to reduce P_{sc} are: β and τ. The power supply scaling greatly affects the reduction of short-circuit power dissipation. Note that this analysis was done for an unloaded inverter. For a loaded gate, if the output signal and input signal have equal rise/fall times, the short-circuit power dissipation will be less than 20% of the total power [5]. So it is very important to keep the edges fast, to have negligible P_{sc} or at least, it is desirable to have equal input and output rise/fall times.

If the load capacitance is high, the output rise/fall times become larger than the input ones. In this case, the input changes completely before the output changes significantly. Therefore, the short-circuit current is near zero. Note that if V_{DD} is approaching $(V_{Tn} + V_{Tp})$ or is less, the short circuit current can be eliminated because both devices can not conduct simultaneously.

4.3.4 Other Power Issues

The total power dissipation of a CMOS gate is given by

$$P_{total} = P_s + P_d + P_{sc} \tag{4.54}$$

It represents the total power of a gate when it is switching at the same rate as the operating frequency. In Chapter 8, we will discuss how to estimate the power dissipation of a complex circuit.

Other power dissipation issues exist, such as: worst case power estimation and temperature effect. These conditions are : maximum V_{DD} and junction temperature, and fast-fast process. Static power dissipation (subthreshold current) is increased by the increased temperature and increased power supply. Dynamic power is not sensitive to the temperature but it is affected greatly by the worst case V_{DD}. Short-circuit power dissipation depends on the temperature just as the short-circuit current does. It is also dependent on the power supply. The mobility and threshold voltage decrease with increasing temperature. Each of these two parameters has an opposite effect on the current. So it is important to consider the worst case power consumption evaluation in any design.

The simulated average total power dissipation can be easily measured by the SPICE simulator using POWER MEASUREMENT commands. However, several papers in the literature have introduced "power meter" in circuit simulation to measure the power dissipation [6, 7, 8].

4.4 CAPACITANCE ESTIMATION

Previously we saw that the speed and power dissipation of CMOS gates depend strongly on the total output load capacitance. This capacitance is the sum of three components as shown in Fig. 4.15.

- Total input capacitances of N driven gates noted C_{in};
- Parasitic output capacitance of the drive gate noted C_p; and
- Wiring capacitance noted C_w.

For simplicity we estimate, in this section, the average value of C_L over the range of the output swing. This approach is used only for initial estimation

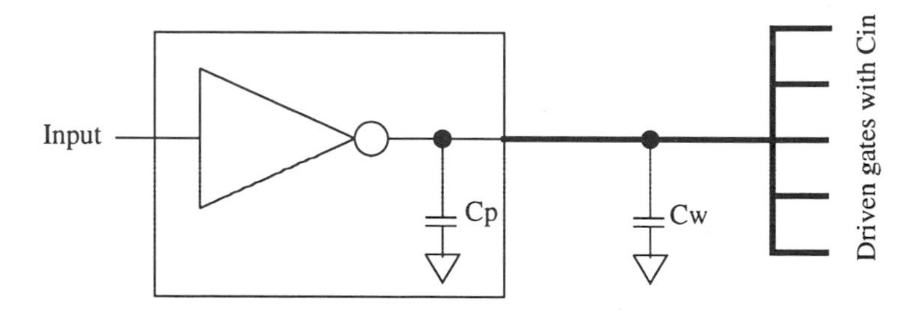

Figure 4.15 Load capacitance in a CMOS gate

of the design. More circuit simulation and layout extraction and post-layout simulation are needed for more accuracy. Moreover, it is sometimes interesting to derive a simple expression for the load capacitance to see the impact of important parameters on the speed and the power dissipation. We first examine the different components of the output load capacitance; then we illustrate by an example the estimation approach.

4.4.1 Estimation of C_{in}

The total capacitance of the driven gates can be evaluated by summing the input capacitance of all the receiving gates and we have

$$C_{in} = \sum_{i=0}^{N} C_{gate,i} \tag{4.55}$$

The gate capacitance of the receiving gate can be approximated by

$$C_{gate} = C_{ox} \sum_{i=1}^{n} (WL)_i \tag{4.56}$$

where n is the number of transistors of the gate. This expression sums the gate capacitances of all the transistors composing the driven circuit. For a CMOS inverter it is given by

$$C_{gate} = C_{ox} \left[(WL)_p + (WL)_n \right] \tag{4.57}$$

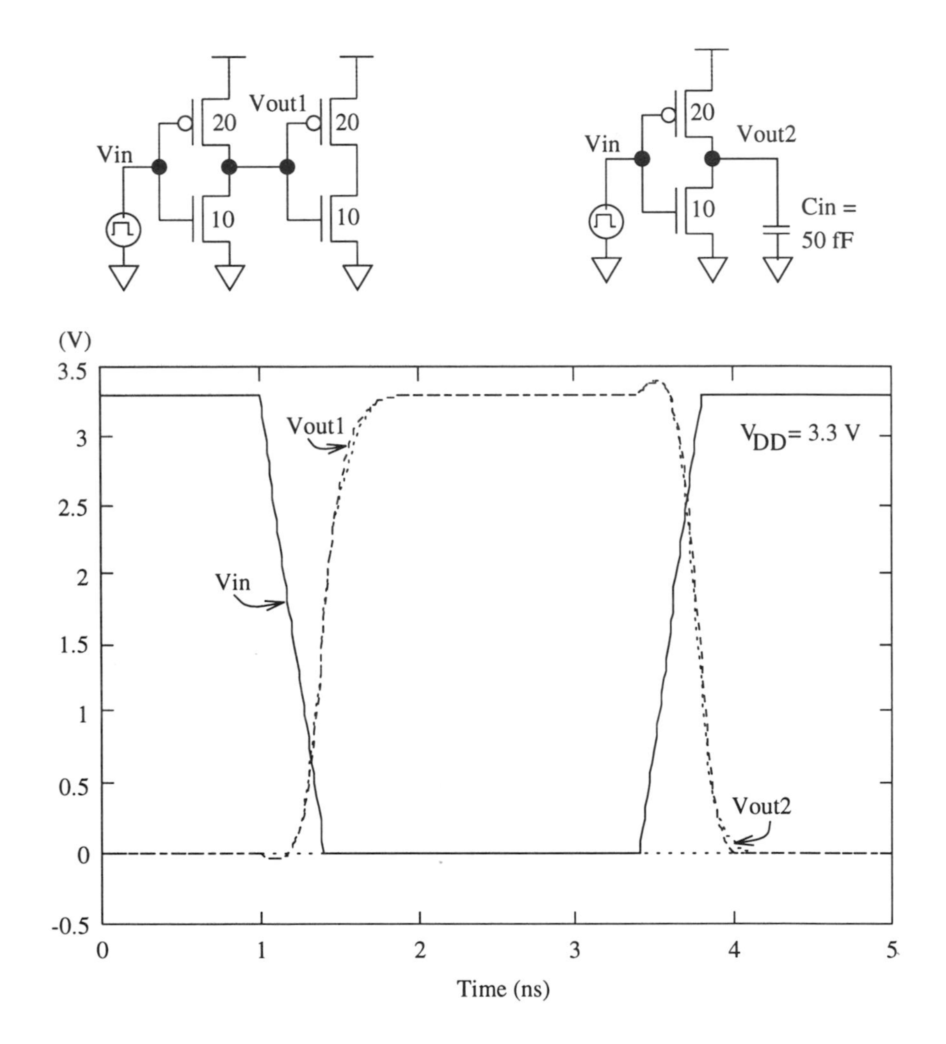

Figure 4.16 Equivalent capacitance of a gate

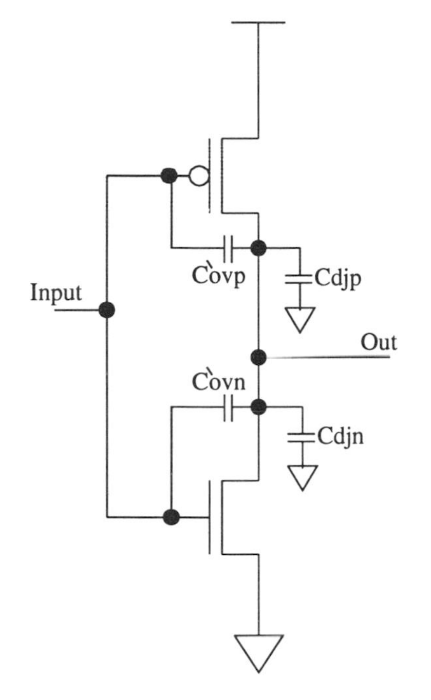

Figure 4.17 Parasitic capacitances at the output of a CMOS inverter

Figure 4.16 shows an example of the equivalent gate capacitance of the receiving gate. The driven inverter has the following drawn sizes : $W_p = W_n = 20\ \mu m$ and $L = 0.8\ \mu m$. This gate can be replaced by an equivalent capacitance $C_{gate} \approx 50\ fF$, which is approximately the same as the one calculated from Equation (4.57).

4.4.2 Parasitic Capacitances

Fig. 4.17 shows the main contributions to the output parasitic capacitances of a CMOS inverter. Thus, it is estimated by

$$C_p = C'_{ovp} + C'_{ovn} + \bar{C}_{djp} + \bar{C}_{djn} \tag{4.58}$$

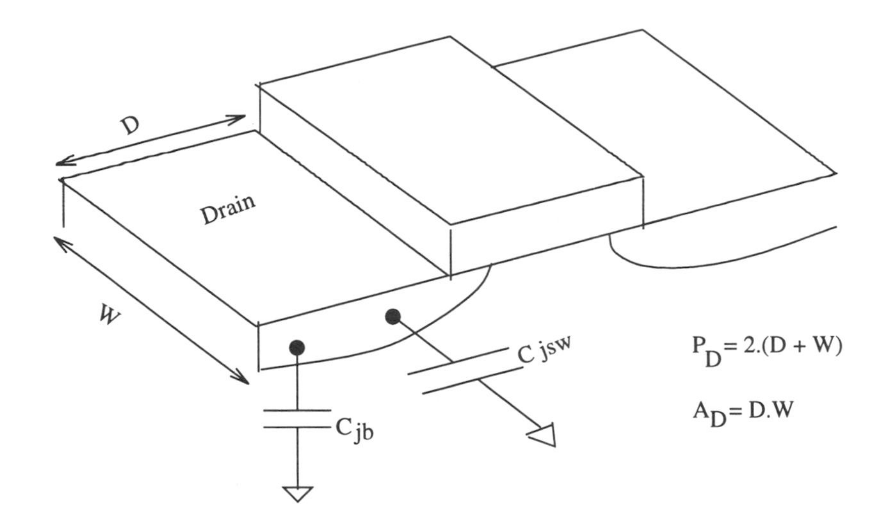

Figure 4.18 Drain junction capacitance components

The drain overlap capacitance for NMOS and PMOS is given by

$$C'_{ov} = C_{ov}W \tag{4.59}$$

C_{ov} is defined in SPICE parameters of Chapter 3 as C_{GDO}. The drain junction capacitance is a function of the reverse applied voltage during the switching of the inverter. The average value of this capacitance over the range of output swing is defined by

$$\bar{C}_{dj} = \bar{C}_{jb}A_D + \bar{C}_{jsw}P_D \tag{4.60}$$

where A_D and P_D are the area and the perimeter of the drain junction as shown in Fig. 4.18. The average bottom junction capacitance is

$$\bar{C}_{jb} = \frac{1}{V_{DD}}\int_0^{V_{DD}} \frac{C_{jb0}}{[1+V_j/V_b]^{mj}}dV_j \tag{4.61}$$

The average side-wall capacitance

$$\bar{C}_{jsw} = \frac{1}{V_{DD}}\int_0^{V_{DD}} \frac{C_{jsw0}}{[1+V_j/V_b]^{mjsw}}dV_j \tag{4.62}$$

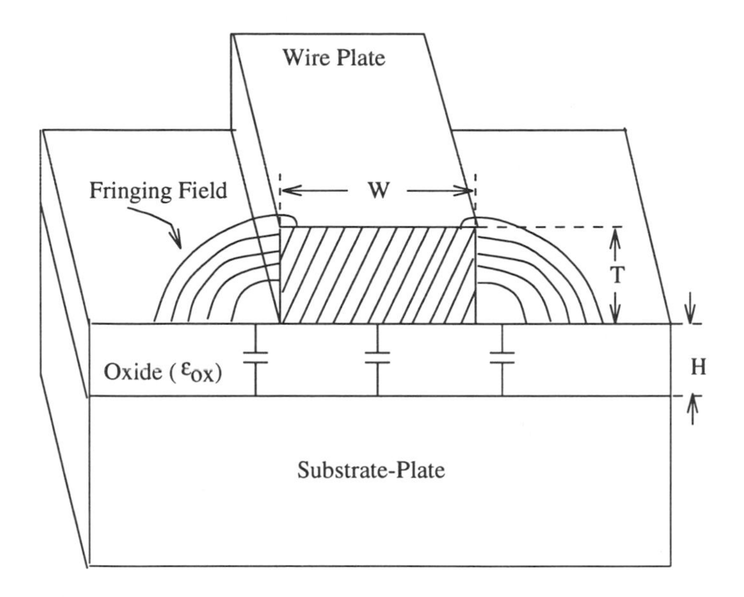

Figure 4.19 Wire capacitance model

4.4.3 Wiring Capacitance

The simple model of wiring capacitance is based on the parallel-plate model [Fig. 4.19] given by

$$C_{wa} = \frac{\epsilon_{ox}}{H} \tag{4.63}$$

where H is the thickness of the insulator layer (oxide), and C_{wa} is the capacitance per area unit. The total capacitance of the wire is

$$C_w = lWC_{wa} \tag{4.64}$$

where W is the width of the wire (metal or poly), and l is the length of the wire. Table 4.1 gives some values of the wiring capacitance per area for the 0.8 μm process presented in Chapter 2. This capacitance can not be known in the early design stage but can be known after layout extraction.

When the thickness of the insulator becomes comparable to that of the wire, T, then the fringing fields at the edge of the wire become important. The effect of the fringing fields is manifested by the increase of the effective area of the plates [Fig. 4.19]. Many approximations have been proposed to compute the

Table 4.1 Typical 0.8-μm CMOS wire capacitance for parallel plate.

Layer	Area Capacitance (10^{-18} $F/\mu m^2$)
Metal2 to Substrate	11
Metal2 to Metal1	25
Metal1 to Substrate	19
Metal1 to poly	28
Metal1 to diffusion	27
Gate poly over field oxide	58

Table 4.2 Typical 0.8-μm CMOS wire fringing capacitance.

Layer	Perimeter Capacitance (10^{-18} $F/\mu m$)
Metal2 to Substrate	38
Metal2 to Metal1	47
Metal1 to Substrate	44
Metal1 to poly	48
Metal1 to diffusion	47
Gate poly over field oxide	44

effect of fringing capacitance. One relatively accurate empirical approximation is given by [9]

$$C_{wl} = \epsilon \left[(\frac{W}{H}) + 0.77 + 1.06(\frac{W}{H})^{0.25} + 1.06(\frac{T}{H})^{0.5} \right] \quad (4.65)$$

where C_{wl} is the total capacitance of the wire per unit length. The contribution of the fringing effect in many cases is important. Table 4.2 shows the fringing capacitance per unit of length.

4.4.4 Example

Consider an inverter with $W_p = 2W_n = 20$ μm with 3 μm length of each drain and source. This inverter is driving a line of metal1 of 100 μm length by 2 μm width and an inverter with $W_p = 2W_n = 20$ μm operating at $V_{DD} = 3.3$ V.

The total load capacitance is computed using the 0.8 μm device parameters presented in Chapter 3 as follows:

- The gate capacitance of the driven inverter is

$$\begin{aligned} C_g &= [W_p L_p + W_n L_n] C_{ox} \\ &= [20 \times 0.8 \ + \ 10 \times 0.8] \times 2fF \ \approx \ 48fF \end{aligned}$$

- The total overlap capacitance at the output is

$$C_{ovt} = C_{GDOp} W_p + C_{GDOn} W_n$$

  Then

$$\begin{aligned} C_{ovt} &= 20 \times 215 \times 10^{-3} + 10 \times 274 \times 10^{-3} \\ &= 4.30 + 2.74 \ \approx \ 7 \ fF \end{aligned}$$

- The total drain junction capacitances can be approximated at mid-voltage of 1.65 V (1/2 of V_{DD}) instead of computing integrals. We have for one drain junction

$$C_{dj} = \frac{A_D C_{jb0}}{[1 + 0.5 V_{DD}/V_b]^{mj}} + \frac{P_D C_{jsw0}}{[1 + 0.5 V_{DD}/V_b]^{mjsw}}$$

  The drain areas are 60 μm^2 and 30 μm^2 for PMOS and NMOS respectively. The drain perimeters are 46 μm and 26 μm for the PMOS and NMOS transistors respectively. The total junction capacitance can be easily calculated and is

$$C_j \ \approx \ 32 \ fF$$

  Note that this capacitance increases with the power supply voltage reduction.

- The wire capacitance is estimated by adding the two components parallel plate and fringing capacitances. The area of the wire is 200 μm^2 while its perimeter is 204 μm. We have

$$\begin{aligned} C_w &= W \times l \times C_w(per \ \ area) \ + \ 2(W + l) \times C_w(per \ \ length) \\ &= 200\mu m^2 \times 19 \times 10^{-3} fF/\mu m^2 + 204\mu m \times 44 \times 10^{-3} fF/\mu m \\ &= 3.8 \ + \ 9.0 \ \approx \ 13 \ fF \end{aligned}$$

  Note that the fringing capacitance is an important portion of the total wire capacitance.

Hence the total capacitance at the output is 100 fF. Note that the contribution of the junction capacitance is important. The contribution of each component varies from one circuit to another and it depends on the layout style used. Before starting any circuit layout, it is important to keep in mind an estimation of capacitances such as the gate and output capacitance of 1 unit size inverter and the wire capacitance of, for example, 100 μm poly line and 100 μm metal1 line. With these data, when starting the design, it is possible to size different transistors correctly.

4.5 CMOS STATIC LOGIC DESIGN

From the CMOS inverter we can realize any static logic function by using the complementary NMOS and PMOS transistors. In this section we present the design of NAND/NOR, complex and transmission gates. The *fanin* of any complex gate is defined as the number of inputs of this gate. The *fanout* of a complex logic gate is the number of driven inputs attached to the output of this gate.

4.5.1 NAND/NOR Gates

Fig. 4.20 shows a 2-input NAND gate (NAND2) and a 2-input NOR gate (NOR2). Each input requires a complementary pair. In the case of the NAND gate, the PMOS transistors are connected in parallel, while the NMOS transistors are connected in series. But in the case of the NOR gate, the NMOS devices are connected in parallel, while the PMOS devices are connected in series. These gates consume only dynamic power while the DC power dissipation is zero (if V_T's are high) because there is no DC path between V_{DD} and ground for any logic combination of the input. For the NAND and NOR gates of Fig. 4.20, any input combination ($AB = 00, 01, 11, or 10$) there is no path between the two rails.

The design of these gates, or any CMOS static gate, follows that of an inverter. As discussed in Sections 4.1 and 4.2, an inverter is designed to meet a given DC and transient performance, then $(W/L)_n$ and $(W/L)_p$ are determined. The $(W/L)_n$ and $(W/L)_p$ of the devices of a logic gate are determined as follows: For example we want to design a 3-input NAND (Fig. 4.21(a)) to have the same DC and transient as that of an inverter driving the same C_{out} (Fig. 4.21(b)).

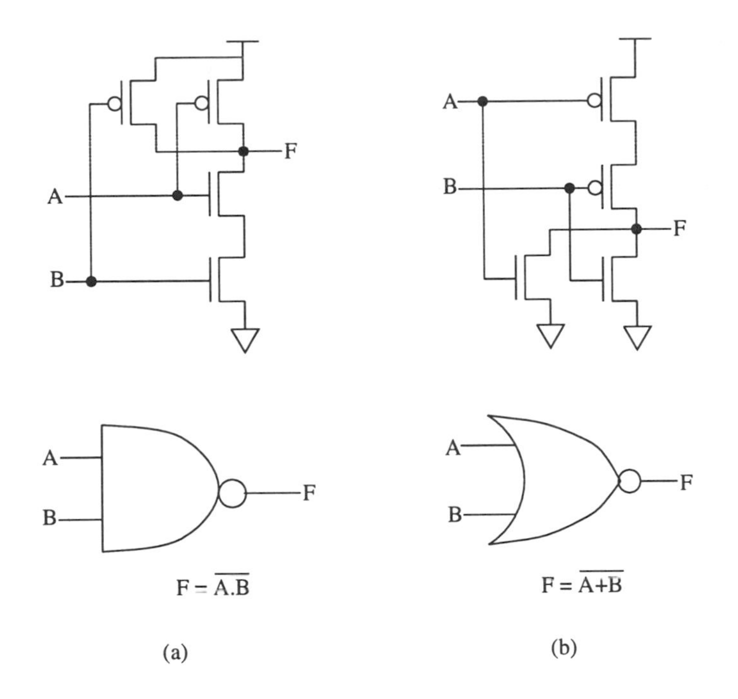

Figure 4.20 CMOS logic gates: (a) 2-input NAND; (b) 2-input NOR.

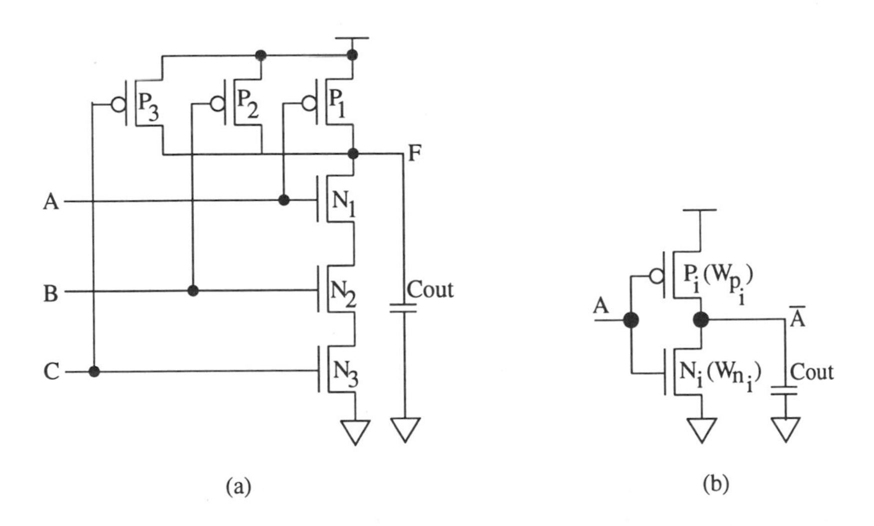

Figure 4.21 Sizing a CMOS gate.

We assume that

$$W_n = W_{n1} = W_{n2} = W_{n3} \tag{4.66}$$

and

$$W_p = W_{p1} = W_{p2} = W_{p3} \tag{4.67}$$

The first thing to do is to approximate the gate by an equivalent inverter where the effective β is given by

$$\frac{1}{\beta_{neff}} = \frac{1}{\beta_{n1}} + \frac{1}{\beta_{n2}} + \frac{1}{\beta_{n3}} = \frac{3}{\beta_n} \tag{4.68}$$

and

$$\beta_{peff} = \beta_p \tag{4.69}$$

To have V_{inv} of the gate in the midway of the power supply in DC characteristics, the following condition should be satisfied for the 3-input NAND gate (see Equation 4.18)

$$\beta_{peff} = \beta_{neff} \tag{4.70}$$

which means that

$$\beta_p = \frac{\beta_n}{3} \tag{4.71}$$

To have the same delay as an inverter with determined sizes, we should have (assuming that L is the same)

$$W_{pi} = W_{peff} = W_p \tag{4.72}$$

and

$$W_{ni} = W_{neff} = \frac{W_n}{3} \tag{4.73}$$

But in practice the size of these transistors, composing the 3-input NAND gate, should be increased because the output parasitic capacitance of the NAND gate (or any complex gate) is larger than that of the inverter. Hence

$$W_p > W_{pi} \tag{4.74}$$

and

$$W_n > 3W_{ni} \tag{4.75}$$

Note that by circuit simulation, we can properly size the transistors. Moreover, it should be noted that the back-gate bias effect has to be taken into consideration in the design of the series NMOS devices in NAND gate (or series PMOS in NOR). The series-connected MOSFETs, during switching, exhibit a threshold voltage increase due to a non-null source-substrate voltage as shown in the simulation example of Fig. 4.22. In Fig. 4.22(a), the transistor N_1 of the first NAND3 gate near the output *out*1, is driven by the latest signal because N_2 and N_3 are already ON. Therefore, the node a_1 is at the ground level and the source of the transistor N_1 is not subject to the body effect. In the other NAND3 gate, the transistor N_4 and N_5 are ON, while N_6 receives the input signal. In this case, the node a_2 and b_2 are at a certain voltage level. Hence, during the discharging period the transistors N_4 and N_5 are subject to the body effect. This effect slows the discharge of the output as shown in Fig. 4.22(b). The output *out*1 is discharged more rapidly than the output *out*2. One way to reduce the body effect at the logic level is to put the transistor, driven by the latest arriving signal, near the output. The early arriving signals should be used to discharge the nodes susceptible to the body effect. For example in an adder circuit, the transistor driven by the carry is placed near the output.

Let us derive the output parasitic capacitance of the m-input NAND gate and compare it to that of the CMOS inverter of Fig. 4.21(b). We have

$$C_p = mW_pC_{ovp} + W_nC_{ovn} + mC_{djp} + C_{djn} \tag{4.76}$$

The C_{djn} of the m-input gate is larger than that of the CMOS inverter by the ratio W_n/W_{ni}. From the above equation it is obvious that C_p of the m-input NAND gate is larger than that of the CMOS inverter.

Note that for the same performance and for the same number of inputs the NAND gate consumes less silicon area than that of a NOR gate because of the smaller area taken by the NMOS devices. Hence, CMOS NAND gates are more widely used than NOR gates. Moreover, the NOR gate consumes more power than the NAND gate.

4.5.2 Complex CMOS Logic Gates

The strategy used to build NAND/NOR gates can be extended to build more complex logic gates. Complex logic functions can be realized by connecting several NAND, NOR and INVERTER gates. However, they can also be efficiently realized using a single CMOS logic gate. Any complex CMOS gate is formed by two N and P logic blocks as shown in Fig. 4.23(a). The two blocks have the same number of transistors. Fig. 4.23(b) shows a three-input complex CMOS gate and its logic equivalent symbol. The topology of the block N is the dual of the block P, i.e., parallel connections become series and vice versa. In either the P or the N logic blocks, the parallel combination is placed far from the output to minimize the output capacitance and hence improves the speed and maybe the dynamic power dissipation. For example, the contribution of

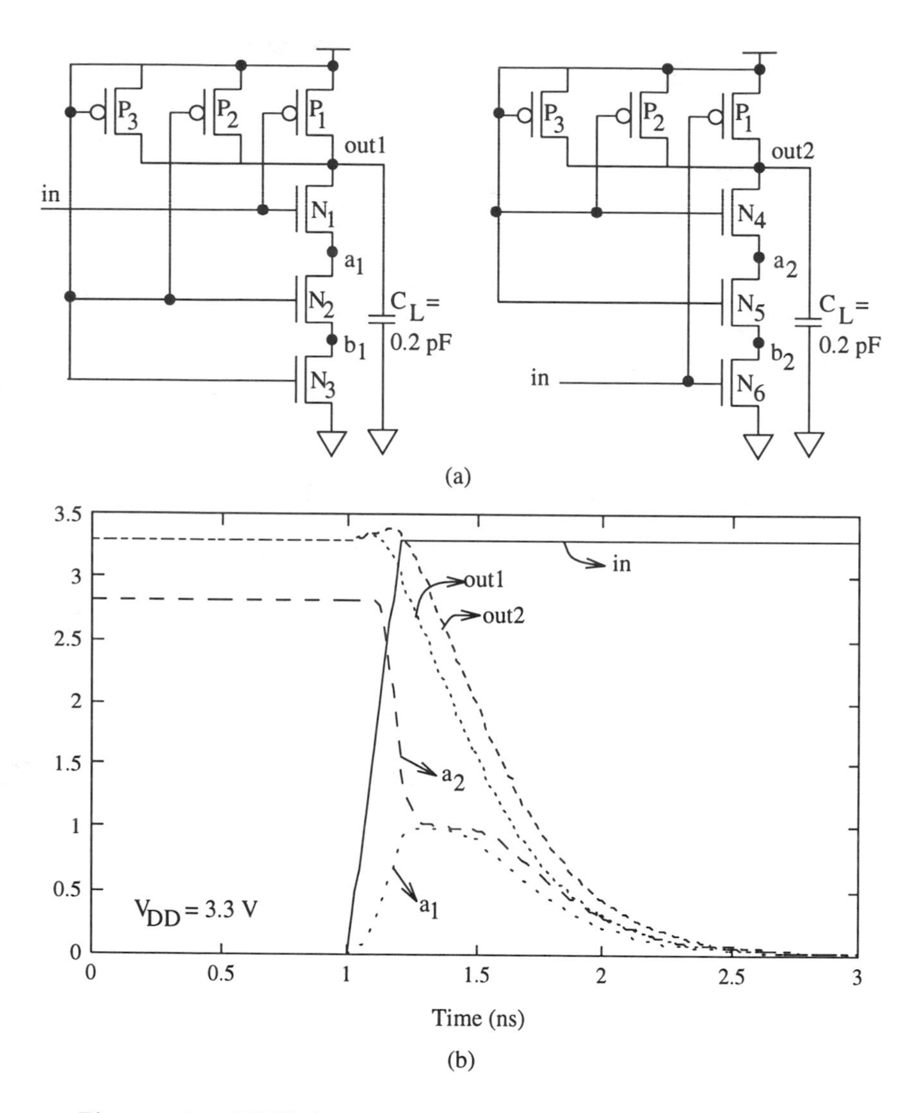

Figure 4.22 SPICE simulation of the body effect on NAND3 gate.

the N block to the output capacitance in Fig. 4.23(b) is less than that of Fig. 4.23(c). There is no direct DC path between V_{DD} and ground for any of the logic input combination. In practice, the complex CMOS gates are used for a maximum fanin of 5-6.

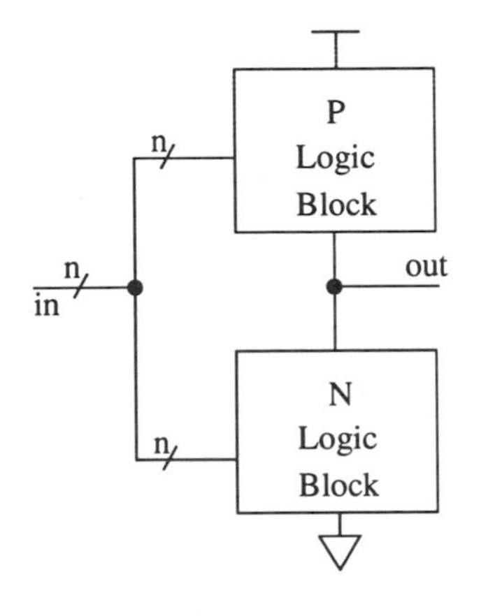

(a)

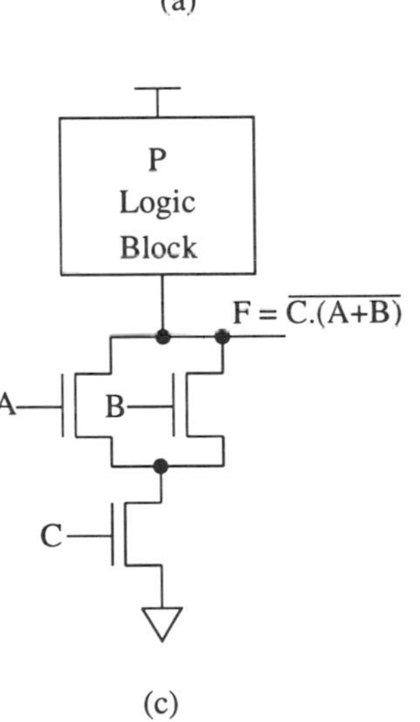

(c)

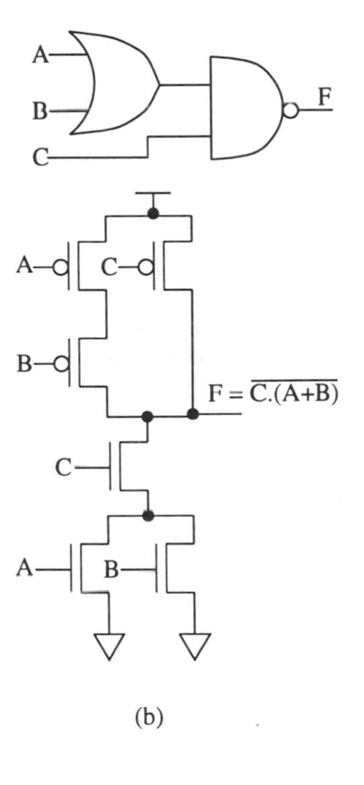

(b)

Figure 4.23 CMOS complex logic gate.

4.5.3 Switching Activity Concept

So far, we have discussed the dynamic power dissipation of an inverter due to the load capacitance. What about a CMOS complex gate driving a load capacitance ? The dynamic power dissipation has two components in a complex gate. The internal cell power, P_{int_dyn}, and the capacitive load power. The internal cell power consists of the power dissipated by of the internal capacitive nodes. Sometimes the internal short-circuit power is added to the internal cell dynamic power.

The dynamic power for a complex gate cannot be estimated by the simple expression $C_L V_{DD}^2 f$, because it might not always switch when the clock is switching. The switching activity determines how often this switching occurs on a capacitive node. For N periods of $0 \rightarrow V_{DD}$ and $V_{DD} \rightarrow 0$ transitions, the switching activity α determines how many $0 \rightarrow V_{DD}$[2] transitions occur at the output. In other words, the activity α represents the probability[3] that a transition $0 \rightarrow V_{DD}$ will occur during the period $T = 1/f$. f is the periodicity of the inputs of the gate. The average dynamic power of a complex gate due to the output load capacitance is

$$P_d = \alpha C_L V_{DD}^2 f \tag{4.77}$$

The internal power dissipation, due to the internal capacitive nodes, can be characterized by simulation. Fig. 4.24 illustrates an example of a complex gate with internal nodes. The internal dynamic power of a cell is given by

$$P_{int\_dyn} = \sum_{i=1}^{n} \alpha_i C_i V_i V_{DD} f \tag{4.78}$$

where n is the number of the internal nodes, α_i is the switching activity of each node i, C_i is the parasitic capacitance of the internal node, and V_i is the internal voltage swing of each node i. The parasitic capacitance at the output is included with the load C_L. Note that internal voltage swing can be different than V_{DD}.

4.5.4 Switching Activity of Static CMOS Gates

In this section we consider the computation of the switching activity of static CMOS gates. We will discuss the case of dynamic gates and other circuit styles

[2] During this transition the energy $C_L V_{DD}^2$ is drawn from the supply.
[3] We assume that the gate does not experience glitching.

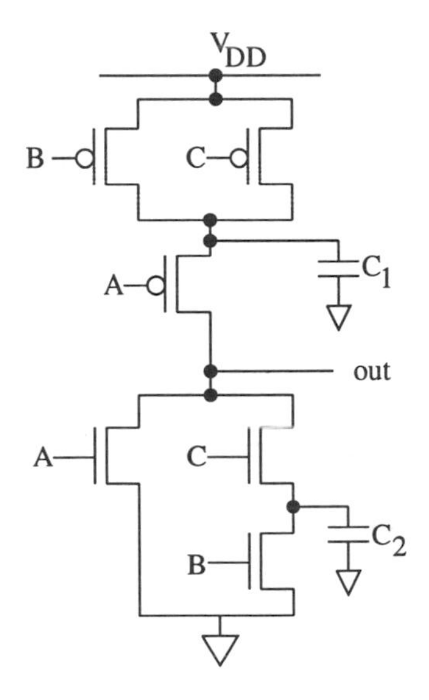

Figure 4.24 Internal capacitive nodes for a complex gate.

in the next sections. First we consider the case of a NOR gate. Then we treat several static gates. Table 4.3 illustrates the truth table of the NOR gate. From the table the probability that the output is at zero is 3/4 and that it is at one is 1/4. The probability for $0 \rightarrow V_{DD}$ transition is computed by multiplying the probability that the output will be at zero, P_0, by the probability it will be at one, P_1.

$$P_{NOR2} = P_0.P_1 = \frac{3}{4} \times \frac{1}{4} = \frac{3}{16} \tag{4.79}$$

We assume that the inputs are uniformly distributed (i.e, the probabilities P(A=1)=P(B=1)=1/2).

We can show that for any boolean function, the activity of a static gate is given by

$$\alpha = P(0 \rightarrow 1) = P_0.P_1 \tag{4.80}$$

where P_0 is computed by dividing the number of zeros by the total number of input combinations ($N = 2^n$ for n-input gate) and P_1 is computed by dividing the number of ones by N. P_0 is also equal to $(1 - P_1)$. Fig. 4.25 shows the probability that the output makes an $0 \rightarrow 1$ transition for several static gates. The probability of transitions at the inputs are assumed uniformly distributed.

Table 4.3 Truth table of a NOR gate. Only for one combination out of four the output is at 1.

A	B	Out
0	0	1
0	1	0
1	0	0
1	1	0

Now we consider the case where the inputs are not equi-probable. In this case, the probability at the output depends on the probabilities of transitions at the primary inputs. As an example of computing the activity, suppose we have a NOR2 gate with P_A and P_B the mutually independent signal probabilities of the inputs A and B to be at "1". The probability of a logical "1" at the output of this gate ($\bar{A}.\bar{B}$) is

$$P_1(NOR2) = (1 - P_A) \,.\, (1 - P_B) \quad (4.81)$$

and the probability of a logical "0" at the output of the NOR2 is

$$P_0(NOR2) = (1 - P_1) \quad (4.82)$$

Hence, the probability of transition $0 \rightarrow 1$ is the product of P_0 by P_1. For a NAND2 gate, the probability of a $0 \rightarrow 1$ transition at the output of this gate is

$$P(0 \rightarrow 1) = (1 - P_A.P_B) \,.\, P_A P_B \quad (4.83)$$

Consider now another example for a gate with the following boolean function

$$G = A.\bar{B}.C \quad (4.84)$$

The probability of the output G being "1" is given by

$$P_1(G) = P_A \,.\, (1 - P_B) \,.\, P_C \quad (4.85)$$

The probability of "0" being the stable value is $(1 - P_1(G))$. Hence the probability of a $0 \rightarrow 1$ transition can easily be computed by the product of $P_0(G)$ and $P_1(G)$.

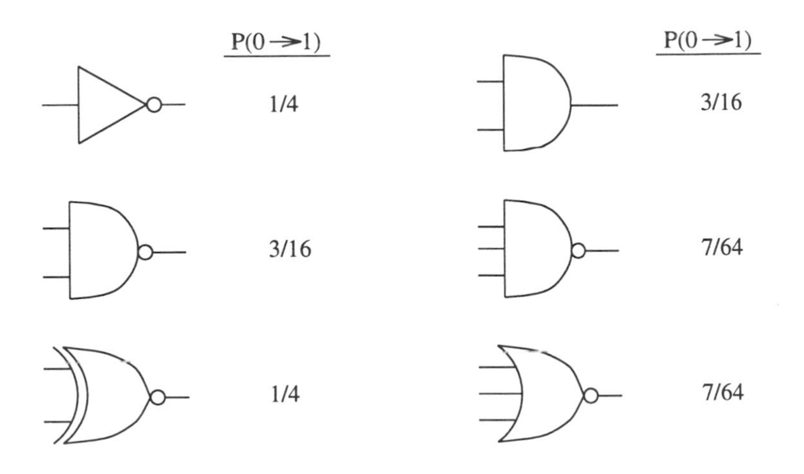

Figure 4.25 Output activities for static logic gates with uniformly distributed inputs.

4.5.4.1 Example

As an example of a logic decision for low-power, consider the different implementation of an 6-input AND gate driving a 0.1 pF load. As shown in Fig. 4.26, we may compare the following implementations:

- Implementation 1 : an 6-input NAND and an inverter.
- Implementation 2 : two 3-input NANDs and one 2-input NOR.
- Implementation 3 : three 2-input NANDs and one 3-input NOR.

The library used of such a comparison is a high-performance standard cell library optimized for speed. Table 4.4 shows some characteristics of the library, where the average delay is reported which is the average value of the rise and fall delay times. $W_p = 2W_n = 10\ \mu m$ is set for all the transistors composing the different gates. The delay is a function of the output load capacitance[4] C_o in pF. The area is a function of a unit area called cell grid. Each unit area for a cell has a certain height and width. Also included in this Table, is the input capacitance of a gate and the output parasitic capacitance in fFs. We make, for this example, the following assumptions:

[4] This capacitance does not include the output parasitic one.

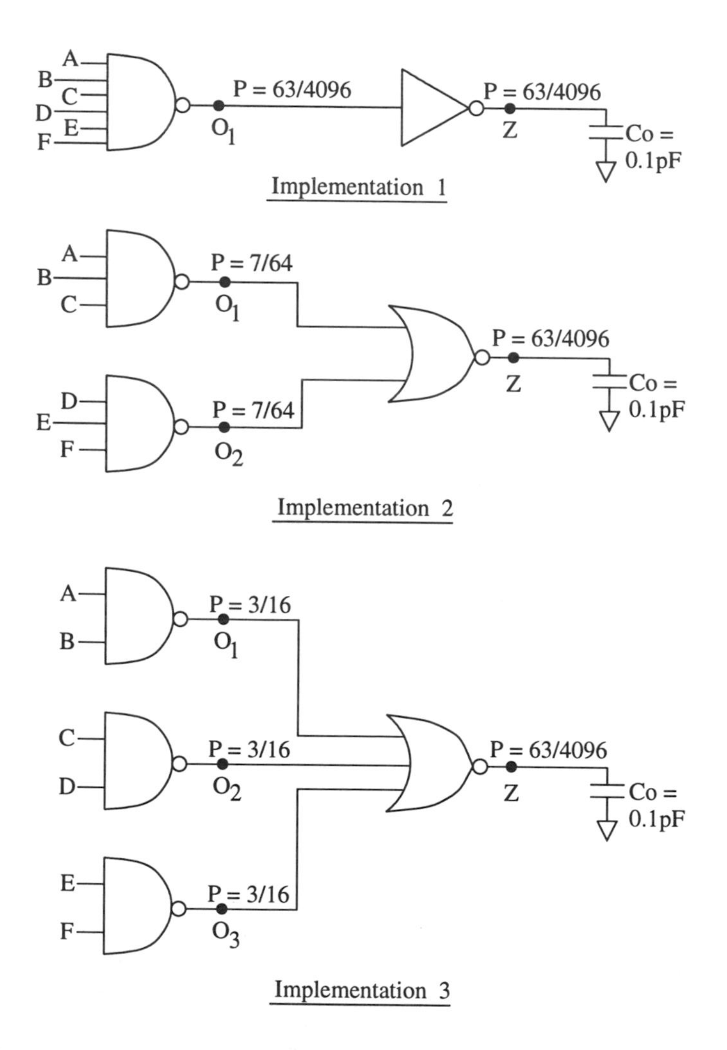

Figure 4.26 Example of 6-input AND function to be implemented from a standard cell library.

- We neglect the wiring capacitance between the different cells; and
- We neglect also the internal power of each gate.

Table 4.4 Characteristics of a high-performance 0.8 μm CMOS library.

Gate type	Area (cell unit)	Output cap. (fF)	Input cap. (fF)	Average delay (ns)
INV	2	85	48	$0.22 + 1.00\,C_o$
NAND2	3	105	48	$0.30 + 1.24\,C_o$
NAND3	4	132	48	$0.37 + 1.50\,C_o$
NAND6	7	200	48	$0.65 + 2.30\,C_o$
NOR2	3	101	48	$0.27 + 1.50\,C_o$
NOR3	4	117	48	$0.31 + 2.00\,C_o$

First we compare the delay and the area of the different implementations. Using the data of Table 4.4, the results are reported in Table 4.5. The delay may be computed or simulated by SPICE as illustrated in Table 4.5. The implementations 2 and 3 offer the best speed compared to the first one. However, they require more area.

Table 4.5 Areas and delay for 6-int AND gate implementations.

	Implem. 1	Implem. 2	Implem. 3
Area (cell unit)	9	11	13
Computed delay (ns)	1.1	0.85	0.87
SPICE delay (ns)	1.1	0.86	0.83

Let us now compare the power dissipation using the power cost function. It is defined by

$$Power\ \ Cost = \sum_i P_{0\to 1,i} C_i \tag{4.86}$$

where $P_{0\to1,i}$ is the probability of transition $0 \to 1$ at each node i and C_i is the total capacitance at each node i. We assume that the inputs A, B, C, D, E, and F are uncorrelated and random (i.e., $P_i = 0.5$). For the implementations of Fig. 4.26, we compute the transition probabilities. Table 4.6 summarizes the procedure of probabilities computation of different nodes in the circuit.

Table 4.6 Transition probabilities:

Implementation 1	O_1	Z
P_1	63/64	1/64
$P_0 = 1 - P_1$	1/64	63/64
$P_{0\to1}$	**63/4096**	**63/4096**

Table 4.6 *(continued)*

Implementation 2	O_1	O_2	Z
P_1	7/8	7/8	1/64
$P_0 = 1 - P_1$	1/8	1/8	63/64
$P_{0\to1}$	**7/64**	**7/64**	**63/4096**

Table 4.6 *(continued)*

Implementation 3	O_1	O_2	O_3	Z
P_1	3/4	3/4	3/4	1/64
$P_0 = 1 - P_1$	1/4	1/4	1/4	63/64
$P_{0\to1}$	**3/16**	**3/16**	**3/16**	**63/4096**

Note that the node O_1, in implemention 1, has a lower switching activity compared to the other two. To compute the power cost function we will not include the primary inputs. Table 4.7 illustrates the results of this calculation. The results indicate that implementation 1 has the lowest power. So technology mapping is important for low-power applications.

We consider now another example using low-area 0.8 μm CMOS standard cell library for the 6-input AND implementation. Some characteristics of this library are shown in Table 4.8. Compared to the library presented in Table 4.4, this library uses small transistors with $W_p = W_n = 4\ \mu m$. Compared to the

Table 4.7 Power cost evaluated for the three implementations using high-performance library.

	Implem. 1	Implem. 2	Implem. 3
Power cost (fF)	6.7	42.5	89.4

case of the high-performance library, the cell area unit, in the low-area case, is smaller by a factor of 1.5. Note that the delays of different gates are higher. However, the input gate and output parasitic capacitances are lower. Thus, this library can be used for low-power function implementation.

Table 4.8 Characteristics of a low-area 0.8 μm CMOS library.

Gate type	Area (cell unit)	Output cap. (fF)	Input cap. (fF)	Average delay (ns)
INV	2	35	13	$0.23 + 3.73\ C_o$
NAND2	3	60	13	$0.28 + 4.40\ C_o$
NAND3	4	65	13	$0.34 + 5.00\ C_o$
NAND6	7	81	13	$0.53 + 7.13\ C_o$
NOR2	3	62	13	$0.35 + 6.27\ C_o$
NOR3	4	69	13	$0.47 + 8.84\ C_o$

Table 4.9 Power cost evaluated for the three implementations using low-area library.

	Implem. 1	Implem. 2	Implem. 3
Power cost (fF)	3.5	19.5	43.7

The delays reported in Table 4.8 do not include the effect of the input voltage slope. The delay, of the different implementations, was simulated with SPICE and it is almost the same for all the configuration. The delay is $\sim$ 1.5 *ns*. Using the same reasoning discussed earlier we can compute the power cost function using this library. The transition probabilities are the same, except the total node capacitances which are different. The results of the power cost evaluation are illustrated in Table 4.9.

The power cost, in the case of low-power library, is almost half of that of high-performance. Still, implementation 1 has a low-power characteristic while the speed is almost the same compared to the others. The area is also lower than the other implementations. This example shows that the power dissipation can be reduced at the gate level. Even if we take into account the wire capacitances between the cells still, the conclusion is valid. The topic of low-power at the gate-level is discussed more in Chapter 8. Keep in mind, that in this comparison, the internal power of the gates has not been considered.

4.5.5 Glitching Power

Note that in the probability discussed so far, we assumed that the gates had zero delay. In that case, we are not taking into account the glitches and we consider only the transitions between stable states. Glitches must be considered if we assume non-zero delay at gates. Thus the total dynamic power of a circuit is the total dynamic power with zero delays power and the glitching power. So what is the glitching phenomenon?

In a static logic gate, the output or internal nodes can switch before the correct logical value is being stable. To illustrate this spurious transition, Fig. 4.27 shows an example of a circuit with a cascaded configuration. When the inputs ABC make the following transition $100 \rightarrow 111$, the output, with zero delay gates, should stay high. However, considering a unit delay for each gate, the output O_1 is delayed compared to the input C and hence causing the output Z to evaluate with the new value of C and the old value of O_1. In that case, the output experiences a dynamic hazard (glitch). This transition increases the dynamic power of the circuit and adds a dynamic component to the switching activity.

Another example is shown in Fig. 4.28(a). The cascaded circuit exhibits a glitching problem. However, the same function can be implemented using balanced delay implementation as shown in Fig. 4.28(b). These are some rules to avoid this problem:

- Balance delay paths; particularly on highly loaded nodes. Insert, if possible, buffers to equalize the fast path; and

- Avoid if possible the cascaded implementation; and
- Redesign the logic when the power due to the glitches is an important component.

4.5.6 Basic Physical Design

To implement simple gates, the physical layout should be performed. It is usually easy to draw a layout of a gate with well arranged transistors. For example, for the inverter, Fig. 4.29(a) shows a possible layout implementation. The metal1 is used for the power lines. Many variations can be drawn, depending on the use of the gate. Fig. 4.29(b) shows another layout variation of the inverter where metal2 is used as the power lines. For clarity the wells and body ties are not shown in these layouts.

Similarly, the schematic of NAND2 and NOR2 gates can be converted to layouts. Fig. 4.30(a) shows one possible layout of a two-input NAND gate. The layout can also be arranged to draw the input poly lines vertically. The layout artist should draw the gate taking into consideration the environment of this cell (the connectivity to others). Fig. 4.30(b) shows the layout of a two-input NOR gate. Note that the junction areas should be optimized during the layout to reduce the power dissipation and improve the speed of the cell. An implementation of a 2-input NOR gate with a high output drain junction capacitance is shown in Fig. 4.31.

To do a layout of a complex gate (i.e, several tens of transistors), the following general layout guidelines can be used :

- Set the sizing of the transistors composing the gate;
- Run V_{DD} and V_{SS} in metal (1 or 2) horizontally. For example, V_{DD} at the top and V_{SS} at the bottom of the cell in semi-rectangular form;
- Define the polysilicon gate lines orientations and order them for maximum active area crossover to form the gate regions;
- Place the N-block (NMOS transistors) near V_{SS} and the P-block (PMOS transistors) near V_{DD}. The PMOS devices should be located in the common N-well if they use the same bulk potential;
- Adhere to the design rules and use if possible an interactive DRC (Design Rule Checker);

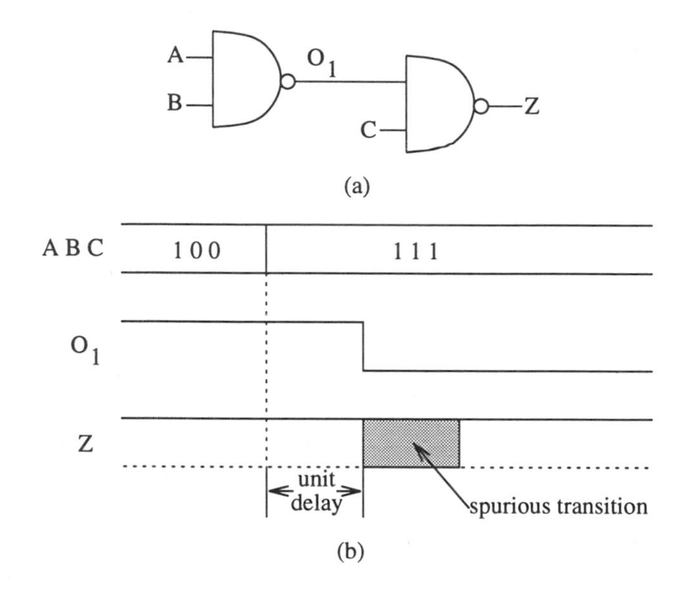

Figure 4.27 Glitching in static CMOS gates: (a) cascaded structure; (b) voltage waveforms.

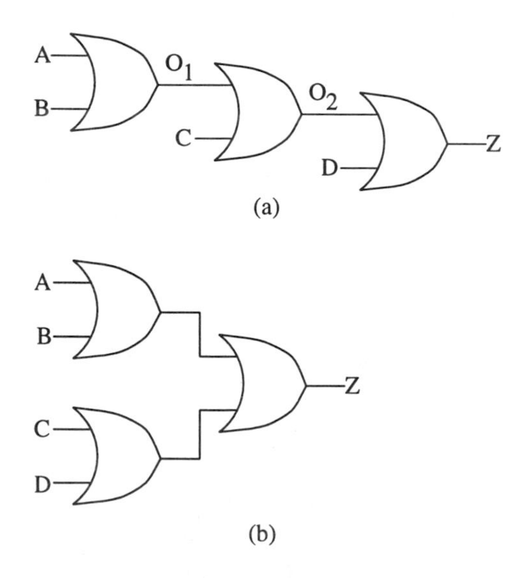

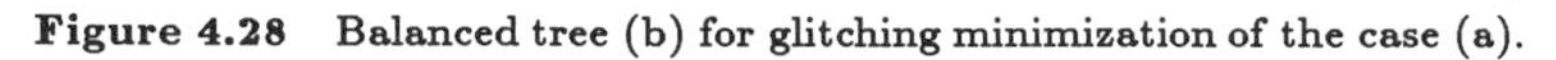

Figure 4.28 Balanced tree (b) for glitching minimization of the case (a).

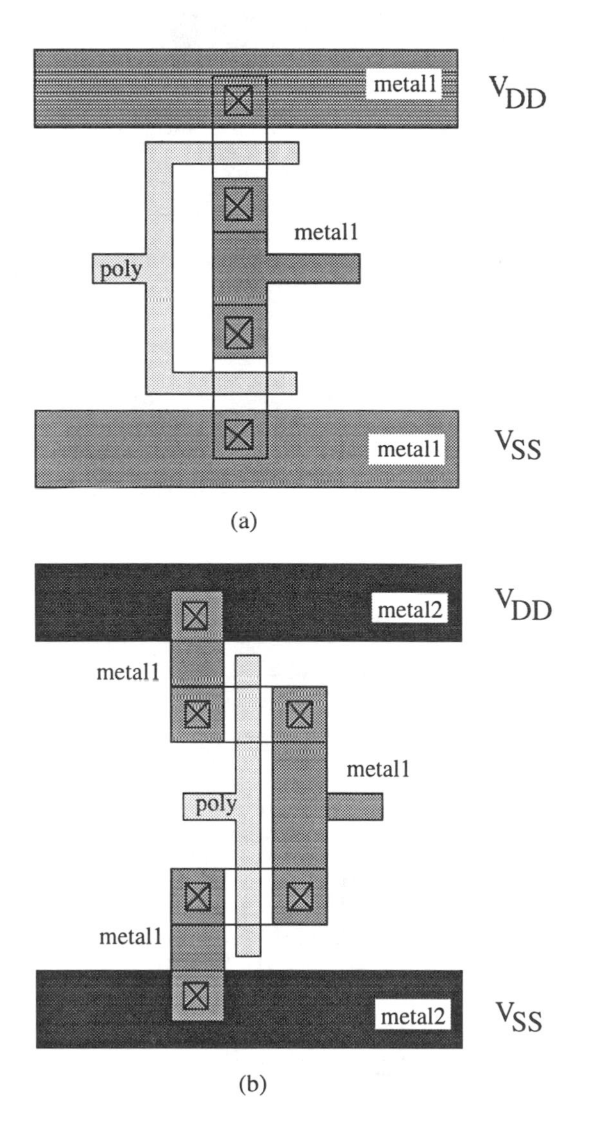

Figure 4.29 Example of inverter layouts.

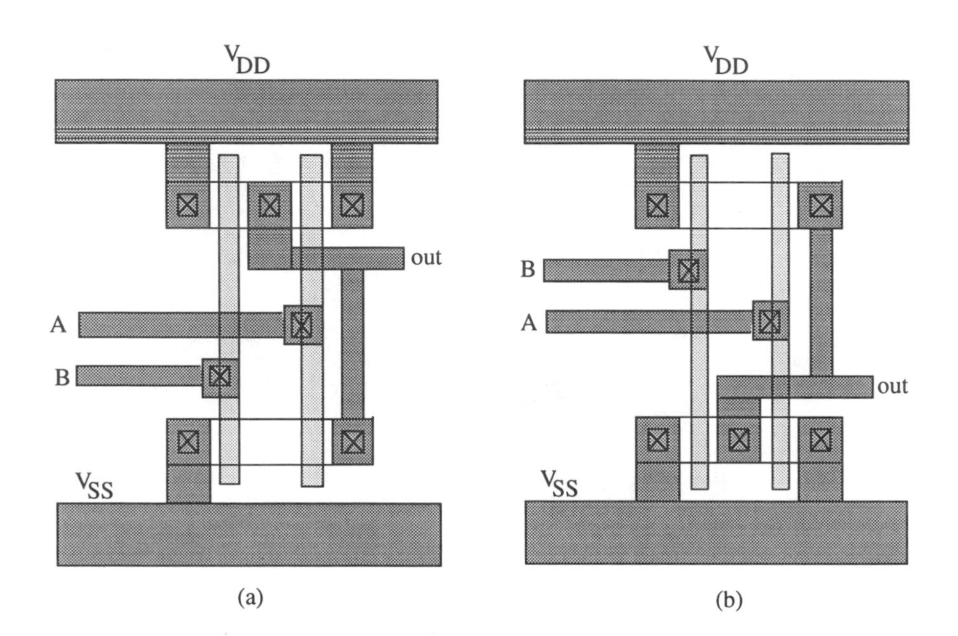

Figure 4.30 Typical: (a) NAND2 gate; (b) NOR2 gate layouts.

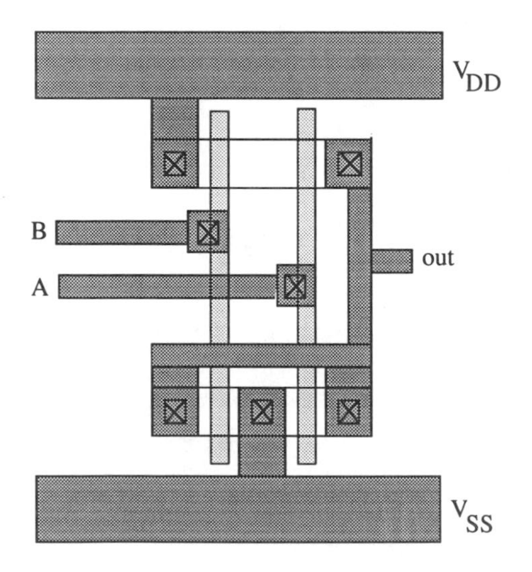

Figure 4.31 Typical NOR2 gate with high output drain capacitance.

- Keep the internal junction and wire capacitances to the minimum to minimize the power and the delay; and
- Complete the connection of different nodes inside the cell using the different layers available (metall, poly, etc.).

Note that the power line widths are drawn taking into consideration the current consumed by the cell because the electromigration phenomena sets the minimum width of conductors.

For low-power design, these are some layout guidelines:

- Identify, in your circuit, the high switching activity nodes;
- Use for these high activity nodes low-capacitance layers such as metal2, metal3, etc.;
- Keep the wires of high activity nodes short;
- Use low-capacitance layers for high capacitive nodes and busses.
- For large width devices, use special layout; such as interdigitated fingers [3] and donut (round transistor); to achieve a low drain junction capacitance; and
- Design complex cells or blocks using, as much as, possible custom approach.

4.5.7 Physical Design Methodologies

There are many layout methodologies to do the physical implementation of a complex circuit. The first methodology is called *full-custom design*, where the layout of each transistor is optimized. The layout of a complex block is performed by custom design for reasons of speed. However, this style leads to low design productivity and is rarely used in ASICs and digital processors. But, when the low-power is an issue the full-custom design can be used to minimize the power of the circuit.

Another design methodology is the standard-cell approach (or semi-custom design) . That is, several gates and functions are created in the library such as:

- NAND, NOR, XOR, AOI, OOAI, latches, buffers, multiplexers, full-adder, flip-flops, etc.;
- Linear cells : low-battery detector, power-up reset, etc.;
- MSI/LSI functions : ALU (Arithmetic and Logic Unit), counters, magnitude comparators, etc.;
- Compiled macrocells : register file, FIFO (First In First Out), ROM (Read Only Memory), parallel multiplier, etc.; and
- Macrocells : 8/16-bit microcontroller, 16-b fixed point DSP, UART (Universal Asynchronous Receiver/Transmitter), etc.

A circuit is designed by capturing the schematic or the functional model (VHDL, Verilog, etc.) of the cells. The layout is generated by an automatic placement and routing. An example of a CMOS standard cell library can be found in [10]. In standard cell approach, the logic cells have the same height and the width is variable. In many libraries, the cells are available in two layout styles. In the area-optimized cell, the cells are made as small as possible. In the performance-optimized style, cells are optimized for high-speed performance and, as a result, occupy more area than the small cells. Even the height of the cells in the two styles is different. A typical standard cell layout for a NAND gate is shown in Fig. 4.32. This methodology provides lower cost and higher productivity than the full-custom one. For low-power applications, the small and large cells for the same function can be carefully chosen to optimize the power in a complex design without degrading the timing requirement.

The third layout methodology is the gate array[5]. The gate arrays consist of implemented cells and need only the personalization steps. Fig. 4.33 illustrates an example of gate-array core using Sea-Of-Gates structure. It consists of I/O and internal cell areas. The I/O cell area contains pads with input/output buffers. The internal cell array contains a continuous array of NMOS and PMOS transistors. Hence, the transistors and interconnects are already predefined. The design of a logic gate consists of wiring the different transistors using metallization and contacts. The isolation of a logic gate is performed by tying the polysilicon gates of the limiting transistors to V_{SS} or V_{DD} depending on the type of gate diffusion. Routing channels are routed over unused transistors. This methodology permits the reduction of the design cost at the expense of area, power and performance. One recent gate array architecture was based on multiplexers with small size transistors to maintain low-power characteristics [11].

[5] Programmable Gate Arrays are not discussed in this book.

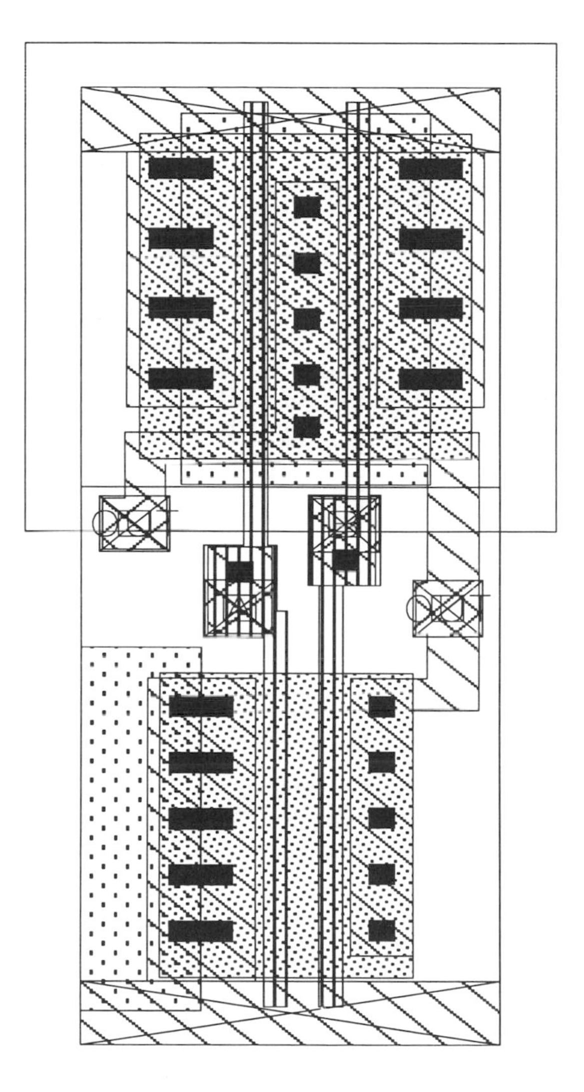

Figure 4.32 An example of standard cell layout (NAND2).

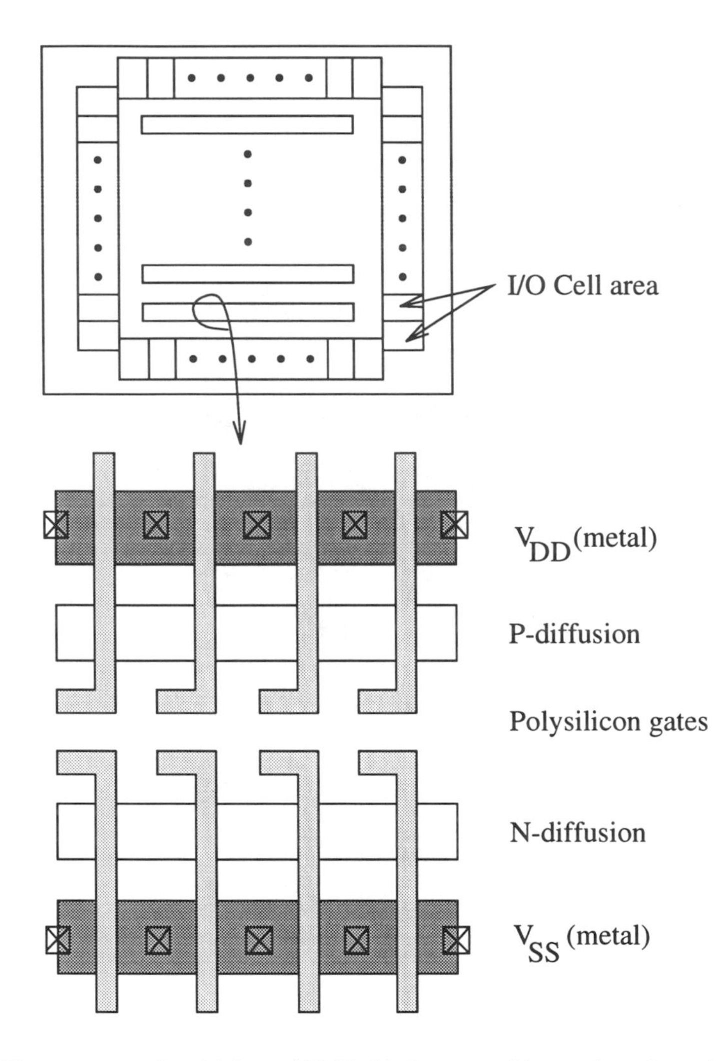

Figure 4.33 Sea-Of-Gates (SOG) chip layout and internal configuration.

Comparing these layout approaches, the full-custom methodology offers the best approach to minimize the power dissipation. However, for a complex design, it is costly to use such a design strategy. The standard cells approach provides good performance and an improved design time. However, in many libraries the devices are oversized for performance purposes and consequently, the power dissipation would be high. To efficiently use the standard cells tech-

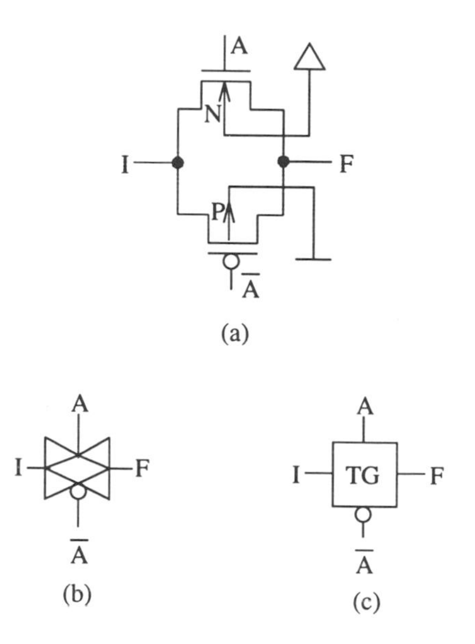

Figure 4.34 (a) CMOS transmission gate; (b) and (c) schematic symbols.

nique for low-power applications, the library should be expanded to include several versions of the same function with different driving capabilities. In that case, powerful synthesis tools are needed to optimize the power while maintaining the timing specifications. Moreover, both the standard cells and gate arrays styles require new place and route tools for low-power design.

4.5.8 Conventional CMOS Pass-Transistor Logic

Another alternative to CMOS static complementary logic is the conventional pass-transistor logic based on MOS switches. Fig. 4.34 shows a CMOS transmission gate (TG) as primitive element. It consists of a complementary pair connected in parallel. It acts as a switch, with the logic variable A as the control input. If A is low, the gate is OFF and presents a high resistance between the terminals. If A is high, the gate is ON and acts as a switch with an on resistance of R_n and R_p in parallel. The equivalent resistance of the TG is $R_{TG} = R_n || R_p$. This resistance is always less than the smallest among R_n and R_p. This permits a fast switching characteristic. When the input I is at V_{DD}, then the output F is quickly charged initially by the NMOS, then at the

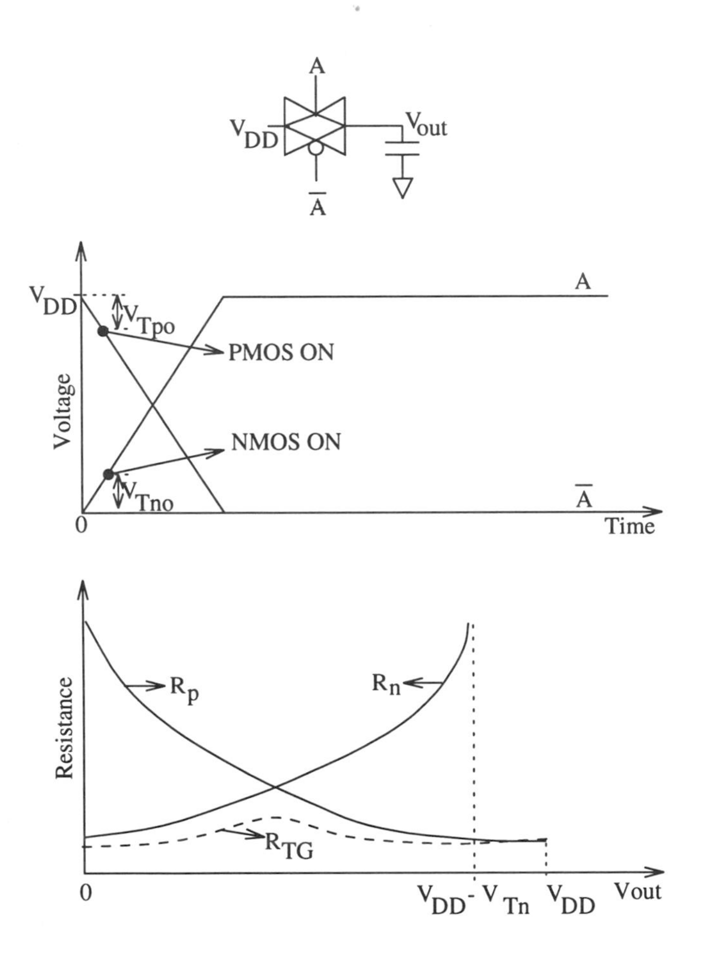

Figure 4.35 Waveforms and equivalent resistance.

end by the PMOS transistor as illustrated by the equivalent resistances of Fig. 4.35. In this figure, we assume that at $V_{out} = 0$, A and $\bar{A}$ are set to their final values. During this transient switching phase the NMOS is subject to the body while the PMOS is not. When a zero, at the input I, is to be transmitted then the PMOS is subject to the body effect. The PMOS and NMOS transistors should be sized such that they charge and discharge the output symmetrically. If $V_{Tn} = |V_{Tp}|$ and the body effect is symmetrical then we can size the devices such as $\beta_n = \beta_p$. Sometimes, equal sized NMOS and PMOS devices can be used. It is easy to see that the delay of the TG gate is approximately independent of the input level. This is not the case if the pass-logic uses a single-channel transistor. A drawback of the CMOS TG is that it consumes more area than a single-channel transmission gate (NMOS TG or PMOS TG). Thus, if the area is of prime concern, NMOS TGs are used.

Any CMOS TG logic (we call it here conventional pass-transistor logic) function can be implemented using the TG primitive element described above. In such implementation the transistor count, hence the silicon area, is low compared to standard static CMOS implementation. This is highlighted in the implementation of such functions as multiplexing, demultiplexing, decoding and addition. Fig. 4.36 shows a 4:1 multiplexer, where the data lines A, B, C and D are controlled by S_1 and S_2 such that

$$F = A.S_1.S_2 + B.S_1.\bar{S_2} + C.\bar{S_1}.S_2 + D.\bar{S_1}.\bar{S_2} \qquad (4.87)$$

This form of logic is used when the inputs and their logic complements are available. The implementation does not need V_{DD} or ground lines. However, the implementation suffers from a number of drawbacks; the driving capability of the circuit is limited and the delay increases with long TG chains. Moreover, the circuit does not provide a restoration of the logic levels, i.e., the logic gates are passive with no gain elements. Fig. 4.37 shows an example on how to restore the voltage levels in chained TGs. When 8 TGs are put in series, the output signal changes very slowly. However, when an inverter stage is added every 4 TG stages, the level is restored as shown in the SPICE voltage waveforms of Fig. 4.37.

The CMOS TG logic can be used in CMOS circuit design offering an extra degree of circuit design freedom. An example is the full-adder. The adder circuits will be discused in detail in Chapter 7. Fig. 4.38 shows the schematic of the XOR gate which is used by the adder. When the input A is low, $\bar{A}$ is high. The transmission gate TG is closed, then the output is equal to B. When A is high, $\bar{A}$ is low. The inverter formed by the transistors N and P is enabled, then the output is equal to A. The TG gate is open in this case. To implement an adder lets first review its functions. The boolean function of a full-adder are:

$$S_{out} = A \oplus B \oplus C_{in} \qquad (4.88)$$

$$C_{out} = A.B + C_{in}(A + B) \qquad (4.89)$$

A and B are the inputs, C_{in} the carry input, S_{out} is the sum output, and C_{out} is the carry output. The truth table of an adder is shown in Table 4.10.

The CMOS implementation of a one-bit full-adder is shown in Fig. 4.39(a). It requires 28 transistors and has two gate delays. In this circuit the transistors

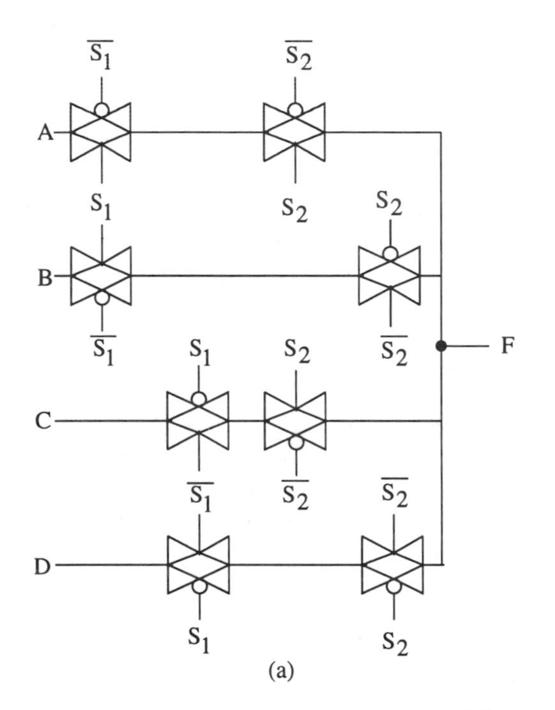

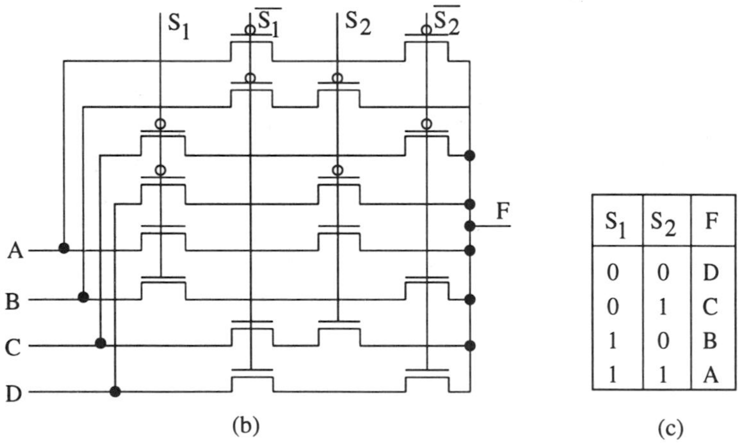

S_1	S_2	F
0	0	D
0	1	C
1	0	B
1	1	A

Figure 4.36 Example of a CMOS multiplexer using TGs: (a) logic diagram; (b) circuit schematic; (c) truth table.

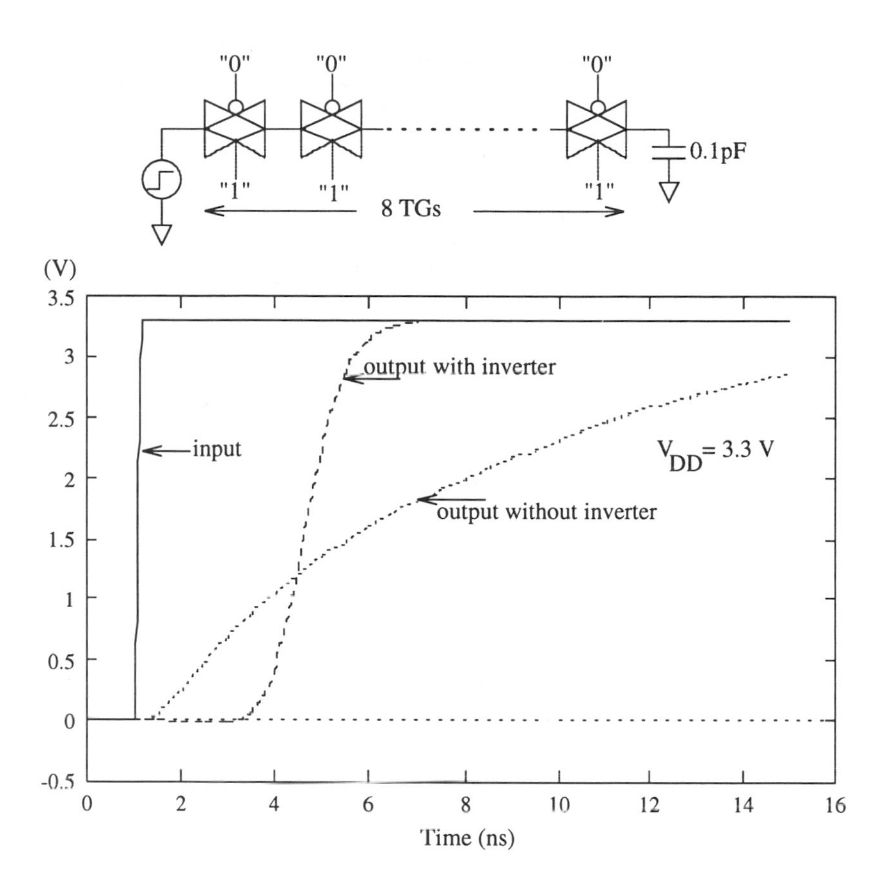

Figure 4.37 Levels restoration in TG's series using one inverter each four TGs.

controlled by the carry signal C_{in} should be placed close to the output. This will offset the body effect problem, since the carry is the latest arriving signal. An optimized implementation of the full-adder is shown in Fig. 4.39(b). It uses only 18 transistors and is based on the XOR function shown in Fig. 4.38 and the TG gates. Hence, this adder is more compact and faster and consumes less power than the complementary static one.

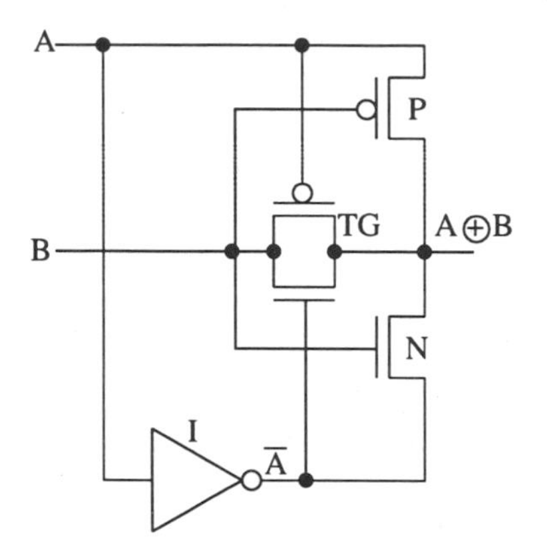

Figure 4.38 TG XOR gate.

A	B	C_{in}	S_{out}	C_{out}
0	0	0	0	0
0	1	0	1	0
1	0	0	1	0
1	1	0	0	1
0	0	1	1	0
0	1	1	0	1
1	0	1	0	1
1	1	1	1	1

Table 4.10 Adder Truth Table

4.5.9 CMOS Static Latch

Fig. 4.40 shows a cross-coupled CMOS static latch. In the storage mode (input $LD = 0$), when the node A is high, B is low, P_1 and N_2 are ON while P_2 and N_1 are OFF. Similarly, when A is low, B is high, P_1 and N_2 are OFF while P_2 and N_1 are ON. The standby power dissipation of the cell is very small. The state of the latch is changed by turning the two transmission gates ON (LD high) and applying the input and its complement.

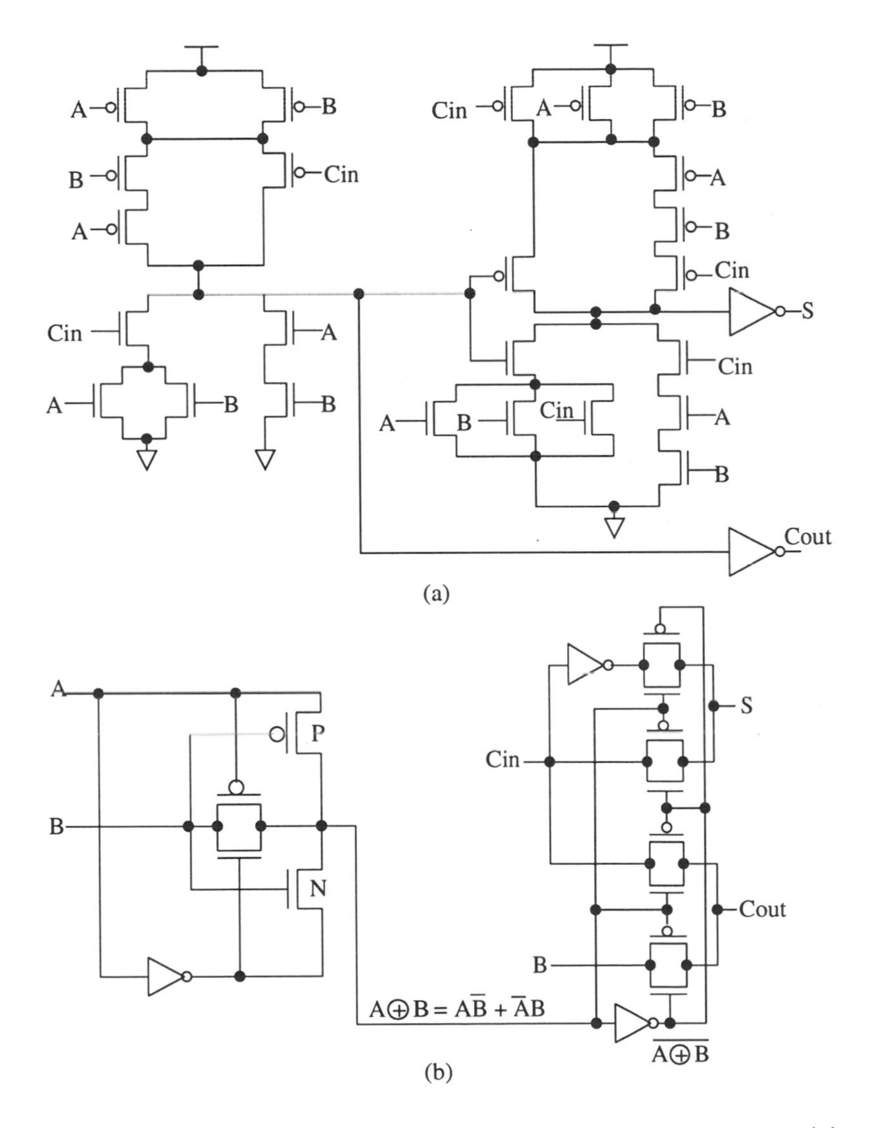

Figure 4.39 CMOS static full adders using: (a) complementary style; (b) TGs.

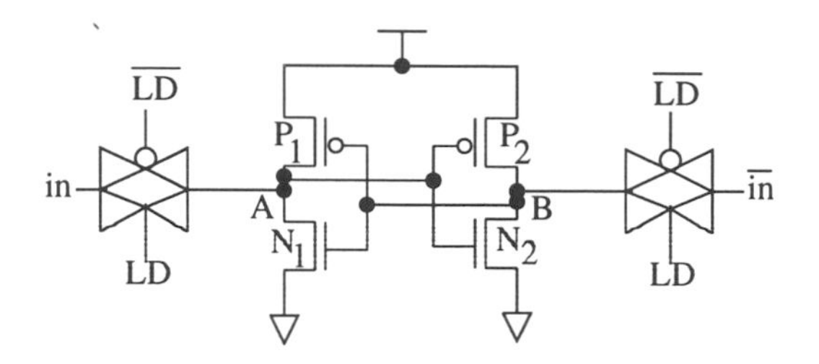

Figure 4.40 CMOS cross-coupled static latch.

4.6 CMOS LOGIC STYLES

CMOS logic has been known to have a negligible static power dissipation. However, this is valid as long as V_T is not too low. However, it has low-speed and consumes large area because for n-input, twice the number of transistors is required. As a result, it is sometimes desirable to have faster and smaller logic gates at the cost maybe of parameters such as : noise margins, power dissipation, etc. This section discusses many CMOS logic alternatives to complementary CMOS and also the clocking issues in a VLSI system.

4.6.1 Pseudo-NMOS CMOS Logic

The gate area of complementary CMOS can be reduced if CMOS circuits are designed in a way similar to NMOS circuit families [12]. A PMOS device is used to replace the depletion-type device in NMOS family. This type of circuit is referred to as pseudo-NMOS, as shown in the inverter of Fig. 4.41. When the input A is low, the output is high and at V_{DD}. When A goes to a high level, N turns ON while P is still ON. In this case, the output never reaches zero and takes a value V_{OL} determined by the ratio β_p/β_n and the logic is called *ratioed.* To examine V_{OL}, we use simple analysis. When A is at V_{DD}, N is in the linear region while P is saturated. By equating the currents using simple models, we have

$$\beta_n(V_{DD} - V_{Tn})V_{OL} = \frac{\beta_p}{2}(V_{DD} - |V_{Tp}|)^2 \tag{4.90}$$

Hence for $V_{Tn} = -V_{Tp} = V_T$

$$V_{OL} = \frac{\beta_p}{2\beta_n}(V_{DD} - V_T) \tag{4.91}$$

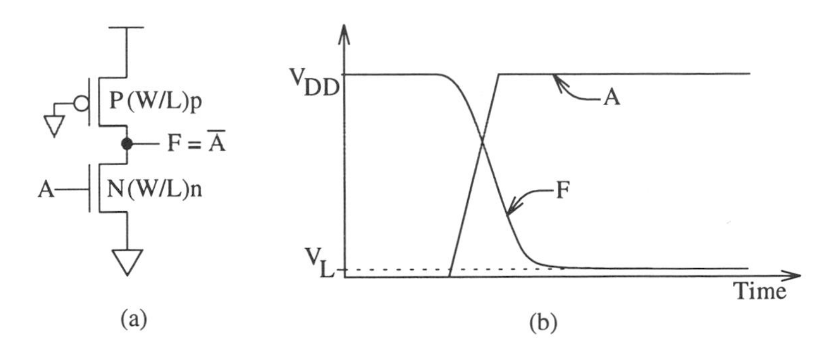

Figure 4.41 Pseudo-NMOS: (a) inverter; (b) voltage waveforms.

Thus V_{OL} depends strongly on the ratio β_p/β_n. For example, if we need a $V_{OL} = 0.04V_{DD}$ and $V_T = 0.2V_{DD}$, then the ratio β_p/β_n should be equal at least to 0.1. If the NMOS transistor is minimum size, the PMOS should be weak to provide adequate noise margins (low V_{OL}). In this case, the rise time of the gate is too slow. If we improve the rise time, the ratio condition tends to increase the gate area and hence the input capacitance.

Although this circuit offers a reduction in total transistor count and ease of layout, it has the disadvantage of non-zero static power dissipation. Since the pull-up PMOS is always ON, a current flows from V_{DD} to ground whenever the pull-down section of the pseudo-NMOS is turned ON. This current is the source of the static power dissipation. When a pseudo-NMOS gate, with output at V_{OL}, is driving another one, the driven gate, with OFF pull-down section, leaks a high subthreshold current but still this current is lower than the one when the pull-down in ON. For n-input pseudo-NMOS gate there are (n+1) transistors. Fig. 4.42 illustrates an example of complex gate implemented in pseudo-NMOS style. This logic has been used in many applications such as decoding logic for memories and PLA. Because of its high static power, it is not suitable for low-power applications.

4.6.2 Dynamic CMOS Logic

To reduce the area and improve the speed of CMOS circuits, another popular style called *dynamic logic* is used. Fig. 4.43 shows a dynamic CMOS gate. This logic is referred to as *domino* CMOS logic [13]. The domino gate shown in Fig. 4.43(a) consists of a dynamic CMOS circuit followed by a static CMOS

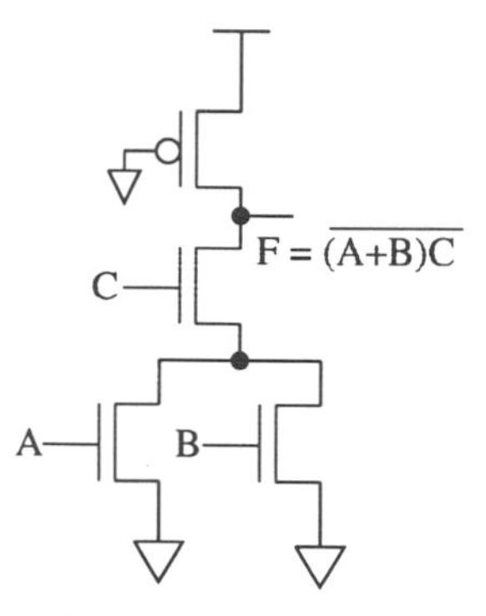

Figure 4.42 Pseudo-NMOS complex logic gate.

buffer. The dynamic circuit consists of a PMOS precharge transistor P_1, an evaluation NMOS transistor N_1, a storage capacitor C, and an N-logic block which is a series-parallel combination of NMOS transistors activated by the inputs and implementing the required logic. The storage capacitance represents the parasitic at node A.

This circuit uses a single clock phase *clk*. During the precharge phase ($clk = 0$), the storage capacitance is charged through the PMOS pull-up P_1 to V_{DD} and the inputs have no effect since there is no path to ground. The output of the buffer is precharged to ground. During the evaluation phase ($clk = 1$), N_1 is ON, and depending on the logic performed by the N-logic block, the node A is either discharged or it will stay precharged.

Fig. 4.43(b) shows an example of complex gate. In a cascaded set of domino logic stages, as shown in Fig. 4.44, the first stage evaluates and causes the next one to evaluate (like domino falls). The number of cascaded stages is limited by the evaluation clock phase.

Compared to pseudo-NMOS, domino logic has the same input capacitance and improved rise time. However the fall time is affected since there is one more transistor in the pull-down section. Also the gate is suitable for high-fanout operation because of the CMOS buffer. Moreover, it is efficient in area for high fanin because $n + 4$ transistors are required compared to $2n$ for CMOS static gate.

Some limitations of the gate are:

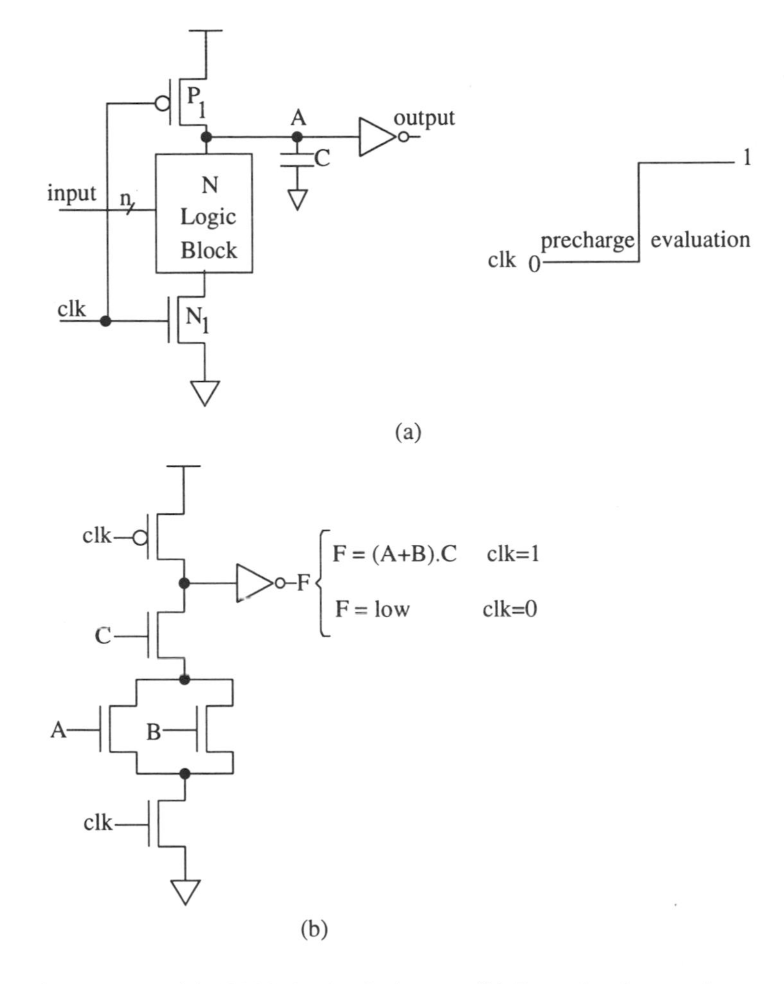

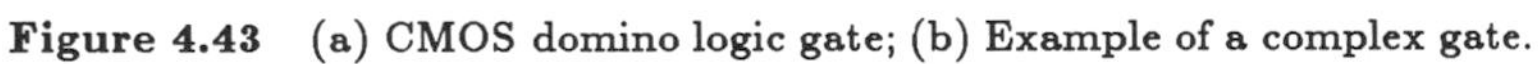
Figure 4.43 (a) CMOS domino logic gate; (b) Example of a complex gate.

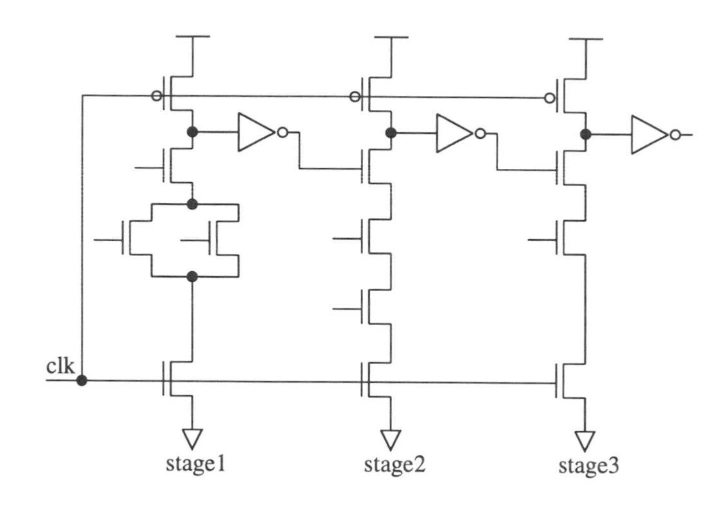

Figure 4.44 Domino logic chain.

- The domino gate has a problem called *charge sharing or redistribution.* Fig. 4.45 gives an example to explain this problem. During the precharge, the node A is at V_{DD} and charge CV_{DD} is stored on the capacitance C. We assume (worst-case) that the parasitic capacitance of nodes B and C, C_1 and C_2 respectively, have zero charges. During the evaluation, the node A should stay at V_{DD}, however, due to C_1 and C_2, charge sharing take place. Using the charge conservation principle before and after redistribution, we have

$$CV_{DD} = (C + C_1 + C_2)V_A \tag{4.92}$$

Hence the final voltage of node A is

$$V_A = \frac{C}{C + C_1 + C_2} V_{DD} \tag{4.93}$$

If for example $C_1 = C_2 = 0.5C$ then this voltage would be $V_{DD}/2$. This voltage can alter the logic and provoke the CMOS buffer to dissipate high static power dissipation.

- If the clock frequency is too low, the node A leaks the charge stored on C due to the leakage currents. The dynamic node can leak its charge in a time of few hundreds of μs to few ms, depending on the temperature, the storage capacitance and the leakage current. When

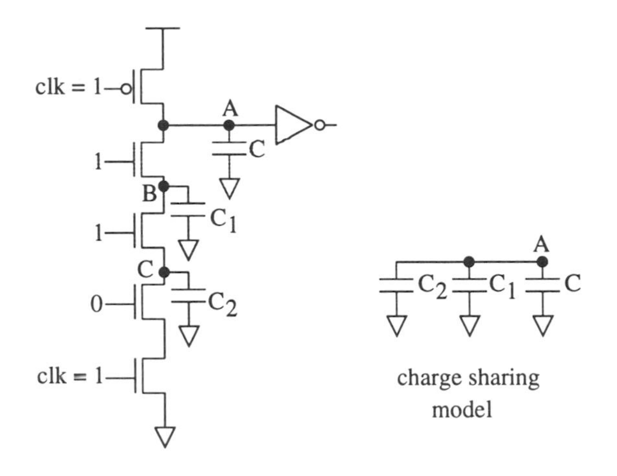

Figure 4.45 Charge sharing in dynamic CMOS logic.

using power-down techniques, the dynamic nodes should not be left floating for a long time. If the leakage is high with low V_T devices, the charge can be deleted in a time as low as 100 ns. This problem is similar to charge sharing. Fig. 4.46 shows two alternates to solve the problems of charge sharing and leakage. In Fig. 4.46(a), a weak PMOS (low W/L) is added as pull-up transistor. This circuit operates like pseudo-NMOS during evaluation phase. Hence it consumes some static power dissipation. If the circuit operates at high-frequency, the added weak PMOS has no role because it does not have enough time to operate. Note that this weak PMOS increases the output capacitance and then it slows this dynamic gate. To eliminate the DC path during evaluation, the gate of the weak PMOS can be driven from the output of CMOS buffer as shown in Fig. 4.46(b). This circuit adds another capacitance at the output of the inverter. A third alternate circuit which solves only the problem of charge sharing is shown in Fig. 4.47. In this circuit configuration, intermediate nodes of complex gate are precharged with additional precharge PMOS devices.

- Another limitation of the domino logic gate is that it implements non-inverting logic functions. However, this is not a serious limitation and can be overcome, if the need arises, by using CMOS static gates. The designer can mix both static and dynamic CMOS logic circuits in a given design to optimize the overall performance.

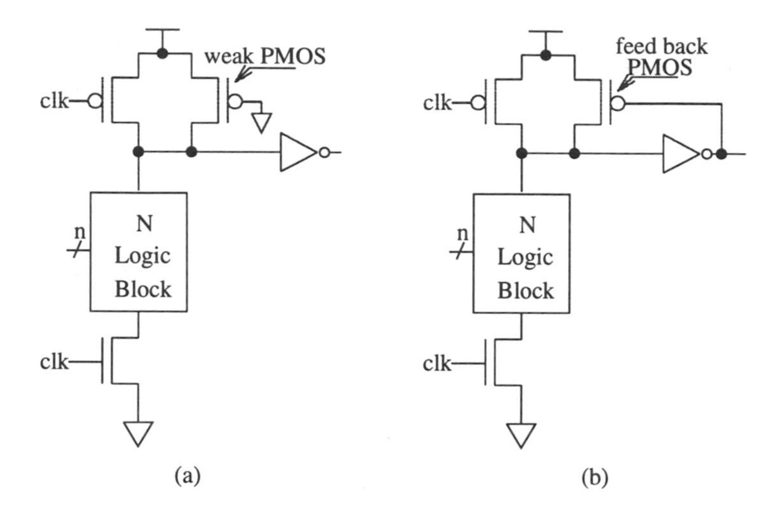

Figure 4.46 CMOS domino logic alternates.

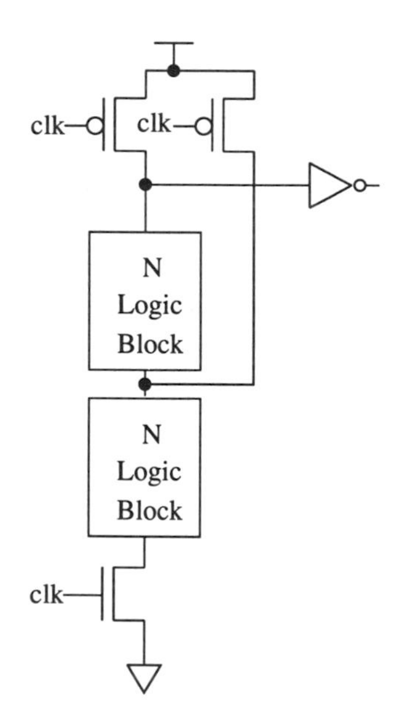

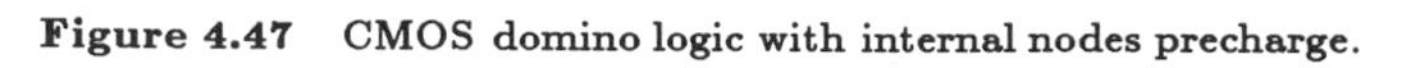

Figure 4.47 CMOS domino logic with internal nodes precharge.

Historically, dynamic design style have been devised for low-power characteristics because of the reduced device count. Moreover, dynamic gates do not experience short-circuit power dissipation and glitching problems as in static circuits. However, to drive the clocked transistors, a large clock distribution network is needed. This highly loaded network consumes a significant amount of dynamic power particularly at high frequency of operation[6]. The switching activities of dynamic gates are higher than those of static gates. In a dynamic gate the output makes a $0 \rightarrow 1$ transition during the precharge cycle only if the N-bloc discharges the output during the evaluation phase. Hence, the probability of $0 \rightarrow 1$ transition is given by

$$P_{0\rightarrow 1} = P_0 \tag{4.94}$$

where P_0 is the probability that the output has a "0" output. For a two-input NAND dynamic gate, the output has only one zero for 4 input states. So,

$$P_{0\rightarrow 1} = P_0 = \frac{1}{2^2} = \frac{1}{4} \tag{4.95}$$

For a NOR2 gate, we have

$$P_{0\rightarrow 1} = P_0 = \frac{3}{2^2} = \frac{3}{4} \tag{4.96}$$

Another refinement of the domino CMOS logic is shown in Fig. 4.48 [14], where the CMOS buffer is removed. N and P logic blocks are alternated and each drive the other. When *clk* is low (0), the first and third stage are precharged high and the second stage is precharged low.

Fig. 4.49 shows another NP domino logic called NORA (No Race) [15]. Two sections *clk* and $\bar{clk}$ are shown in Fig. 4.49. It is constructed by cascading N and P blocks followed by C^2MOS (clocked CMOS) latch. CMOS buffers (inverters) are used to provide logic inversion. When $clk = 1$ (evaluation phase in section *clk*), the C^2MOS latchs operates like an inverter. When $clk = 0$, the latch moves into *hold* state because the output NMOS and PMOS transistors are OFF. In this case, the old data is latched at the output. This latch is used to avoid signal races. A NORA pipeline is shown in Fig. 4.50 and it consists of alternating *clk* and $\bar{clk}$ sections. Signal races do not occur in this structure because of the use of C^2MOS. Another logic has been proposed to overcome charge sharing by using additional clocking signals. It is called Zipper CMOS logic. For more details refer to [16].

[6] See the example of the DEC Alpha chip in Section 4.9.4.

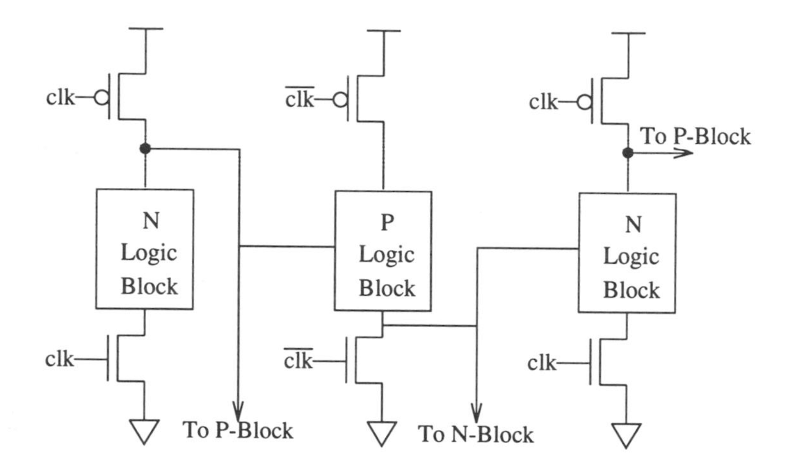

Figure 4.48 NP domino logic.

An example of a pipelined full-adder (FA) NORA circuit is shown in Fig. 4.51. This cell can be used in many designs such as a pipelined multiplier. The output C^2MOS latches can only use three transistors rather then four. The NMOS and PMOS transistor P_s and N_c respectively, can be removed from the output C^2MOS latches. The reason is that during precharge phase ($clk = 0$), the output nodes A and B are set to ground and V_{DD} respectively. Thus, the transistors P_1 and N_2 are turned OFF. Hence, the clocked transistors P_s and N_c can be removed and the FA cell is isolated from other sections during precharge.

4.6.3 Design Style Comparison

If we compare the above discussed design styles, static CMOS logic is the slowest circuit, but the power efficiency is the best, particularly if minimum size devices are used. Hence, it is suitable for low-power, medium speed applications. Note that the static CMOS logic occupies the largest chip area because complementary functions are needed. The circuit designer can include, in static logic, pass-transistor logic to improve the speed and area. Pseudo-NMOS logic style can be faster than static CMOS logic, however its rise time is long. This is limited by the low output logic level. Moreover, the most serious drawback of pseudo-NMOS logic is the high power dissipation in the standby mode. N-P domino logic is fast, because it has small input capacitance like pseudo-NMOS

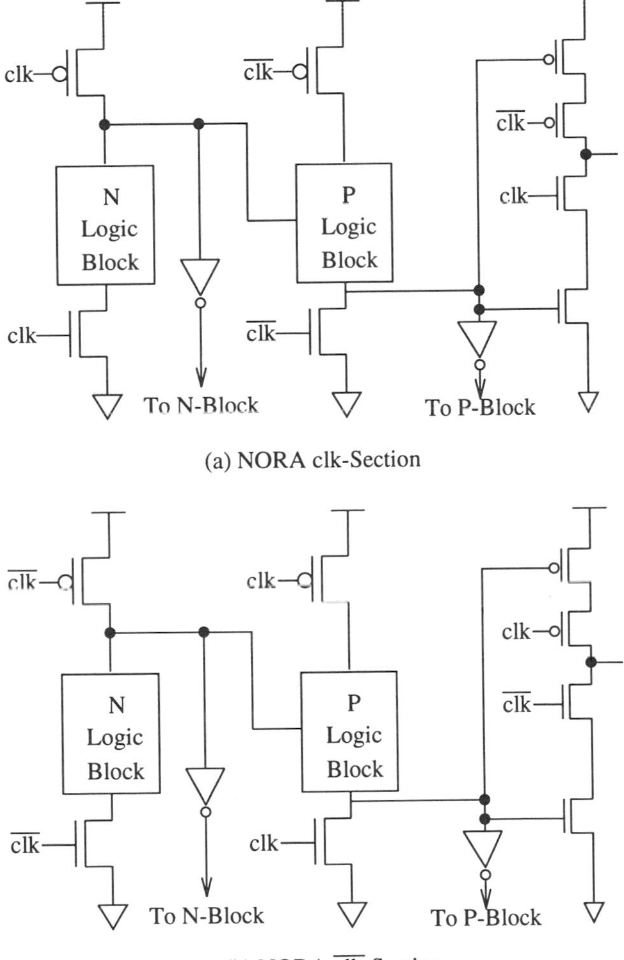

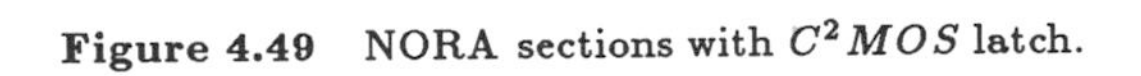

Figure 4.49 NORA sections with C^2MOS latch.

Figure 4.50 NORA pipeline logic.

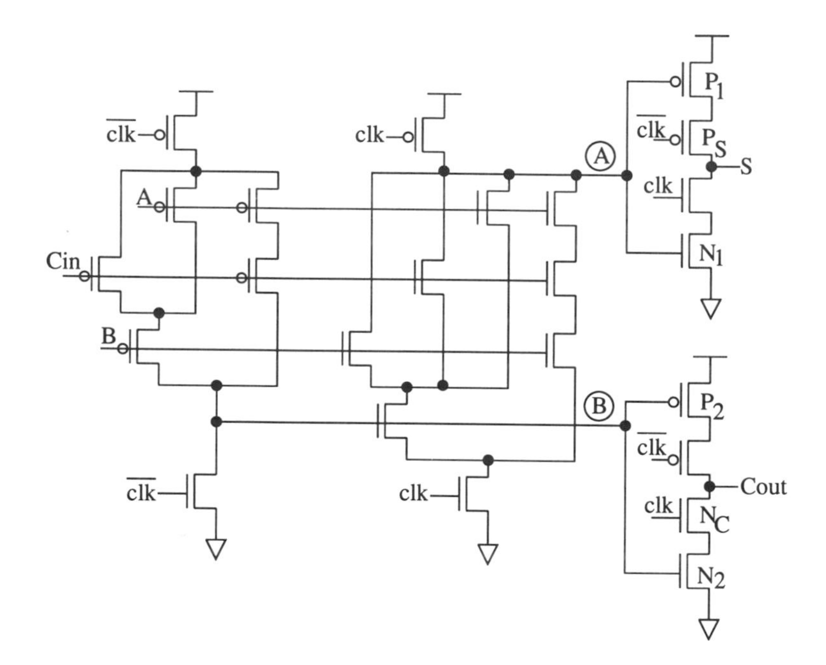

Figure 4.51 Pipelined full-adder NORA circuit.

logic and improved rise time. The power dissipation consumed by this logic is high due to the high switching activity of the clock even if the circuit is not used. However, power-down techniques can be used to control the clock of the logic. Using this style, requires from the designer to spend more design effort than the static style to solve all the problems of dynamic logic such as: charge sharing, clock skew, precharging, etc. Finally, we note that pass-transistor logic is very promising for high-performance low-voltage low-power applications.

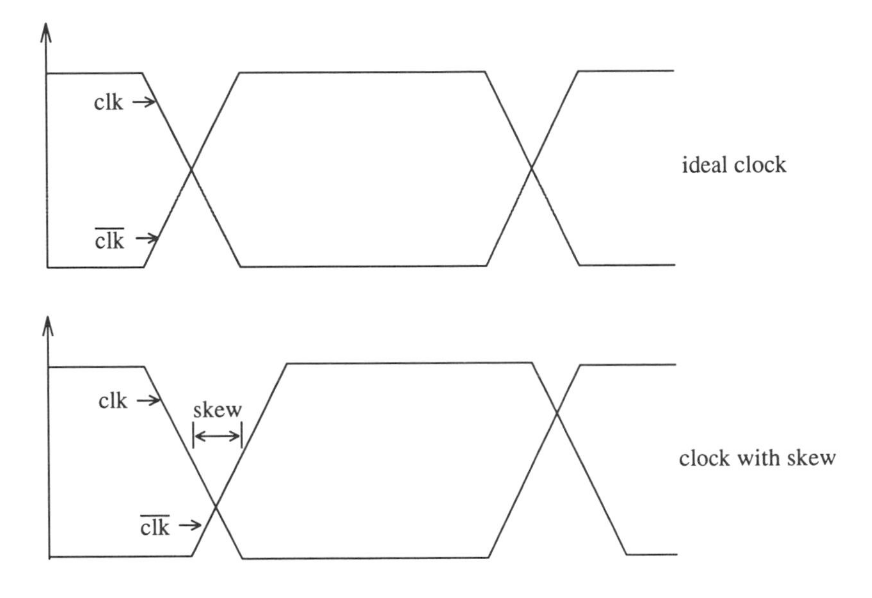

Figure 4.52 Clock skew.

4.6.4 Clock Skew in Dynamic Logic

Clock skew is a critical design parameter in high-speed circuits. Fig. 4.52 shows the clock skew in single complementary-phase clock signals. If $c\bar{l}k$ is generated from clk, clock skew is possible. The time skew is measured between the half-V_{DD} points of clk and $c\bar{l}k$ signals. In the presence of clock skew, a glitch can be transmitted from one section to another as illustrated in the example of Fig. 4.53(b). This structure contains one stage between the two C^2MOS latches, and a glitch can be transmitted to the last C^2MOS latch. The example of Fig. 4.53(c) does not have this problem. It has been shown that to eliminate the signal race in N-P domino logic, an even number of inversions should be used between stages [17]. Moreover, the clock skew problem should be minimized to improve the speed of dynamic circuits. One possible solution of single complementary-phase clock generation, with minimal skew and process-insensitive, is the one shown in Fig. 4.54 [18]. The delays $clk_i \rightarrow clk$ and $clk_i \rightarrow c\bar{l}k$ are equalized with special buffer sizing.

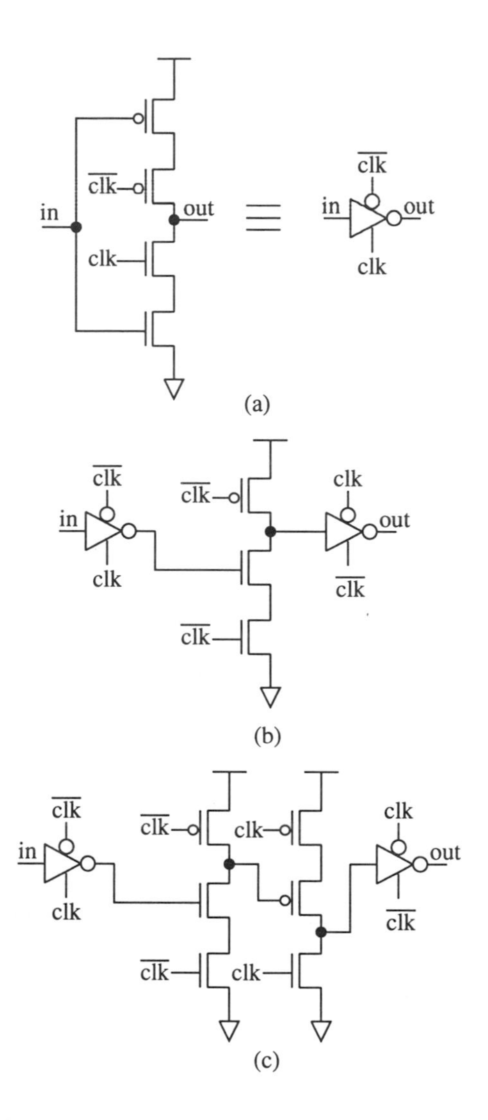

Figure 4.53 N-P dynamic logic circuits: (a) clocked inverter; (b) signal race; (c) no signal race.

4.7 CLOCKING

One way to synchronize thousands of signals in a VLSI system is to employ a clocking strategy. The clock controls the flow of data in the digital system and reduces the complexity of design.

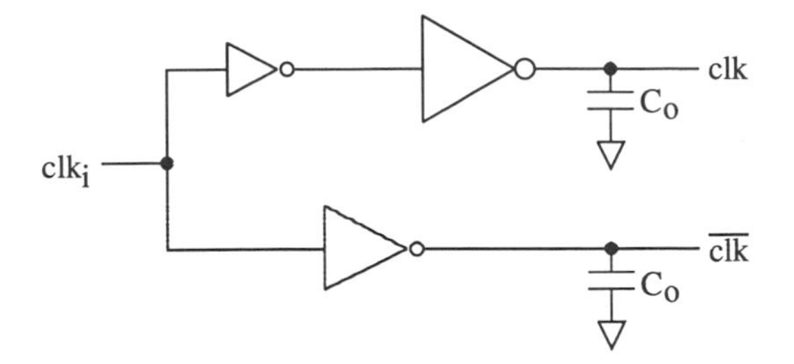

Figure 4.54 Improved delay-equalizing buffer design.

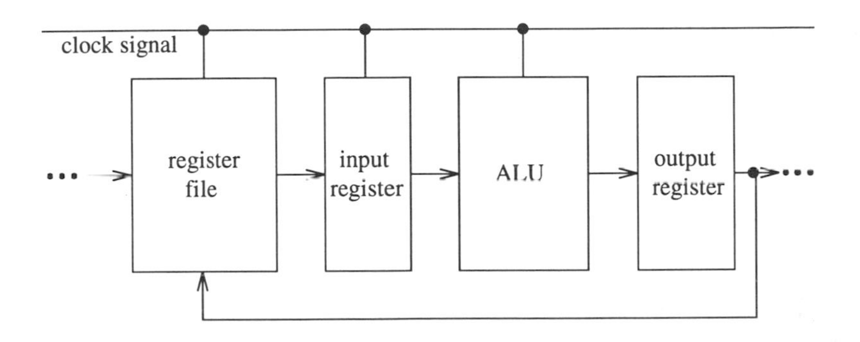

Figure 4.55 clocked pipeline system.

Most VLSI processors are constructed using a set of functional blocks (ALU, shifter, register file, etc.) connected via pipeline registers as shown in the example of Fig. 4.55. The clock signal can be split to one, two, three or four phases. Typically the phases are non-overlapping.

First we present the different storage elements (latches, registers), then we treat two clocking strategies : single-phase and two-phase with emphasis on the former which is usually the main option available in standard cell and gate-array approaches. The clock distribution issues are discussed in Section 4.9.4.

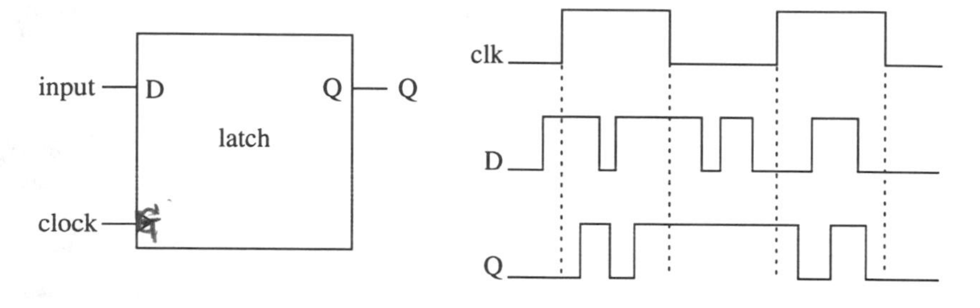

Figure 4.56 D-latch operation.

4.7.1 Storage Elements

There are many types of storage elements. Some of the ones used in VLSI design are the following:

4.7.1.1 D-Latch

Sometimes called level-sensitive latch. Its operation is shown in Fig. 4.56. The output changes with the input when the clock is high (case of positive level-sensitive latch). The D input must be stable within a time window around the positive transition of the clock (Fig. 4.57). The input data is passed to the output within a delay t_l. The time window is defined by two times; called setup time t_s, and hold time t_h. Setup time, t_s, is the time needed for the D input to be stable, prior to the clock edge. More specifically, it is the delay between the input of the latch and the storage node. Hold time, t_h is the time needed for the D input to be stable after the clock edge. This time relates to the delay between the clock input and the storage point.

There are a variety of implementations for this D-latch. Fig. 4.58 reviews some of the static versions. The circuit of Fig. 4.58(a) has a weak inverter used as feedback path for latch mode. The voltage at node A is not changed by noise or leakage because the feedback inverter would keep the level. The feedback inverter should have low (W/L) for NMOS and PMOS (weak inverter) compared to the transmission gate and forward inverter. This assures that the transmission gate is capable of overdriving the feedback inverter when data is being written to the latch. The feedback inverter should be carefully sized to guarantee switching for all process corners and maximum fanout condition.

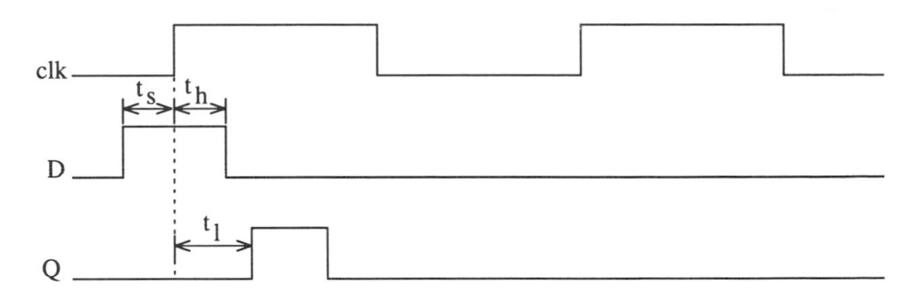

Figure 4.57 D-latch time parameters.

The problem of ratioed design in Fig. 4.58(a) can be avoided by using the modified version in Fig. 4.58(b), where a transmission gate is added in the feedback path. When $clk = 1$, the data is passed to the storage node and the feedback node is disconnected. When $clk = 0$, the feedback loop is closed, and the latch is in store (latch) mode. Fig. 4.58(c) shows another version of Fig. 4.58(b), where the outputs are buffered. This latter latch is found in the cells library of standard-cell and gate-array. All these described static latches store their state even if the clock is stopped. Note that these latches do not dissipate any DC power.

To reduce the size of the static latches, dynamic versions can be used as illustrated in Fig. 4.59, Fig. 4.60 and Fig. 4.61. Fig. 4.59 shows a simple dynamic latch, where the storage node A, temporarily stores the data. Note that latches have a property called "transparency": output follows the input when the clock is asserted. Otherwise they are "opaque". Fig. 4.60 shows two other latches [19]. The circuits of Fig. 4.60(a) is transparent when the clock clk, is high and latches the data (opaque) when the clock is low. This latch is positive level-sensitive. The negative level-sensitive is shown in Fig. 4.60(b). Note that these latches use one clock line (clk).

The circuits of Fig. 4.60 have reduced noise immunity. For example, for the circuit of Fig. 4.60(a), when the latch is opaque ($clk = 0$), the node A may be tristated high with Q tristated low. The node A is isolated and may be susceptible to noise which reduces its voltage. The reduced voltage of node A can cause the PMOS P_2 leaking current, thereby destroying the output Q. This problem was addressed with latches designed in DEC Alpha microprocessor [21]. For example the circuit of Fig. 4.61 is an improved version of Yuan and Svensson [19]. A weak PMOS device P_3 is added to solve the problem of noise in positive level-sensitive latch. The operation of this latch follows. When clk

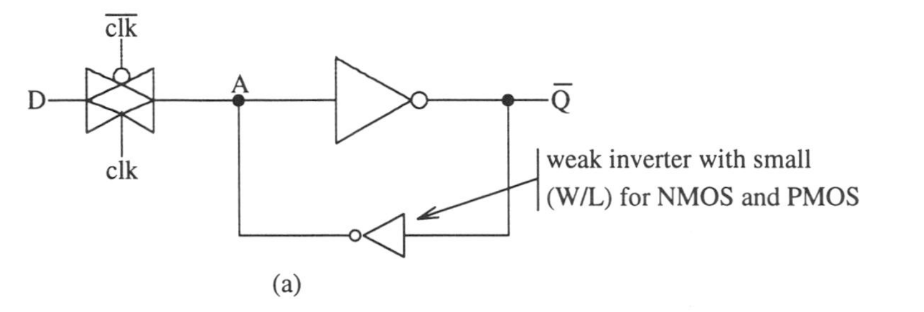

(a)

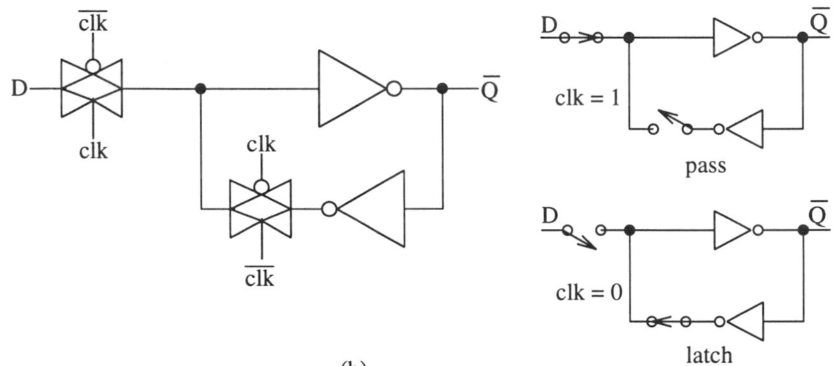

(b)

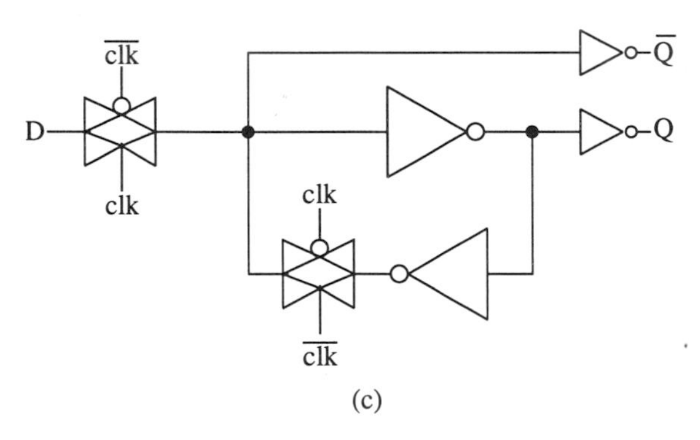

(c)

Figure 4.58 Static CMOS single clock latch circuits: (a) cross-coupled inverters latch; (b) transmission-gate latch; (c) buffered version of (b).

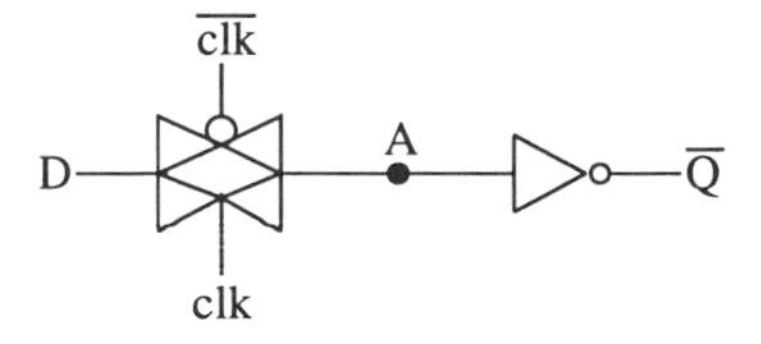

Figure 4.59 Simple dynamic CMOS single-clock latch.

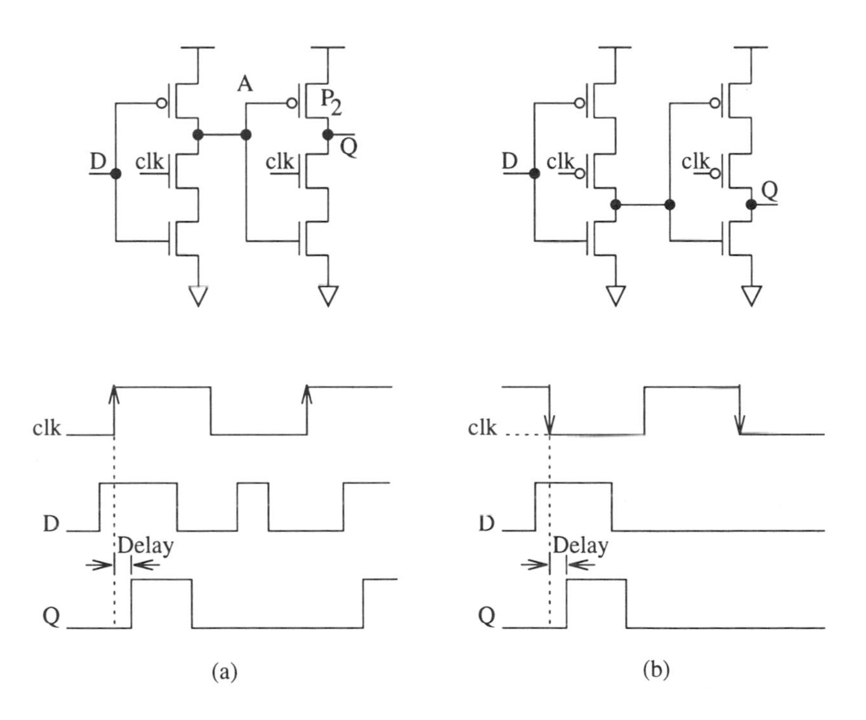

Figure 4.60 single clock dynamic latches: (a) positive level-sensitive; (b) negative level-sensitive.

is high, P_1, N_1 and N_3 function like an inverter. P_2, N_2 and N_4 function also like an inverter. Therefore the latch passes the input D to the output Q. If D falls to low, then A is high and Q is low. When *clk* is low, N_3 and N_4 are OFF. If D goes to high, P_1 is OFF, while the nodes A and Q are tristated high and low respectively. The added P_3, in this case, is ON and holds P_2 OFF. This device supplies current to node A and counters any noise.

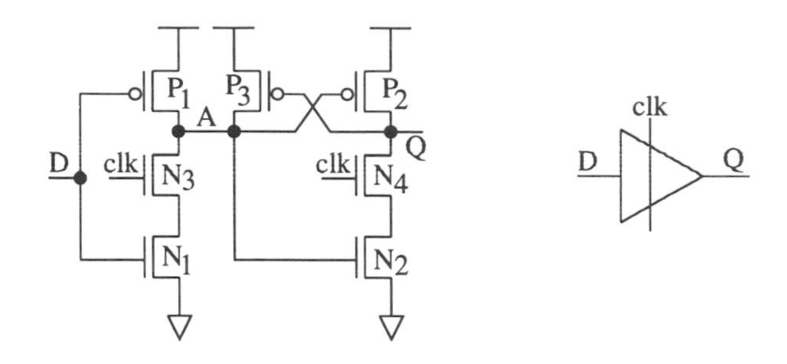

Figure 4.61 Non-inverting dynamic latch with improved noise immunity.

For flexibility reason many latches have been designed for DEC Alpha chip [21]. Some are illustrated in Fig. 4.62. These latches have been designed for all process corners and circuit conditions (supply voltage, temperature, rise/fall times, etc.). The results showed no appreciable evidence of race-through for *clk* rise/fall times at or below 0.8 *ns*. With 1-ns rise/fall times, the latches showed some signs of failure. A 0.5 *ns* for rise/fall times was set for the clock in this chip.

4.7.1.2 Edge-Triggered D-flip-flop, (ETDFF)

Sometimes this flip-flop is called edge-triggered register. Fig. 4.63 shows a static version (buffered) of the D flip-flop with positive edge-triggered, and the voltage waveforms. It is constructed by using two latches. The first one called *master*, is positive level-sensitive. The second one called *slave*, is negative level-sensitive. When the clock is low, the storage node A follows the input, while the node B stores the old data and is disconnected. Then, when the clock makes a transition from 0 to 1, the node A stores the input value during the transition, then ceases to sample any input data. When $clk = 1$, the master is in the the hold mode and the node A passes the data to storage node B of the slave latch which is then passed to the output Q and $\bar{Q}$. In this case, the output is disconnected from the input D. Hence, the flip-flop does not have the transparency property of the latch. When the clock returns back to 0, the slave is in hold mode. By reversing the two latches, a negative edge-triggered flip-flop can be constructed. This circuit can be found in standard-cell and gate-array libraries and represents an important cell in synchronized design. With high operating frequency, it is desirable to balance the delay of *clk* and

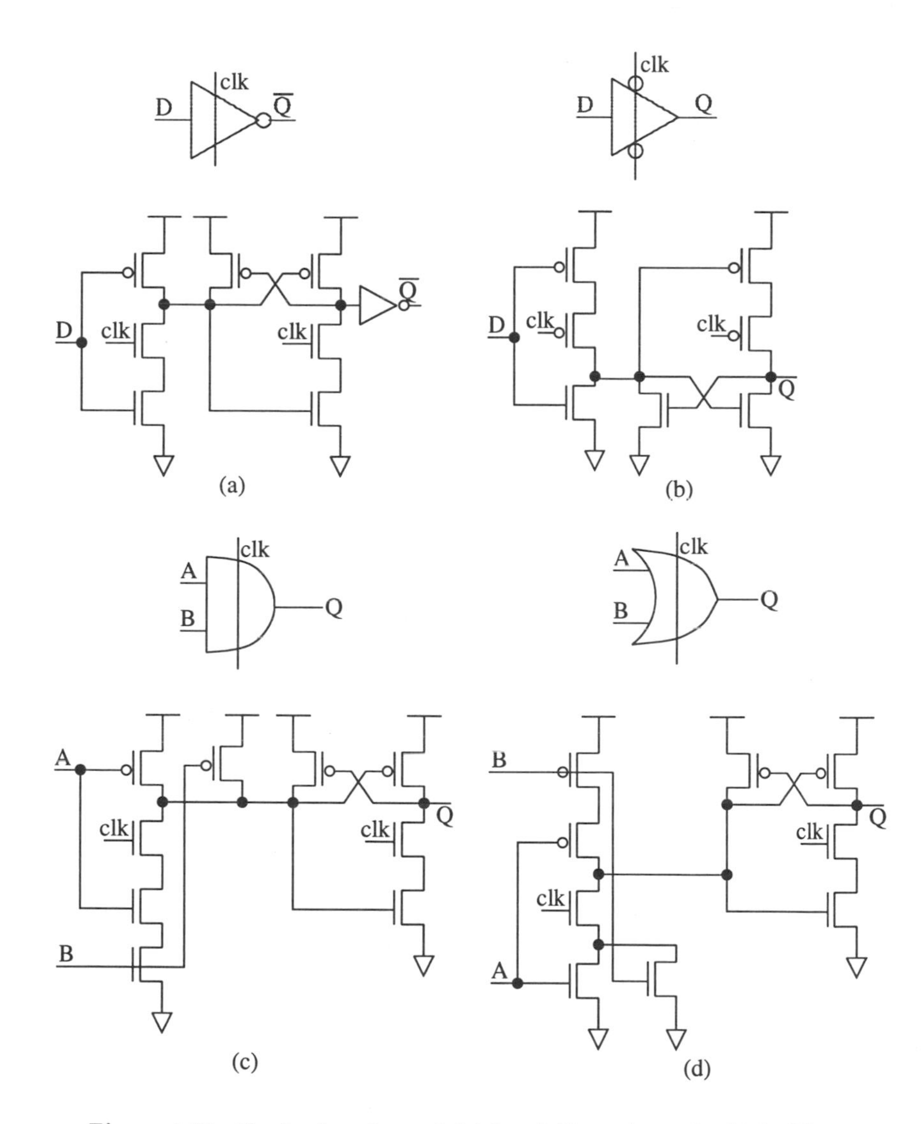

Figure 4.62 Single-phase dynamic latches: (a) inverting active-high; (b) non-inverting active-low; (c) two-input AND active-high; (d) two-input OR active-high.

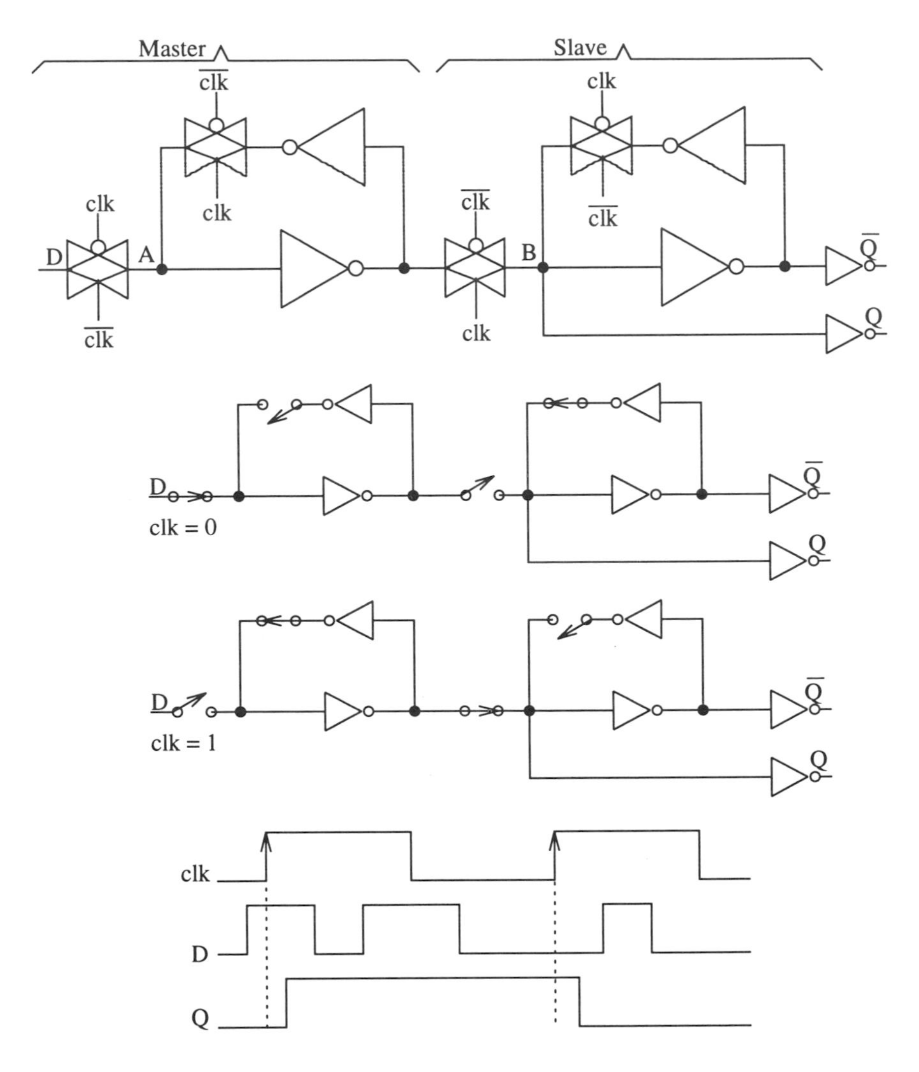

Figure 4.63 Positive edge-triggered flip-flop.

$c\bar{l}k$ locally, to reduce the clock skew problem. The clock skew, in single-phase strategy can lead to invalid data storage.

A dynamic version of the positive ETDFF is shown in Fig. 4.64 [19]. The operation of this circuit is illustrated by the voltage waveforms. The value

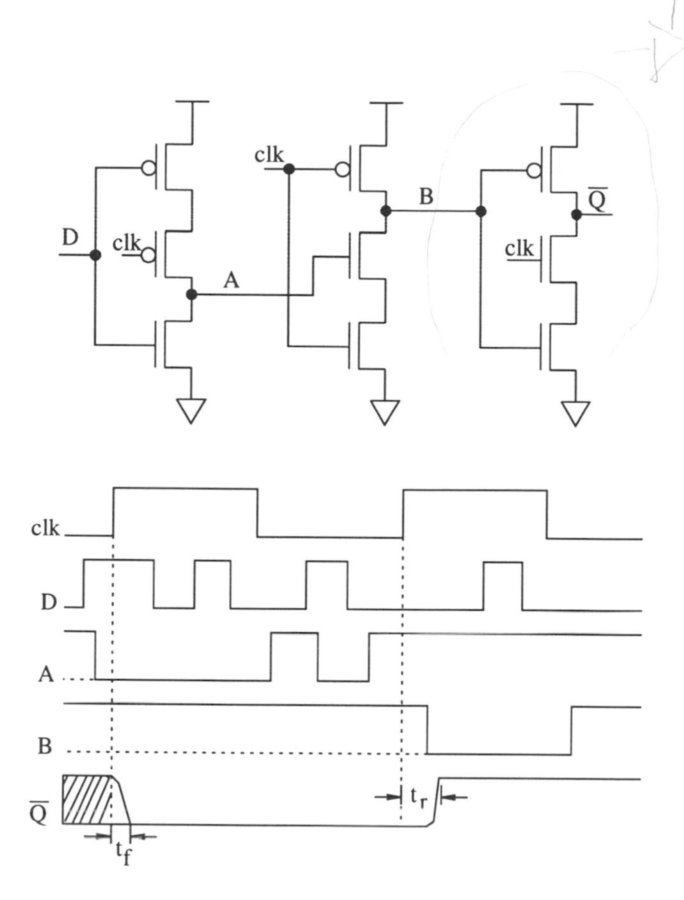

Figure 4.64 Single-phase positive ETDFF.

of the hold time of this flip-flop is close to zero [20]. This dynamic flip-flop, compared to the static one, needs only 9 transistors and one clock line. The negative ETDFF is shown in Fig. 4.65.

4.7.1.3 Miscellaneous

Many other latches and flip-flops are available; for example in gate-array libraries such as the JK flip-flop and the toggle (T) flip-flop. Fig. 4.66 shows the T flip-flop with reset control. When $clk = 1$, the output Q is complemented, whereas when $clk = 0$, Q keeps its old state.This T flip-flop provides divide-by-2 operation. A JK flip-flop is shown in Fig. 4.67. When J and K inputs are low, the outputs are maintained on the positive edge of the clock. If

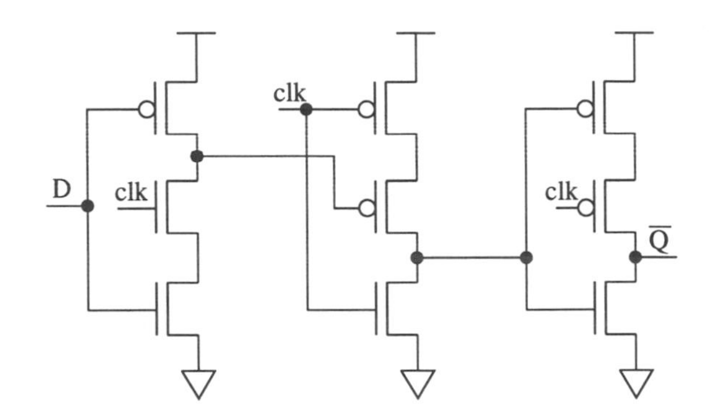

Figure 4.65 Single-phase negative ETDFF.

$J = 0$ and $K = 1$, the output Q is set to 0, whereas when $J = 1$ and $K = 0$, the output Q is set to 1. When both J and K are high then the output are complemented.

4.7.2 Single-Phase Clocking

Generic single-phase finite-state-machine (FSM) is shown in Fig. 4.68. The storage element can be either a latch or a register (flip-flop). For the latch case, it demands more constrained design because of the transparency property of the latch. When the latch is transparent, the state-signals can pass the logic block more than once during one clock cycle. To avoid race condition in this FSM, the clock width (of transparency) has to satisfy a two sided-constraint [22]. Hence, single-phase with latches, in the case of FSM, is insidiously complex.

To reduce the complexity of timing constraint, single-phase ETDFFs can be used. The flip-flop is never transparent. At the clock edge, the state is stored and it cannot pass the logic more than once during one clock cycle. Designing and synchronizing VLSI circuits with ETDFFs is rather simple and straightforward particularly when using static flip-flops.

For high-speed CMOS applications it is necessary that the storage elements should be carefully designed with minimum delay, setup time and clock skew. In this case, tristate dynamic latches can be used efficiently. Fig. 4.69 shows an example of using dynamic latches [21]. Notice that $L1$ and $L2$ are transparent latches separated by random logic and are not simultaneously active. When

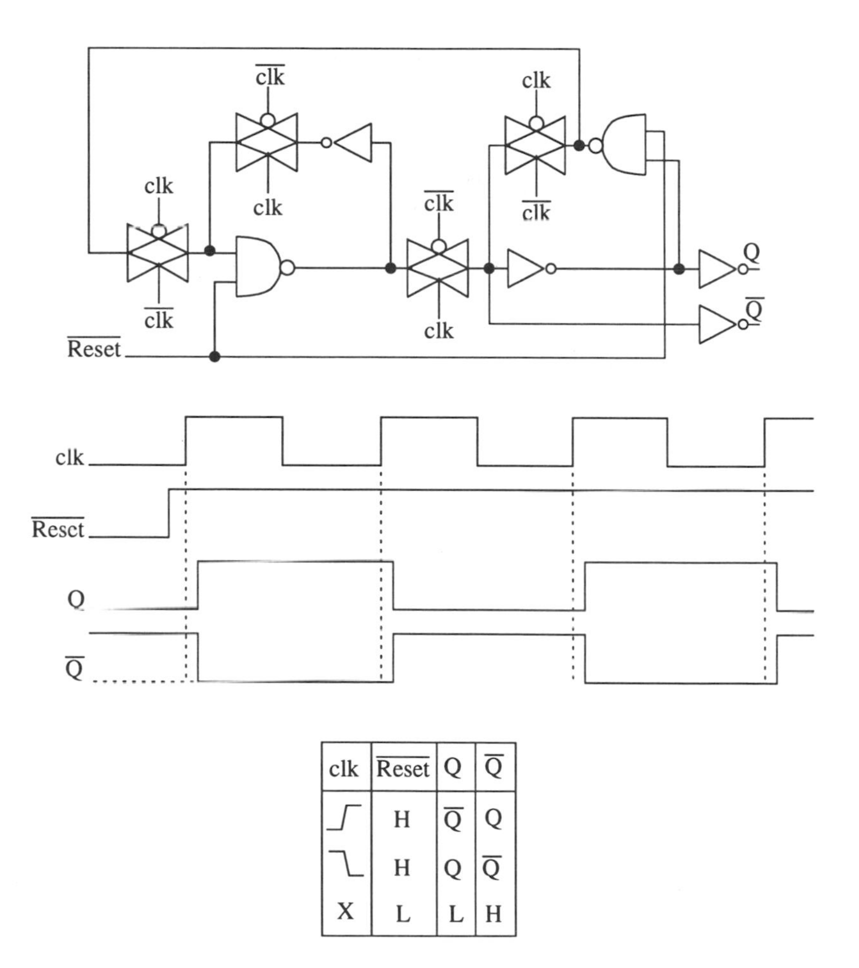

clk	$\overline{\text{Reset}}$	Q	$\overline{\text{Q}}$
rising edge	H	$\overline{\text{Q}}$	Q
falling edge	H	Q	$\overline{\text{Q}}$
X	L	L	H

Figure 4.66 Toggle flip-flop with reset (buffered version).

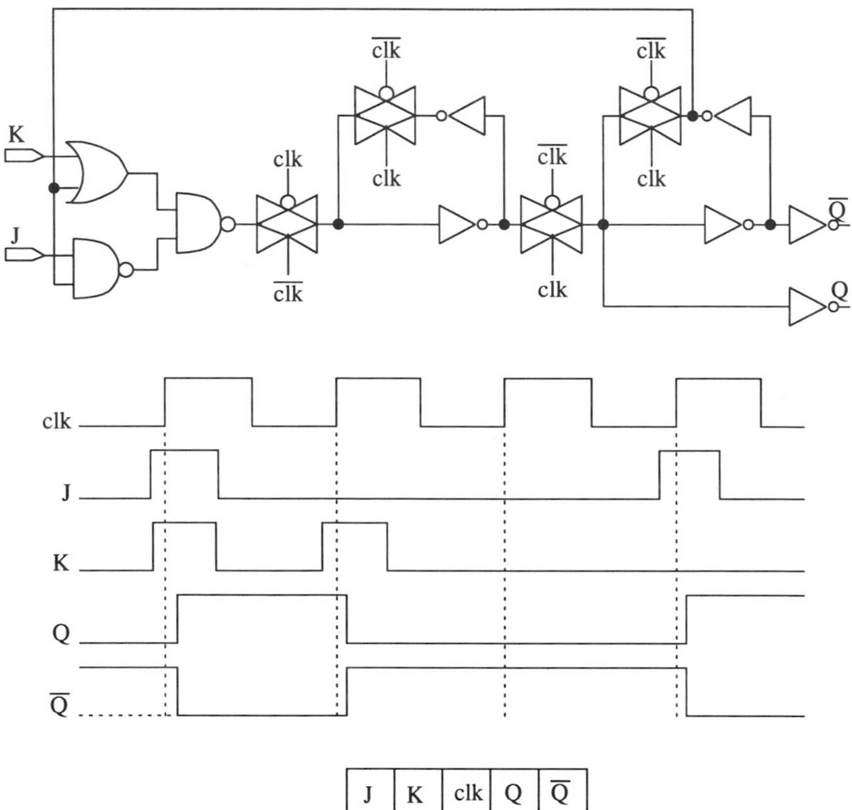

J	K	clk	Q	$\overline{Q}$
L	L	$\uparrow$	Q	$\overline{Q}$
L	H	$\uparrow$	L	H
H	L	$\uparrow$	H	L
H	H	$\uparrow$	$\overline{Q}$	Q

Figure 4.67 JK flip-flop.

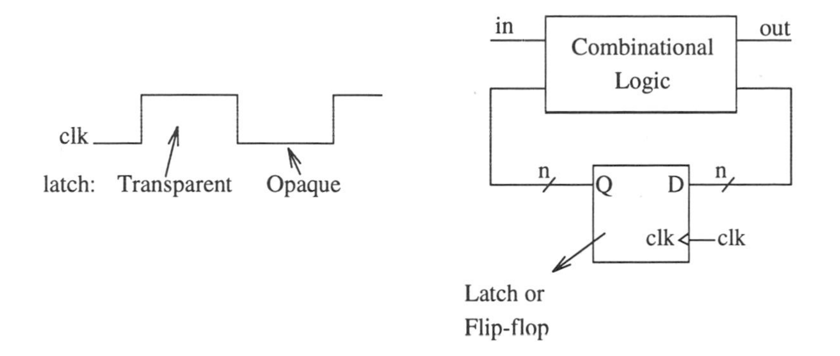

Figure 4.68 Single-phase finite state machine.

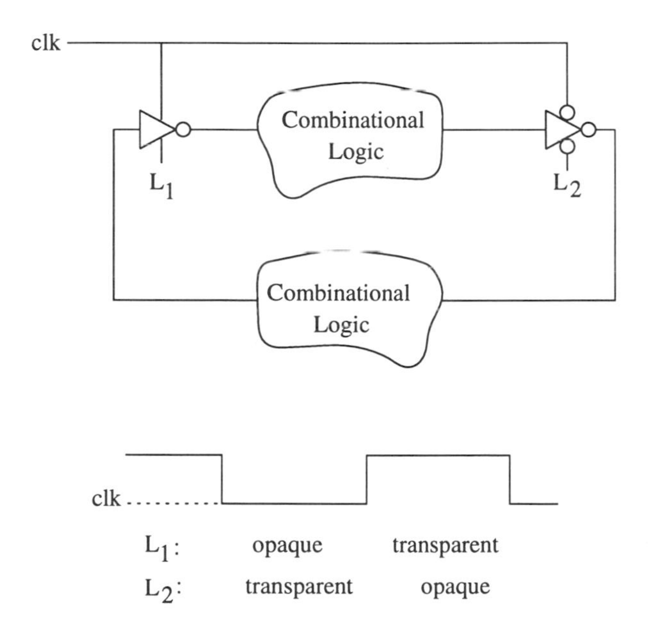

Figure 4.69 Single-phase systems with dynamic latches.

clk is high, $L1$ is transparent, whereas when *clk* is low, $L2$ is transparent. The minimum number of logic gates between latches can be zero and the maximum is constrained by the cycle time.

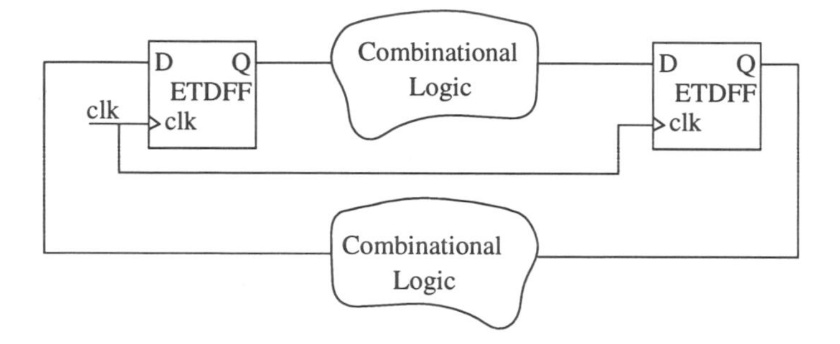

Figure 4.70 Single-phase system with ETDFFs.

Fig. 4.70 shows another example of single-phase system using ETDFFs. This system is edge based and the minimum cycle time is given by [22]

$$t_{cycle,min} = t_{ff,max} + t_{logic,max} + t_{setup,max} + t_{skew,max} \tag{4.97}$$

where t_{ff}, $t_{logic,max}$, $t_{setup,max}$ and $t_{skew,max}$ are worst case delays of the flip-flop, combinational logic block, setup time and clock skew. When designing with gate-array and/or standard cell approaches, the single-phase clocking scheme using static ETDFFs is the only option available for the designer.

4.7.3 Two-Phase Clocking

Two-phase non-overlapping clocking strategy removes many constraints existing in single-phase discipline. However, the use of two-phase (or multiple-phase) non-overlapping clock structures becomes more difficult as clock frequencies and chip size increase. This is because of the increase in clock skew and clock interconnect wiring. For high-speed applications, single-phase strategy is preferred and tends to be widely used in many VLSI systems' designs.

Fig. 4.71 shows an example of two-phase non-overlapping clocking scheme. The first latch $L1$ is transparent when the clock clk_1 is high, whereas $L2$ is transparent when clk_2 is high. The example of Fig. 4.71 is not the only way to build a two-phase system. Latches can be replaced by two-phase master-slave flip-flops where the master latch is clocked by clk_1 and the slave latch by clk_2. This latter structure does not have transparency property.

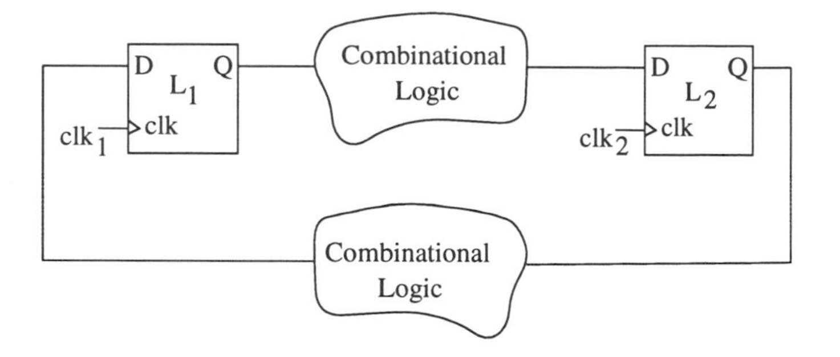

Figure 4.71 Two-phase system with latches.

4.8 PASS-TRANSISTOR LOGIC FAMILIES

Several pass-transistor logic families, for logic circuit design, have been proposed for improving the speed of CMOS circuits. Such families are: the conventional CMOS pass-transistor logic, the Complementary Pass-transistor Logic (CPL) [23], the Dual Pass-transistor Logic (DPL) [24], and the Swing Restored Pass-transistor Logic (SRPL) [25]. In this section, CPL, DPL, and SRPL logics are presented and compared.

4.8.1 CPL

The main concept behind CPL is shown in the block diagram of Fig. 4.72. It consists of NMOS pass transistor logic network driven by two sets of complementary inputs and two CMOS inverters used as buffers.

Fig. 4.73 illustrates an example of AND/NAND gate built in CPL logic. At the node Q for example we have

$$Q = A.B + B.\bar{B} = A.B \tag{4.98}$$

At the output of the corresponding inverter we have NAND function. The NMOS pass-transistor logic network forms pull-up and pull-down functions. When the inputs (AB) have the following combination (11), the voltage of the node Q is at a voltage given by

$$V_Q = V_{DD} - V_{Tn}(V_Q) \tag{4.99}$$

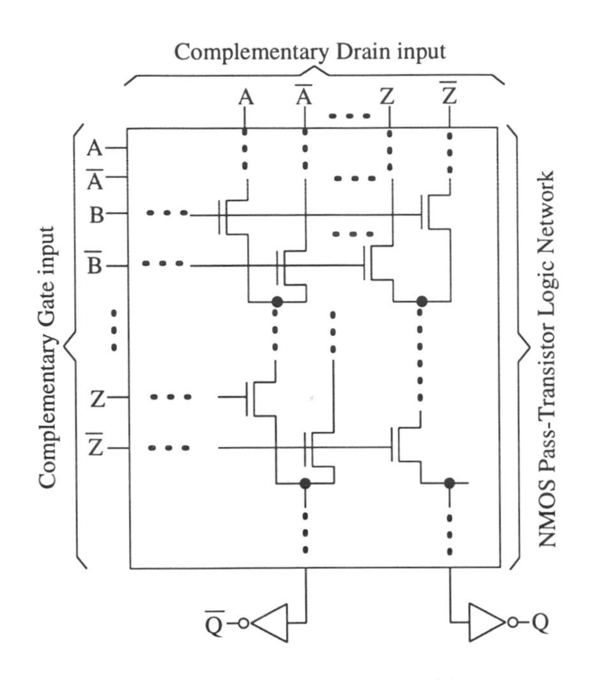

Figure 4.72 Basic CPL logic circuit.

where V_{Tn} is the threshold voltage subject to the body effect. So the inverting buffers translate the swing of the output from ground to $V_{DD} - V_{Tn}$, to a full-rail logic swing (ground to V_{DD}). The logic threshold voltage of the inverting buffers should be shifted to lower voltage than $V_{DD}/2$. Hence the β ratio of the inverter in this case should be higher than unity. This inverting buffer permits also to drive large load capacitance efficiently. When the output of logic networks are at $V_{DD} - V_{Tn}$ then all the output inverters are driven by reduced swing, as shown in Fig. 4.74. Hence, the DC power of the inverter increases because the pull-up PMOS device is not completely OFF. The V_{GS} of the pull-up PMOS is equal to $-V_{Tn}$. Moreover, the drive capability of the pull-down NMOS transistor is reduced particularly if the power supply voltage is reduced. The noise margins are also affected. To solve the problem of DC power dissipation we can design NMOS transistors with lower V_T than that of the PMOS transistor. Also, the body effect should be controlled. Another way to solve all the problems associated with the reduced high-level is to add to the CPL a PMOS latch as shown in the case of the AND/NAND circuit of Fig. 4.75. In this case, the two added PMOS transistors can be sized to be

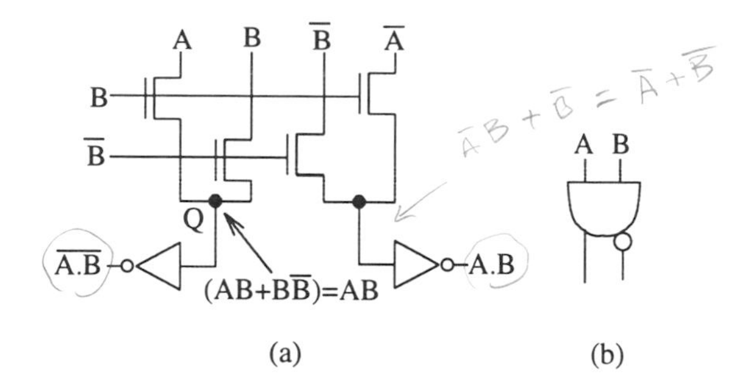

Figure 4.73 CPL AND/NAND circuit.

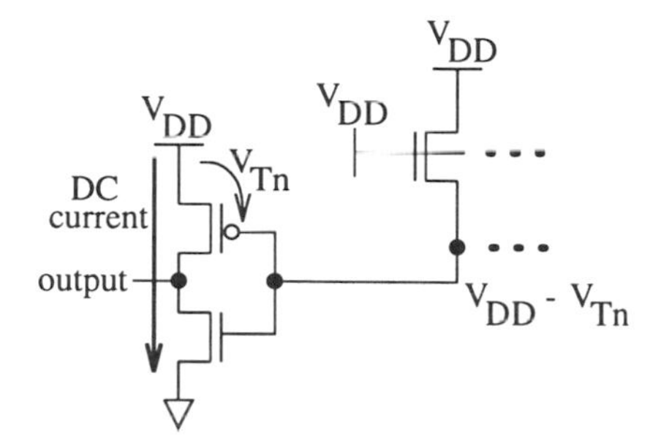

Figure 4.74 High level degradation in CPL.

minimum, as long as the high-level reaches V_{DD} in the given cycle time. We call this style PMOS latch CPL. Careful design should be considered when the NMOS network has minimum size devices. Otherwise the high-level stored in the latch cannot be discharged.

Fig. 4.76 shows examples of CPL arrays for OR/NOR and XOR/XNOR functions. With only 4 transistors we can produce many two-input functions with their complement. More examples are shown in Fig. 4.77 for 3-input AND/NAND and OR/NOR gates. In these examples 8 NMOS transistors are needed to generate the 3-input functions. Any complex logic function can be constructed easily using this principle of NMOS network transistors. For example the full-adder circuit can be constructed using wired CPL as shown in Fig. 4.78. The circuit is constructed using basic CPL primitives discussed before.

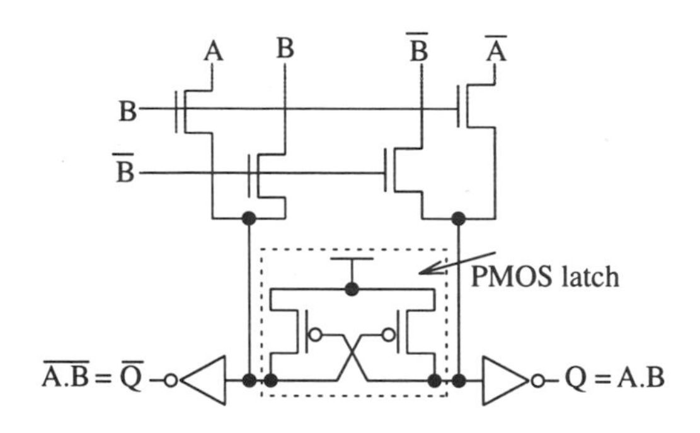

Figure 4.75 CPL AND/NAND circuit with PMOS latch.

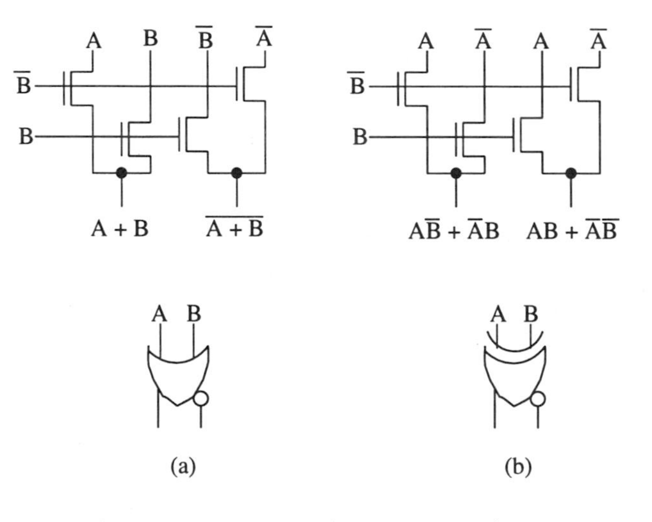

Figure 4.76 CPL OR/NOR and XOR/XNOR.

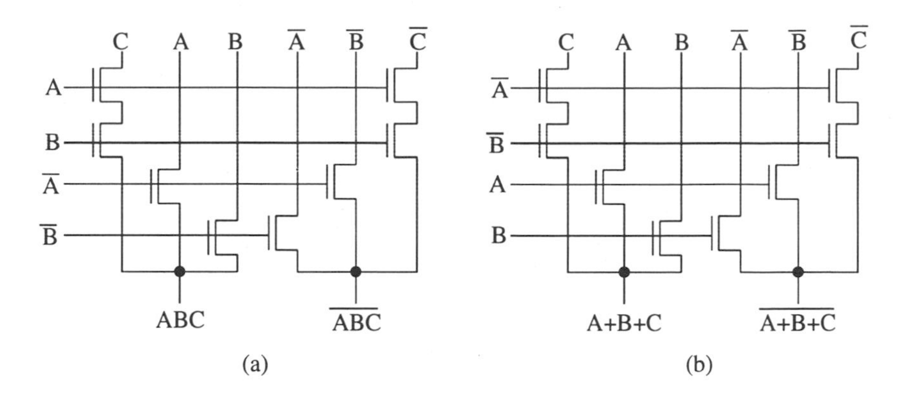

Figure 4.77 CPL 3-input: (a) AND/NAND; (b) OR/NOR logic arrays.

Also the sizes of the transistors are shown in this figure for fast operation. The transistors of the NMOS network, far from the output, have larger size than those closer to the output. This is because the NMOS devices, closer to the output, pass a reduced swing. The sizing of the transistors depends on the circuit type, layout and device's parameters. Compared to full-adder implemented in standard static CMOS style, the adder of Fig. 4.78 is much faster and dissipates less power due to the low internal swing. Also the schematic of this CPL adder is structured resulting in simplified layout.

One drawback associated with the CPL logic is the driving capability which is limited and the delay increases with long pass-transistor chains. So buffering is needed to restore the transmitted level and improve the driving capability.

4.8.2 DPL

The DPL is a modified version of CPL suitable for low-voltage applications. It alleviates the problems of CPL associated with the reduced high level. Example for AND/NAND gate is illustrated in the schematic of Fig. 4.79. It consists of NMOS and PMOS pass transistors in contrast to CPL gate, where only NMOS devices are used. In the example of AND/NAND gate, the NMOS transistor are used to pass the ground while the PMOS transistors are used to pass the high level (V_{DD}). The output of the DPL is full rail-to-rail swing owing to the addition of PMOS. However, this addition results in increased

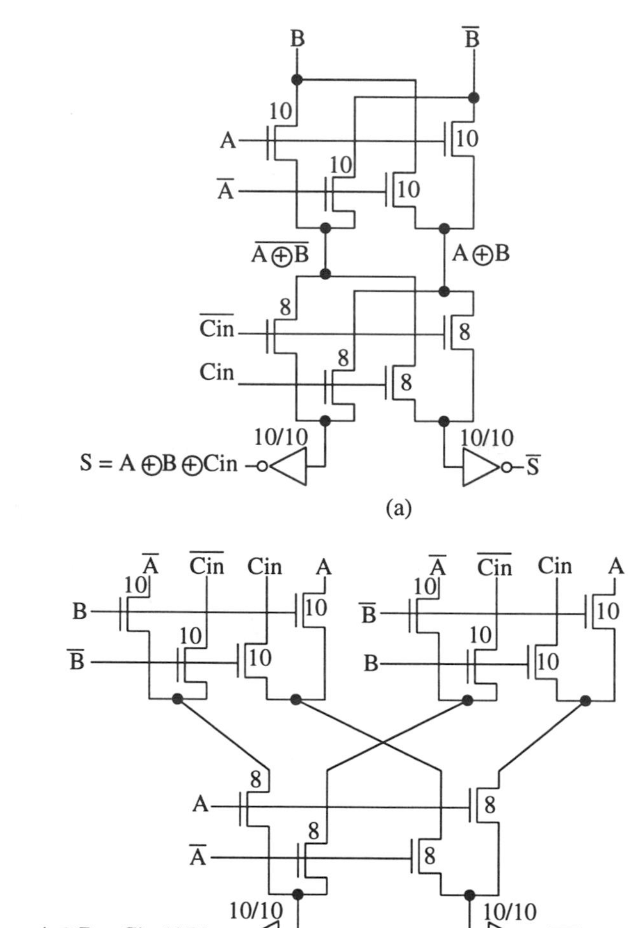

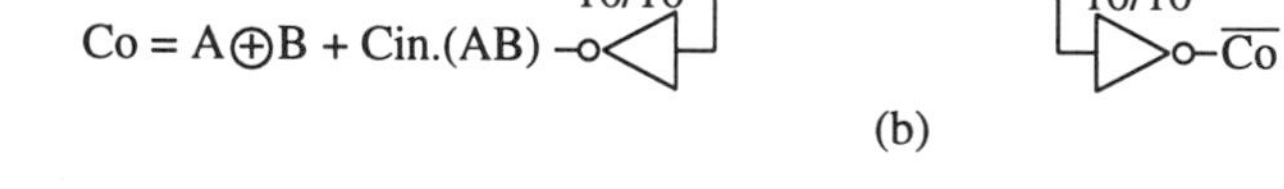

Figure 4.78 CPL full-adder: (a) sum block; (b) carry block.

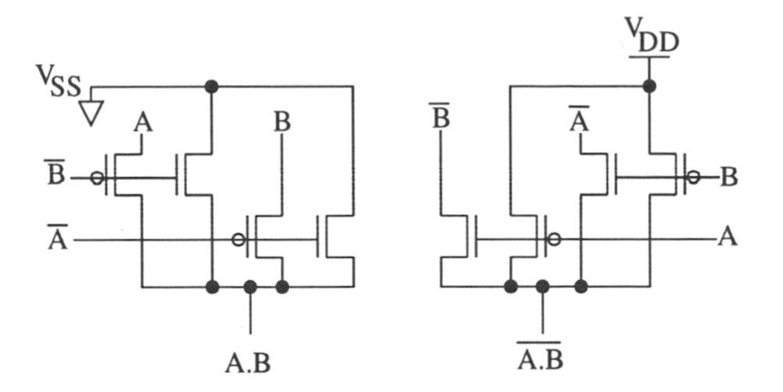

Figure 4.79 DPL AND/NAND gate.

input capacitance compared to CPL. This will not limit the performance of DPL as will be explained.

Fig. 4.80 shows a comparison between the switching characteristics of CPL, conventional pass-transistor CMOS and DPL XOR gates. In the truth tables, the column labeled "Pass" shows which signals are passed and perform the XOR function. These are some features of DPL

- The DPL gate has a balanced input capacitance. This reduces the dependence of the delay on the input data, contrary to the CPL and conventional CMOS pass-transistor logic where the input capacitances for the signals A and B are not the same.

- In DPL, for any input combination, there are always two current paths driving the output. This compensates for any reduction in speed due to the additional PMOS. For example, when the inputs A and B are low, A is passed by a PMOS while B is passed by an NMOS.

A DPL full-adder implementation is shown in Fig. 4.81. When all the input A, B and C are low, for example, there are two current paths to the output buffer. This implementation uses DPL primitives such as AND/NAND, OR/NOR, XOR/XNOR and MUX to generate the carry and sum signals.

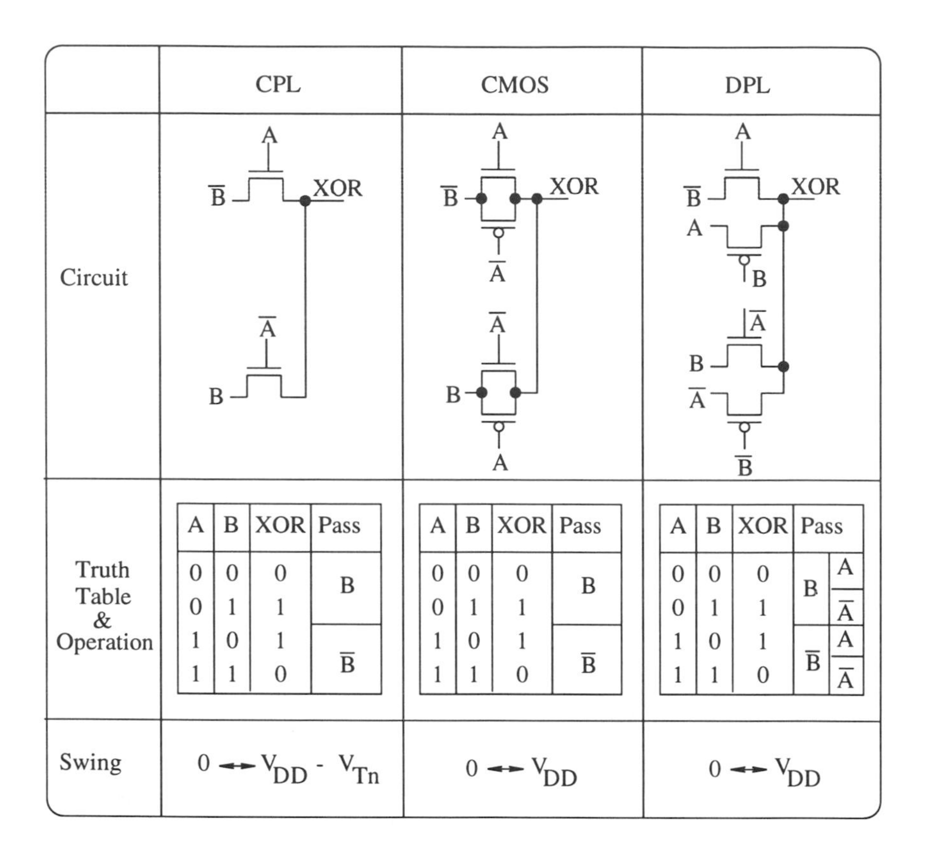

Figure 4.80 Comparison of CPL, conventional CMOS TG and DPL pass-transistor logics for XOR gate.

4.8.3 Modified CPL

Another technique which uses CPL-like style suitable for low-power/low-voltage is the Swing Restored Pass-transistor Logic (SRPL) [25]. Figure 4.82 shows the basis of SRPL logic gate. One part is the NMOS network with the CPL style discussed previously and the second part, is a CMOS latch. The cross-coupled CMOS inverters (latch) permit to restore the logic levels. So, any logic function in SRPL can be implemented using CPL network and a CMOS latch at the output. The sizing of such a logic is critical for speed and power dissipation issues. Fig. 4.83 shows an example of AND/NAND gate using SRPL. Increasing the size of the NMOS transistors in the network, $W_{network}$

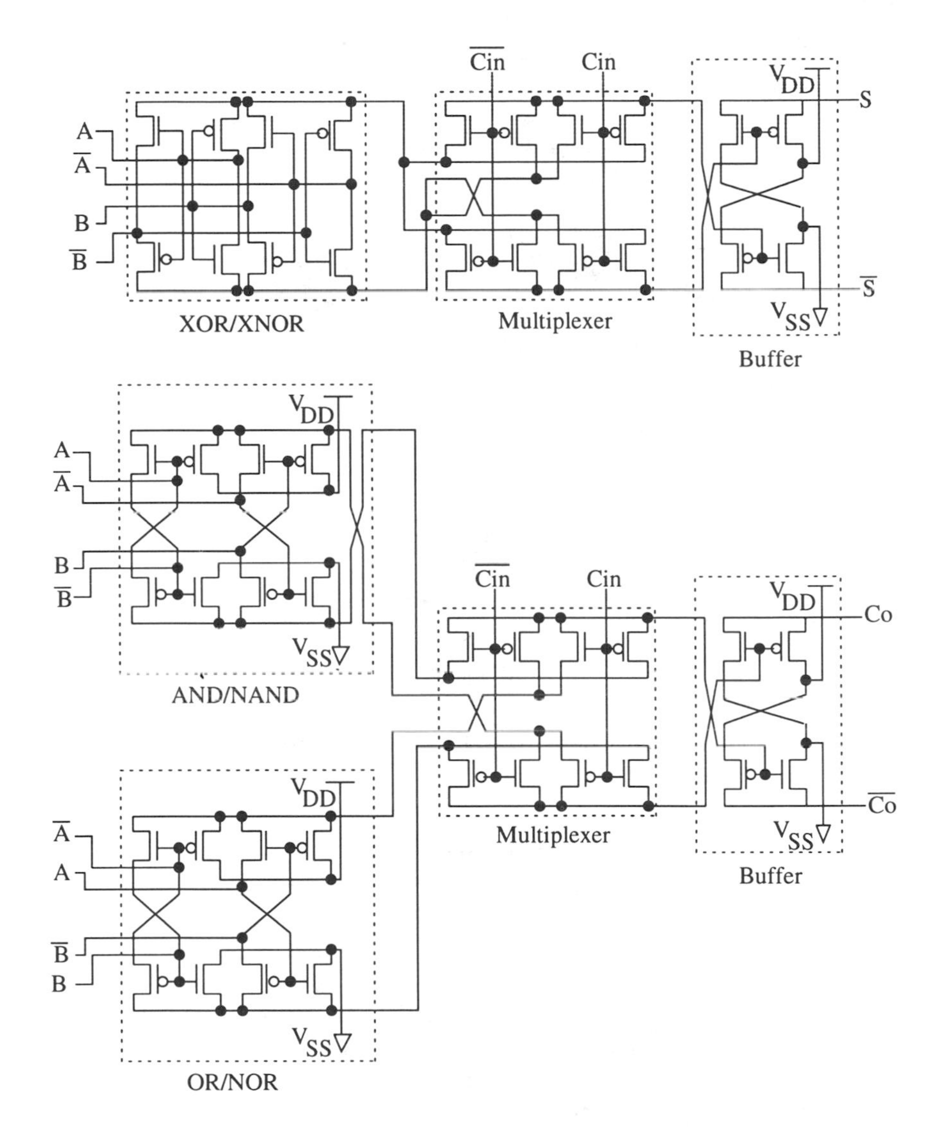

Figure 4.81 DPL full-adder.

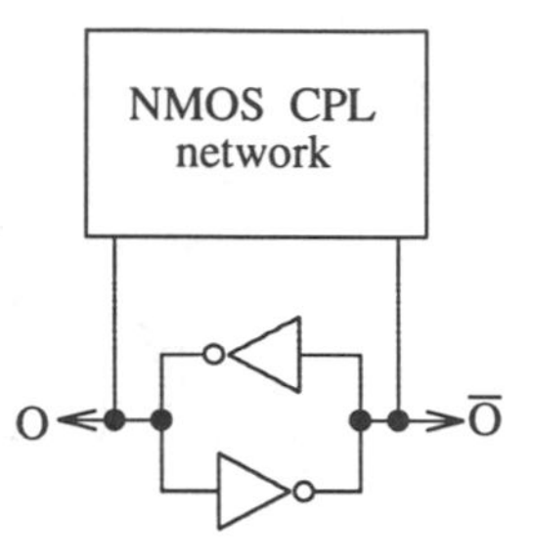

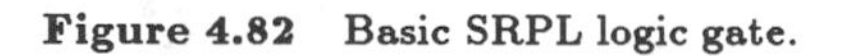
Figure 4.82 Basic SRPL logic gate.

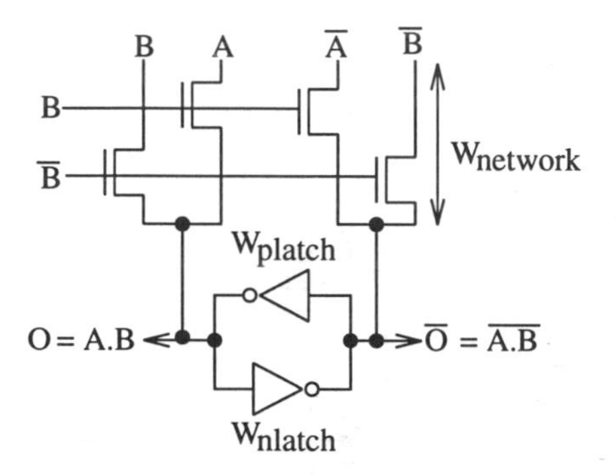

Figure 4.83 AND/NAND gate in SRPL.

improves the speed as shown in the simulation curve of Fig. 4.84. It has been found that the size of the latch should be minimum, for a fast operation, using the 0.8 μm device parameters of Chapter 3. If the size of the NMOS transistors in the network is small, the output of the SRPL gate fails to switch to ground because the equivalent impedance of the network is lower than the one seen by the output to V_{DD}. This problem becomes worse when many gates are cascaded. Fig. 4.85 illustrates this problem in 2 AND/NAND cascaded gates. When the input goes from V_{DD} to ground, the nodes A and B, initially at V_{DD}, cannot be completely discharged.

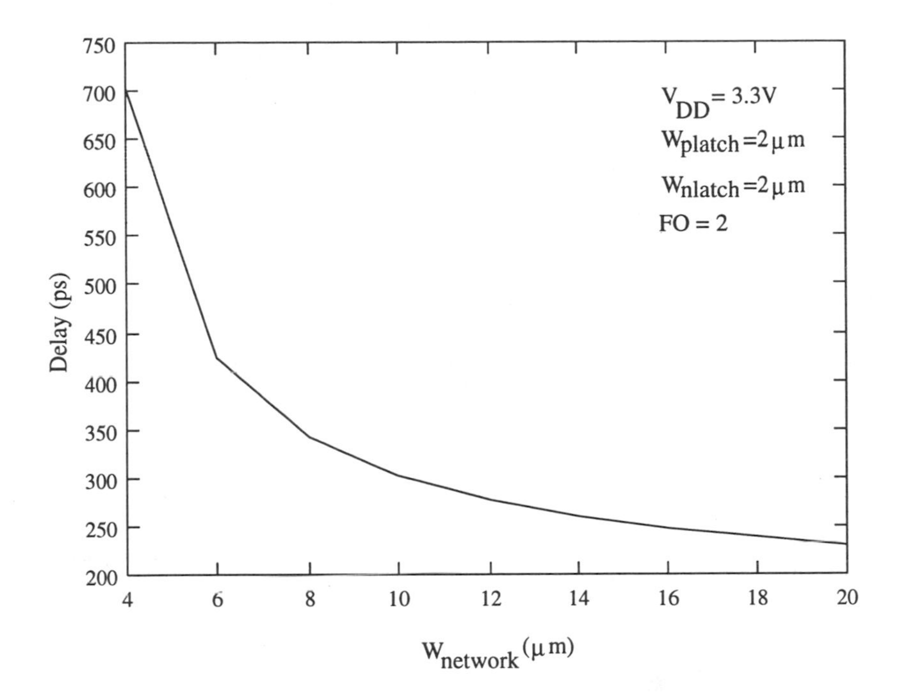

Figure 4.84 Delay of AND/NAND gate versus NMOS size.

4.8.4 Pass-Transistor Logics Comparison

The speed and power dissipation of the different pass-logic styles, so far presented, depend on the circuit type and the application of the circuit (cascaded gates, driving a fixed load, etc.). For the case of a full-adder, used in a multiplier array, a comparison is given in Chapter 7. In general, SRPL has the lowest power dissipation but careful design is needed when small device sizes are used. The DPL consumes more power than SRPL and PMOS latch CPL, because of the higher transistor count. Both CPL and SRPL circuits have the smallest area and the fastest speed. In summary, CPL-like styles are promising, for low-power and high-speed applications.

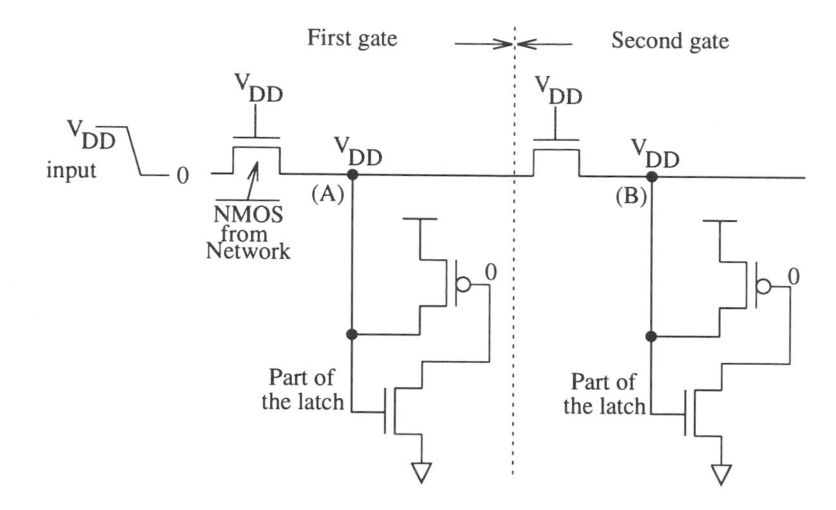

Figure 4.85 Example of AND/NAND gates malfunction when minimum size NMOS is used in the network.

4.9 I/O CIRCUITS

I/O circuits connect the on-chip logic circuitry to the external world. They play an important role in the limitation of speed and power dissipation of the whole chip. In this section many I/O circuits are discussed such as input and output buffers, clock distribution, clock buffering and low-swing I/O. The power dissipation issues related to these circuits are also studied. Layout techniques for I/O circuits are not covered in this chapter.

4.9.1 Input Circuits

To distribute an input signal to the internal circuitry of a chip, an input buffer is needed. It has its gate connected to the input pad. Excessive electrostatic charge, on the input pad, can break down the oxide and destroy the transistors of the input buffer. For an oxide thickness of 100 A, the breakdown voltage is $\approx 7\ V$. The voltage build on the gate, from the electrostatic charge, can be as high 300 V [26]. Fig. 4.86 shows an example of electrostatic discharge protection. If the voltage, at the node N, goes above V_{DD} or below ground, then the coupling diodes D_1 and D_2 limit the voltage excursion of the node N within $-V_{BE}$ and $V_{DD} + V_{BE}$. The role of the resistance R, is to limit the

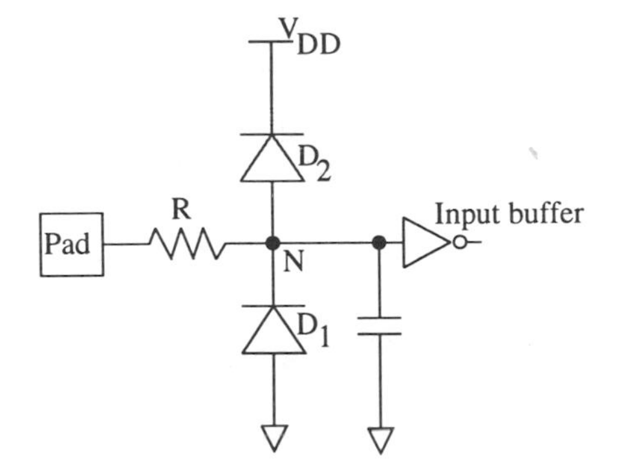

Figure 4.86 Electrostatic Discharge (ESD) protection in the input pad.

peak current that flows in the diodes. Typical values of R are few a hundred of Ω and are realized using the diffusion layers. The input protection circuit has a parasitic RC time constant which can limit high-speed operation. It ranges from a few tens of *ps* to a few hundreds of *ps*.

The input buffer, connected to this input pad, consists in general of a number of inverter stages to drive the internal circuitry. The input buffer, for clock distribution, needs special care and design and is discussed in Section 4.9.4.

4.9.1.1 Static Power Dissipation

When the input signal has TTL (Transistor-Transistor Logic) levels, the conventional CMOS buffer is used to translate these levels to CMOS levels. The TTL interface has historically specified input voltage levels of 0.8 V for the low-level input maximum, and 2.0 V for the high-level input minimum. The recently passed 3.3 V "Low-Voltage TTL (LVTTL)" standard is shown in Table 4.11.

The individual input inverters are designed by setting their W/L ratio such that the switching point of the buffer is near 1.4 V (middle of V_{IL} and V_{IH}). To have this switching point of 1.4 V at 5 V power supply voltage, the ratio W_n/W_p of the input inverter of the buffer should be at 2.9 using 0.8 μm CMOS technology. At 3.3 V, this ratio should only be equal to 0.7. However, since the TTL voltage swing is limited to 1.2 V, the input buffer is always dissipating

Table 4.11 LVTTL levels for $V_{DD} = 3.3 \pm 10\%$.

Minimum high output	Minimum high input	Maximum low output	Maximum low input
2.4 V	2.0 V	0.4 V	0.8 V

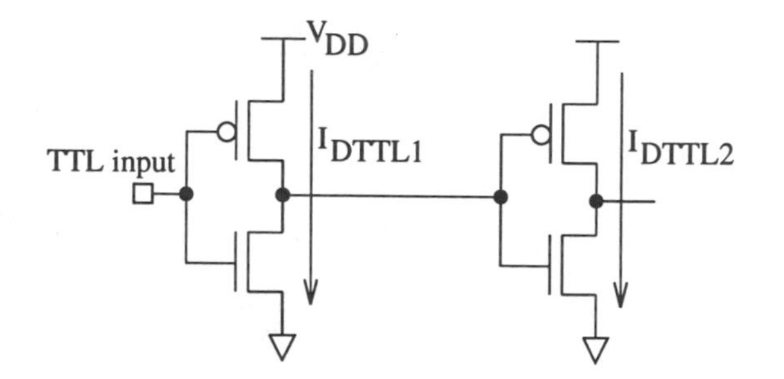

Figure 4.87 TTL input buffer.

DC power, as shown in Fig. 4.87, particularly if the V_T of the devices is low. If the first inverter does not fully translate the input TTL levels then the second stage dissipates some DC power. The static power dissipated by a TTL input buffer is

$$P_{TTL} = V_{DD} I_{DTTL} \tag{4.100}$$

where

$$I_{DTTL} = I_{DTTL1} + I_{DTTL2} \tag{4.101}$$

I_{DTTL} is the average dissipated current for the cases when the input is at low and high levels. At $V_{DD} = 3.3\ V$, the input buffer dissipates more static power when the input is high than when it is low. Fig. 4.88 shows the characteristics of the static power dissipation of the input buffer. Note that when V_{DD} is scaled down the DC current is reduced because the V_{GS} of the pull-up PMOS of the input buffer is reduced. If the number of TTL input pads is large, then the DC power of the input buffers could be an important and limiting factor. A static power-saving input buffer for reducing I_{DTTL} for 5 V power supply voltage has been proposed in [27].

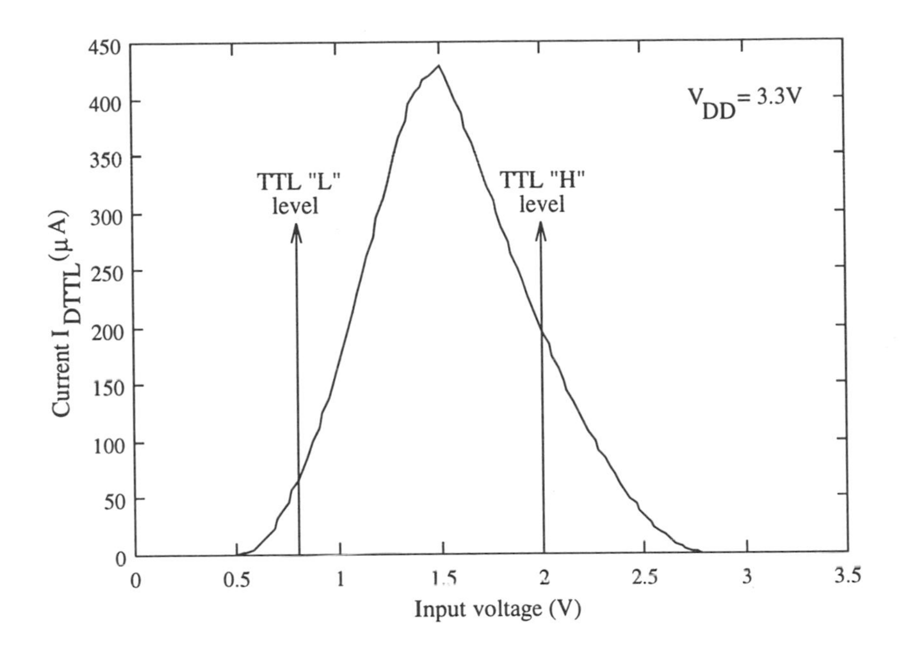

Figure 4.88 Simulated static power dissipation of input buffer.

4.9.1.2 Dynamic Power Dissipation

The dynamic power dissipation of the input pad is mainly internal power. The total dynamic power of all the input pads (of the same type of example) is

$$P_I = AN_iE_{ii}f \tag{4.102}$$

where A is the switching activity, N_i the number of the input pads and E_{ii} is internal energy of the input pad in Watt/Hz.

When the input signal has ECL levels, then an ECL input buffer, with ECL-to-CMOS converter are used. In general they are implemented in BiCMOS technology and consume a DC power. An ECL-CMOS converter can be designed in full CMOS [28].

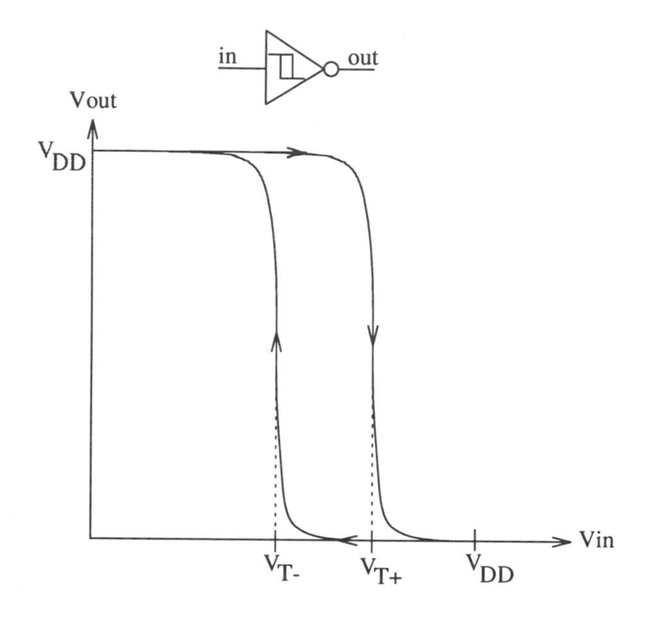

Figure 4.89 Transfer characteristic of Schmitt trigger.

4.9.2 Schmitt Trigger

When the input signal to a chip is slowly varying, a hysteresis circuit is needed at the input pad to generate a clean edge. A circuit called Schmitt trigger can be used for this function. They are often found at the on-chip inputs. Fig. 4.89 illustrates the transfer characteristic of an ideal Schmitt inverter with hysteresis voltage $V_H = V_{T+} - V_{T-}$. For 3.3 V power supply with 3.6 V for fast process and 3.0 for slow process, typical values are : $V_{T+,max} = 1.7\ V$ and $V_{T-,min} = 1.0\ V$. The Schmitt circuit switches at different thresholds. When the input is rising, it switches when $V_{in} = V_{T+}$ and when the input is falling, it switches when $V_{in} = V_{T-}$. Fig. 4.90 shows an example of how the Schmitt trigger turns a signal with a very slow transition into a signal with a sharp transition.

A CMOS version of the Schmitt trigger is shown in Fig. 4.91. When the input is rising, initially the NMOS transistors are OFF. The V_{GS} of the transistor N_2 is given by

$$V_{GS2} = V_{in} - V_{FN} \tag{4.103}$$

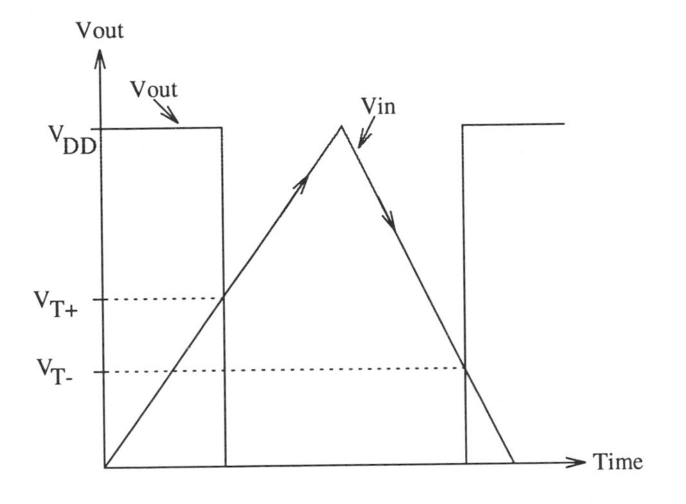

Figure 4.90 Voltage waveforms for slow input.

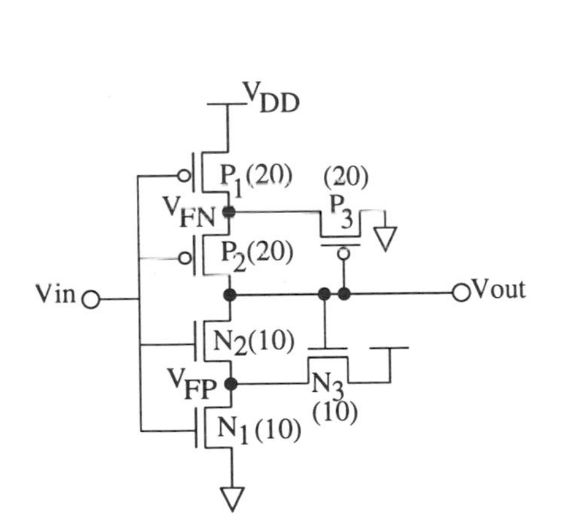

Figure 4.91 The CMOS Schmitt trigger schematic.

When $V_{in} = V_{T+}$, N_2 enters in conduction mode which means $V_{GS2} = V_{Tn}$ then[7]

$$V_{FN} = V_{T+} - V_{Tn} \tag{4.104}$$

[7] We neglect the body effect of N_2

The voltage V_{FN} is controlled by N_1 and N_3. These transistors operate in saturation because

$$V_{GS1} = V_{T+} \tag{4.105}$$

$$V_{DS1} = V_{FN} = V_{T+} - V_{Tn} \tag{4.106}$$

and

$$V_{GS3} = V_{DD} - V_{FN} \tag{4.107}$$

$$V_{DS3} = V_{DD} - V_{FN} \tag{4.108}$$

The drain currents flowing in N_1 and N_3 are equal. Then using a simple MOS model we have

$$\frac{\beta_{n1}}{2}(V_{T+} - V_{Tn})^2 = \frac{\beta_{n3}}{2}(V_{DD} - V_{T+})^2 \tag{4.109}$$

We have

$$V_{T+} = \frac{V_{DD} + \sqrt{\beta_n}V_{Tn}}{1 + \sqrt{\beta_n}} \tag{4.110}$$

where

$$\beta_n = \frac{\beta_{n1}}{\beta_{n3}} = \left\{\frac{W_{n1}/L_{n1}}{W_{n3}/L_{n3}}\right\}_{eff} \tag{4.111}$$

This equation shows that the trigger point is independent of the process parameters except for V_{Tn}. By symmetry, the trigger point for falling transition, can be deduced from the pull-up section. We have

$$V_{T-} = \frac{\sqrt{\beta_p}(V_{DD} + V_{Tp})}{1 + \sqrt{\beta_p}} \tag{4.112}$$

where

$$\beta_p = \frac{\beta_{p1}}{\beta_{p3}} = \left\{\frac{W_{p1}/L_{p1}}{W_{p3}/L_{p3}}\right\}_{eff} \tag{4.113}$$

If $\beta_n = \beta_p$ and $V_{Tn} = -V_{Tp} = V_T$, then

$$V_{T+} = \frac{V_{DD}}{2} + \frac{V_T}{2} \tag{4.114}$$

$$V_{T-} = \frac{V_{DD}}{2} - \frac{V_T}{2} \tag{4.115}$$

$$V_H = V_{T+} - V_{T-} = V_T \tag{4.116}$$

In this case the hysteresis voltage can be made equal to V_T. The short-circuit power dissipation of the Schmitt trigger can be very important since the rise/fall times of the input signal is very long.

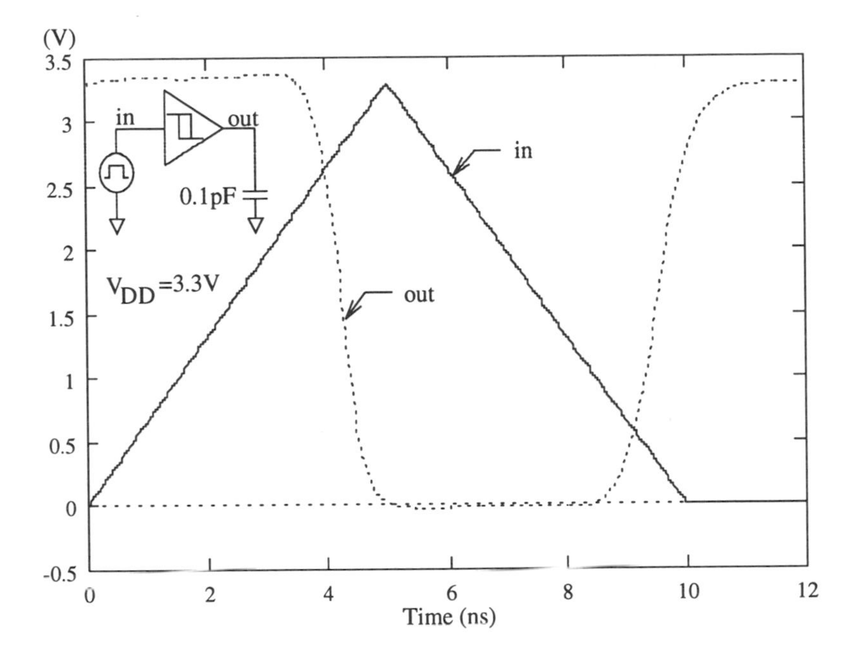

Figure 4.92 Input and output voltage waveforms of the Figure 4.91.

Fig. 4.92 shows SPICE simulation of the circuit of Fig. 4.91 in 0.8 μm technology. In this example, the load capacitance is 0.1 pF and the total power dissipation is 0.85 mW. The dynamic power dissipation, due to the load and parasitic capacitances, is 0.40 mW. Therefore, the power due to the short-circuit is 0.45 mW, which represents $\sim$ 53 % of the total power dissipation.

4.9.3 CMOS Buffer Sizing

When the gate is intended to drive a large load capacitance (larger than the input capacitance of the gate), the driving capability is limited and the delay is large. If we increase the size of the gate (driver configuration), we improve the rise/fall times but still the delay can be improved by putting several stages of buffering between the first gate and the load. The objective in a buffer configuration is to get the input signal to the load as quickly as possible. Each stage in the buffer chain should have its transistor widths larger than the previous

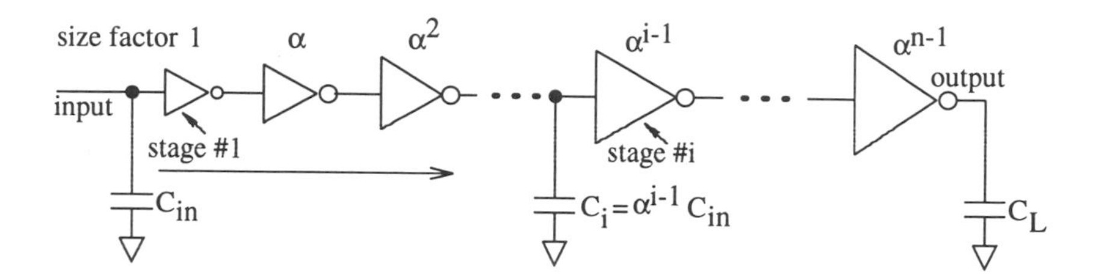

Figure 4.93 Buffer chain.

one by a factor α. This is illustrated in Fig. 4.93. We use a similar simple analysis for minimum delay as the one proposed by Mead and Conway [29].

The first inverter has small size devices and has an input capacitance C_{in}. The input capacitance for stage i is given by

$$C_i = \alpha^{i-1} C_{in} \quad i = 1, 2, ..n \tag{4.117}$$

The number of stages needed, n, is deduced from

$$C_L = \alpha \cdot C_n \tag{4.118}$$

which gives; using Equation (4.117)

$$\alpha^n = \frac{C_L}{C_{in}} \tag{4.119}$$

Hence

$$n = \frac{\ln(C_L/C_{in})}{\ln \alpha} \tag{4.120}$$

Now we want to find the ratio α. The delay of the first stage, driving an identical one, is τ_0. The delay of the ith stage is

$$t_{di} = \alpha \tau_0 \quad i = 1, 2, ..n \tag{4.121}$$

For n-stages, the total delay is

$$t_d = \sum_{i=1}^{n} t_{di} = n \alpha \tau_0 \tag{4.122}$$

The delay is then

$$t_d = \ln(C_L/C_{in}) \frac{\alpha}{\ln \alpha} \tau_0 \tag{4.123}$$

Question : What are the values of the size ratio α and the number of stages n to optimize the delay ?

By differentiating t_d equation with respect to α and then setting it equal to zero, we have

$$\alpha_{opt} = e \approx 2.7 \tag{4.124}$$

The optimum number of stages is

$$n_{opt} = \ln(C_L/C_{in}) \tag{4.125}$$

In this analysis, we have neglected the parasitic output capacitance of each stage. Other studies [30, 31, 32, 33] illustrate that the size ratio α depends on the ratio of the parasitic output capacitance and load capacitance. In [34] a new approach for CMOS tapered buffers, with large C_L/C_{in} ratio, was proposed. It uses a variable size ratio between the stages.

The power dissipation of a CMOS buffer is mainly dominated by dynamic power dissipation for large V_T. The short-circuit power dissipation can be neglected as first-order analysis [34]. If we include the parasitic output capacitance. So stage i, has a total output capacitance

$$C_{oi} = \alpha^i C_{in} + \alpha^{i-1} C_p \tag{4.126}$$

we assume that the parasitic capacitance of stage i is proportional to the size ratio α. The dynamic power dissipation at the output of gate i is

$$P_i = C_{oi} V_{DD}^2 f = V_{DD}^2 f(\alpha^i C_{in} + \alpha^{i-1} C_p) \tag{4.127}$$

or

$$P_i = V_{DD}^2 f \alpha^{i-1}(\alpha C_{in} + C_p) \tag{4.128}$$

The total power is

$$P_T = \sum_{i=1}^{n} P_i = V_{DD}^2 f(\alpha C_{in} + C_p) \sum_{i=1}^{n} \alpha^{i-1} \tag{4.129}$$

Hence

$$P_T = V_{DD}^2 f(\alpha C_{in} + C_p) \frac{\alpha^n - 1}{\alpha - 1} \tag{4.130}$$

The power efficiency of the buffer can then be defined as

$$\eta = \frac{P_L}{P_T} \tag{4.131}$$

where P_L is the power dissipated, due to the load C_L, which is simply $C_L V_{DD}^2 f$. P_T is the total power dissipated given by Equation (4.130). This power efficiency, for a given C_L, C_{in} and C_p, is a function of only the factor α. The term $1 - \eta$ characterizes the additional power dissipation overhead, needed by the buffer chain to drive the load C_L. For high values of α, the power efficiency of the buffer increases. In practice α can be in the range of 2-to-10. This value of α can be set depending on speed, delay and power dissipation constraints.

4.9.4 Clock Drivers and Clock Distribution

Usually when the clock is to be distributed on-chip, input buffers are needed. The clock circuit has to drive very high internal load with extremely fast fall/rise times. For example, in the case of DEC Alpha chip [21] the clock load is 3.2 nF. If this load has to be driven by a large driver, in rise/fall times of 0.5 ns when the clock frequency is 200 MHz [$T_{clock} = 5\ ns$], then the average transient current would be

$$I_{avr} = C\frac{\Delta V}{\Delta t} = \frac{3.2 \times 10^{-9} \times 3.3}{0.5 \times 10^{-9}} = 21\ A \tag{4.132}$$

@V_{DD} = 3.3V power supply. The corresponding dynamic power dissipation due to this clock loading is

$$P = CV_{DD}^2 f = 3.2 \times 10^{-9} \times 3.3^2 \times 200 \times 10^6 \approx 7\ W \tag{4.133}$$

This example shows how the clocking is an important design issue. A clocking strategy should be used to distribute the clock to the different functional blocks of chip with minimum clock skew and low-power dissipation.

The clock skew problem is due mainly to two issues

- The difference in RC interconnect time constants: For example in Fig. 4.94 node A and node B have two different branch lengths to node C. In this case, the delays of the signals at node A and node B vis a vis node C are different. Therefore, the clock skew is equal to the time difference between these two signals.

- Unbalanced loads at different nodes: As shown in the example of Fig. 4.95, if the loads at the nodes A and B, C_A and C_B respectively, are different. Then the skew between the signals at these nodes exists.

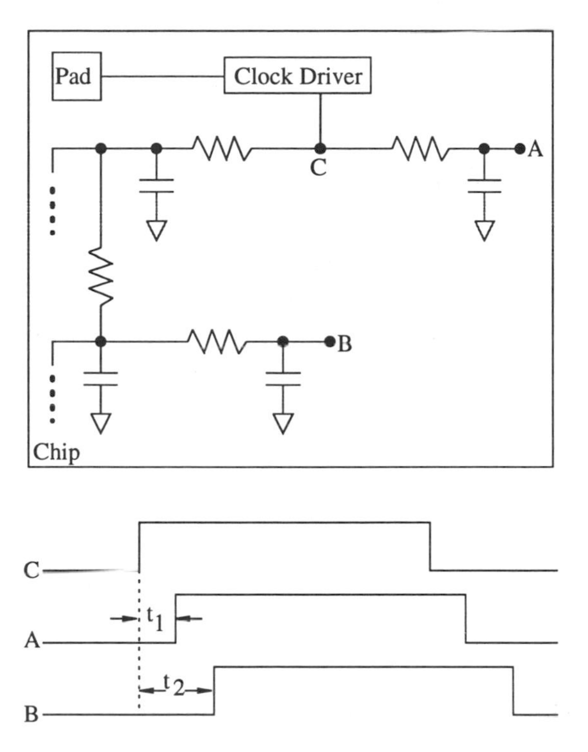

Figure 4.94 Clock skew due to the difference in RC delays in branches A and B.

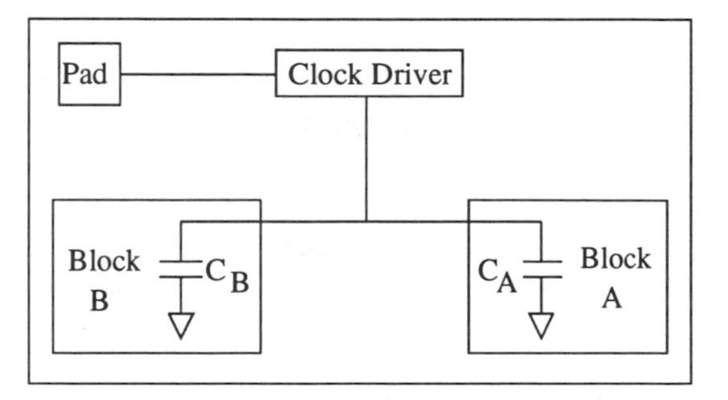

Figure 4.95 Clock skew due to the unbalanced loads at block A and block B.

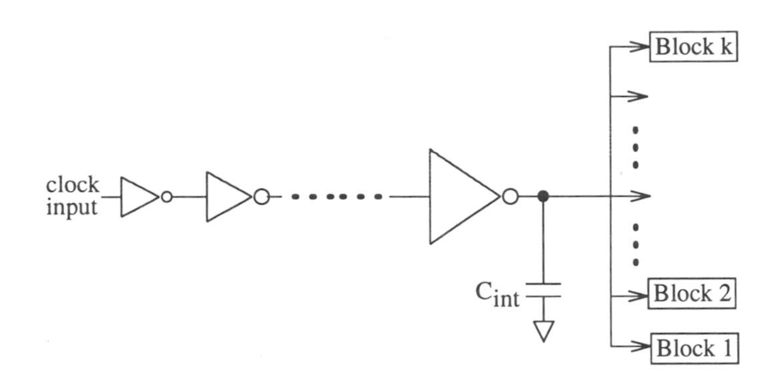

Figure 4.96 Cascaded inverters for global clock distribution.

Several strategies have been proposed to minimize clock skew. The first approach is to use cascaded inverters (buffer) to drive a large load and feed all blocks as shown in Fig. 4.96. The buffer chain is designed by the approach presented in Section 4.9.3. In another approach, the clock distribution is accomplished by using a tree of clock buffers well sized as illustrated by Fig. 4.97. Identical buffers are used in each level and each buffer sees the same load capacitance. Equalizing clock buffer loads is possible by : 1) equalizing the interconnect lengths between the buffers of different levels, and 2) the addition of dummy buffers at the slightly loaded buffer output. The last distribution level has buffers which drive the functional elements such as registers. This structure results in very reduced skew and the only skew that exists is the one produced by variations in process parameters. To further minimize the skew, identical layout for all the buffers, should be used. As an example of tree approach is the following case. To distribute the clock signal to 64 elements (for example registers), 3 stages (levels) of buffering with 1-to-4 tree structure are required. A variety of software packages have been developed for clock tree synthesis [35, 36].

To reduce the high dynamic power dissipation (few Watts) in clock distribution at a fixed power supply, many techniques can be used such as:

1. Using a low capacitance clock routing line such as metal3. This layer of metal can be, for example, dedicated to clock distribution only.
2. Using low-swing drivers at the top level of the tree or in intermediate levels.

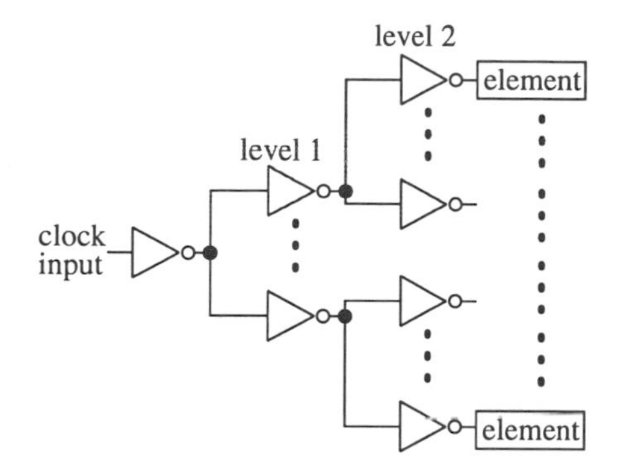

Figure 4.97 Clock tree distribution.

For the second approach, a half-swing clocking scheme has been proposed [37]. Fig. 4.98 shows the half-swing clock driver which generate half V_{DD} clock signals (four phases) to the elements (eg., latches). Using the charge sharing principle, the node of half-V_{DD} can be expressed by

$$H - V_{DD} = \frac{C_1 + C_A}{C_1 + C_4 + C_A + C_B} V_{DD} \quad \textit{when clk is low} \tag{4.134}$$

$$H - V_{DD} = \frac{C_2 + C_A}{C_2 + C_3 + C_A + C_B} V_{DD} \quad \textit{when clk is high} \tag{4.135}$$

where C_A and C_B are added capacitors to the power lines. C_1 through C_4 are the load capacitances of the driver. When C_A is equal to C_B and both are large enough, compared to C_1-C_4, then H-V_{DD} node is stabilized at $V_{DD}/2$.

Fig. 4.99 shows the clocking schemes of the latches driven by the clock driver. Compared to the conventional scheme which uses two clock phases, the half-swing scheme requires four clock phases. Two phases are for PMOSs and two are for NMOSs as shown in Fig. 4.99(b). This scheme reduces the power by 75%. However, the delay of the latch is increased by the new clocking scheme, which can be acceptable [37].

4.9.5 Output Circuits

To drive the output pad, a high drive capability driver is needed to achieve adequate rise and fall times. In this case, inverter chain is used to handle the

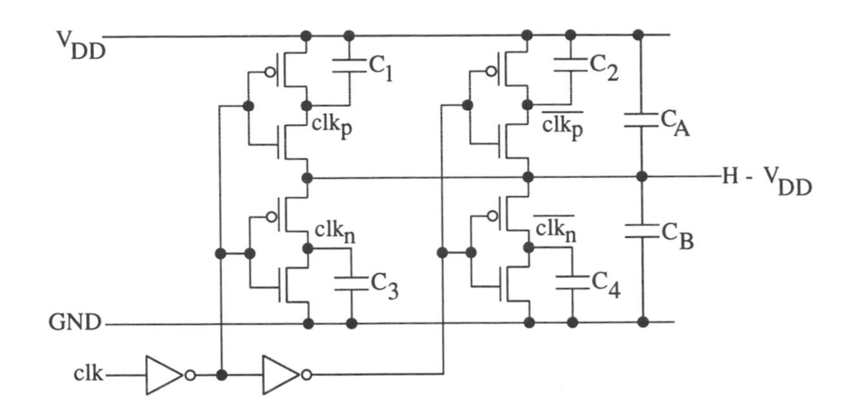

Figure 4.98 Half-swing clock driver.

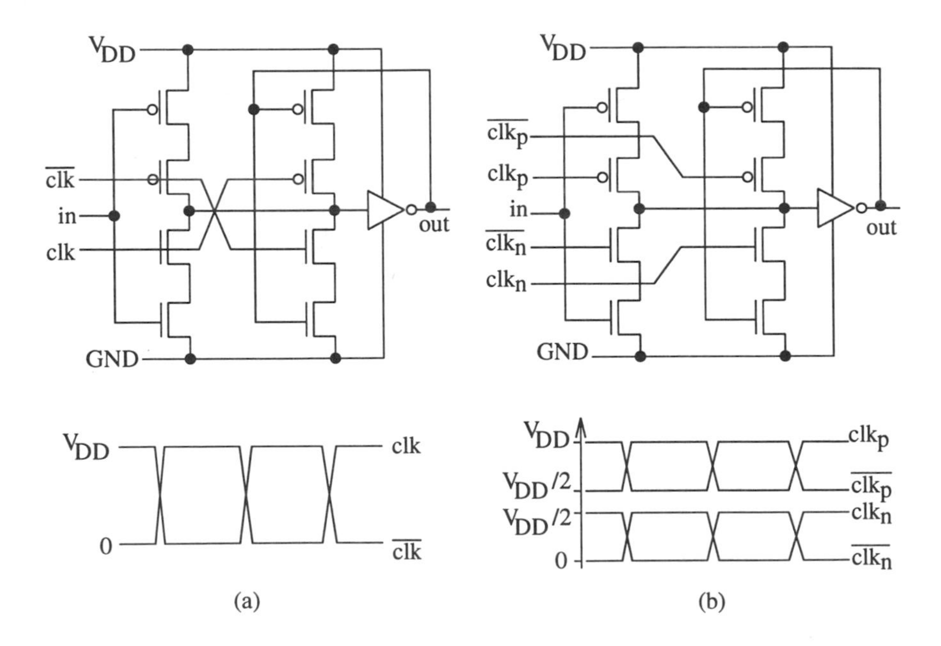

Figure 4.99 (a) Conventional; (b) half-swing clocking schemes for a latch.

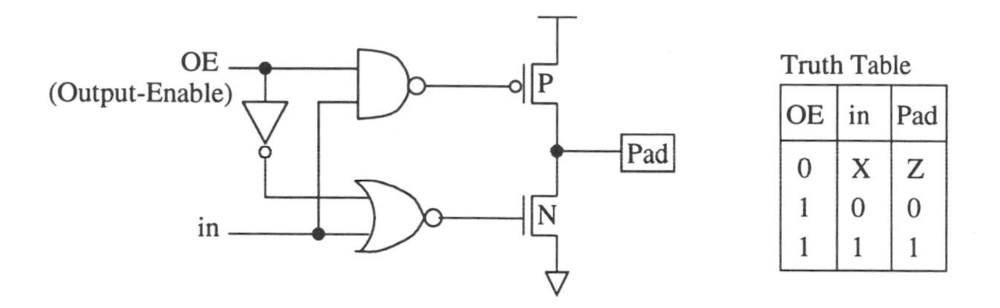

Figure 4.100 Tristate output buffer.

large load of the pad, package wiring, and off-chip load. This capacitance can be few tens of pF. A typical value of this capacitance is 50 pF. There are many types of output pads such as tristate, bidirectional, low-V_{DD} (3.3 V) to high-V_{DD} (5 V) output buffer and low-swing output.

4.9.5.1 Tristate and Bidirectional Circuits

Fig. 4.100 shows a tristate circuit to drive large pad capacitance. When the output enable signal is high, the output data is the same as the input data. When the output enable signal is low, then the output of the pad is in high impedance state (Z). Both the output NMOS and PMOS transistors are cutoff. Fig. 4.101 shows the bidirectional I/O circuit which is quite useful when we need to save the number of I/O pads. Sometimes an input buffer is included in the bidirectional pad. The operation of this circuit is obvious.

4.9.5.2 Power Dissipation of Output Circuit

The total power dissipation at the output pads can be divided into the static power dissipation and the dynamic power dissipation. The static power dissipation is due mainly to the leakage currents (junction and subthreshold) if the output pads are driving CMOS logic. If the V_T of the devices is large enough, then the static power dissipation of the output pads is neglected. However if V_T is small, then the DC power, due to the subthreshold current, for the output pads is

$$P_s = N_o I_{DS,mean} V_{DD} \tag{4.136}$$

where N_o is the number of output pads and $I_{DS,mean}$ is the average subthreshold current for both cases when the input is low and high. For low V_T the

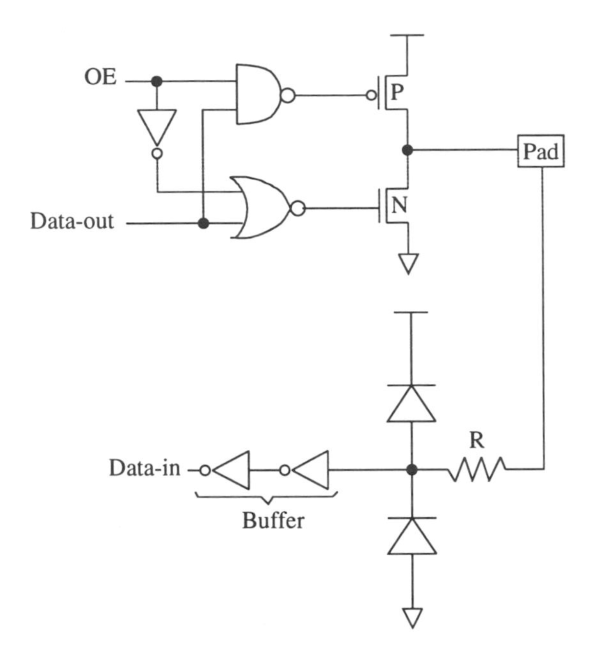

Figure 4.101 Bidirectional pad.

$I_{DS,mean}$ value would be important, because the devices in the output buffer have large size particularly the output transistors. $I_{DS,mean}$ should be computed in worse case where the V_T has its minimum value. Thus for future technologies where the threshold voltage is low and the number of output pads is large, this static power dissipation would be very important and can be a limiting factor for low-power applications. Hence low-power circuit techniques are needed for output buffers.

If the CMOS output buffer is intended to drive bipolar TTL inputs (not CMOS TTL inputs), then important current is sunk. Fig. 4.102 shows the final stage of the buffer driving a TTL logic. Since, bipolar TTL inputs can source significant amounts of current, a CMOS output buffer must sink this current. For 3.3 V power supply, this current can be in the range of 1 mA to 12 mA depending on the strength of the output driver. The static power dissipated by the one output pad driving bipolar TTL inputs is

$$P_{TTLol} = V_{OL} I_{OL} \tag{4.137}$$

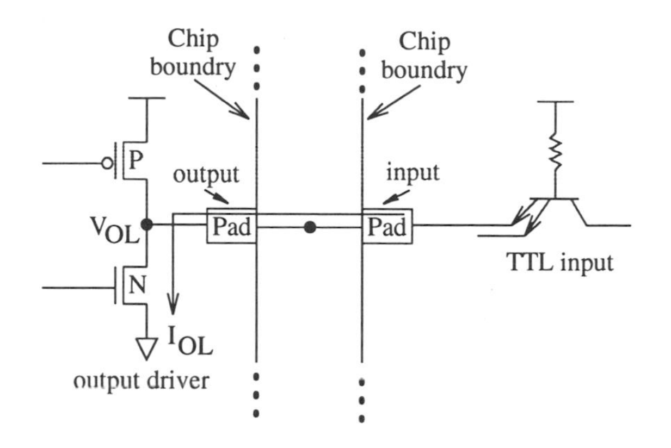

Figure 4.102 TTL output buffer.

where I_{OL} is the current sunk by the output buffer and is equal to the sum of the current from all the bipolar TTL inputs. $V_{OL} = 0.4\ V$ for low TTL output. This disspated power is due to the output NMOS pull-down transistor and can be an important issue as far as the chip heat is concerned. Note that the corresponding energy is not drawn from the internal power supply.

Another component of the total power dissipated at the output pads is the dynamic power. It is given by

$$P_{dyn} = A(N_o E_{io} + N_o C_o V_{DD}^2) f \tag{4.138}$$

where E_{io} is the internal switching energy of the output pad, and C_o is the average output load capacitance (including the pad load). As an example, 64 output pads switching with an activity of 10% at 200 MHz dissipate 0.8 W (@$V_{DD} = 3.3\ V$, $E_{io} = 70\ \mu W/MHZ$ and $C_o = 50\ pF$). This value is very important to take into account.

The total power dissipation of the bidirectional pads can be evaluated using the approaches developed for the input and output circuits.

4.9.5.3 3.3-to-5 V Output Interface

When a 3.3 V chip is connected to a 5 V chip, zero DC power dissipation interfaces are needed. If the conventional CMOS is used to interface the 3.3 V logic to 5 V logic, the DC power would be large. Fig. 4.103 illustrates this

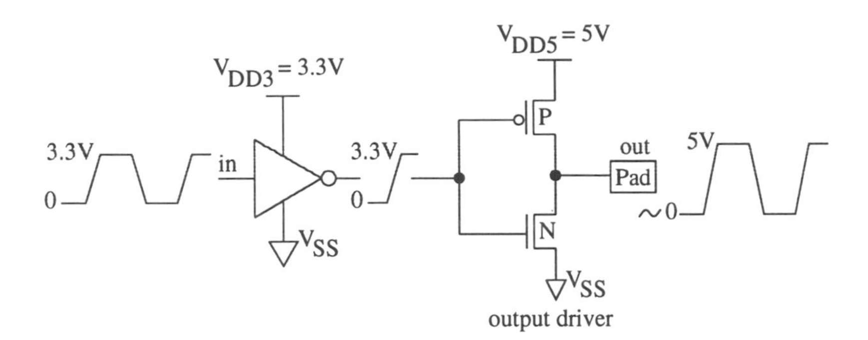

Figure 4.103 Conventional CMOS as 3.3-to-5 V interface.

problem. For example, if the 3.3 V inverter drives high into the 5 V inverter, the V_{GS} of the PMOS transistor P, is equal to 1.7 V. This value is larger than V_T of the device and thus results in large DC power dissipation in the range of milliwatts. Since this power is for every I/O, then for a whole ASIC chip it could be hundreds of mW. This situation is unacceptable for low-power applications.

The circuit of Fig. 4.104 defines a solution to the problem of DC power dissipation [38]. The circuit has two power supplies, denoted V_{DDL} and V_{DDH} corresponding to low-V_{DD} (example 3.3 V) and high-V_{DD} (example 5 V), respectively. For low input data, node A is at V_{DDL} and node B is at zero. The NMOS transistor N is conducting and the output is at V_{SS}. Since the output is zero, the feedback PMOS transistor, P_f, is also conducting. The pass NMOS transistor N_p is cutoff, thus the node C is pulled up to V_{DDH}. Then the PMOS transistor P is completely OFF. Hence no leakage is in this state except the junction leakage currents and the subthreshold currents. For high input data, node A is at zero and node B is at V_{DDL}. In this case the NMOS transistor N is OFF and the pass transistor N_p is conducting. Initially the feedback PMOS transistor P_f is ON and since N_p is conducting, then proper sizing of P_f and N_p (higher conductance of N_p) will permit node C to be discharged through N_p. This causes P to conduct, which in turn charges the output to V_{DDH}. Then the feedback device P_f is completely OFF. Thus this interface results in very limited leakage current and solves the problem of interface.

As mentioned, the transistors P_f and N_p should be sized properly so that the circuit does not latch the previous data. P_f should be much smaller than

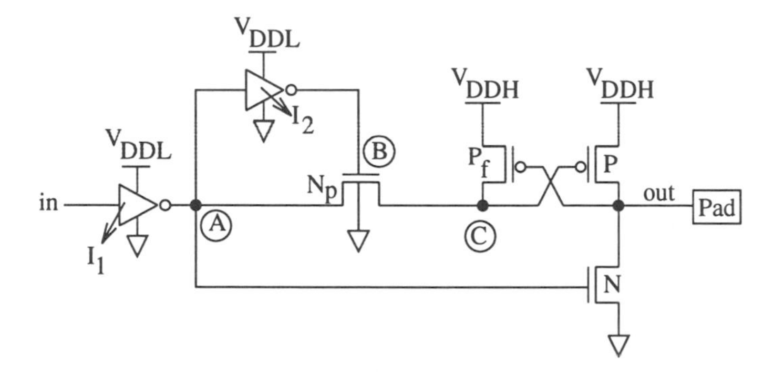

Figure 4.104 CMOS low-to-high voltage interface.

N_p. We use simple analysis to find the relationship between the sizes of the two transistors. For high input data, initially the node C is at V_{DDH}. Thus the NMOS N_p is in saturation and the PMOS P_f is in the linear region. By assuming that the drain current of N_p is much higher than that of P_f, we have

$$\frac{\beta_{np}}{2}(V_{DDL} - V_{Tn})^2 > \beta_{pf}(V_{DDH} - |V_{Tp}|)V_{DDH} \tag{4.139}$$

which yields the following relation

$$\frac{\beta_{np}}{\beta_{pf}} > 2\frac{(V_{DDH} - |V_{Tp}|)V_{DDH}}{(V_{DDL} - V_{Tn})^2} \tag{4.140}$$

where β_{np} and β_{pf} are the βs of the NMOS transistor N_p and the PMOS transistor P_f, respectively. The low-to-high voltage converter has a negligible DC current when the input is stable since all the devices are completely OFF. This technique can be used to interface any low-voltage to higher voltage.

4.9.6 Ground Bounce

When a high drive current CMOS driver switches, it generates high current spikes. This current can generate noise, as shown in Fig. 4.105. The current flows through the impedance between the pad and supply node and produces a voltage noise. This noise is often called $L\frac{di}{dt}$ or *ground bounce*. The L is due to the package inductance. The ground bounce is given by

$$V_L = L\frac{di}{dt} \tag{4.141}$$

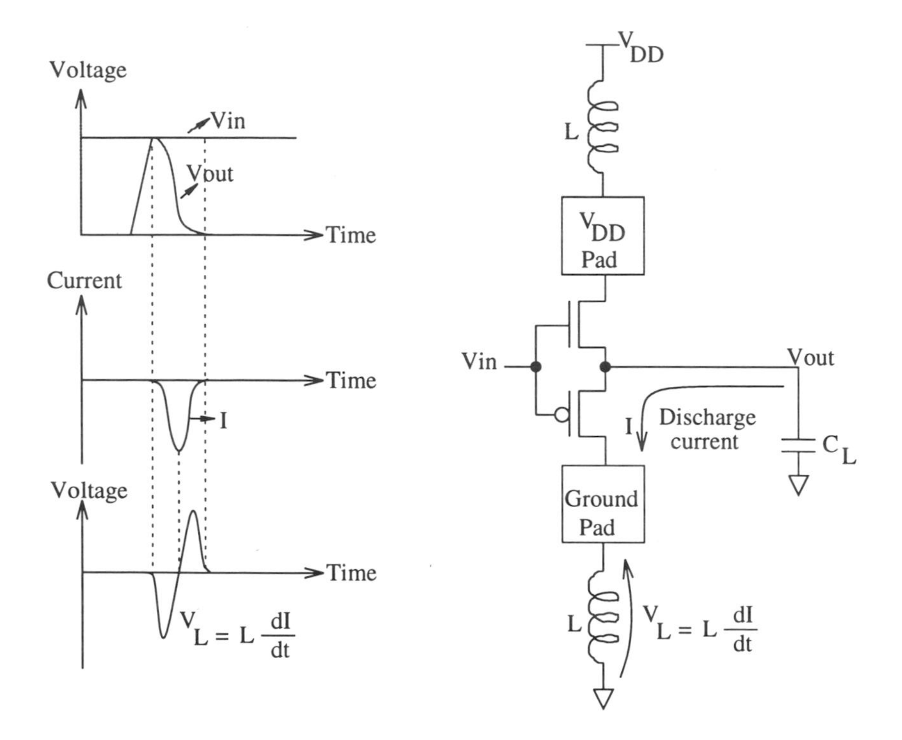

Figure 4.105 CMOS driver with parasitic inductance.

This noise problem can occur on power lead and is termed *power bounce.* We will use only one name to refer to this problem. Consider a CMOS output driver driving the output pad of 50 pF at 3.3 V in 2 ns rise/fall times. It can be shown [39] that $\frac{di}{dt}$ is related to the fall/rise times by

$$\left|\frac{di}{dt}\right|_{max} \geq 4\frac{C_o V_{DD}}{t_{r/f}^2} \tag{4.142}$$

The di/dt can be as high as 165 mA/ns. If for example 8 drivers are allowed to switch simultaneously per each V_{DD}/V_{SS} pads pair, the resulting ground bounce for $L = 1\ nH$ is 1320 mV. This value can be a problem, particularly for low-voltage applications, since this ground bounce consumes a large fraction of the digital noise margins. Some of the problems encountered are 1) false triggering, 2) double clocking, and/or 3) missing clocked pulses.

I/O buffers are not the only source of ground bounce in CMOS circuits. Clock buffers and slightly the core logic can also cause serious ground bounce in the supply leads when driving large loads. Careful power supply routing should be taken when we power large buffers. The resistance of the metal should be minimized so the voltage drop, due to the current spike, is reduced.

There are many techniques to reduce the ground bounce. One simple approach is to use separate supply pins for the output buffers. Some approaches, based on reducing L and di/dt, are the following:

- Multiple supply pads and pins is one way to reduce the inductance of the supply. A recent chip uses 127 power/ground pins out of a total of 293 pins [40].
- Placement of power and ground pins, adjacent one to the other reduces the effective inductance of power and ground pins by mutual inductance. This approach causes an increase in chip size and cost.
- Circuit techniques to reduce the di/dt of the output and clock buffers, while maintaining adequate performance. The simplest way is to control the rise/fall times while maintaining the timing requirement. However, this approach has a serious problem, since worst-case-slow process dictates the buffer sizing (worse-case delay), while best-case-fast process dictates the ground bounce level. Hence the buffer design is constrained by the two extremes of process variations. Once the buffer is sized to satisfy the worse-case delay, the worse-case ground bounce may exceed the fixed level. This problem can be solved by controlling the signal slope at the input of the output transistors of the buffer [41].
- For clock buffers, and in high-performance design, on-chip by-pass capacitance are added between the power bus and the substrate as shown in Fig. 4.106. This capacitance lowers the impedance of the power supply. On-chip bypass capacitance does not reduce the noise produced by output buffers.
- Another approach is to reduce the output voltage swing of the large buffer.

In conclusion, to reduce the ground bounce, all the techniques can be combined to reduce L and di/dt. The reader can refer to many other techniques to reduce the ground bounce [42, 43, 44, 45].

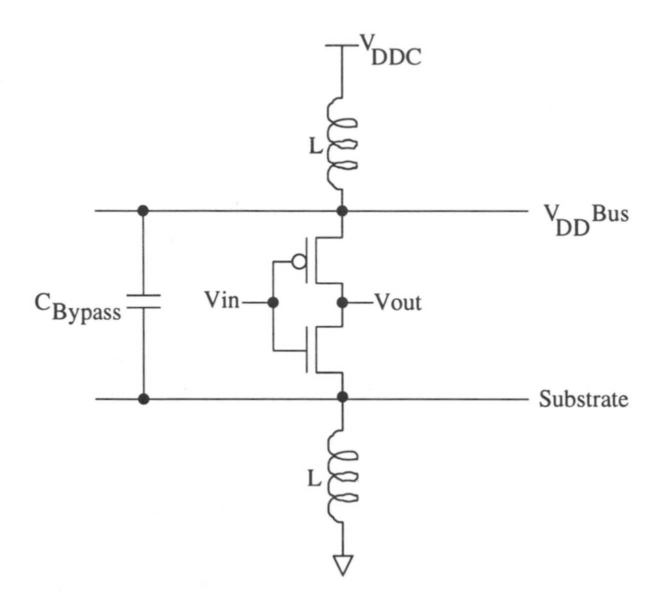

Figure 4.106 On-Chip bypass capacitance.

4.9.7 Low-Swing Output Circuit

With the advent of high-performance VLSI chips, which operate beyond 100 MHz and have over 100 I/Os on the same chip, high data rate CMOS I/O interfaces with low-swing signals are needed such as ECL (Emitter Coupled Logic) [46, 47, 48], BTL [49], GTL [50], and CMTL (Current Mode Transceiver Logic) [51]. Conventional unterminated interconnects (between VLSI chips) for CMOS-level signals usually have poor signal quality with severe overshoot and ringing, accompanied by EMI (electromagnetic interference) and the possibility to trigger the latch-up.

Fig. 4.107 shows two chips connected to the bidirectional transmission line (50 Ω termination resistors) through GTL I/O (Gunning I/O) transceivers. Both ends of the transmission line are terminated to prevent reflections. The load seen by each driver is 25 Ω. The termination voltage V_{TM} is about 1.2 V. The output driver is an open-drain NMOS pull-down transistor and when it is inactive the output is at high-level signal V_{OH} equal to V_{TM}. The input receiver uses a differential comparator with external reference voltage $V_{ref} = 0.8\ V$.

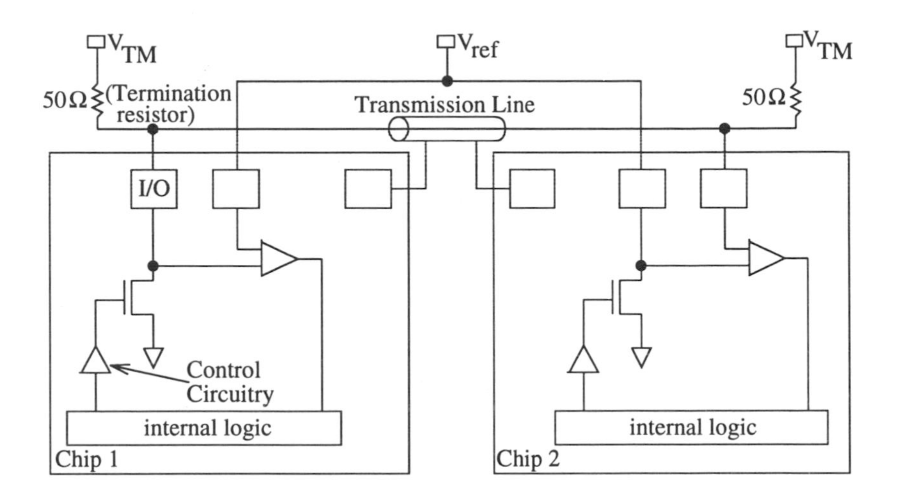

Figure 4.107 CTL I/O with two chips connected to transmission line.

Fig. 4.108 shows an output driver in open-drain configuration which includes circuitry to reduce overshoot and the turn-off di/dt. When V_{in} is low, P_1 turns ON which itself turns N_3 and N_4 ON. In this case,, the maximum output voltage is $V_{OL,max} = 0.4\ V$. The power dissipated by the pull-down NMOS is maximum and mainly static. The static current is equal to $(V_{TM} - V_{OL})/R = 0.8/25 = 32\ mA$[8]. Hence, the maximum static power dissipated on-chip is $P = 32\ mA \times 0.4 = 12.8\ mW$ for each I/O. Typical value of V_{OL} is 0.24 V, thus the nominal power dissipated by each active driver is 9.2 mW. When the input goes from low to high, N_2 turns ON and N_3 is still ON because the signal through the two inverters I_1 and I_2 is delayed by about 1 ns. The transistor N_1 is weak, hence the output discharge is controlled by N_2 and N_3. These transistors let the drain of N_4 connected to its gate as long as V_{DS4} is higher than V_T. When N_3 turns OFF, then N_1 discharges the gate of N_4 to the ground. Thus, the turn-off of N_4 is controlled. In this case, there is no DC power dissipated.

Fig. 4.109 shows the input buffer which employs a differential comparator. This circuit switches to high (low) V_{out} when $V_{in} - V_{ref} > 50\ mV$ ($< -50\ mV$), respectively over process, power supply and junction temperature variations.

[8] Note that this current is supplied by V_{TM} and not V_{DD}

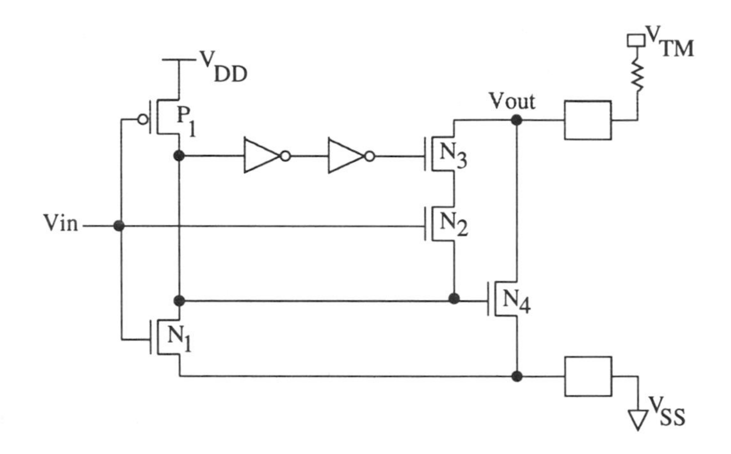

Figure 4.108 CMOS output driver in GTL I/O.

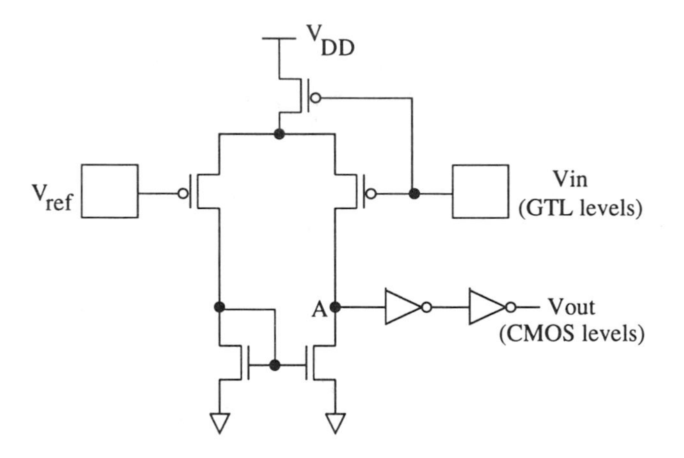

Figure 4.109 CMOS input receiver in GTL I/O.

The average power dissipated by this input receiver is 5.5 mW at 5 V power supply.

4.10 LOW-POWER CIRCUIT TECHNIQUES

Remember that the total power dissipated by a circuit has three components. Two of them which are very important are : 1) the static power (P_s), and 2) the dynamic power (P_d). This section treats some of the circuit techniques for achieving low-power while maintaining performance. Techniques to reduce the power at subsystem/system and architecture levels will be discussed in Chapters 6, 7 and 8.

4.10.1 Low Static Power Techniques

One important source of static power dissipation is the use of low threshold voltage. With device scaling, the power supply voltage is scaled. If the threshold voltage is not scaled, and is equal or greater than one half V_{DD}, the gate delay increases drastically [52]. The threshold voltage should be less than 20% of V_{DD}, in order to maintain performance at low supply voltage. At 1 V power supply, the threshold voltage[9] can be as low as 0.1 V. However, reducing V_T causes serious standby subthreshold current increase, due to the exponential relation between the current and V_T. With low V_T the process fluctuation can increase this current more. For VLSI integration and future ULSI, the total standby current can be high and not acceptable for low-power applications.

To reduce this subthreshold current, associated with low V_T devices, there are many techniques. These techniques are based on the principle to reverse bias the V_{GS} voltage of the MOS device (in the case of NMOS) in the standby mode of operation, as shown in Fig. 4.110. With $V_{GS} = V_{GR}$, where V_{GR} is negative, the standby state of the device moves from state α to state β. We cite two techniques using this principle:

4.10.1.1 Self-Reverse Biasing

This technique has been used mainly to reduce the static power dissipation in standby mode of the memory decoded-driver [53]. The drivers, in memory, have a large number of circuits, arranged repeatedly, but only a few of them operate simultaneously. The circuit of Fig. 4.111 can drastically reduce the subthreshold current of the drivers. The technique simply consists of inserting a PMOS transistor P_c with a size W_c between the power supply V_{DD} and the common source node A. All the PMOS transistors ($P_{d1}, P_{d2}, ..., P_{dn}$) of the

[9] Constant current threshold voltage.

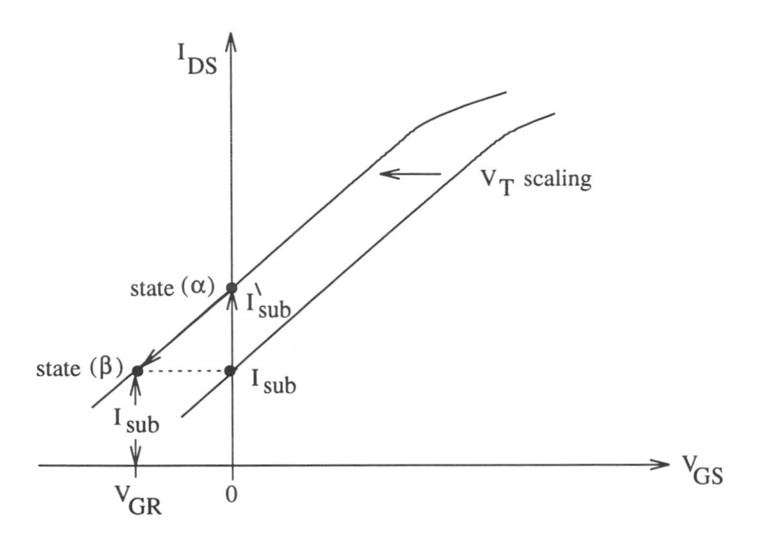

Figure 4.110 Principle of subthreshold reduction.

drivers have, in this example, the same size W_d and common source (node A). The number of drivers n can be between a few hundreds to a few thousands. The MOS transistors in the drivers have low $|V_{Td}|$ (e.g., 0.1 V). The PMOS transistor P_c have a threshold voltage $|V_{Tc}|$ slightly higher than $|V_{Td}|$ (e.g., $0.2 - 0.4$ V).

In active mode, the input S is low and the transistor P_c is ON. For the drivers only one circuit is ON. In order that the PMOS transistor P_c does not affect the drive current of the drivers, its size W_c should be larger than W_d, depending on the capacitance of the common source, which is huge for high n. In standby mode, the input S is high and the PMOS transistor P_c is OFF. The inputs of all drivers are set to high (V_{DD}). Without the PMOS transistor P_c, the total subthreshold current would be n times the current of each driver. This makes this current very high. Hence P_c reduces and limits the subthreshold current. The voltage of the common source node A, is reduced by an amount ΔV_{SRB} (a few hundreds of mV). This causes the PMOS transistors of all drivers to have self-reverse-biasing gate-source voltage, which drastically reduces the subthreshold current. The time needed for the node to stabilize to $V_{DD} - \Delta V_{SRB}$ (or the time needed to switch from the active to standby mode) is called evolution time and can be very high (order of 1 ms) compared to the delay of the driver. The reason is that only the leakage and subthreshold currents which

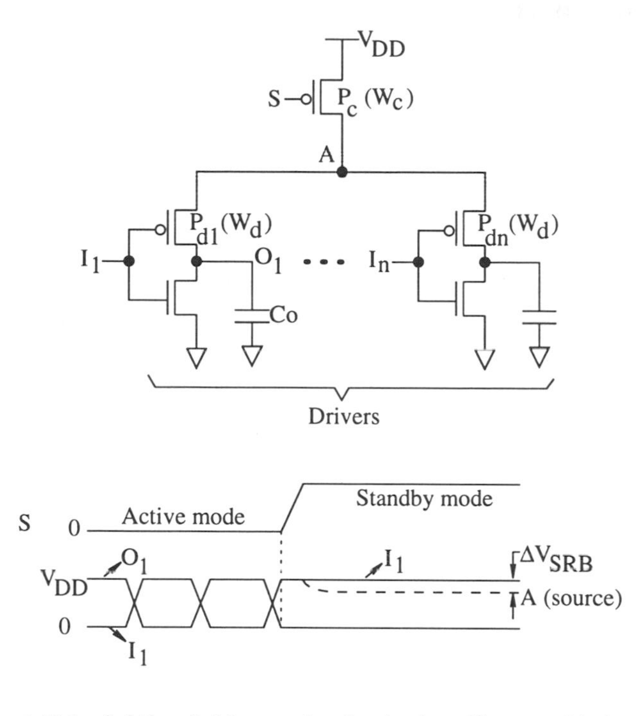

Figure 4.111 Subthreshold current reduction by self-reverse biasing.

discharge the node A in this mode. This time can be unsignificant to low-power operation if the standby mode time is large enough as in the case of many low-power applications. When the input S is turned low (active mode), the time needed for the common source A to recover (reaches almost V_{DD}) is too low and can be lower than the delay time. Hence, it does not interrupt the start of normal operation.

Lets derive now the subthreshold current expressions before and after reduction by SRB technique. The total subthreshold current without the self-reverse-biasing technique is given by

$$I_{sub1} = n.I_0 \frac{W_d}{W_0} \exp \frac{-|V_{Td}|}{S/ln10} \tag{4.143}$$

With the transistor P_c, the subthreshold current is given by

$$I_{sub2} = I_0 \frac{W_c}{W_0} \exp \frac{-|V_{Tc}|}{S/ln10} \tag{4.144}$$

We assume that the devices have the same I_0, W_0 and S. By dividing the current equations (4.143) and (4.144), we have, for the subthreshold current, a reduction factor γ

$$\gamma = \frac{I_{sub1}}{I_{sub2}} = n.\frac{W_d}{W_c} \exp \frac{(|V_{Tc}| - |V_{Td}|)}{S/ln10} \quad (4.145)$$

For example for $n = 512$, $W_c = 10W_d$, (with this ratio the speed is not affected), $V_{Tc} = 0.3\ V$, $V_{Td} = 0.1\ V$ and $S = 90\ mV/decade$, the factor $\gamma = 8.5 \times 10^3$. So, the saving, in subthreshold current, is sufficient. The parameter ΔV_{SRB} can be easily deduced. Note that this technique needs multi-V_T technology.

4.10.1.2 *Multi-V_T Technique*

This technique is similar to the one discussed above, but it can be applied to any CMOS logic [54, 55]. The basic idea is shown in the example of the NAND gate of Fig. 4.112. Here the MOS transistors P and N have high V_T (e.g., 0.6 V extrapolated) for 1 V power supply applications. Also the logic gate has MOSFETs with low V_T ($\leq 0.3\ V$). The signal SL is used to switch the gate in active or sleep (standby) mode. The virtual supply lines V_{DDV} and V_{SSV} are common for many gates. We call this logic multi-threshold CMOS logic (MT-CMOS).

In the active mode, the signal SL is low, P and N are ON, so the virtual supply lines V_{DDV} and V_{SSV} can be set to almost V_{DD} and ground, respectively. Hence, the low-V_T logic can switch efficiently, but care should be taken in the sizing of the P/N devices compared to the logic. Fig. 4.113 shows the effect of sizing the high-V_T devices on the delay of the gate. The width of P/N should be at least 10 times larger than that of logic cells. This condition depends greatly on the parasitic capacitances of the virtual supply lines C_1 and C_2 [see Fig. 4.112]. If C_1 and C_2 are large then the width of P and N transistors can be reduced, because these capacitances tend to suppress the bouncing of V_{DDV} and V_{SSV} and hence improve the speed. The high-V_T MOSFETs can be common for several logic gates (e.g, 10).

In the standby (sleep) mode, the signal SL is high, then P and N are OFF. Hence, the subthreshold current is limited by that of these high-V_T devices. In this case, the static power dissipation is dramatically reduced in the sleep mode. The subthreshold reduction factor can be deduced using the analysis presented in the previous section. One problem associated with this MT logic is that the evolution and recovery times can be large.

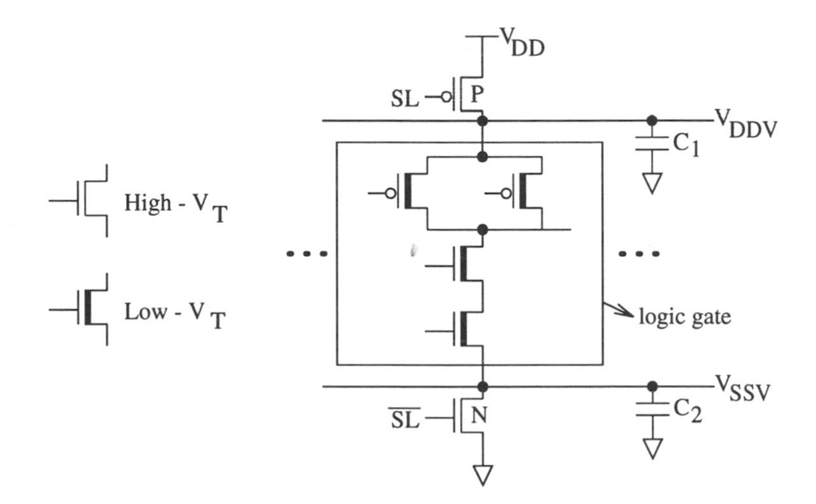

Figure 4.112 Multi-threshold voltage CMOS circuit.

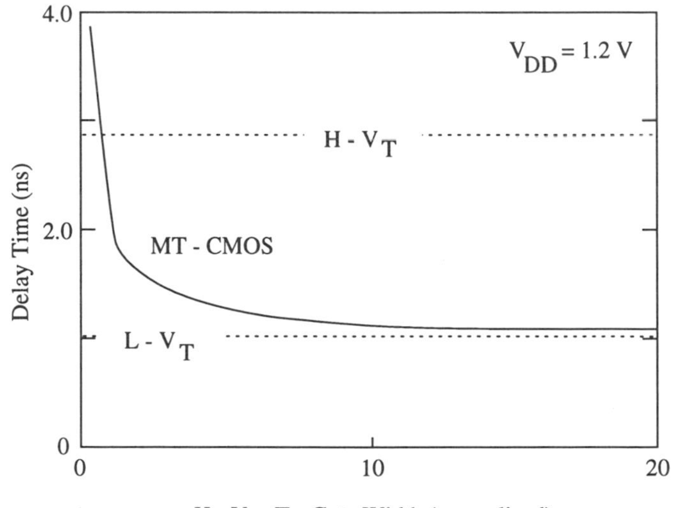

Figure 4.113 Effect of high-V_T MOS width on the performance of MT-CMOS.

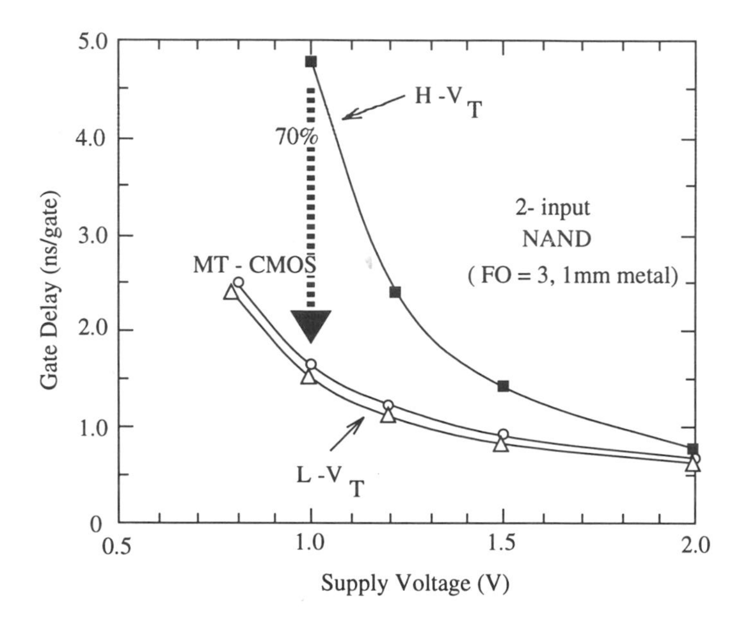

Figure 4.114 2-input NAND gate delays.

The measured delay, as a function of the supply voltage for 2-input NAND gate with FO= 3 and wiring load of 1 mm (0.25 pF), is shown in Fig. 4.114. The technology is 0.5-μm CMOS with low $V_{Tn} = 0.25\ V$, low $V_{Tp} = -0.35\ V$, high $V_{Tn} = 0.55\ V$ and high $V_{Tp} = -0.65\ V$. The MT-CMOS logic has almost the same speed as the full low-V_T logic. The logic delay time is reduced by 70% at 1 V as compared with that of the high-V_T one.

For holding the level of the output during the sleep mode, a level holder is necessary as shown in Fig. 4.115. It consists only of cross-coupled inverters with high-V_T devices powered from the power supply V_{DD}.

The source of the static power dissipation is not only low V_T devices. Several other issues contribute to static power increase. These are some circuit design guidelines to reduce the static power dissipation :

- Avoid the use of pseudo-NMOS circuits in your design.

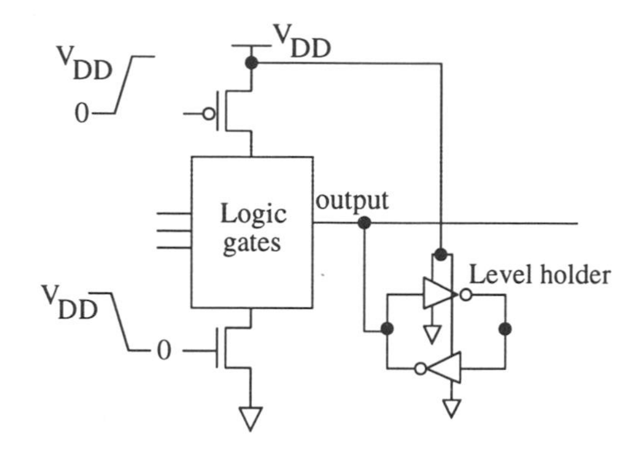

Figure 4.115 CMOS gate with level holder.

- Avoid the use of TTL-compatible I/O or devise low-DC current level converters.
- Do not use low V_T devices in the I/O buffers, otherwise the DC power increases remarkably because the MOS transistors of the I/O buffers have large sizes. If you do not have any option, then use the subthreshold reduction techniques.

4.10.2 Low Dynamic Power Techniques

ASICs and VLSI processor clocks are improving rapidly, reaching the sub-GHz range [21, 56]. The power dissipation of CMOS digital circuits, operating at these high-frequencies, increases drastically and it can be the main performance limiting factor. Therefore, low-power circuit techniques are needed to reduce the dynamic power of digital circuits. Moreover, low-power chip consumption is extremely important in order to extend the battery life of portable systems [57].

In general the dynamic power dissipation of a gate (i) is given by:

$$P_{di} = \alpha_i C_i V_i V_{DD} f \qquad (4.146)$$

where α_i is the gate activity, V_i is the voltage swing, C_i is the load and parasitic capacitances and f is the operating frequency of the system. Equation (4.146) demonstrates that there are several ways to reduce P_{di}:

1. Reduce the power supply voltage. Scaling V_{DD} from 3.3 V to 1 V results in a power reduction factor of 11. However, this approach leads to speed degradation for a given technology. But if device scaling is applied, in a next generation technology, the delay will improve and hence the operating frequency. In a complex digital system local supply reductions can be used for non-critical circuits.

2. Reduce, temporarily, the clock frequency of unused blocks on a VLSI chip using an on-chip power management unit or reduce the gate activity. These can be done at the architectural level.

3. Reduce the output capacitance C_i. As a first order approximation this capacitance is composed of the interconnect capacitance C_{int} and the total input capacitances of the driven gates C_{inp}. The latter can be reduced using low input capacitance logic family [60] such as CPL-like. Also using minimum size logic gates in non critical parts of the design can reduce the dynamic power significantly.

When C_{int} dominates, as in busses and high-capacitance interconnections (interblock wires), then circuit techniques, based on low-swing signal, while maintaining the power supply voltage, can lead to power dissipation reduction [58, 59]. With increasing chip dimensions and integration density, the capacitances of wires will dominate. It is expected that the power dissipation associated with the busses and the interconnections in future ULSI chips will reach half of the total power dissipation [58].

These are some guidelines for the design of low-dynamic power circuits :

- Choose the technology that has low junction and oxide capacitances for the same performance.
- Avoid, if possible, the use of dynamic logic design style.
- For any logic design, reduce the switching activity, by logic reordering and balanced delays through gate tree to avoid glitching problem.
- Use low-input capacitance logic family.
- In non-critical paths, use minimum size devices whenever it is possible without degrading the overall performance requirements.
- If pass-transistor logic style is used, careful design should be considered.

4.11 ADIABATIC COMPUTING

As discussed in Section 4.3.2, the energy provided by the supply to charge a load C_L of a driver during charging and discharging is

$$E = C_L V^2 \tag{4.147}$$

where V is the power supply voltage as shown in Fig. 4.116(a). Half of the energy is dissipated by the resistor of the pull-up PMOS device during the charging phase. A similar argument applies to the discharge resistor of the pull-down NMOS transistor. This analysis is valid even if a step power supply voltage, V, is applied to the network. From Fig. 4.116(b), the voltage drop across the resistor, R_p varies from V (supply voltage) to zero. Hence, the energy dissipated by R_p is given by

$$E_R = \int_V^0 V_R dQ = \int_V^0 V_R C_L d(V - V_R) \tag{4.148}$$

then

$$E_R = \frac{1}{2} C_L V^2 \tag{4.149}$$

or

$$E_R = C_L V \bar{V_R} \tag{4.150}$$

where $\bar{V_R}$ is the average voltage drop across the resistor of the pull-up PMOS.

If the power supply voltage has two half steps, as shown in Fig. 4.116(c), the energy dissipated by the resistor is

$$E_R = \frac{1}{4} C_L V^2 \tag{4.151}$$

So less energy is dissipated by the resistor, when the average voltage is reduced, while keeping the swing and load capacitance constant. This is the principle of *Adiabatic Switching* [61, 62, 63].

For multi-steps power supply voltage, as shown in Fig. 4.116(d), the total energy dissipated is given by [61]

$$E = C_L \frac{V^2}{N} = \frac{E_{conventional}}{N} \tag{4.152}$$

and the one dissipated by the resistor is

$$E_R = \frac{1}{2} C_L \frac{V^2}{N} \tag{4.153}$$

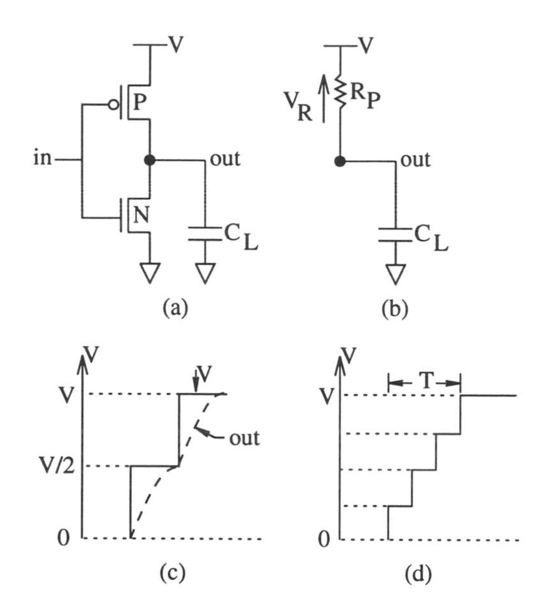

Figure 4.116 Adiabatic switching of an inverter (driver): (a) inverter; (b) equivalent charging circuit; (c) two-step power supply; (d) multi-step power supply.

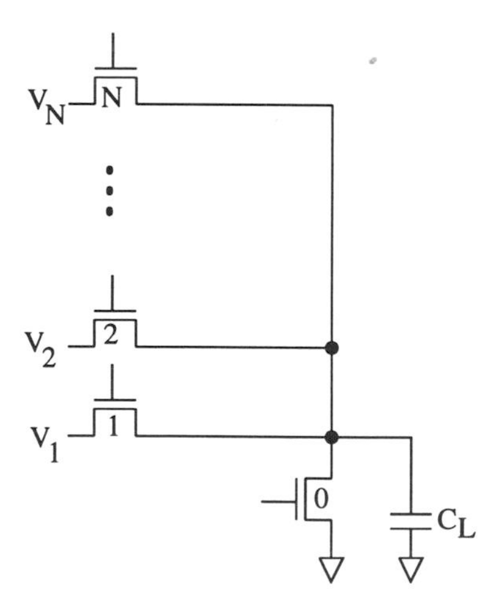

Figure 4.117 A stepwise driver for a capacitive load.

where N is the number of voltage steps uniformly distributed. Fig. 4.117 shows an example of a driver with uniformly distributed supplies which are switched in successively. The voltage V_i is given by

$$V_i = i\frac{V}{N} \tag{4.154}$$

To charge the load, V_1 through V_N are connected to the load in succession (by closing switch 1, opening switch 1, closing switch 2, etc.). To discharge the load, V_{N-1} through V_1 are switched in the same way, and the switch 0 is closed, connecting the output to ground. Note that the supply voltage, with multi-steps, needs a longer time period than the conventional case to charge up the load capacitance. This technique has been used for large loads.

Another variation is to use a supply voltage with a ramp form[10] [62]. In this case, the energy is drastically reduced if a long time period is used. For the inverter for example, pulsed power supplies (PPS) are applied to the circuit.

The adiabatic computing becomes attractive only when the delay is not critical, because in that technique the energy is traded for delay. The energy-delay product of the adiabatic circuit is much worse than the conventional CMOS gates [64].

4.12 CHAPTER SUMMARY

This chapter has provided an introduction to low-power CMOS design. The power dissipation components of a CMOS gate have been discussed. Techniques to reduce the different components, at physical and circuit levels, were presented. Novel CMOS design styles such as CPL, DPL, and SRPL were examined. Several issues in CMOS circuit design, such as clock distribution, ground bouncing, etc., were reviewed. This chapter represents a base, for Chapters 6, 7, and 8, where subsystems and low-power architectures are discussed.

[10] The multi-steps supply, if further partitioned, it reaches the ramp waveform.

REFERENCES

[1] N. H. E. Weste and K. Eshraghian, "Principles of CMOS VLSI Design : A Systems Perspective," second edition, Addison-Wesley, Reading, MA, 1993.

[2] J. P. Uyemura, "Circuit Design for CMOS VLSI," Kluwer Academic Publishers, Norwell, MA, 1992.

[3] M. I. Elmasry, "Digital MOS Integrated Circuits II", IEEE Press Book, 1993.

[4] R. M. Swanson and J. D. Meindl, "Ion-Implanted Complementary MOS Transistors in Low-Voltage Circuits", IEEE J. Solid-State Circuits, vol. 7, no. 2, pp. 146-153, April 1972.

[5] H. J. M. Veendrick, "Short-Circuit Dissipation of Static CMOS Circuitry and Its Impact on the Design of Buffer Circuits," IEEE J. Solid-State Circuits, vol. 19, no. 4, pp. 468-473, August 1984.

[6] S. M. Kang, "Accurate Simulation of Power Dissipation in VLSI Circuits," IEEE J. Solid-State Circuits, vol. 21, no. 5, pp. 889-891, October 1986.

[7] G. J. Fisher, "An Enhanced Power Meter for SPICE2 Circuit Simulation," IEEE Trans. Computer-Aided Design, vol. 7, pp. 641-643, May 1988.

[8] G. Y. Yacoub and W. H. Ku, "An Enhanced Technique for Simulating Short-Circuit Power Dissipation," IEEE J. Solid-State Circuits, vol. 24, no. 3, pp. 844-847, June 1989.

[9] N. Meijs, and J. T. Fokkema, "VLSI Circuit Reconstruction From Mask Topology," Integration, vol. 2, no. 2, pp. 85-119, 1984.

[10] D. V. Heinbruch, "CMOS3 Cell Library," Addison-Wesley, Reading, MA, 1988.

[11] R. J. Landers, and S. Mahant-Shetti, "Multiplexer-Based Architecture for High-Density, Low-Power Gate Arrays," in Symposium on VLSI Circuits, Tech. Dig., Honolulu, pp. 33-34, June 1994.

[12] M. I. Elmasry, "Digital MOS Integrated Circuits I", IEEE Press Book, 1981.

[13] R. H. Krambeck, C. M. Lee and H-F S. Law, "High Speed Compact Circuits with CMOS", IEEE J. Solid-State Circuits, vol. 17, no. 3, pp. 614-619, June 1982.

[14] V. Friedman and S. Liu, "Dynamic Logic CMOS Circuits", IEEE J. Solid-State Circuits, vol. 19, no. 2, pp. 263-266, April 1984.

[15] N. F. Gonclaves and H. J. DeMan, "NORA: a Race Free Dynamic CMOS Technique for Pipelined Logic Structures" IEEE J. Solid-State Circuits, vol. 18, no. 3, pp. 261-266, June 1983.

[16] C. M. Lee and E. W. Szeto, "Zipper CMOS," IEEE Circuits and Devices Mag., vol. 2, no. 3, pp. 10-17, May 1986.

[17] N. Weste and K. Eshraghian, "Principles of CMOS VLSI Design : A Systems Perspective," Addison-Wesley, Reading, MA, 1985.

[18] F. Lu and H. Samueli, "A 200-MHz CMOS Pipelined Multiplier-Accumulator Using a Quasi-Domino Dynamic Full-Adder Cell Design," IEEE J. Solid-State Circuits, vol. 28, no. 2, pp. 123-132, February 1993.

[19] J. Yuan and C. Svenson, "High-Speed CMOS Circuit Technique," IEEE J. Solid-State Circuits, vol. 24, no. 1, pp. 62-71, February 1989.

[20] M. Afghahi and C. Svensson, "A Unified Single-Phase Clocking Scheme for VLSI Systems," IEEE J. Solid-State Circuits, vol. 25, no. 1, pp. 225-233, February 1990.

[21] D. W. Dobberpuhl et al., "A 200-MHz 64-b Dual-Issue CMOS Microprocessor", IEEE J. Solid-State Circuits, vol. 27, no. 11, pp. 1555-1567, November 1992.

[22] H. B. Bakoglu, "Circuits, Interconnects, and Packaging for VLSI," Addison Wesley, Reading, MA, 1990.

[23] K. Yano, et al., "A 3.8-ns CMOS 16x16 Multiplier Using Complementary Pass-Transistor Logic", IEEE J. Solid-State Circuits, vol. SC-25, no. 2, pp. 388-394, April 1990.

[24] M. Suziki, et al., "A 1.5-ns 32-b CMOS ALU in Double Pass-Transistor Logic", IEEE J. Solid-State Circuits, vol. SC-28, no. 11, pp. 1145-1151, November 1993.

[25] A. Parameswar, H. Hara, and T. Sakurai, "A High-Speed, Low-Power, Swing Restored Pass-Transistor Logic Based Multiply and Accumulate Circuit for Multimedia Applications," IEEE Custom Integrated Circuits Conference, Tech. Dig., San Diego, CA, pp. 278-281, May 1994.

[26] L. A. Glasser and D. W. Dobberpuhl, "The Design and Analysis of VLSI Circuits", Addison-Wesley, Reading, MA, 1985.

[27] T. Kobayashi et al., "A Current-Controlled Latch Sense Amplifier and a Static Power-Saving Input Buffer for Low-Power Architecture", IEEE J. Solid-State Circuits, vol. SC-28, no. 4, pp. 523-527, April 1993.

[28] M. S. J. Steyaert, et al., "ECL-CMOS and CMOS-ECL Interface in 1.2-μm CMOS for 150-MHz Digital ECL Data Transmission Systems", IEEE J. Solid-State Circuits, vol. SC-26, no. 1, pp. 18-24, January 1991.

[29] C. Mead and L. Conway, "Introduction to VLSI Systems", Addison-Wesley, Reading, MA, 1980.

[30] N. C. Li, G. L. Haviland and A. A. Tuszynski, "CMOS Tapered Buffer", IEEE J. Solid-State Circuits, vol. SC-25, no. 4, pp. 1005-1008, August 1990.

[31] M. Nemes, "Driving Large Capacitances in MOS LSI Systems", IEEE J. Solid-State Circuits, vol. SC-19, no. 1, pp. 159-161, February 1984.

[32] N. Hedenstierna and K. O. Jeppson, "CMOS Circuit Speed and Buffer Optimization", IEEE Trans. Computer-Aided Design, vol. CAD-6, no. 2, pp. 276-281, March 1987.

[33] A. J. Al-Khalili, Y. Zhu and D. Al-Khalili, "A Module Generator for Optimized CMOS Buffer", IEEE Trans. Computer-Aided Design, vol. CAD-9, no. 10, pp. 1028-1046, October 1990.

[34] S. R. Vemuru and A. R. Thorbjornsen, "Variable-Taper CMOS Buffer", IEEE J. Solid-State Circuits, vol. SC-26, no. 9, pp.1265-1269, September 1991.

[35] J. Burkis, "Clock Tree Synthesis for High Performance ASICs", in IEEE ASIC Intern. Conf. and Exhibit, Rochester, NY, pp. P9-8.1-P9-8.3, September 1991.

[36] P. D. Ta and K. Do, "A Low-Power Clock Distribution Scheme for Complex IC System", in IEEE ASIC Intern. Conf. and Exhibit, Rochester, NY, pp. P1-5.1-P1-5.4, September 1991.

[37] H. Kojima, S. Tanaka, and K. Sasaki, " Half-Swing Clocking Scheme for 75% Power Saving in Clocking Circuitry," Symposium on VLSI Circuits, Tech. Dig., Honolulu, pp. 23-24, June 1994.

[38] J. S. Caravella and J. H. Quigley, "Three Volt to Five Volt Interface Circuit with Device Leakage Limited DC Power Dissipation", in IEEE ASIC Intern. Conf. and Exhibit, Rochester, NY, pp. 448-451, September 1993.

[39] M. Shoji, "CMOS Digital Circuit Technology", Prentice Hall Inc., Englewood Cliffs, NJ., 1988.

[40] F. Abu-Nofal et al., "A Three-Million Transistor Microprocessor", in IEEE International Solid-State Circuits Conf., pp. 108-109, February 1992.

[41] T. Gabara and D. Thompson, "Ground Bounce Control in CMOS Integrated Circuits", in IEEE International Solid-State Circuits Conf., pp. 88-89, February 1988.

[42] T. Gabara, "Ground Bounce Control and Improved Latch-up Suppression Through Substrate Conduction", IEEE J. Solid-State Circuits, vol. 23, no. 5, pp. 1224-1232, October 1988.

[43] M. Hashimoto and O-K Kwon, "Low dI/dt Noise and Refletion Free CMOS Signal Driver", in IEEE Custom Integrated Circuits Conf., Tech. Dig., pp. 14.4.1-14.4.4, 1989.

[44] T. Wada, M. Eino and K. Anami, " Simple Noise Model and Low-Noise Data-Output Buffer for Ultra-High-Speed Memories", IEEE J. Solid-State Circuits, vol. 25, no. 6, pp. 1586-1588, December 1990.

[45] R. Senthinathan and J. L. Prince, "Application Specific CMOS Output Driver Circuit Design Techniques to Reduce Simultaneous Switching Noise", IEEE J. Solid-State Circuit, vol. 28, no. 12, pp. 1383-1388, December 1993.

[46] T. Knight and A. Krymm, "A Self-Terminating Low-Voltage-Swing CMOS Output Driver", IEEE J. Solid-State Circuits, vol. 23, no. 2, pp. 457-464, April 1988.

[47] H-J Schumacher, J. Dikken and E. Seevinck, "CMOS Subnanosecond True-ECL Output Buffer", IEEE J. Solid-State Circuits, vol. 25, no. 1, pp. 150-154, February 1990.

[48] M. Pedersen and P. Metz, "A CMOS to 100K ECL Interface Circuit", in IEEE International Solid-State Circuits Conf., Tech. Dig., pp. 226-227, February 1989.

[49] J. Martinez, "BTL Transceivers Enable High-Speed Bus Design", EDN, August 1992.

[50] B. Gunning, L. Yuan, T. Nguyen and T. Wong, "A CMOS Low-Voltage-Swing Transmission-Line Transceiver", in IEEE International Solid-State Circuits Conf., Tech. Dig., pp. 58-59, February 1992.

[51] J. H. Quigley, J. S. Caravella and W. J. Neil, "Current Mode Transceiver Logic (CMTL) for Reduced Swing CMOS, Chip to Chip Communication", in IEEE International ASIC Conference and Exhibit, Rochester, NY, Tech. Dig., pp. 452-457, September 1993.

[52] M. Kakumu, "Process and Device Technologies of CMOS Devices for Low-Voltage Operation," IEICE Trans. Electron., Vol. E76-C, No. 5, pp. 672-680, May 1993.

[53] T. Kawahara et al., "Subthreshold Current Reduction for Decoded-Driver by Self-Reverse-Biasing," IEEE J. Solid-State Circuits, vol. 28, no. 11, pp. 1136-1144, November 1993.

[54] S. Mutoh et al., "1 V High-Speed Digital Circuit Technology with 0.5-μm Multi-Threshold CMOS," in IEEE International ASIC Conference and Exhibit, Rochester, NY, Tech. Dig., pp. 186-189, September 1993.

[55] M. Horiguchi et al., "SSI CMOS Circuit for Low-Standby Subthreshold Current Giga-Scale LSI's", IEEE J. of Solid-State Circuits, Vol. 28, No. 11, pp. 1131-1135 November 1993.

[56] R. W. Badeau et al., "A 100-MHz Macropipelined VAX Microprocessor," IEEE J. Solid-State Circuits, vol. 27, no. 11, pp. 1585-1597, November 1992.

[57] R. Brodersen, A. Chandrakasan and S. Sheng, "Design Techniques for Portable Systems", in IEEE International Solid-State Circuits Conf., Tech. Dig., pp. 168-169, February 1993.

[58] Y. Nakagome et al., "Sub-1-V Swing Internal Architecture for Future Low-Power ULSI's," IEEE J. Solid-State Circuits, vol. 28, no. 4, pp. 414-419, April 1993.

[59] A. Bellaouar, I. S. Abu-Khater, and M. I. Elmasry, "Low-Power CMOS/BiCMOS Drivers and Receivers for On-Chip Interconnects," IEEE J. Solid-State Circuits, vol. 30, no.1, May 1995.

[60] A. Chandrakasan et al., "Low-Power CMOS Digital Design", IEEE J. Solid-State Circuits, vol. 2, no. 4, pp. 473-484, April 1992.

[61] L. J. Svensson, and J. G. Koller, "Driving a Capacitive Load without Dissipating fCV^2," IEEE Symposium on Low Power Electronics, Tech. Dig., San-Diego, pp. 100-101, October 1994.

[62] T. Gabara, "Pulsed Power Supply CMOS - PPS CMOS," IEEE Symposium on Low Power Electronics, Tech. Dig., San-Diego, pp. 98-99, October 1994.

[63] J. S. Denker, "A Review of Adiabatic Computing," IEEE Symposium on Low Power Electronics, Tech. Dig., San-Diego, pp. 94-97, October 1994.

[64] M. Horowitz, T. Indermaur, and R. Gonzalez, "Low-Power Digital Design," IEEE Symposium on Low Power Electronics, Tech. Dig., San-Diego, pp. 8-11, October 1994.

5

LOW-VOLTAGE VLSI BICMOS CIRCUIT DESIGN

BiCMOS technology offers enhanced performance compared to CMOS at 5 V power supply voltage. Many high-speed BiCMOS SRAMs, gate arrays, ASICs, etc. have been fabricated [1]. In this chapter, we present a variety of BiCMOS logic circuits suitable for 3.3 and sub-3.3 V. The potential gates for digital applications are identified. The chapter starts with the introduction of the conventional BiCMOS (totem-pole) gate which is used in 5 V applications. The degradation of this gate, with supply voltage scaling, is demonstrated. In Section 5.2, we introduce the BiNMOS family suitable for low-voltage applications. Other logic families, for low power supply voltage operation, are discussed in Section 5.3. Low-voltage digital applications of BiCMOS are identified. The reader is referred to BiCMOS books [2, 3] to get more familiar with BiCMOS circuits.

5.1 CONVENTIONAL BICMOS LOGIC

In this section, the conventional BiCMOS logic family is introduced. This family has been used successfully in many applications at 5 V power supply voltage. The reason for the speed advantage of BiCMOS compared to CMOS is explained. At low-voltage, the performance degradation of conventional BiCMOS is shown.

The CMOS inverter of Fig. 5.1 suffers from the limited current drive when the load capacitance is large. To increase the drive capability of CMOS, a bipolar driver can be added at the output of the CMOS inverter. Fig. 5.2 shows one possible configuration to construct what is called a conventional BiCMOS

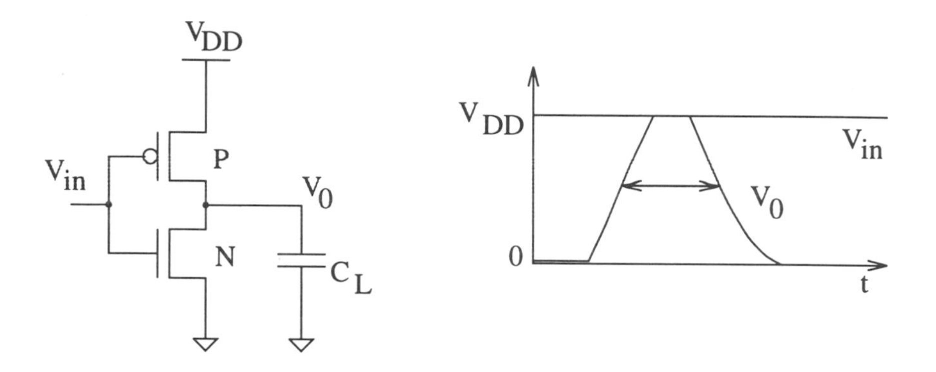

Figure 5.1 CMOS inverter driving a large load capacitance.

inverter. The addition of the bipolar driver stage to the basic CMOS inverter is responsible for the high current driving capability of BiCMOS over CMOS. As a result BiCMOS offers lower delay compared to that of CMOS especially at high loading capacitance.

The operation of this gate is straightforward. When the input is low, the PMOS P is ON and its drain current turns the transistor Q_1 ON. The collector current of Q_1 charges the output load capacitance. As the output reaches $V_{DD} - V_{BEon}$, where V_{BEon} is the turn-on voltage of the bipolar transistor and is about 0.7 V, Q_1 gradually turns OFF. During this period, the NMOS transistor N_{d2} is ON. Since N_{d2} is conducting, Q_2 is in the cutoff region. Transistor N_{d2} can also be controlled by the output node. However, using the base node results in faster operation because the base of Q_1 is pulled up faster than the output node and because the voltage level of the base node is larger. If the input is high, the NMOS transistors N and N_{d1} are ON. Q_1 is OFF while Q_2 turns ON to discharge the output node. As a result, the load capacitance is pulled down. As the output V_o reaches V_{BEon} transistor Q_2 turns OFF and the output stays at this level. The conventional BiCMOS gate provides high drive capability, zero static power dissipation and high input impedance. More discussions on this gate are given in the following sections.

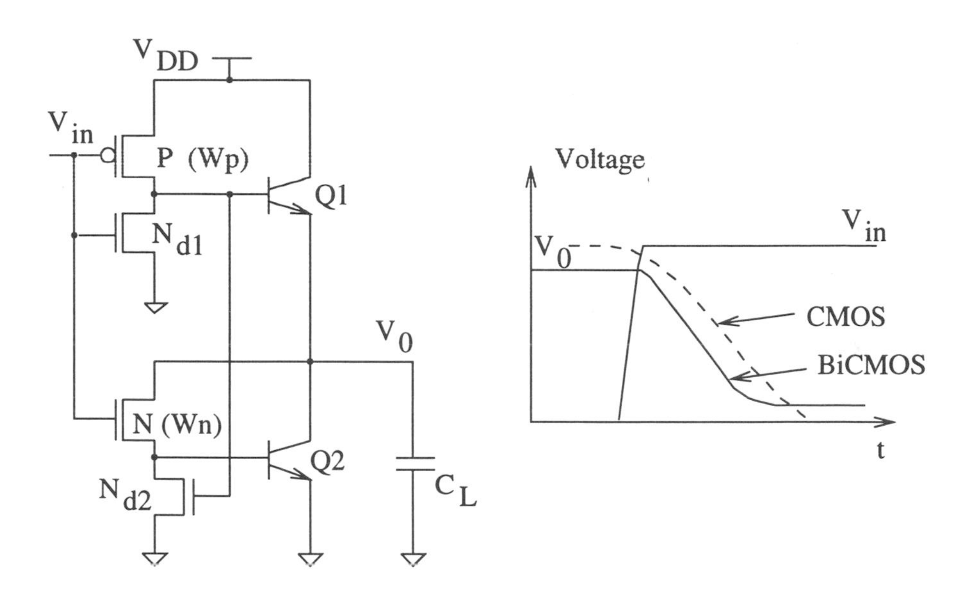

Figure 5.2 Conventional BiCMOS inverter.

5.1.1 DC Characteristics

Fig. 5.3 shows the DC transfer characteristic of the conventional BiCMOS inverter of Fig. 5.2. When the input voltage to the BiCMOS inverter is zero, both the bipolar transistors are OFF. The PMOS device P operates in the linear region with zero drain-source voltage. Due to the subthreshold current of the transistor N ($\sim$ 10 pA), the base-emitter voltage of Q_1 is around 0.45 V. As a result, the output voltage V_o = 4.55 V (@ $V_{DD} = 5\ V$). The base of the bipolar transistor Q_2 is at zero voltage because N_{d2} is ON.

As the input voltage increases, the subthreshold current of N increases causing V_{BE,Q_1} to rise and the output voltage to fall. When the input voltage is around the mid-V_{DD}, both the P and N MOSFETs are ON and operate in the saturation region. Also the bipolar devices are ON. At this point, the BiCMOS inverter is in the high gain region and the output voltage drops sharply towards its low level.

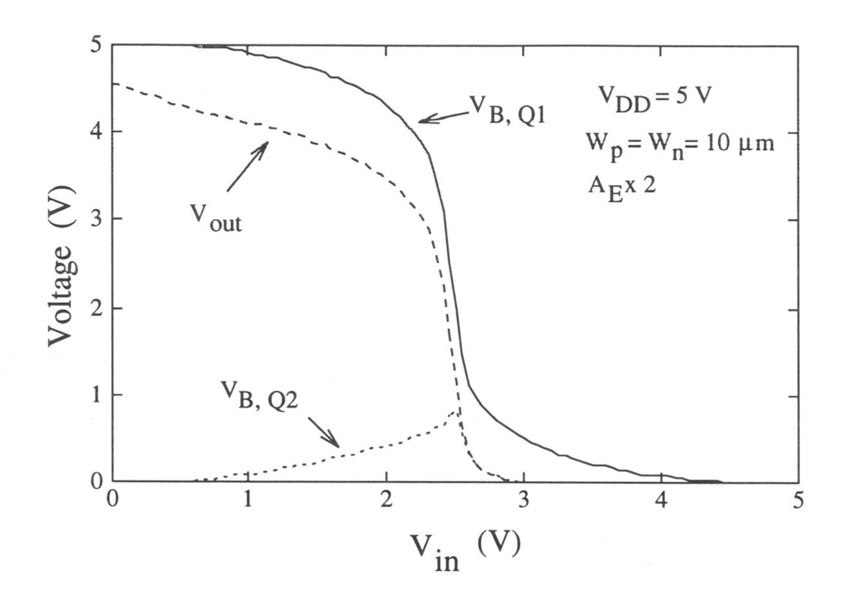

Figure 5.3 The DC transfer characteristic of the conventional BiCMOS at 5 V.

As the input voltage increases again, the base of Q_2 follows the voltage of the output since N is ON. When the input voltage reaches V_{DD}, the PMOS P is OFF. The discharge device, N_{d1} is ON and the base of Q_1 is at zero. Also, the output is completely discharged and N is ON. Then, the base of Q_2 is at zero. In this case, the output voltage is zero and both the base-emitter voltages are zero.

5.1.2 Transient Switching Characteristics

In this section we study the transient behavior of the conventional inverter of Fig. 5.2. The purpose of this analysis is three-fold; i) it serves to understand the transient switching behavior of the gate, ii) to develop a simple analytic model, and iii) also to show the superiority of BiCMOS compared to CMOS. The objective of delay analysis is to point out the important device and circuit parameters that affect the response of the gate. The developed model is very simple and can be used as a first order approximation. We start with the

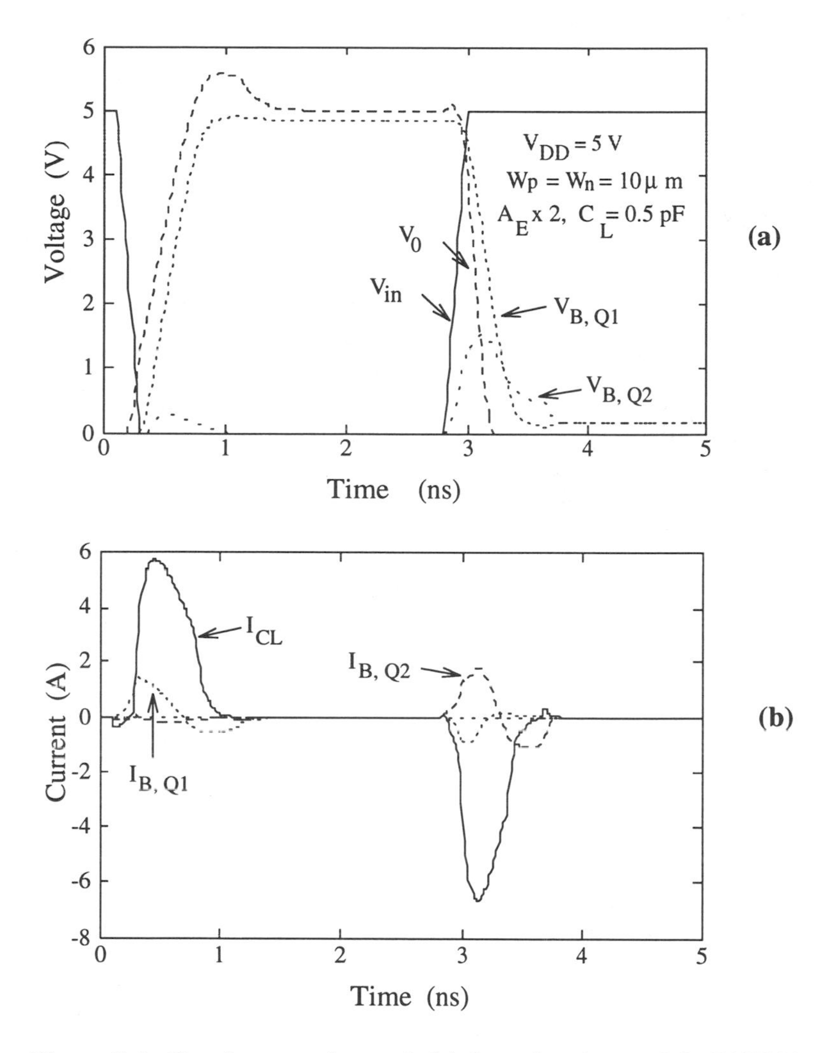

Figure 5.4 Transient wave forms of: (a) the node voltages of the BiCMOS inverter; (b) the base's current and the load C_L current.

analysis of the pull-up section. Then we show the difference in the case of the pull-down section. We assume a step input.

5.1.2.1 Transient Behavior

Fig. 5.4 shows the transient behavior of the BiCMOS inverter of Fig. 5.2. When the input falls to ground, transistor P turns ON and operates initially in the saturation region. Its drain charges the parasitic capacitances at the base and when $V_{BE,Q_1} = V_{BEon}$, Q_1 turns ON. The emitter current increases in a relatively short time to its peak to charge the output load C_L as shown in Fig. 5.4(b). The output voltage is pulled-up following the base voltage of Q_1 as shown in Fig. 5.4(a). As the base of Q_1 exceeds V_{Tn}, N_{d2} turns ON to discharge the base of Q_2 to ground. But due to capacitive coupling, V_{B,Q_2} tends to be pulled-up. When the base voltage is higher than $V_{DD} - V_{DSsat}$, where V_{DSsat} is the saturation voltage of P, the PMOS transistor P enters the linear region and the drain (base) current drops gradually. Consequently, the emitter current of Q_1 starts falling. As the output voltage V_o approaches the theoretical limit of $V_{DD} - V_{BEon}$, Q_1 is expected to turn gradually OFF. However, due to the capacitive coupling between the base and the output node, V_o exceeds this limit as shown in Fig. 5.4(a). The same reasoning can be applied when the input rises to V_{DD}.

5.1.2.2 Analytic Delay Model

A simple delay analysis is carried out in this section. The reader can refer to [4, 5, 6] for other detailed models. We take into account the parasitic capacitances and the bipolar high current effects. We do not take into account the parasitic resistances since they have no appreciable effect with advanced bipolar technology. This model is based on Raje model [7].

Fig. 5.5 illustrates the transient equivalent circuit of the pull-up section (Fig. 5.2) of the conventional BiCMOS gate driving a load capacitance C_L. As we are interested in 50% rise time, the PMOS current can be modeled by the saturation current of the device. This current is given by Equation (3.82) in Chapter 3

$$I_{DSsat} = K_p C_{ox} v_{sat,p} W_p(|V_{GS}| - |V_{Tp}|) \tag{5.1}$$

where V_{GS} is equal to $(V_{in,ol} - V_{DD})$, where $V_{in,ol}$ is the low level of the input. The capacitance C_{pm} accounts for the parasitic capacitances of the MOS devices P, N_{d1} and N_{d2} at the base of the pull-up bipolar transistor. Therefore, it is given by

$$C_{pm} = C_{d,P} + C_{d,N_{d1}} + C_{ox,N_{d2}} \tag{5.2}$$

where $C_{d,P}$ and $C_{d,N_{d1}}$ are the drain junction capacitances of P and N_{d1} and $C_{ox,N_{d2}}$ is the gate oxide capacitance of N_{d2}. The overlap capacitances of P

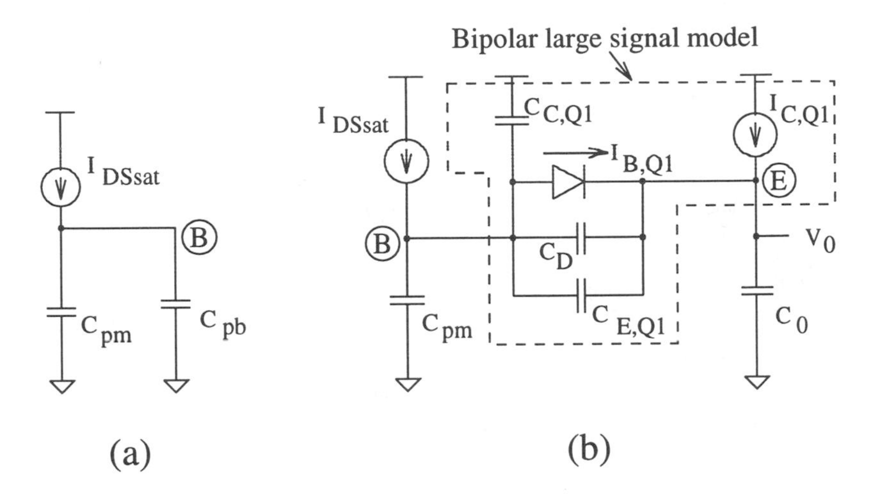

Figure 5.5 Equivalent circuit of the BiCMOS gate when: (a) Q_1 is OFF; (b) Q_1 turns on.

and N_{d1} are assumed negligible. The bipolar parasitic capacitance C_{pb} of Fig. 5.5(a) is given by

$$C_{pb} = C_{C,Q_1} + C_{E,Q_1} \tag{5.3}$$

The total load capacitance, C_o, shown in Fig. 5.5(b), is given by

$$C_o = C_L + C_{S,Q_2} + C_{C,Q_2} \tag{5.4}$$

where C_L is the external load capacitance, C_{S,Q_2} is the average collector-substrate capacitance of Q_2 and C_{C,Q_2} is the average base-collector capacitance of Q_2. Recall from Section 3.5.3 that the base-emitter diffusion capacitance is given by

$$C_D = \tau_f \frac{dI_{C,Q_1}}{dV_{BE}} \tag{5.5}$$

where the τ_f is the forward transit time subject to high-level effects.

The delay can be divided into three components :

1. The first component, t_1, is defined as the time required to turn Q_1 ON. The model of Fig. 5.5(a) can be used in this case. Writing the current equation at the base node of Q_1, we have

$$I_{DSsat} = (C_{pm} + C_{bp}) \frac{dV_{B,Q_1}}{dt} \tag{5.6}$$

Solving that equation and assuming that initially the base-emitter of Q_1 is zero, we have

$$t_1 = (C_{pm} + C_{bp})\frac{V_{BE,on}}{I_{DSsat}} \tag{5.7}$$

If the initial V_{BE} is not zero then the above expression should be corrected. Typical value of t_1 is 17.5 ps for a total parasitic capacitance at the base node of 50 fF, $V_{BE,on} = 0.7\ V$, and $I_{DSsat} = 2\ mA$.

2. The second component, t_2, is defined as the time required to charge the diffusion capacitance, C_{D,Q_1}. Starting from t_1, the collector current begins to quickly rise and then reaches its peak value, I_{CP}. The output voltage changes slowly (see waveforms of Fig. 5.4). So t_2 is then defined as the time required for the collector current to reach its peak. This delay component is given by

$$t_2 I_{DSsat} = \tau_f I_{CP} \tag{5.8}$$

which means that the charge furnished by the PMOS is needed to charge diffusion capacitance. Therefore,

$$t_2 = \tau_f \frac{I_{CP}}{I_{DSsat}} \tag{5.9}$$

The peak collector current of Q_1 can be approximated using Equation (3.111) [Section 3.5.2]. So we have

$$I_{CP} = \sqrt{\beta_0 I_{Kf} I_{DSsat}} \tag{5.10}$$

where β_0 is the value of the gain for low-level injection and I_{Kf} is the forward knee current. Note that τ_f is increased by the collector current [see equation (3.127) Section 3.5.3]. Hence, an average value of the forward transit time should be used in the above delay expression. The initial value of τ_f is 12 ps and it can reach 50 ps when the collector current reaches, for example, 5 mA. For $I_{DSsat} = 2$ mA, typical value for t_2 is 78 ps (average forward transit time is 31 ps).

3. The third component, t_3, is defined as the time required to charge the total load capacitance to the middle point of the output swing. If we assume that the voltage across the base-emitter of Q_1 is almost constant, then we have the following approximation

$$\frac{dV_o}{dt} = \frac{I_{DSsat} + I_{C,Q_1}}{C_{C,Q_1} + C_o} \tag{5.11}$$

If we assume that I_{C,Q_1} is constant during this time [see Fig. 5.4], and the mid-point of the output is $V_{DD}/2$, then we have

$$t_3 = \frac{(C_{C,Q_1} + C_o)V_{DD}}{2(I_{DSsat} + I_{CP})} \tag{5.12}$$

The value of this delay varies by more than an order of magnitude depending on the device's size and the load capacitance. For example, for a load C_L of 1 pF, this delay, t_3, has a typical value at 5 V power supply voltage of 400 ps, while for a load of 100 fF a typical value is 70 ps.

Hence, the total delay t_d can be written as

$$t_d = t_1 + t_2 + t_3 \tag{5.13}$$

The first delay is associated with the parasitics at the base, the second one with the forward transit time and the last one is a function of the load capacitance. For small loads, t_2 and t_3 dominate. However, for large output loads, the third delay term, t_3 dominates.

The expression of the pull-down time is similar to that of the pull-up time except for the value of the drain current of the transistor N [see Fig. 5.2]. The saturation current of this device is given by

$$I_{DSsat} = K_n C_{ox} v_{sat,n} W_n (V_{GS} - V_{Tn}) \tag{5.14}$$

The V_{GS} for the NMOS during the switching is affected by V_{BE} drop while the one of the PMOS is not. This voltage is given by

$$V_{GS} = V_{in,hi} - V_{BE} \tag{5.15}$$

So the effective gate-source voltage of the NMOS is lower than that of PMOS. The sizing of the NMOS and PMOS devices does not follow the rule used for CMOS. It can only be determined from circuit simulation to get symmetrical rise/fall delay times.

The slope of the characteristic delay-load of the BiCMOS gate is larger than that of CMOS, since it is equal to $V_{DD}/2(I_{DSsat}+I_{CP})$. For a CMOS gate, the slope is simply $V_{DD}/2(I_{DSsat})$. The saturation current in the CMOS is slightly higher than that of BiCMOS because the CMOS inverter has a PMOS with slightly wider device [see next Section]. However, the slope of the BiCMOS inverter is larger due to large I_{CP}. Therefore, the BiCMOS gate has a higher drivability than CMOS.

5.1.3 CMOS and BiCMOS Comparison

Lets compare the delay of BiCMOS gate to CMOS gate, having both of them the same input capacitances. We consider the case of inverters with the following sizes. For the BiCMOS inverter, we have : $W_p = W_n = 10\ \mu m$, $W_{N_{d1}} = W_{N_{d2}} = 2\ \mu m$, and the emitter area is $\times 2$ the minimum area. For the CMOS inverter, we have $W_p = 15\ \mu m$ and $W_n = 7\ \mu m$. For unloaded inverters and from the delay expression of the BiCMOS inverter discussed above, $t_{d,CMOS} < t_{d,BiCMOS}$ because the BiCMOS circuit has more parasitics and requires an initial delay to turn ON the bipolar device. For large loads, $t_{d,CMOS} > t_{d,BiCMOS}$, as explained previously. Fig. 5.6 shows the simulated delays of the CMOS and BiCMOS inverters function of the fanout. Fanout is defined here as the ratio of the load seen by the gate to the input capacitance. In other words, fanout is equal to the number of the gates connected to the output of the driving gate, all having the same input capacitance. The inputs are driven by a small size inverter of the same type to have typical input waveform fall/rise times. For low fanout, 1-to-2, CMOS outperforms BiCMOS at 5 V power supply voltage. However, when the fanout is greater than 3, BiCMOS outperforms CMOS; particularly for high loads. In Fig. 5.6, the crossover capacitance (or fanout), denoted C_x, is typically in the order of 100 fF. This crossover value is critical for the performance of BiCMOS; particularly when the supply voltage is scaled down.

5.1.4 Power Dissipation

As discussed, the BiCMOS gate of Fig. 5.2 has no DC current path from V_{DD} to V_{SS} if the input has rail-to-rail swing. Hence the static power dissipation is negligible if V_T of the MOS devices is high. The dynamic power dissipation of the gate can be estimated from the circuit diagram of Fig. 5.7.

It is estimated by

$$P_d = C_{p1}V_{DD}^2 f + C_{p2}V_{BEmax}^2 f + C_o V_{DD}(V_H - V_L)f \tag{5.16}$$

The first term is due to the total parasitic capacitance at the base node of Q_1 where the swing is $\sim V_{DD}$. The second term is also due to the parasitic capacitance at the base node of Q_2. The swing at this node is limited to V_{BEmax} when the collector current reaches its peak. Finally the third term is related to the output load capacitance, C_L, and the parasitic capacitance at the output. The swing is only $V_H - V_L$, where V_H and V_L are the high-level and the low-level of output, respectively. These levels are affected by the output load.

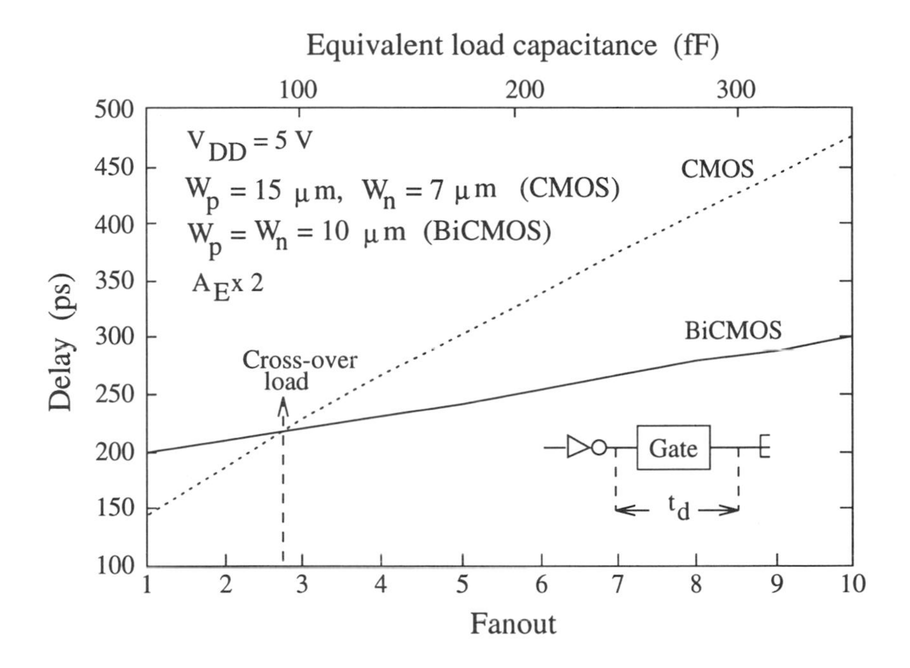

Figure 5.6 Delay's comparison of CMOS and conventional BiCMOS inverters with the same input capacitance.

For small loads the power of BiCMOS is greater than that of CMOS, while for large loads, they have almost the same dynamic power. Table 5.1 shows the simulation results of the power dissipation for both gates at 5 V power supply. At a fanout of 1, CMOS consumes much lower power than BiCMOS and it is faster. However at a fanout of 10, the BiCMOS is faster (37.5% delay reduction) and it dissipates only 24% power more than CMOS.

When a BiCMOS gate is driving another BiCMOS, or a CMOS gate, the driven gate exhibits a DC power dissipation. This DC current is not acceptable, particularly when the circuit is in standby mode. This is due to the reduced swing at the output of the first gate. Fig. 5.8 shows an example of BiCMOS gate driving a CMOS gate. If for example the output of the first gate (BiCMOS) is V_{BE}, the V_{GS} of the driven NMOS would be higher than zero and around the V_{Tn} resulting in appreciable DC power. Furthermore, the drive current of the driven gate would be reduced; particularly at low power supply voltage. Another disadvantage of the reduced swing is the noise margin reduction.

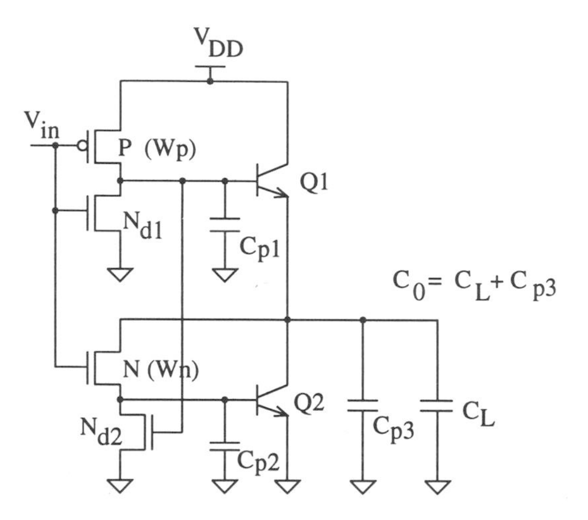

Figure 5.7 Equivalent circuit for power dissipation estimation.

Table 5.1 CMOS/BiCMOS power dissipation versus load @$V_{DD} = 5\ V$ and f = 100 MHz.

Driver	Fanout=1	Fanout=5	Fanout=10
CMOS (mW)	0.67	0.83	1.26
BiCMOS (mW)	0.23	0.58	1.02

5.1.5 Full-Swing with Shunting Devices

Previously we have seen that BiCMOS circuits exhibit reduced output swing. To overcome these shortcomings, various types of BiCMOS gates have been devised. These are based on the conventional BiCMOS circuits with base-emitter or collector-emitter shunting techniques or on other logic circuits which will be discussed in the following sections. Figure 5.9 shows some of the circuits based on shunting devices. Fig. 5.9(a) illustrated one full-swing (FS) configuration called "FS type" gate [8] which uses MOS devices to achieve full-swing. For the charging phase, as the output exceeds V_H, Q_1 ceases to source current to the load, and the load capacitance is charged through the shunting PMOS transistor P_s. When the input goes to HIGH, the load is discharged through

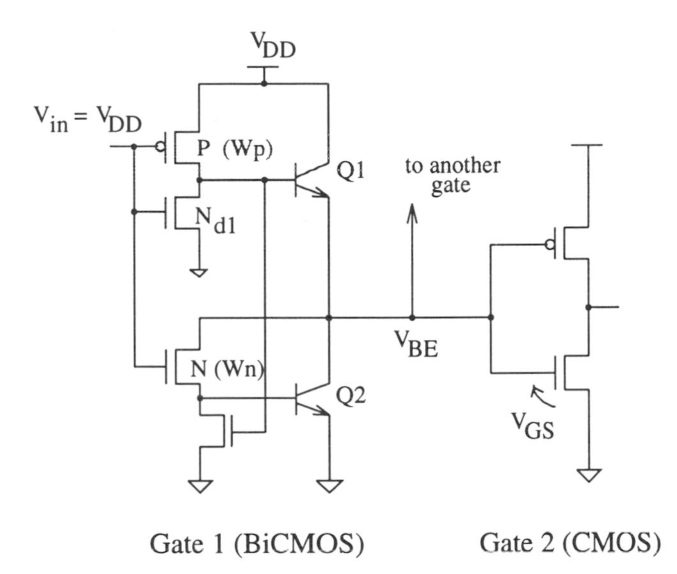

Figure 5.8 DC power dissipation of the driving gate.

N and N_s. When V_o falls below V_L, Q_2 ceases to sink current from the load capacitance. Then the output is discharged to the ground through only the MOS transistors N and N_s. The final charging and discharging phases occurs through the shunting devices. Hence, these phases can be slow because the MOS shunting devices have low drive capabilities. When this FS BiCMOS gate is operating under high frequency, the output swing can be reduced. Another drawback of this circuit is that part of the current supplied by P (N) is wasted through the shunting transistors which weakens the bipolar drive. The shunting transistors P_s and N_s can be minimum size.

The problem of the base drive inherent in the "FS type" BiCMOS gate can be overcome by using feedback (FB) from the output through an inverter as shown in Fig 5.9(b). This circuit is called "FB type" [9]. During the pull-up transition, the shunting device P_s is initially OFF and the PMOS transistor P supplied all its current to the base of Q_1. When V_o is approaching its high level, the inverter I turns ON P_s which itself charges the output node to V_{DD}. The pull-down transition can be explained similarly. The shunting devices P_s and N_s and the inverter I can be sized properly to achieve greater speed than the other configurations, even the conventional BiCMOS gate.

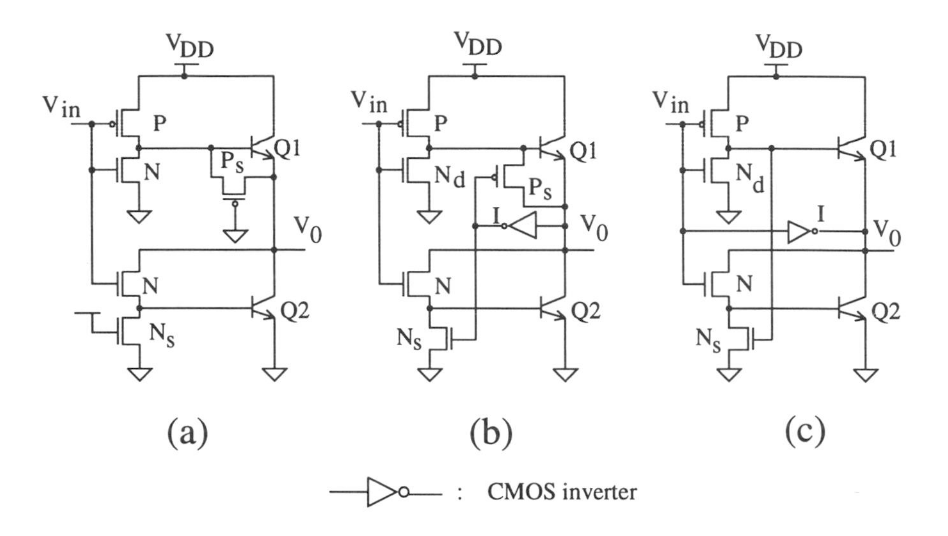

Figure 5.9 Full-swing BiCMOS gate types: (a) "FS type"; (b) "FB type"; (c) "CE-shunting type.

Another full-swing configuration is the one shown in Fig. 5.9(c). It uses a parallel inverter from the input to shunt the collector-emitter (CE) of Q_1 and Q_2 outputs. The disadvantage of this gate is the increased input capacitance.

5.1.6 Power Supply Voltage Scaling

The output bipolar stage introduces V_{BE} voltage losses at the output node as discussed earlier. When a BiCMOS gate is driving another BiCMOS gate, the conventional BiCMOS gate loses its superior performance over CMOS at lower power supply voltage. The major cause of this problem is the pull-down section of the BiCMOS gate. The V_{GS} voltage of the driving NMOS transistor of the pull-down section is equal to $V_{DD} - 2V_{BE}$. As V_{DD} is reduced, V_{GS} is significantly reduced, resulting in degradation of drain current, hence the driving capability of the conventional BiCMOS gate. Fig. 5.10 shows the delay of a BiCMOS inverter in comparison to that of a CMOS as the supply voltage is scaled down. The reported delay times were extracted from SPICE simulation by measuring the delay of the second gate in a chain of identical inverters. All gates were equally loaded by a load $C_L = 0.25\ pF$ and one fanout. All the circuits have the same input capacitance. The BiCMOS inverter fails to

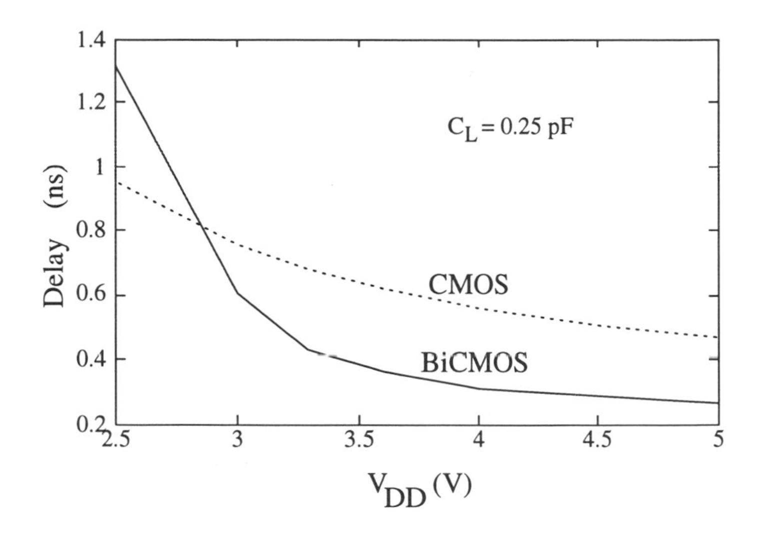

Figure 5.10 CMOS and conventional BiCMOS inverters delay vs the power supply voltage.

operate at 2 V power supply. The BiCMOS outperforms CMOS but for 3 and sub-3 V it looses its superior performance.

The limit of operation of the conventional BiCMOS gate with the power supply voltage is determined by the NMOS device of the pull-down section. The drive current of this NMOS device is $(V_{DD} - 2V_{BE} - V_{Tn})$. Hence, $V_{DD,min} \sim 2.2\ V$. Therefore, high-performance BiCMOS circuits, at low-voltage, are needed that minimize

- Technology/process complexity;
- Circuit complexity by using less device count;
- Area occupied by the gate; and
- Power dissipation.

5.2 BINMOS LOGIC FAMILY

BiCMOS technology can gain much of its performance edge over CMOS with circuit techniques that minimize or eliminate the effects of V_{BE} loses. To overcome the problem of delay degradation in conventional BiCMOS with supply voltage, many novel circuits were proposed. In this section, a practical family suitable for 3.3 V and sub-3.3 V operation regime is outlined.

Fig. 5.11 shows the BiNMOS family of BiCMOS circuits. The basic circuit technique used in BiNMOS [10] is the use of the NPN bipolar transistor only in the pull-up section of the output stage [Fig. 5.11(a)]. The pull-down section is kept as CMOS. In CMOS circuits, the PMOS transistor is two-to-three times slower than an NMOS transistor, when same sizes are compared. In the BiNMOS circuit, the use of the PMOS, with the bipolar driver in the pull-up section, will balance the unsymmetrical response of CMOS.

In the basic circuit of Fig. 5.11(a), the output reachs only $V_{DD} - V_{BE}$ level. This increases the delay and power dissipation of the subsequent gates. If a resistor (in this case the gate is called BiRNMOS) or a grounded gate PMOS transistor is inserted between the emitter and the base of the pull-up bipolar transistor, the output achieves full-swing. However, this will degrade the speed of the gate because the base current is bypassed by the inserted element and hence is reduced.

Many alternatives have been proposed such as BiPNMOS [11], and PBiNMOS [12] to realize full-swing output. The BiPNMOS is shown in Fig. 5.11(c). A small size PMOS transistor and an inverter are added to the basic BiNMOS gate. The PMOS device realizes full-swing output when the output changes from low to high. The added PMOS, P_s turns ON only when the output reaches the threshold voltage of the feedback inverter. Hence, the base current supplied by the pull-up PMOS transistor is not affected by this added PMOS transistor. Consequently, the BiPNMOS gate has higher performance than conventional BiNMOS and BiRNMOS. One drawback of the BiPNMOS is the increased output load capacitance due to the inverter I.

The PBiNMOS gate configuration shown, in Fig. 5.11(d), uses a small size PMOS device in parallel with the bipolar pull-up transistor to realize full-swing output. This configuration results in better performance compared to the other circuit structures but slightly increases the input capacitance of the gate. In this section, we show that a properly optimized PBiNMOS gate is faster than CMOS, even at low power supply and load.

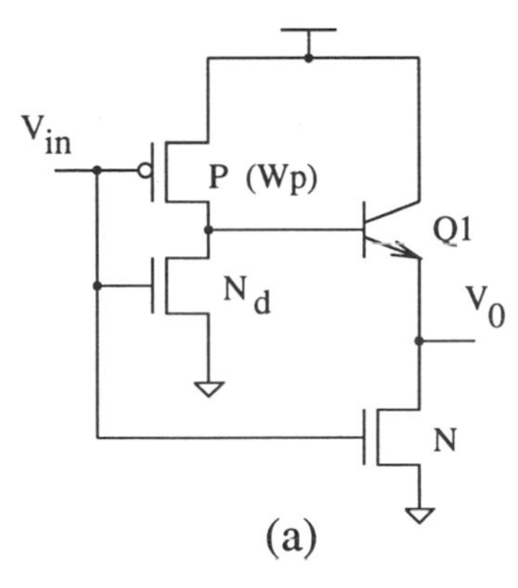

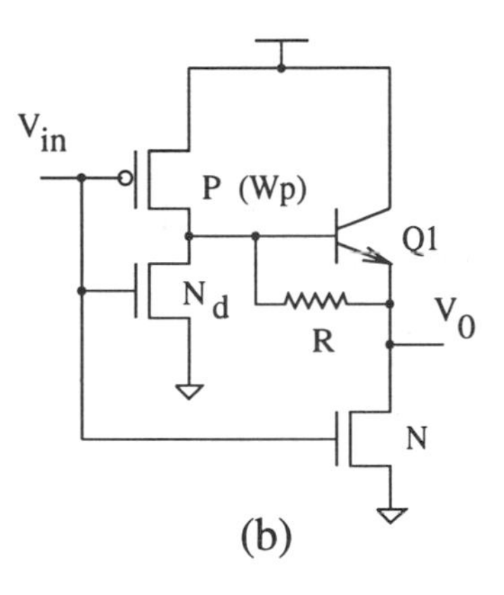

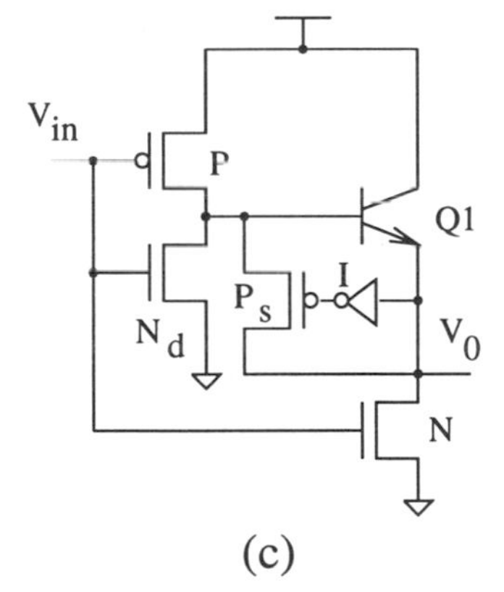

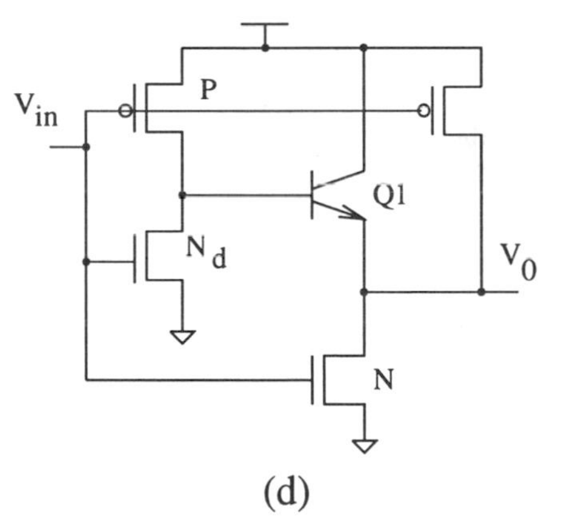

Figure 5.11 BiNMOS logic circuit families: (a) BiNMOS; (b) BiRMOS; (c) BiPNMOS; (d) PBiNMOS.

5.2.1 BiNMOS Gate Design

In this section we discuss the effect of the circuit parameters available to the designer to optimize the PBiNMOS gate for low fanout fast operation using the 0.8 μm BiCMOS device parameters discussed in Chapter 3. We optimize the design of the inverter. Then, the technique can be extended to more complex gates.

Finding the proper sizing of the input MOSFET's P and N (W_p and W_n respectively) is not trivial. The sizing of N_d and P_s [see Fig. 5.11(d)] is not critical. For typical applications, it is enough to use near minimum size devices. When the delay of the PBiNMOS is plotted versus the width of one of the devices P or N, for different fanouts, a common optimum width exits as shown in Fig. 5.12(a) with a flattened region. This optimum is due to the fact that when increasing the size, the drivability of the gate increases. However, the equivalent output load also increases. Then at a certain size, an optimum delay exits. From this figure, the optimum W_p is 9 μm and $W_n = 11$ μm (particularly for low-fanout). Note that in Fig. 5.12(a), we have chosen $W_p \approx 0.8 W_n$. This is explained in more detail below.

When the BiNMOS inverter is used as a driver of a fixed load (e.g., bus), instead of driving gates, then we should consider the delay of the driver, including the delay of the stage that drives it. In Fig. 5.12(b), the total delay of the PBiNMOS driver and the CMOS inverter that drives it is plotted for two fixed loads: 0.2 pF and 0.5 pF. The CMOS stage has a minimum size. The minimum delay is around the point determined previously for the fanout case.

The choice of the emitter area in this gate depends on the technology and the load. For the 0.8 μm BiCMOS at 3.3 V power supply voltage, it was found that using the minimum emitter area ($A_E \times 1 = 0.8 \times 4\ \mu m^2$) gives the minimum delay for the range of loads $\leq 1pF$.

Fig. 5.13 shows that the optimal W_p/W_n ratio is the same for different fanouts and is equal to 0.8. This point also gives almost symmetrical fall/rise delays. So even if the fanout is unknown, the optimum gate is fixed and the sizes depend only on the device parameters. This result is very important for standard cells and gate arrays where the cells are designed with unknown loads.

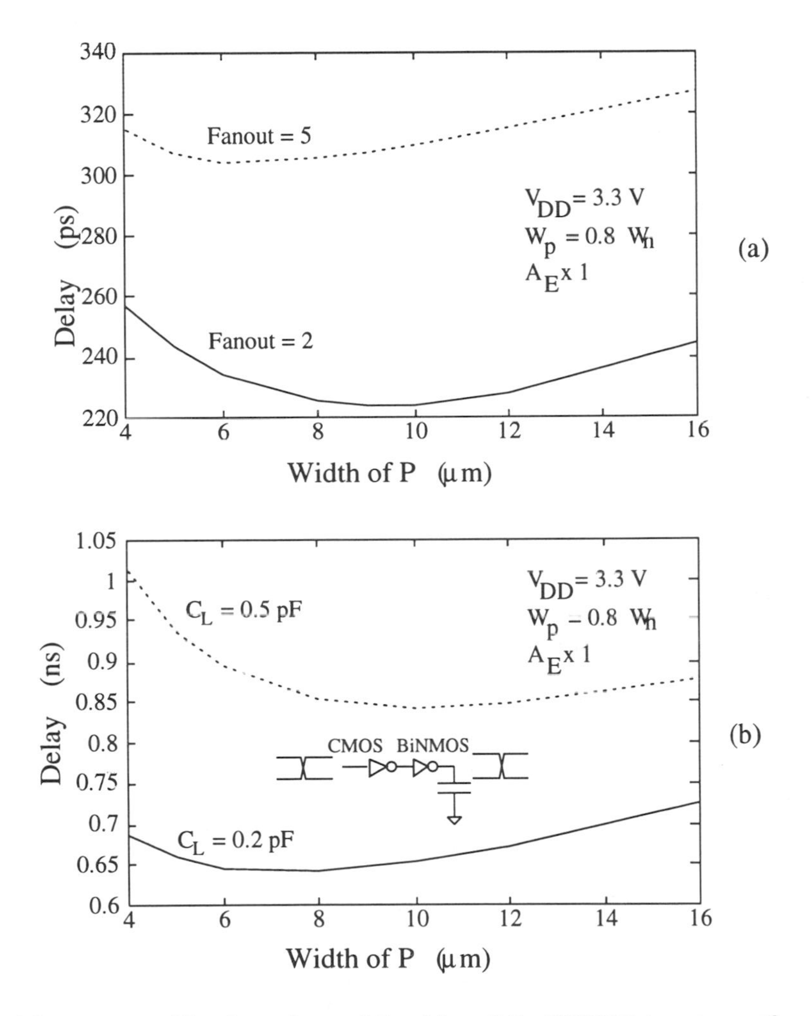

Figure 5.12 The dependence of the delay of the BiNMOS inverter on the size of the input MOS transistors: (a) fanout; (b) fixed load.

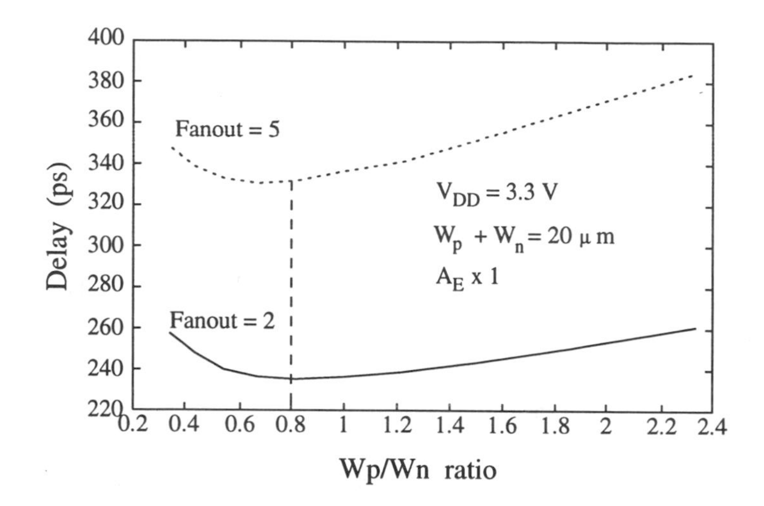

Figure 5.13 The delay of PBiNMOS inverter versus the ratio of W_p/W_n for a fixed input capacitance.

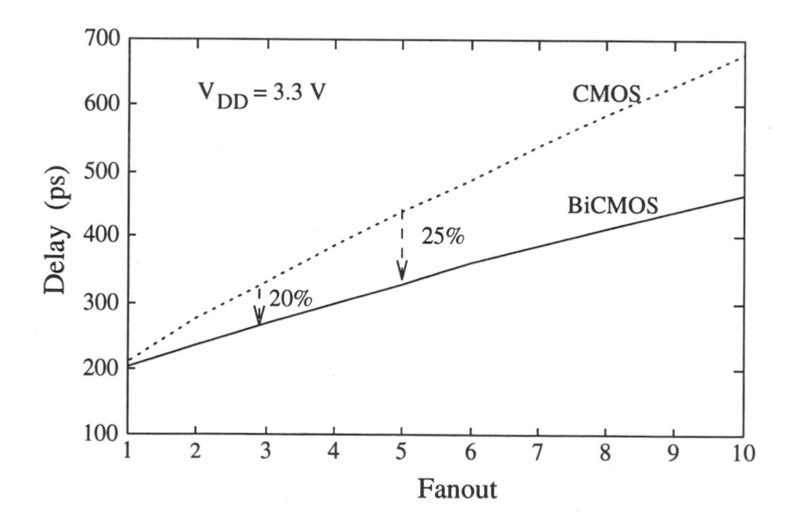

Figure 5.14 Comparison of the CMOS and PBiNMOS delays for the same input capacitance function of the fanout.

5.2.2 CMOS and BiNMOS Comparison

Fig. 5.14 shows the delay of CMOS and PBiNMOS inverters function of the fanout. Both gates have the same input capacitance. The important result of this plot, is that the PBiNMOS gate is always faster than CMOS, except for a fanout of 1, where PBiNMOS is slightly faster. For a fanout of 3, which is a typical value in many designs, the delay is reduced by 20%. For a higher fanout, the delay is reduced by 25-40%. This result is quite different from the case of conventional BiCMOS where a high fanout (or load) is required for BiCMOS.

Let us compare the power dissipation of the gates for different fanout. Table 5.2 shows this comparison for small fanouts. The power dissipations of both gates are comparable and are the same for a fanout (≥ 3). The small size additional bipolar in the BiNMOS gate does not result in significant power dissipation overhead. This result shows that the BiNMOS family is an excellent choice for low-power and high-speed operation. However for a fanout 1-2, still the CMOS can be used.

Table 5.2 CMOS/PBiNMOS power dissipation versus fanout @$V_{DD} = 3.3\ V$ f = 100 MHz.

Driver	Fanout=2	Fanout=3	Fanout=5
CMOS (μW)	149	192	277
PBiNMOS (μW)	171	203	287

5.2.3 BiNMOS Logic Gates

Since the PBiNMOS is used extensively in 3.3 V digital integrated circuits, some logic gates are presented. Combinational PBiNMOS logic circuits are easily constructed using the basic PBiNMOS inverter of Fig. 5.11(d). Two-input NOR and NAND gates are shown in Fig. 5.15(a) and Fig. 5.15(b). The logic function is implemented using the PMOS and NMOS blocks as in CMOS technology. The bipolar device Q_1 is used as a current drive. More complex functions can be implemented using standard CMOS gate formation theory. The layout of the PBiNMOS inverter is shown in Fig. 5.16. The BJT consumes area in the PBiNMOS gate. However, when complex gates are implemented with more MOS devices, the extra area of the BJT is reduced.

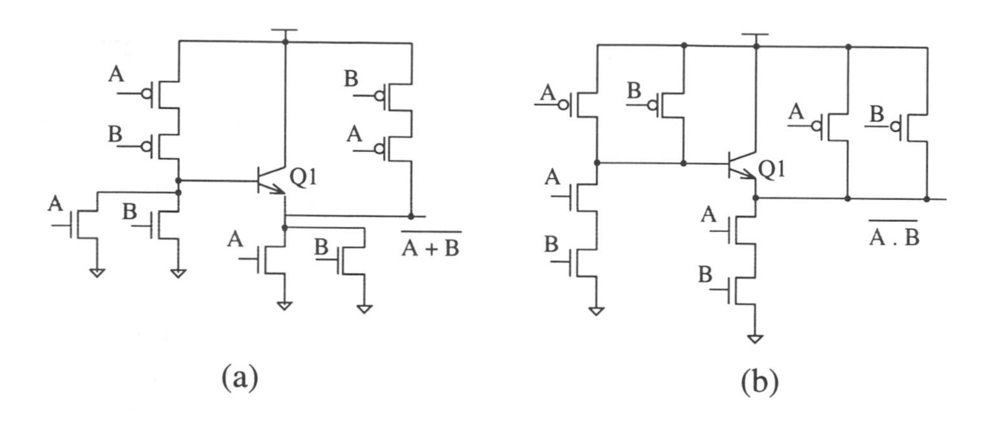

Figure 5.15 Circuit schematics of : (a) PBiNMOS NOR2 ; (b) PBiNMOS NAND2.

One technique to reduce the area penalty of the BJT is to use merged N-well bipolar and PMOS devices.

5.2.4 Power Supply Voltage Scaling

For future technologies, the power supply voltage will be scaled below 3.3 V. Fig. 5.17 shows the delay of PBiNMOS and CMOS inverters for a fanout=3 versus the power supply voltage scaling. The reported delay times were extracted from SPICE simulation by measuring the delay of the second gate in a chain of identical inverters. In this case, the full-swing operation, at the input of a PBiNMOS inverter, is provided by an identical gate, where a shunting PMOS is used. Fig. 5.17, shows that PBiNMOS is faster than CMOS down to 2.5 V. At 2.5 V the delay reduction is 15%. The crossover power supply voltage between PBiNMOS and CMOS is around 2.15 V. Note that in this comparison we used a 0.8 μm BiCMOS technology optimized for 5 V operation. In this case, to compare the BiNMOS to CMOS at low-voltage, deep-submicron technology should be used. From the device level point of view, scaled technology is expected to improve the performance of BiNMOS at low-voltage. However, 2 V is the limit of the use of BiNMOS, since almost half of the swing at sub-2 V is provided by the poor shunting PMOS device.

In summary, BiNMOS family provides the following advantage:

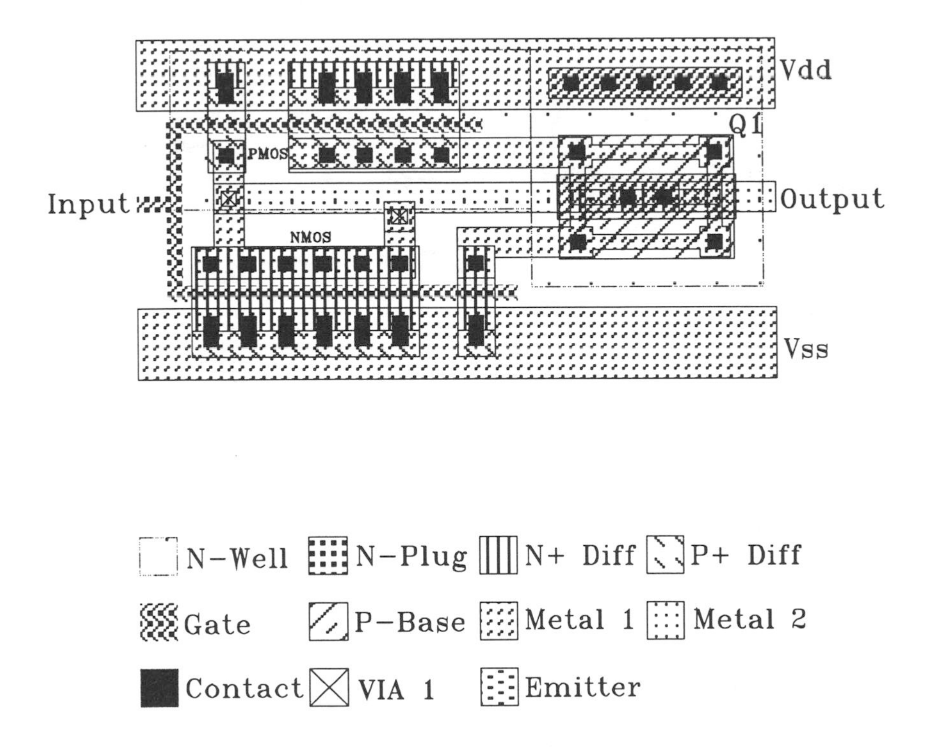

Figure 5.16 The layout of a PBiNMOS inverter.

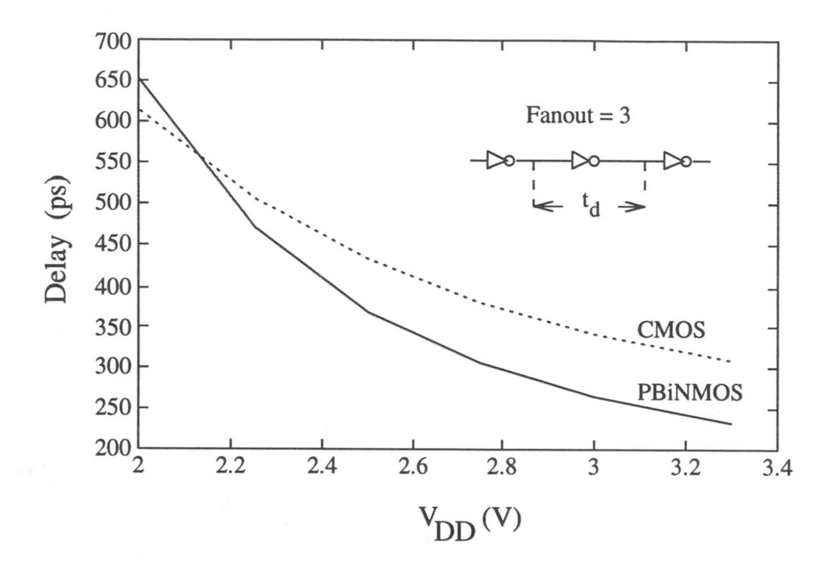

Figure 5.17 Effect of V_{DD} scaling on th PBiNMOS and CMOS inverter's delay.

- Simple gate compared to other BiCMOS logic circuits;
- Good performance at 3.3 and 2.5 V power supply voltage generations even at low-fanout; and
- Needs simple BiCMOS process.

The only disadvantage of BiNMOS is its poor performance for sub-2 V operation. The small area penalty of BiNMOS is not a problem since for complex gates the overhead of the bipolar device is minimized.

5.3 LOW-VOLTAGE BICMOS FAMILIES

In this section, several BiCMOS logic circuits proposed for low-voltage high-speed digital applications are reviewed [13]. Many of these circuits have not been widely used in BiCMOS products. However, some of the logic circuits presented in this section exhibit high-performance at low-voltage down to $\sim 1\ V$.

For fast operation at low-voltage the full-swing operation should be realized with bipolar devices. Otherwise, the techniques based on shunting devices do not provide high drivability.

5.3.1 Merged and Quasi-Complementary BiCMOS Logic

In this section two circuit techniques to overcome the shortcomings of the conventional BiCMOS gate are discussed and compared. These gates are intended to be used for sub-3.3 V operation. Also they are devised to solve the problem of using PNP transistor (see next section on Complementary BiCMOS). In all these circuits, the improvement is done mainly on the pull-down section of the conventional BiCMOS, since it is the major cause of speed degradation at low-voltage.

5.3.1.1 Merged BiCMOS (MBiCMOS)

To improve the performance of the pull-down section of the conventional BiCMOS circuit, with power supply scaling, PMOS/NPN pull-down BiCMOS gate has been proposed [14] as shown in Fig. 5.18. In this pull-down configuration, a PMOS transistor P_2, is used to drive the NPN bipolar transistor, Q_2. The gate of the PMOS P_2 is tied to the base of Q_1. The CMOS inverter formed by the transistors P_1 and N_{d1} supplies rail-to-rail voltage swing to the pull-down PMOS. Hence, the V_{GS} voltage of the driving PMOS transistor is not affected by V_{BE} loss as in the case of conventional BiCMOS. This gate is called Merged BiCMOS (MBiCMOS) because of the advantage of the gate for possible PMOS/NPN device's merging.

The pull-up section is similar to the one in conventional BiCMOS. The operation of the pull-down sections is as follows. When the input is high, N_{d1} pulls the base of Q_1 down to ground and P_2 turns ON. The transistor P_2 supplies the base current to Q_2. The bipolar transistor Q_2 discharges the load capacitance to lower voltage equal or less than V_{BEon}.

Still this structure suffers from the 2 V_{BE} losses. The only improvement in MBiCMOS, compared to conventional BiCMOS, is the higher drive current of the pull-down section. If the N-well of the pull-down PMOS transistor is tied to the V_{DD} rail, its threshold voltage will experience a degradation due to the body effect during the pull-down transient. As a result, the drivability of the pull-down PMOS transistor is degraded. A simple solution to eliminate this problem is to shunt the source and the substrate of the PMOS transistor, P_2.

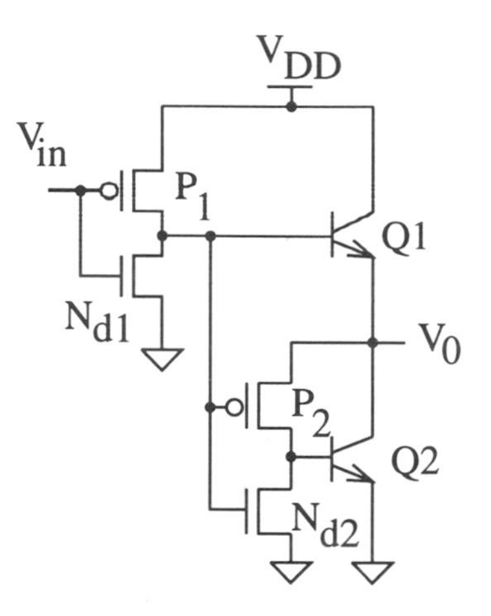

Figure 5.18 The MBiCMOS gate.

It was shown that this configuration (with shunted source/substrate) is faster than its CMOS counterpart down to 2.2 V supply voltage using sub-0.5 μm BiCMOS technology [15, 16].

5.3.1.2 *Quasi-Complementary BiCMOS*

Another variation of the MBiCMOS is called "Quasi-complementary BiCMOS" [17]. A "quasi-PNP" connection is generated in the pull-down section of the conventional BiCMOS as shown in Fig. 5.19. It consists of PMOS and NPN transistors (Fig. 5.19(b)). This configuration resembles the MBiCMOS gate of Fig. 5.18. The QCBiCMOS has two attractive features. The first one is that the drain current of the pull-down section does not suffer the $2V_{BE}$ losses as in the case of conventional BiCMOS. The second one is that the pull-down waveform is steep, due to the good charge retention capability of the bipolar transistor. The feedback circuit formed by the two cross-coupled inverters, I_1 and I_2, permits the discharge of the base of the pull-down transistor immediately after the pull-down transition.

The QCBiCMOS gate keeps its superiority over CMOS down to 2 V. At 2 V it has better performance than BiNMOS logic circuit. However for sub-2 V, it looses its performance. Furthermore, it consumes large area and needs a relatively large fanout to outperform CMOS.

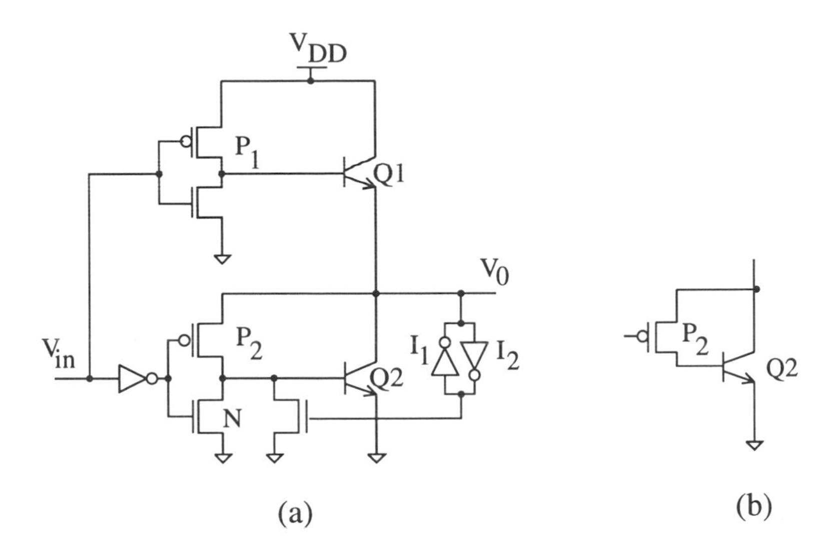

Figure 5.19 (a) The QCBiCMOS gate; (b) the quasi-PNP configuration.

5.3.2 Emitter Follower Complementary BiCMOS Circuits

Full-swing operation can also be achieved by using what is called the Complementary BiCMOS (CBiCMOS). The use of complementary BiCMOS has been encouraged by the recent advances in bipolar technology, which led to high-performance PNP transistors. It is expected that the NPN and PNP transistors will exhibit close performance when the devices are scaled down and the base doping increases. In this section, we study the emitter-follower (EF) CBiCMOS.

Fig. 5.20 shows the use of complementary bipolar output stage to form the basic complementary BiCMOS circuits [18, 19]. The pull-up section is similar to the conventional BiCMOS. The pull-down section is symmetrical to the pull-up. The current of the NMOS transistor N does not suffer of V_{BE} reduction due to Q_2 as in conventional BiCMOS. The static swing varies between V_{BEon} and $V_{DD} - V_{BEon}$. However, as explained in Section 5.1.2, the actual swing might be larger than the static design. The balanced transconductance of the PMOS/NPN and NMOS/PNP makes it easier to obtain symmetrical fall and rise time. Hence this circuit eliminates the degradation of the pull-down delay with power supply voltage of the conventional BiCMOS.

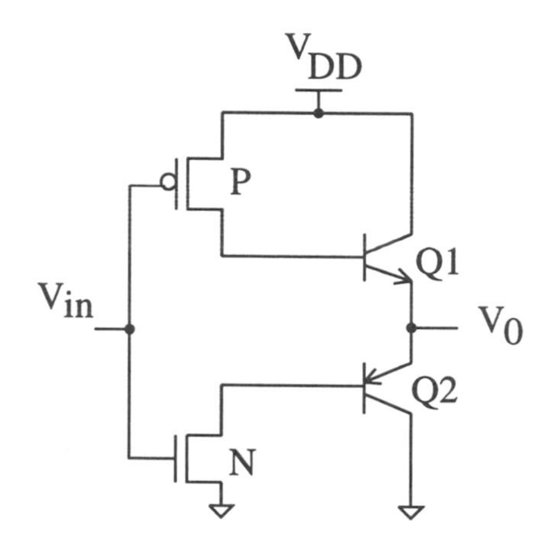

Figure 5.20 Schematic of the basic CBiCMOS.

The gate of Fig. 5.20 can be modified to achieve full-swing operation by using emitter-base shunting devices. Fig. 5.21(a) shows EF CBiCMOS with shunting technique. The shunting MOS transistors of the base-emitters permit restoration of the full logic level of the output. But still the full-swing is achieved with the two slow MOS devices. Some of the base current can be consumed by the shunting devices which weakens the drive of Q_1 and Q_2. To overcome this problem, the feedback technique can be used as shown in the circuit of Fig. 5.21(b). The turn ON of the shunting devices is delayed by the feedback inverter, I.

These CBiCMOS circuits have two drawbacks: poor performance at 2 V power supply voltage and less, and high processing cost because of the high performance PNP device needed. This low performance, at low voltage, is due mainly to the fact that $2V_{BE}$ output swing is generated by the two shunting transistors.

5.3.3 Full-Swing Common-Emitter Complementary BiCMOS Circuits

So far all the presented full-swing circuits, such as PBiNMOS, CBiCMOS, MBiCMOS and QCBiCMOS, achieve the rail-to-rail swing by using resistors or MOSFETs that operate in the linear region. These techniques are effective

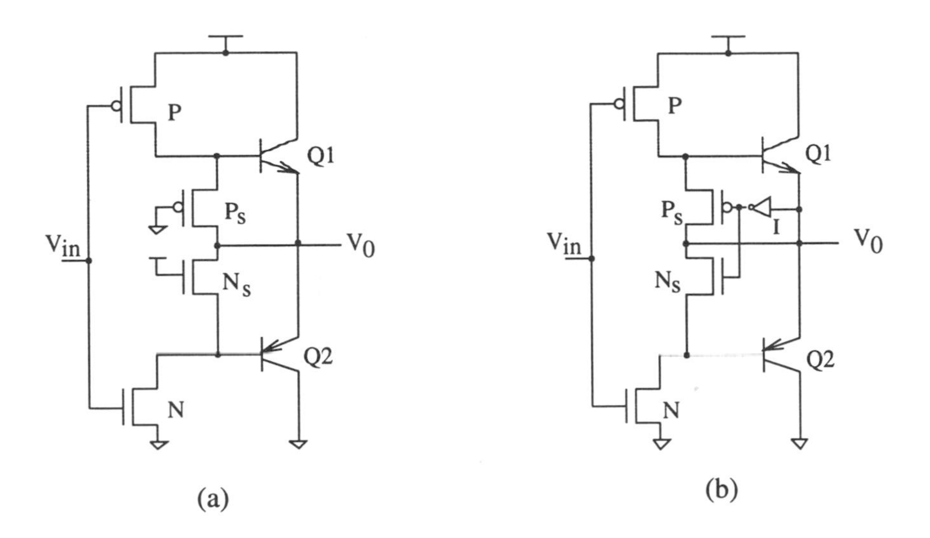

Figure 5.21 Schematic of EF CBiCMOS gates with shunting devices.

only when the operating frequency is low, where the gate can complete its full-swing operation and/or when the load capacitance is small [20]. Full-swing circuits with full bipolar drive are needed. In this section, CBiCMOS variation suitable for sub-2 V operation, called Transient Saturation (TS) is presented.

Fig. 5.22 shows the basic common-emitter complementary BiCMOS (CE-CBiCMOS) circuit. The circuit is symmetrical and has symmetrical fall and rise times. When the input goes to high, N turns ON to sink the current from the base of the PNP transistor Q_2. When the base voltage of Q_2 falls to $V_{DD}-V_{BEon}$, Q_2 turns ON to source the current to the output load capacitance. Q_2 eventually saturates and the output node is pulled-up to $V_{DD} - V_{CEsat}$. At the end of charging the MOS device is still consuming current. The operation of the pull-down section can be explained similarly. Hence, the operation of CE-CBiCMOS is non-inverting and the gate needs an extra CMOS inverter at the input to achieve complement function. In this circuit, the MOS transistors operate in saturation, hence they supply high current for the bipolar transistors. Furthermore, the output swing has near rail-to-rail swing (V_{CEsat} to $V_{DD} - V_{CEsat}$). This circuit offers high-speed at low-voltage, but has two drawbacks; (i) the high-static power dissipation, due to the DC current flowing through the base of either Q_1 or Q_2, and (ii) the excess delay due to the slow process of turning the saturated BJTs OFF.

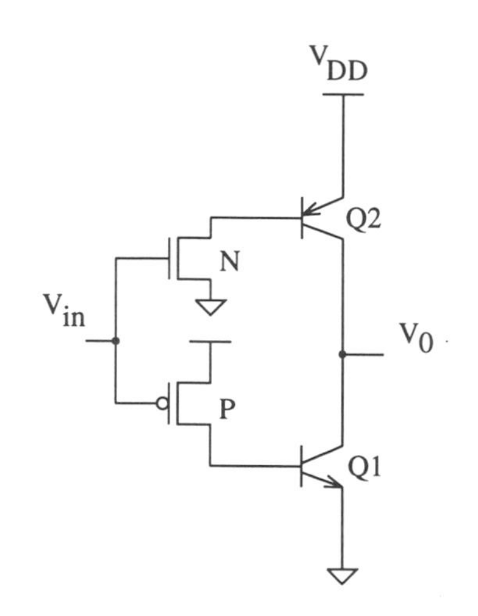

Figure 5.22 Common-emitter CBiCMOS gate.

These two problems have been solved with several implementations [21, 22]. One possible implementation is shown in Fig. 5.23. It is called Transient Saturation Full-Swing (TS-FS) BiCMOS. This logic uses the principle of CE CBiCMOS described in Fig. 5.22. When the input falls, we assume that the output is charged high, then P_3 is ON. P_2 turns ON and the base of Q_1 is charged through P_3 and P_2 [Fig. 5.23(b)]. Consequently, Q_1 discharges the output (load) down. When the output voltage approachs zero, the inverter I_1 turns P_3 OFF and N_4 ON [Fig. 522(c)]. The base voltage of Q_1 falls below V_{BEon} causing it to turn OFF. Although Q_1 saturates, this does not slow the next pull-up transition because the excess minority carriers of Q_1 are discharged immediately after the pull-down operation. Thus, the bipolar transistor saturates transiently. The circuit is symmetrical, hence the operation of the pull-up section can be explained similarly. The PMOS transistor, P_3, cuts off the the DC current path during the pull-down transition to avoid any static power dissipation. The small size output latch, composed of the inverters I_1 and I_2, holds the output level because in steady state there is no path between the output and the supply lines.

Compared to the BiCMOS logic circuits so far presented, TS-FS is faster below 2 V supply, when the load is relatively large ($\sim 1\ pF$). At 1.5 V it is twice as fast as CMOS for large loads. Although this circuit solves the problem of speed degradation of BiCMOS @1.5 V power supply, it still has several drawbacks:

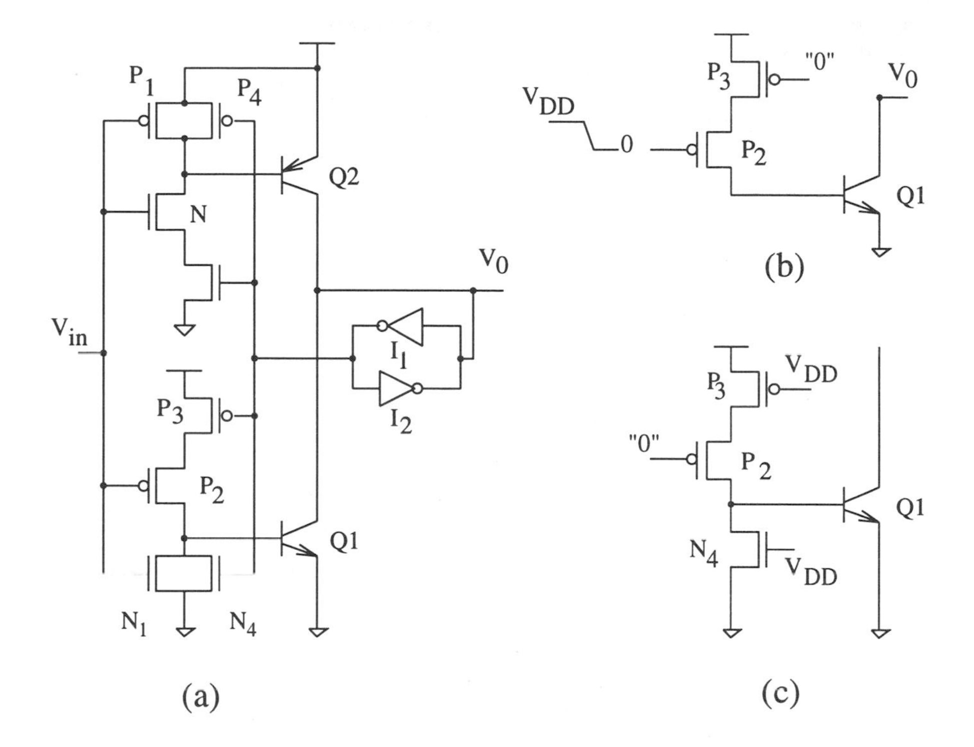

Figure 5.23 (a) Circuit configuration of TS-FS BiCMOS; (b) and (c) transient saturation operation for the pull-down section.

process complexity due to the PNP bipolar transistor; large area; relatively high crossover point with CMOS ($\sim$ 0.4 pF); and it is a noninverting circuit.

5.3.4 Bootstrapped BiCMOS

An alternate way to avoid the negative effect of V_{BE} loss in BiCMOS is simply to use a second supply voltage equal to $(V_{DD} + V_{BE})$. However, this approach is costly because of the additional wires needed to distribute across the chip and the need for the second supply voltage. Another approach is to use bootstrapping technique to pull-up the base of the pull-up bipolar transistor to $(V_{DD} + V_{BE})$ and hence the output to V_{DD}. The generation of voltages higher than the power supply at the gate level adds an extra degree of freedom to BiCMOS. Schottky BiNMOS/BiCMOS circuit configurations using the bootstrapping have been proposed to overcome the negative effect of V_{BE} loss [20]. The full-swing operation is performed by saturating the bipolar transistor of the pull-up section with a base current pulse. After which, the base is isolated and bootstrapped to a voltage higher than V_{DD}. These Schottky circuits outperform all existing BiCMOS families in sub-3V regime down to 2 V, but they need a BiCMOS technology with good integrated Schottky diode. Other examples of a such technique are the bootstrapped BiCMOS circuits published by [23, 24, 25]. The main advantage of the bootstrapped circuits is that they can be realized in conventional BiCMOS process with CMOS and NPN transistor only. In this section, we present one bootstrapped circuit which overcomes many drawbacks of the BiCMOS logic families discussed previously.

5.3.4.1 Basic Concept of Operation

The Bootstrapped Full-swing BiCMOS (BFBiCMOS) inverter is shown in Fig. 5.24. It consists only of CMOS and NPN transistors. Hence, it can be built in a non-complementary BiCMOS technology. The pull-down circuitry is identical to that of TS-FS and was explained previously. The operating principle of the pull-up section can be explained as follows. When the input is high and the output is low, the PMOS transistor P_d is ON. In this case, the base voltage of Q_1 is precharged to V_{TP} which is less than V_{BEon} but close to it. The precharge PMOS transistor MP_p is ON to charge the bootstrapped capacitor C_{boot} to the level V_{DD} (precharge cycle). When the input goes to low and P_1 turns ON, the bipolar transistor Q_1 turns ON almost instantaneously because its base-emitter junction is precharged near V_{BEon}. Consequently the initial turn-on delay of the pull-up section is reduced. This has an impact on the minimum fanout required by BFBiCMOS to outperform CMOS. Once Q_1 turns on, the output node starts to charge the load capacitor C_L toward V_{DD}. Since P_p is OFF, the node n_1 is disconnected from V_{DD} and is floating. Thus as the output voltage V_o rises to V_{DD}, the voltage at node n_1 also rises towards $V_{DD} + V_{BEon}$ (bootstrapping cycle).

When the input is low, the gate of the PMOS transistor P_p turns OFF (almost instantaneously) during the bootstrapping cycle to prevent discharging the bootstrapped node through reverse current from n_1 to V_{DD}. This is achieved through the use of the pseudo-inverter formed by P_i and N_i. During the bootstrapping cycle (the input is low), P_i turns ON and the gate of the precharge transistor P_p is pulled up towards the voltage of n_1. Thus, P_p is completely OFF when the voltage at n_1 exceeds V_{DD}. Furthermore, the PMOS transistor P_d is OFF completely because its gate is driven by the boosted voltage through P_i.

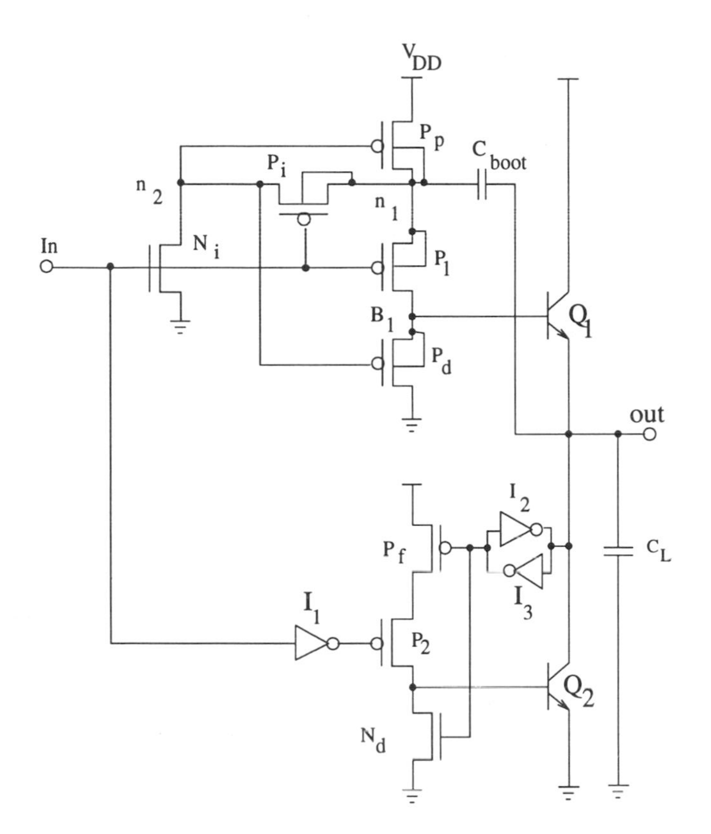

Figure 5.24 The bootstrapped full-swing BiCMOS inverter (BFBiCMOS).

Compared to the Bootstrapped BiCMOS (BS-BiCMOS) [23] of Fig. 5.25, the BFBiCMOS has several advantages. First, the bootstrapped capacitor is driven by the output rather the input as in the BS-BiCMOS. In BS-BiCMOS, the gate of precharge transistor, P_p is driven to V_{DD} and the node n_1 to $V_{DD} + V_{BE}$. Hence, when V_T is lower than V_{BE}, the bootstrapped node leaks its charge and results in less efficient bootstrapping. Third, a PMOS transistor P_d is used to discharge the base to a precharged level V_T, resulting in improved performance. Furthermore, it has a high crossover capacitance and less performance than the BFBiCMOS.

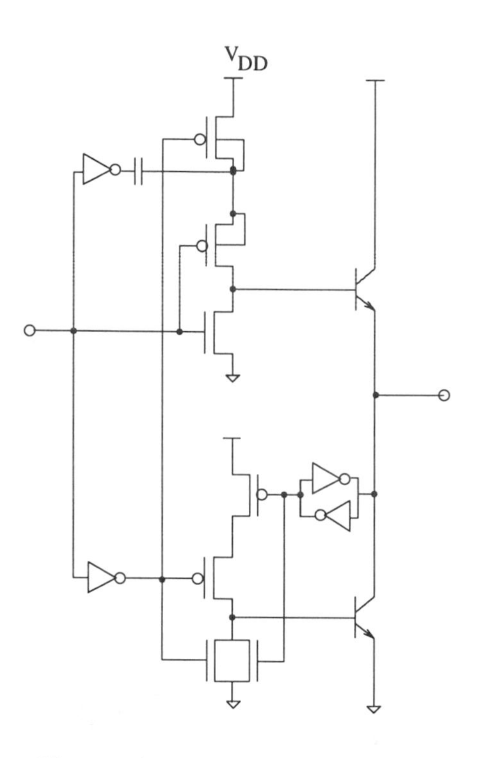

Figure 5.25 The BS-BiCMOS inverter.

The simulated waveforms at 1.5 V power supply of the BFBiCMOS inverter are shown in Fig. 5.26. The base of Q_1 goes to $(V_{DD} + V_{BE})$ when the input is low. Note that when the input is high the base voltage falls to V_T.

5.3.4.2 *Design Issues*

As a first order analysis, the minimum value of C_{boot}, necessary for the bootstrapping condition, can be obtained as follows. During the precharge cycle, the charge of the bootstrapped capacitor is $V_{DD}C_{boot}$ and the charge on C_p, the parasitic capacitance on the node n_1, is $V_{DD}C_p$. The total charge on n_1 during the precharge cycle is

$$Q_{n1} = V_{DD}C_{boot} + V_{DD}C_p \tag{5.17}$$

In order for V_{out} to reach V_{DD}, V_{n1} must reach $V_{DD} + V_{BEon}$ (during the bootstrapping cycle). Thus the charge on C_p is $(V_{DD} + V_{BEon})C_p$ and the

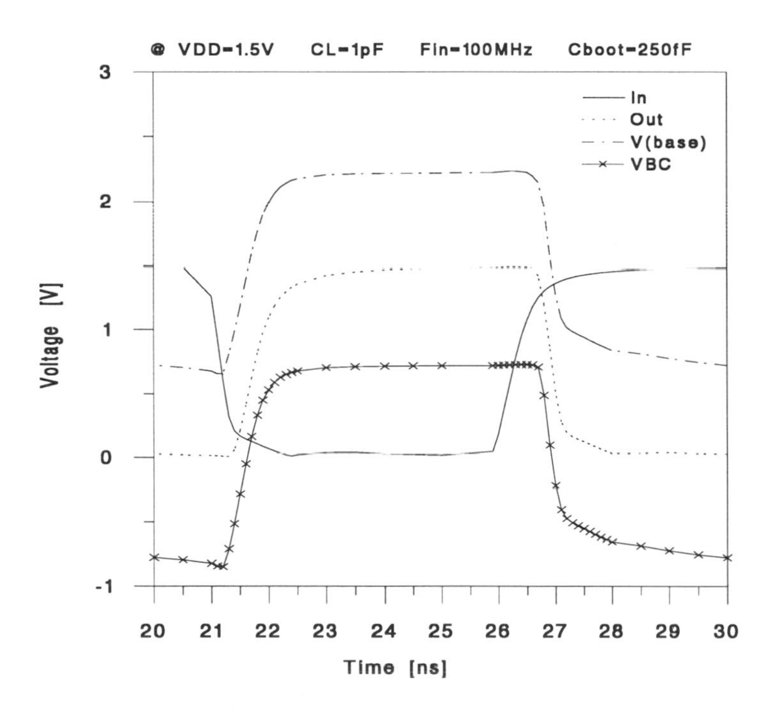

Figure 5.26 Voltage waveforms of the input (in), the output (out) and the base-collector of Q_1 for the BFBiCMOS inverter.

charge on C_{boot} is $V_{BEon}C_{boot}$. The new charge is given by

$$Q'_{n1} = V_{BEon}C_{boot} + (V_{DD} + V_{BEon})C_p \tag{5.18}$$

The charge necessary for the base is

$$Q_b = Q_{n1} - Q'_{n1} \tag{5.19}$$

As an approximation Q_b can be given by

$$Q_b = I_b t_r \tag{5.20}$$

where I_b is the average base current of $Q1$ and t_r is the rise time of the output. From Equations (5.17-5.20) we find that

$$C_{boot} = \frac{V_{BEon}}{V_{DD} - V_{BEon}} C_p + \frac{I_b t_r}{V_{DD} - V_{BEon}} \tag{5.21}$$

This equation indicates that C_{boot} has to be increased as the power supply is scaled down. When power supply scaling is accompanied with device scaling, t_r improves and as a result C_{boot} can be kept small. At 3.3 V, a typical value of C_{boot} is 100 fF, while at 1.5 V, without technology scaling, it is equal to $250fF$.

The bootstrapped capacitance can be implemented using a NMOS transistor with its source and drain connected together. In this case, the capacitance is related to the area and gate oxide thickness of the MOS transistor. Simulations have shown that for 1.5 V power supply voltage, the width and length of this bootstrapped NMOS are equal to 13 μm and 6 μm, respectively. A typical area increase for a two-input NAND gate due to C_{boot} is 10%.

As shown in Fig. 5.24 of the BFBiCMOS inverter, the N-well of the PMOS devices P_p, P_1 and P_i is connected to the bootstrapped node n_1. This prevents their source/drain-well junctions to turn ON during the bootstrapping cycle. Also, it prevents any latch-up which might be caused by the parasitic SRC when the drain/source-well voltages are forward-biased. The PMOS transistor P_d also has its well connected to its source. This eliminates the body effect of the transistor and prevents any leakage during the bootstrapping.

5.3.4.3 BiNMOS Configuration

Fig. 5.27 shows the BiNMOS version of the bootstrapped circuit. The pull-down section uses an NMOS transistor (N_1) as CMOS.

The pull-up section of this BFBiNMOS configuration is slightly different than BFBiCMOS, where a small-size PMOS transistor (P_f) is added. Without this PMOS device, the base-emitter voltage of Q_1 would be equal to V_{BEon} when the output reachs V_{DD}. For low output load, if the input goes to high, the pull-down NMOS device, N_1, discharges the output faster than the PMOS transistor P_d does for the base. Thus, the bipolar transistor Q_1 can turn ON to supply the output. This results in a high fall time delay. The added small-size PMOS transistor, P_f, in the pull-up section solves this problem. It permits, through the use of inverter I_3, to set the voltage of nodes n_1 and B_1 to V_{DD} at the end of the bootstrapping. Hence, the base-emitter voltage of Q_1 is almost equal to zero at the end of the bootstrapping. When the base is discharged from

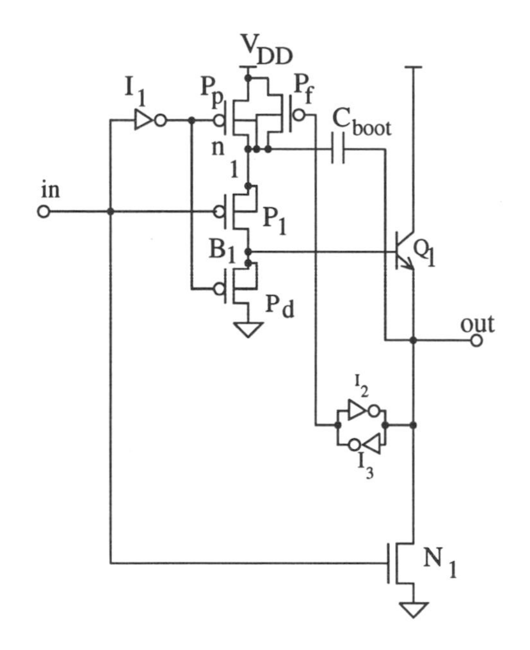

Figure 5.27 Bootstrapped full-swing BiNMOS inverter (BFBiNMOS).

$(V_{DD} + V_{BEon})$ to V_{DD} by the PMOS P_f, inverter I_2 holds the output level at V_{DD}. Without this inverter, the output falls down to a level equal to (V_{DD} - V_{BE}) due to the base-emitter coupling capacitance. The simulated waveforms of the different voltages are shown in Fig. 5.28.

For an n-input gate implementation, the BFBiNMOS requires $4n$ input transistors. Whereas, the BFBiCMOS and the BS-BiCMOS require $5n$ and $6n$ input transistors, respectively. The crossover load capacitance represents one of the important parameters in circuit comparison. It is a measure of the load where BiCMOS circuits start to have speed advantage over that of CMOS. In the range 1.2-3.3 V, BFBiCMOS/BFBiNMOS circuits require almost an equivalent minimum fanout of 5. The BS-BiCMOS have a higher crossover capacitance.

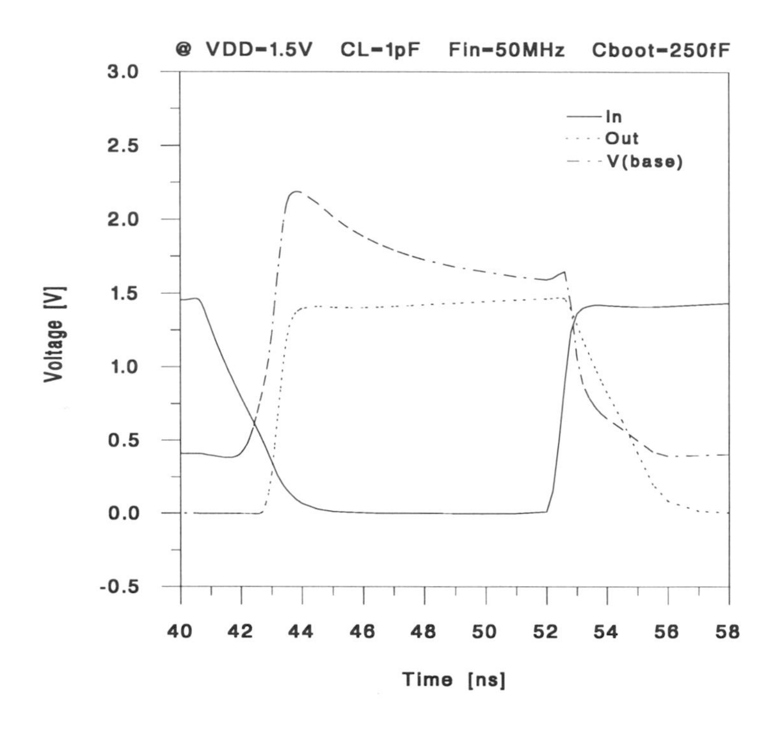

Figure 5.28 Voltage waveforms of the input (in), the output (out), and the base of Q_1.

5.3.5 Comparison of BiCMOS Logic Circuits

In this section, a brief comparison of several BiCMOS logic circuits is presented using a generic 0.35 μm BiCMOS technology given in Table 5.3. For more detailed comparison, the reader can refer to [25].

Two-inputs NAND gate configuration was chosen to evaluate and compare the performance of the circuits shown in Fig 5.29. The logic families compared are: CMOS [Fig. 5.29(a)], PBiNMOS [Fig. 5.29(b)], TS-FS [Fig. 5.29(c)], BS-BiCMOS [Fig. 5.29(d)], BFBiNMOS [Fig. 5.29(e)], and BFBiCMOS [Fig.

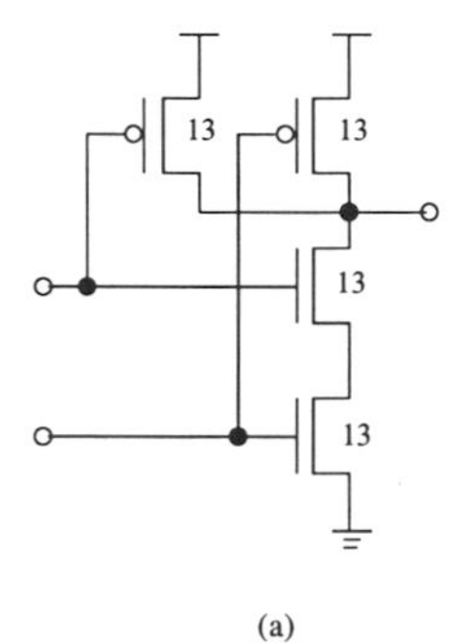

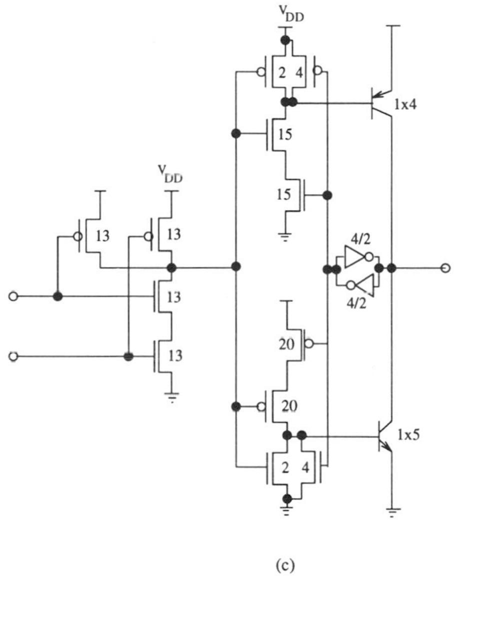

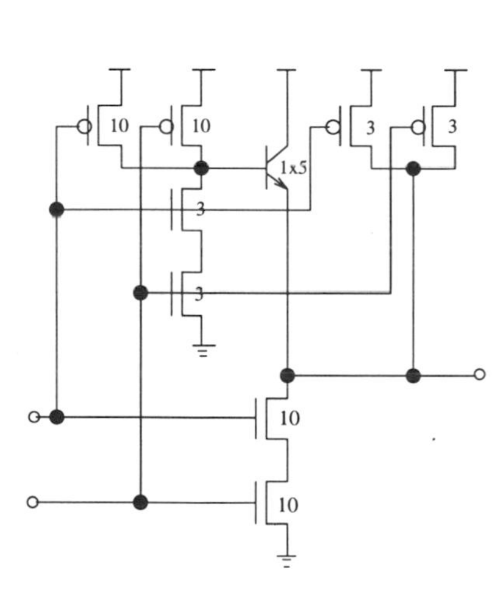

Figure 5.29 Two-input NAND gates: (a) CMOS; (b) PBiNMOS; (c) TS-FS; (d) BS-BiCMOS; (e) BFBiNMOS; (f) BFBiCMOS.

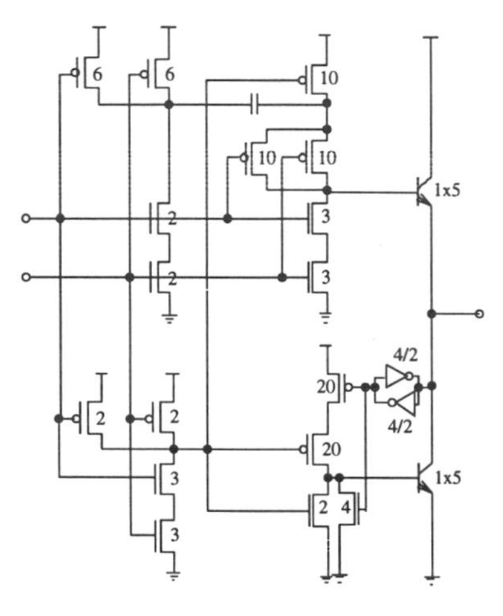

(d)

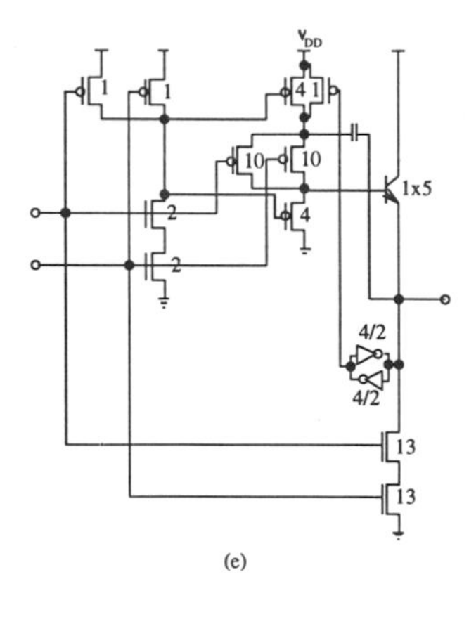

(e)

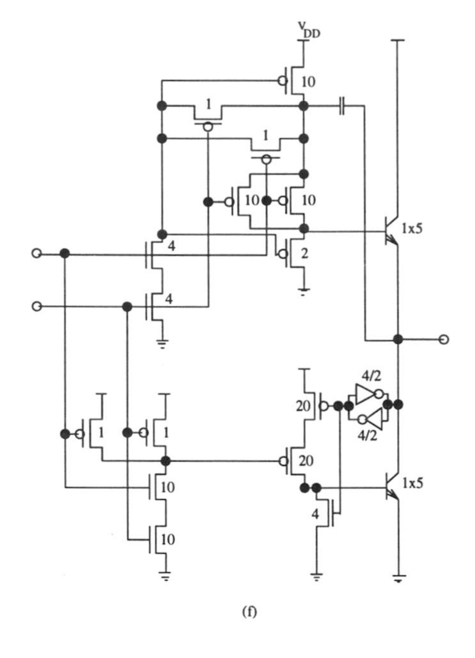

(f)

Figure 5.29 *(continued)*

Table 5.3 Key device parameters for 0.35 μm BiCMOS PROCESS.

MOS	NMOS	PMOS
L_G	0.35μm	0.35μm
L_{eff}	0.23μm	0.34μm
I_{ds}	4.9 mA	2.4 mA
	@ $V_{DS} = V_{GS}$	= 3.3V, W = 10 μm
T_{ox}	90 nm	90 nm
V_T	0.55V	-0.65V
Bipolar	NPN	PNP
A_E	1x5 μm^2	1x4 μm^2
β_f	90	90
τ_f	7 ps	21 ps
C_{je}	16 fF	28 fF
C_{jc}	17 fF	28 fF
C_{js}	52 fF	73 fF
R_c	30 Ω	37 Ω
R_c	28 Ω	31 Ω
R_b	265 Ω	260 Ω

5.29(f)]. The simulations were carried out using a chain of gates. The reported 50% delay times are those of an intermediate gate.

Table 5.4 shows the delay, the average power dissipation and the power-delay product of the different NAND gates at two supplies; 3.3 and 1.5V. The simulation was carried out at a typical load capacitance of 1 pF.

The bootstrapped family consumes more power than CMOS because of the higher internal node capacitance. However, they provide a high speed of operation, particularly the BFBiCMOS, where it has a factor of 3 speed advantage compared to CMOS at 1.5 V. Moreover, the delay-power product of the bootstrapped family is lower than that of CMOS. Notice that at 3.3 V, PBiNMOS has the lowest delay-power product and less delay than CMOS. BiNMOS at 1.5 V is slower than CMOS and is not reported in the table. These results also indicate that the use of the bootstrapped BiCMOS/BiNMOS gate would improve the delay-power product when V_{DD} is scaled down to 1.5 V.

Table 5.4 Delay, power and delay-power product of different BiCMOS NAND2 gates @ 3.3 V and $C_L = 1\ pF$

Logic Type	Delay (ps)	Power ($\mu W/MHz$)	Delay×Power (fJ/MHz)
CMOS	754	12.8	9.7
PBiNMOS	512	12.9	6.6
TS-FS	365	28.2	10.3
BS-BiCMOS	373	20.0	7.5
BFBiNMOS	481	18.5	8.9
BFBiCMOS	289	26.4	7.6

Table 5.4 *(continued)*
Delay, power and delay-power product of different BiCMOS NAND2 gates @ 1.5 V and $C_L = 1\ pF$.

Logic Type	Delay (ps)	Power ($\mu W/MHz$)	Delay×Power (fJ/MHz)
CMOS	1994	2.50	5.0
TS-FS	793	4.94	3.9
BS-BiCMOS	962	3.84	3.7
BFBiNMOS	1175	4.60	3.2
BFBiCMOS	686	3.50	4.1

5.3.6 Conclusion

We have demonstrated, during all the previous sections, that the best family to use for a fanout higher than 5, is the bootstrapped BiCMOS for the range of power supply 1-to-3.3 V. However, due to its higher area occupied, it can be used mainly in high-speed digital applications. Note, when the load is large, in the range of $\sim 1\ pF$, the bootstrapped family provides a high speed and a good delay-power product. One drawback of this family, beside the large area, is that the bootstrapping is sensitive to the shape of the input voltage. One practical gate which can be used in several applications, even when the fanout is low, is the BiNMOS family. It has good performance for 3.3 and 2.5 V power supplies. Also it provides a better delay-product than CMOS. In the next section, many digital applications based on BiNMOS family are outlined.

5.4 LOW-VOLTAGE BICMOS APPLICATIONS

In this section, we present the applications of BiCMOS digital circuits in the implementation of digital building blocks, microprocessors, memories, digital signal processors, and gate arrays. BiNMOS family and its utilization in practical design at 3.3 V is emphasized. Many of the circuits cited are discussed in detail in Chapters 4, 6 and 7.

5.4.1 Microprocessors and Logic Circuits

BiNMOS logic have been used in several microprocessors [26, 27]. In this application, BiNMOS can be used in critical path delay reduction without increasing chip area since BiNMOS needs a low-fanout to outperform CMOS. Among the critical paths, we cite

- Decoders in the register file and the cache memory;
- Sense amplifiers and output buffers in the register file and the cache;
- Booth's encoder, Wallace tree, and the final adder in a multiplier;
- Arithmetic and logic unit in a microprocessor data path; and
- Critical path of the control unit.

In the microprocessor of [26], the PBiNMOS logic family is used at 3.3 V power supply. The critical path of the control unit is reduced by 35% over CMOS. The BiNMOS gates keep their speed advantage even in the worst case ($V_{DD} = 2.7\ V$ and $T = 125\ C$).

BiCMOS logic is not only limited to conventional gates, but many other logics can be devised. One such example is the pass-transistor BiNMOS used in the design of a 64-bit adder [28] similar to the CMOS CPL logic family discussed in Chapter 4. Fig. 5.30 shows an exclusive OR/NOR gate using the pass-transistor BiNMOS gate (abbreviated PT-BiCMOS) using double rail. The outputs of the pass-transistor network are connected to the bases of the bipolar transistors Q_1 and Q_2 to reduce the intrinsic delay. The PMOS transistors P_1 and P_3 are cross-coupled to restore the high level of the pass logic to full V_{DD}. The PMOS transistors, P_2 and P_4, charge the output to full-swing. These transistors are subject to body effect, hence they turn ON later during transitions.

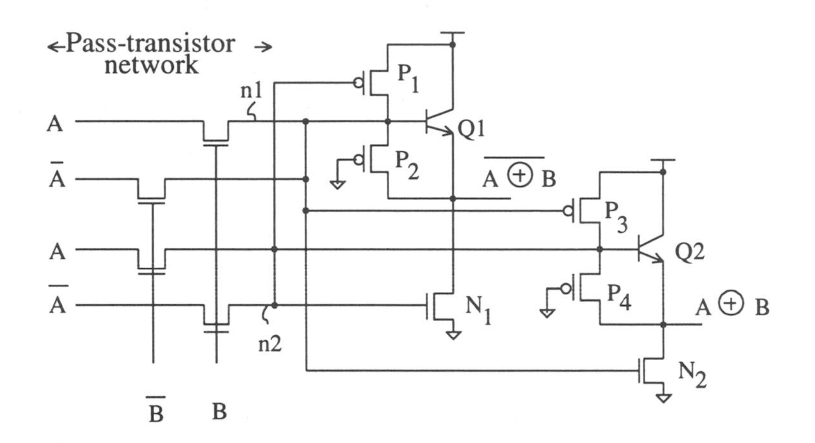

Figure 5.30 Exclusive OR and NOR gate using pass-transistor BiCMOS.

Fig. 5.31(a) compares the delay of exclusive OR and NOR gates using PT-BiCMOS, TG-type CMOS, and CPL-type CMOS using 0.5 μm BiCMOS process at 3.3 V power supply voltage. The fanout=1 is equivalent to a capacitance of 35 fF. The PT-BiCMOS gate is faster than the CMOS gates for any fanout. The power-delay product is also shown in Fig. 5.31(b). The TG gate has the best delay-power product for a fanout lower than 3. However, for a fanout greater than 3, the PT-BiCMOS gate is better.

This PT-BiCMOS has been used in the design of a 64-bit adder [28]. It is used mainly in the P, sum and carry blocks. A delay time of 3.5 ns was obtained for the 64-bit adder at 3.3 V, which is 25% better than the CMOS version. The area and power dissipation penalties of the PT-BiCMOS adder, compared to the CMOS, were 13% and 14% respectively. The speed advantage is kept down to almost 2 V.

5.4.2 Random Access Memories (RAMs)

One of the largest applications of BiCMOS is in RAM design, particularly Static RAMs (SRAMs). The first BiCMOS SRAM was proposed in 1985 [29], then many BiCMOS SRAMs were reported [30, 31, 32, 33, 34, 35, 36, 37]. The major applications of fast BiCMOS SRAMs are cache for workstations and main memory for super computers. Many BiCMOS SRAMs are in production

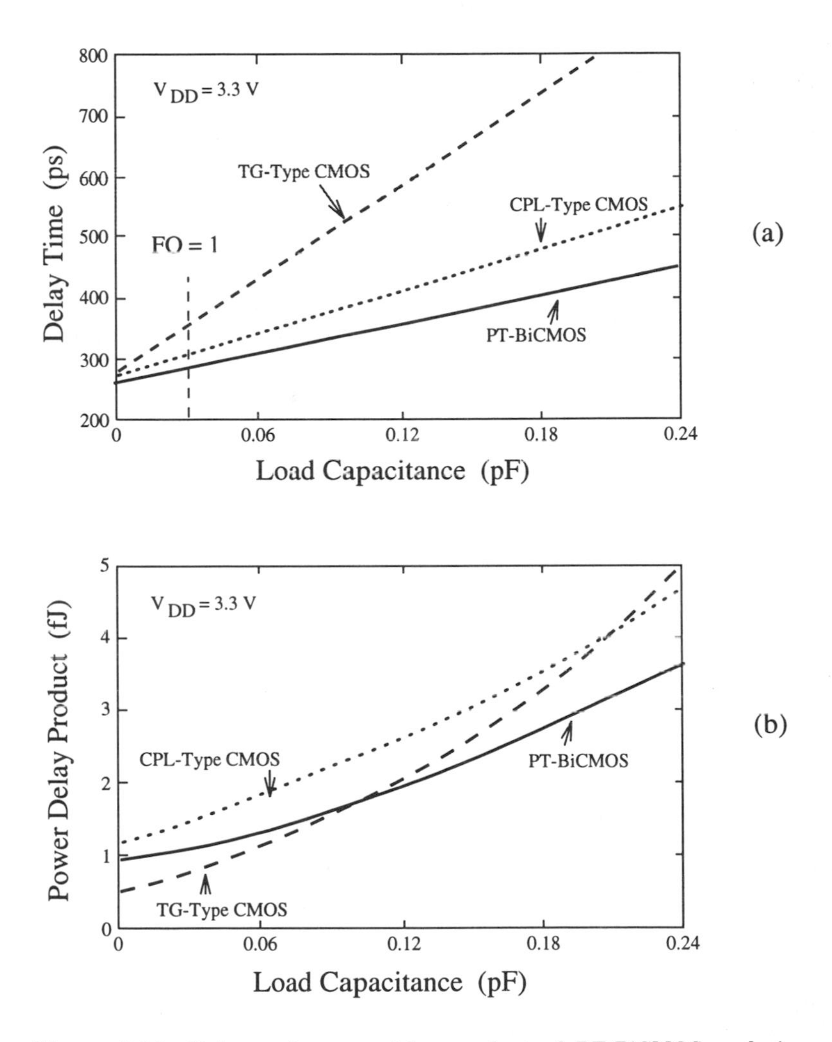

Figure 5.31 Delay and power delay product of PT-BiCMOS exclusive OR/NOR gate [28].

for capacities ranging from 64-Kb to 4-Mb. The BiCMOS SRAMs result in fast speed of operation, compared to CMOS SRAMs, however, they consume a high power dissipation due to the use of some ECL circuits. The BiCMOS SRAMs also provide ECL interfaces. A typical SRAM architecture with ECL interfaces is shown in Fig. 5.32. The memory array is implemented in CMOS for high-density purposes.

In general the BiCMOS circuit techniques are used in the periphery of the memory, such as:

- ECL interface I/O buffers for address and data;
- Input address predecoder. In general, the predecoder is implemented in ECL circuit for the input address interface;
- ECL-CMOS level converters for address and data;
- Word-line decoder and driver;
- Column decoder and driver;
- Sense amplifier in the read path;
- Main amplifier in the read path; and
- Write circuitry.

The above cited circuits represent the major BiCMOS parts in a ECL SRAM. Even if active pull-down techniques are used in the state of the art SRAMs, for high-speed and low-power dissipation, still the power of SRAM's is relatively high.

While most of the SRAMs products are asynchronous, it is clear that synchronous SRAM's development is growing fast. Recently, a 220-MHz pipelined 16-Mb SRAM has been reported in 0.4 μm BiCMOS technology [37]. It uses the BiNMOS logic family at 3.3 V supply. The power dissipation is 1 W at 220 MHz with GTL I/O interfaces. The application of such an SRAM is for synchronous secondary cache memory.

For DRAMs, the application of BiCMOS technology has allowed higher speed of operation [38, 39, 40]. However, the application of BiCMOS to DRAMs was very limited due to several factors: i) low-power requirements, and ii) process

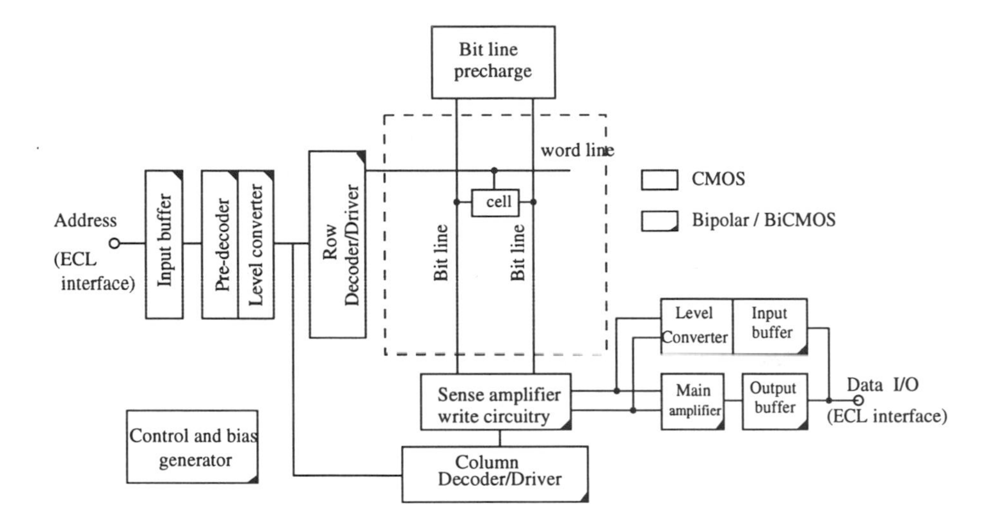

Figure 5.32 State-of-the-art BiCMOS SRAM architecture.

complexity. BiCMOS was limited to some periphery circuits due to layout-pitch matching. It was used in the I/O buffers, decoder and drivers, main sense amplifier and voltage down converter. In general BiCMOS SRAMs and DRAMs are not suitable for low-power applications.

5.4.3 Digital Signal Processors

High-performance DSPs are needed in many applications such as video signal processors, convolvers, filters, etc. BiCMOS technology has been used successfully in DSPs operating at a frequency of 300 MHz [41, 42]. These DSPs operate at 3.3 V power supply voltage using BiNMOS logic family. Among the characteristics of these BiCMOS DSPs, we cite:

- Parallel, pipelined architecture;
- High-performance and high density of integration; In this case, critical data-path functional blocks are customized; and

- BiNMOS is used in the blocks such as: SRAM, ROM (Read Only Memory), ALU (Arithmetic Logic Unit), multiplier, and clock driver, etc.

Fig. 5.33 shows a block diagram of a DSP [41]. This architecture can process any signal processing operation. The BiNMOS inverters are used as clock buffers to reduce the clock skew at 300 MHz clock frequency. The clock is distributed to about 1000 registers. High clock frequency increases drastically power and reduces the power supply voltage due to the power noise (effect of high dissipated current). The BiNMOS inverter, used in the clock distribution, is the conventional one which has a high level of $V_{DD} - V_{BE}$. Hence, the dynamic power of the clock network is reduced by 17% compared to CMOS when using BiNMOS.

Also the BiNMOS logic is used as:

- Output buffer of the Booth encoder of the convolver/multiplier block;
- Decoder driver of the register file; and
- Other drivers.

5.4.4 Gate Arrays

Gate arrays became very popular for a wide spectrum of applications because of their low cost and short turn-around time. Gate array chips consist of a large number of identical sites or basic cells which are usually placed in rows. The rows are separated by routing channels. The core of rows and channels is surrounded by I/O cells at the chip periphery as illustrated in Fig. 5.34.

Each of the basic cells is typically made up of a number of transistors which can be connected to form a two input NAND or NOR gate or a simple latch. The only processing step that can be customized is the metallization. The user of a gate array can implement the system by specifying the required connections between the devices in each cell and then the connection between the various cells. This is done automatically using CAD tools. The number of metal levels used for wiring varies from 2 to 4. The first one or two levels are used for internal wiring of the cell and the upper levels (e.g. third and fourth) for wiring between the cells in the horizontal and vertical directions [43].

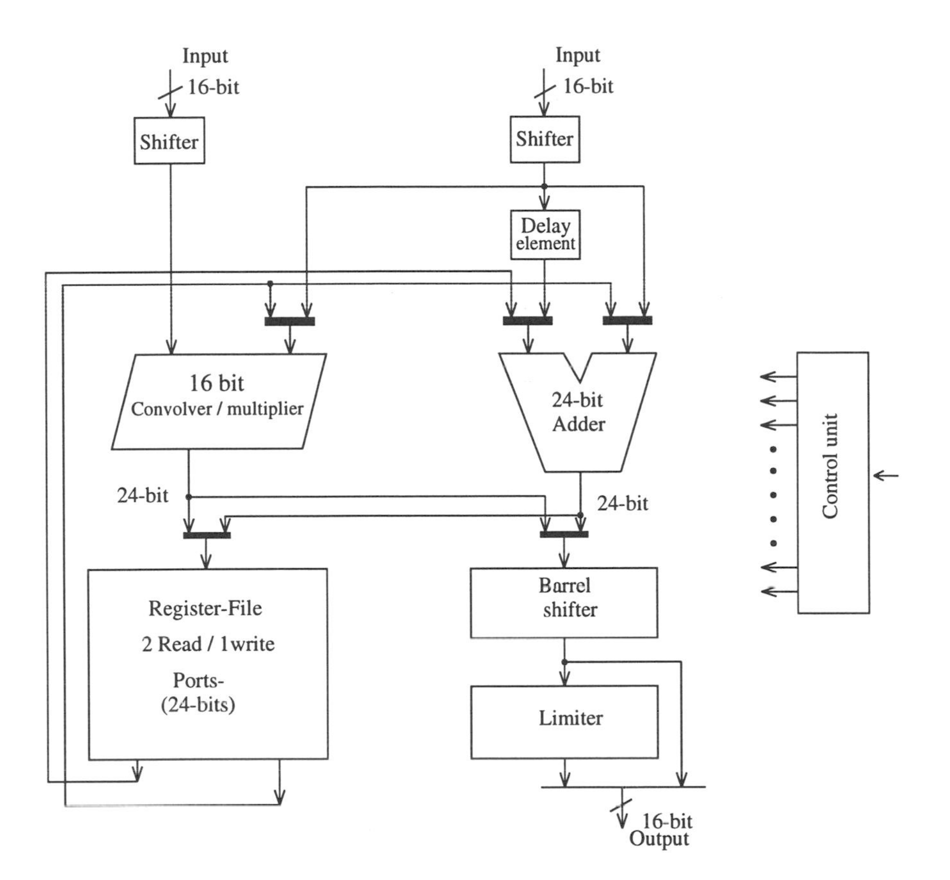

Figure 5.33 Block diagram of a DSP.

BiCMOS technology has been used extensively for building gate arrays and channelless gate arrays (sea-of-gates) [43, 44, 45, 46]. At 3.3 V power supply voltage, BiNMOS logic family has been used [10, 11]. In [11], BiPNMOS logic gate has been proposed for the Channelless gate array. Fig. 5.35 shows a layout of a BiPNMOS basic cell on 0.5 μm BiCMOS technology. A bipolar transistor and a small size MOS transistor are added to the pure CMOS basic cell. These transistors are not only used to implement BiPNMOS gates but also flip-flops, memory macros (RAM, ROM, and CAM), etc. A BiPNMOS two-input NAND gate has 35% delay reduction compared to a similar CMOS gate for a fanout of 7. The speed advantage is maintained down to 2.5 V.

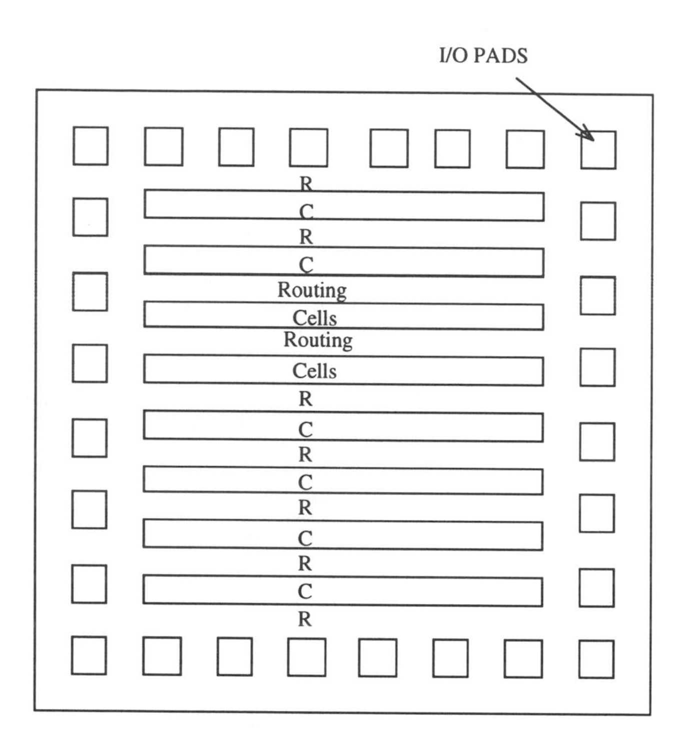

Figure 5.34 Gate Array chip floorplan.

5.4.5 Application Specific ICs (ASICs)

In order to realize high-performance ASICs, fast standard cell library macros for rapid design are important. This library contains custom functional macros such as: adder, Programmable Logic Array (PLA), register file, RAM, cache, Table Look-aside Buffer (TLB), and controller, etc. PBiNMOS logic has been used for such a standard cell library [12]. The cells of logic gates are designed in CMOS and PBiNMOS for the same logic functions. The PBiNMOS gates are used for a relatively high fanout and load, whereas CMOS gates are used for a small fanout. A CAD tool can be utilized to choose the most appropriate cells in the design.

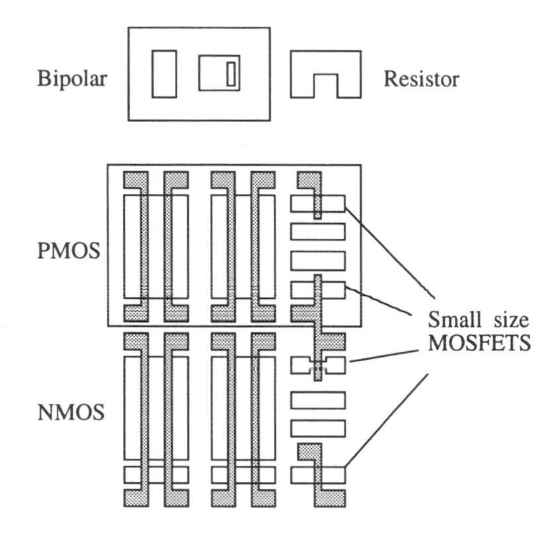

Figure 5.35 Layout of a BiNMOS gate array cell.

5.5 CHAPTER SUMMARY

In this chapter, we have demonstrated the advantage of using BiCMOS over CMOS in terms of speed. We have shown the historical evolution of the different BiCMOS logic families. A variety of alternative circuit techniques for low-voltage operation have been outlined and compared to the conventional BiCMOS. Also we have shown how optimized BiNMOS are faster than CMOS even if the fanout is low (greater than 1). The design techniques can be extended to more complex gates and building blocks such as flip-flops, and adders, etc. Variety of applications where BiCMOS, particularly BiNMOS can be used at low-voltage are reviewed. The addition of the bipolar to CMOS to devise new structures enhances the performance of ICs. This feature improves the access time of memories, register files, ALUs, DSPs, etc. Notice that a large portion of a BiCMOS IC is implemented in CMOS, while bipolar transistors represent a small portion (0.5-4%) for driving or sensing purposes. The power dissipation of BiCMOS circuits, compared to their CMOS counterparts, increases drastically if ECL is used because of the DC current. However, if only BiCMOS logic gates are used, the power increase is not significant compared to speed enhancement. In some cases, like clock distribution network, the power dissipation is reduced when using BiNMOS.

REFERENCES

[1] A. R. Alvarez, "BiCMOS Technology and Applications," Kluwer Academic Pub., MA, Second Edition, 1993.

[2] S. H. K. Embabi, A. Bellaouar and M. I. Elmasry, "BiCMOS Digital Integrated Circuit Design", Kluwer Academic Pub., MA, 1993.

[3] M. I. Elmasry, "Design and Analysis of BiCMOS ICs", IEEE Press, 1994.

[4] G. P. Rosseel, and R. W. Dutton, "Influence of Device Parameters on the Switching Speed of BiCMOS Buffers," IEEE Journal of Solid-State Circuits, vol. 24, no. 1, pp. 90-99, February 1989.

[5] P. Raje, K. Chan, and K. Saraswat, "BiCMOS Gate Performance Optimization using Unified Delay Model," Symposium on VLSI Technology, Tech. Dig., pp. 91-92, 1990.

[6] S. H. K. Embabi, A. Bellaouar, and M. I. Elmasry, "Analysis and Optimization of BiCMOS Digital Circuit Structures," IEEE Journal of Solid-State Circuits, vol. 26, no. 4, pp. 676-679, April 1991.

[7] P. A. Raje, K. C. Saraswat and K. M. Cham, "Performance-driven Scaling of BiCMOS Technology", IEEE Trans. on Electron Devices, ED-39, no. 3, pp. 685-693, March 1992.

[8] J. Gallia, et al., "High-Performance BiCMOS 100K-Gate Array," IEEE Journal of Solid-State Circuits, vol. 25, no. 1, pp. 142-149, February 1990.

[9] Y. Nishio, et al., "A BiCMOS Logic Gate with Positive Feedback," International Solid-State Circuits Conference, Tech. Dig., pp. 116-117, February 1989.

[10] A. E. Gamal et al., "BiNMOS a Basic Cell for BiCMOS Logic Circuits", in Custom Integrated Circuits Conf., Tech. Dig., pp. 8.3.1-8.3.4., 1989.

[11] H. Hara et al., "0.5-um 2M-Transistor BiPNMOS Channelless Gate Array", IEEE Journal Solid-State Circuits, vol. 26, no. 11, pp. 1615-1620, November 1991.

[12] H. Hara et al., "0.5-um 3.3-V BiCMOS Standard Cells with 32-kb Cache and Ten-Port Register File", IEEE Journal Solid-State Circuits, vol. 27, no. 11, pp. 1579-1584, November 1992.

[13] M. I. Elmasry, and A. Bellaouar, "BiCMOS at Low-Supply Voltage," in IEEE Bipolar/BiCMOS Circuits and Technology Meeting, pp. 89-96, October 1993.

[14] P. Raje, et al., "MBiCMOS: A Device and Circuit Technique for Submicron, sub-2 V Regime," International Solid-State Circuits Conference, Tech. Dig., pp. 150-151, 1991.

[15] P. G. Y. Tsui et al., "Study of BiCMOS Logic Gate Configurations for Improved Low-Voltage Performance", IEEE Journal Solid-State Circuits, vol. 28, no. 3, pp. 371-374, March 1993.

[16] S. W. Sun et al., "A Fully Complementary BiCMOS Technology for Sub-Half-Micrometer Microprocessor Applications", IEEE Trans. Electron Devices, vol. 39, no. 12, pp. 2733-2739, December 1992.

[17] K. Yano et al., "Quasi-Complementary BiCMOS for Sub-3V Digital Circuits", IEEE Journal Solid-State Circuits, vol. 26, no. 11, pp. 1708-1719, November 1991.

[18] A. Watanabe et al., "Future BiCMOS Technologies for Scaled Supply Voltage", International Electron Devices Meeting, Tech. Dig., pp. 429-433, December 1989.

[19] H. J. Shin et al., "Full-swing CBiCMOS Logic Circuits", in IEEE Bipolar/BiCMOS Circuits and Technology Meeting, Tech. Dig. pp. 229-233, September 1989.

[20] A. Bellaouar, I. S. Abu-Khater, M. I. Elmasry, and A. Chekima, "Full-Swing Schottky BiCMOS/BiNMOS and the Effects of Operating Frequency and Supply Voltage Scaling," IEEE Journal of Solid-State Circuits, vol. 29, no. 6, pp. 693-700, June 1994.

[21] S. H. K. Embabi, A. Bellaouar, M. I. Elmasry, and R. A. Hadaway, "New Full-Voltage-Swing BiCMOS Buffers", IEEE Journal Solid-State Circuits, vol. 26, no. 2, pp. 150-153, February 1991.

[22] M. Hiraki et al., "A 1.5-V Full-Swing BiCMOS Logic Circuit", IEEE Journal Solid-State Circuits, vol. 27, no. 11, pp. 1568-1574, November 1992.

[23] R. Y. V. Chik and C. A. T. Salama, "1.5 V Bootstrapped BiCMOS Logic Gate", IEE Electronic Letters, Vol. 29, No. 3, pp. 307-309, February 1993.

[24] S. H. K. Embabi, A. Bellaouar, and K. Islam, "A Bootstrapped Bipolar CMOS (B^2CMOS) Gate for Low Voltage Applications," IEEE Journal of Solid-State Circuits, vol. 30, no. 1, pp. 47-53, January 1995.

[25] A. Bellaouar, M. I. Elmasry, and S. H. K. Embabi, " Bootstrapped Full-Swing BiCMOS/BiNMOS Logic Circuits for 1.2-3.3 V Supply Voltage Regime," IEEE Journal of Solid-State Circuits, vol. 30, no. 6, June 1995.

[26] J. Shutz, "A 3.3 V 0.6 μm BiCMOS Superscalar Microprocessor," IEEE International Solid-State Circuits Conference, Tech. Dig., pp. 202-203, 1994.

[27] F. Murabayashi, et al., "3.3 V, Novel Circuit Techniques for a 2.8-Million-Transistor BiCMOS RISC Microprocessor," IEEE Custom Integrated Circuit Conference, Tech. Dig., pp. 12.1.1-12.1.4, May 1993.

[28] K. Ueda, H. Suziki, K. Suda, Y. Tasujihashi, H. Shinohara, "A 64-bit Adder By Pass Transistor BiCMOS Circuit," IEEE Custom Integrated Circuit Conference, Tech. Dig., pp. 12.2.1-12.2.4, May 1993.

[29] K. Ogiue, et al., "A 15 ns/250 mW 64K Static RAM," in ICCD, Tech. Dig., pp. 17-20, 1985.

[30] H. Tran et al., "An 8-ns 1-Mb ECL BiCMOS SRAM with a Configurable Memory Array Size," International Solid-State Circuits Conf. Tech. Dig., pp. 36-37, February 1989.

[31] M. Matsui et al., "An 8-ns 1-Mb ECL BiCMOS SRAM," International Solid-State Circuits Conf., Tech. Dig., pp. 38-39, February 1989.

[32] Y. Maki et al., "A 6.5-ns 1 Mb BiCMOS ECL SRAM," International Solid-State Circuits Conf. Tech. Dig., pp. 136-137, February 1990.

[33] M. Takada et al., "A 5-ns 1-Mb ECL BiCMOS SRAM," IEEE Journal of Solid State Circuits, vol. 25, no. 5, pp. 1057-1062, October 1990.

[34] A. Ohba et al., "A 7-ns 1-Mb BiCMOS ECL SRAM with Program-Free Redundancy," in Symp. VLSI Circuits Conf. Tech. Dig., pp. 41-42, May 1990.

[35] Y. Okajima et al., "A 7-ns 4-Mb BiCMOS SRAM with a Parallel Testing Circuit," International Solid-State Circuits Conf. Tech. Dig., pp. 54-55, February 1991.

[36] N. Tamba et al., "A 1.5 ns 256Kb BiCMOS SRAM with 11K 60 ps Logic Gates," International Solid-State Circuits Conf., Tech. Dig., pp. 246-247, February 1993.

[37] K. Nakamura et al., "A 200-MHz Pipelined 16-Mb BiCMOS SRAM with PLL Proportional Self-Timing Generator," IEEE Journal of Solid-State Circuits, vol. 29, no. 11, pp. 1317-1322, November 1994.

[38] G. Kitsukawa, et al., "An Experimental 1-Mb BiCMOS DRAM," IEEE Journal of Solid-State Circuits, vol. SC-22, no. 5, pp. 657-662, October 1987.

[39] S. Watanabe, et al., "BiCMOS Circuit Technology for High Speed DRAMs," Symposium on VLSI Circuits, Tech. Dig., pp. 79-80, 1987.

[40] G. Kitsukawa, et al., "Design of ECL 1-Mb BiCMOS DRAM," Electronics and Communications in Japan, Part 2, vol. 75, no. 5, pp. 89-102, 1992.

[41] M. Nomura et al., "A 300-MHz, 16-bit, 0.5-μm BiCMOS Digital Signal Processor Core LSI," IEEE Custom Integrated Circuits Conference, Tech. Dig., pp. 12.6.1-12.6.4, May 1993.

[42] T. Inoue, et al., "A 300-MHz 16-bit BiCMOS Video Signal Processor,", IEEE Journal of Solid-State Circuits, vol. 28, no. 12, pp. 1321-1329, December 1993.

[43] F. Murabayshi, et al., "A 0.5 micron BiCMOS Channelless Gate Array," IEEE Custom Integrated Circuits Conference, Tech. Dig., pp. 8.7.1-8.7.4, May 1989.

[44] H. Hara, et al., "A 350 ps 50K 0.8 micron BiCMOS Gate Array with Shared Bipolar Cell Structure," IEEE Custom Integrated Circuits Conference, Tech. Dig., pp. 8.5.1-8.5.4, May 1989.

[45] J. D. Gallia, et al., "High-Performance BiCMOS 100K-Gate Array," IEEE Journal of Solid-State Circuits, vol. 25, no. 1, pp. 142-149, February 1990.

[46] T. Hanibuchi, et al., "A Bipolar-PMOS Merged Basic Cell for 0.8 micron BiCMOS Sea of Gates," IEEE Journal of Solid-State Circuits, vol. 26, no. 3, pp. 427-431, March 1991.

6

LOW-POWER CMOS RANDOM ACCESS MEMORY CIRCUITS

Low-power Random Access Memory (RAM) has seen a remarkable and rapid progress in power reduction. Many circuits techniques for active and standby power reduction in static and dynamic RAMs have been devised. In this chapter we study low-power memory circuit techniques which are very interesting for several other applications. Among these circuits, we examine memory cells, sense amplifiers, precharging circuits, etc. Circuit techniques for 1.x V power supply are also discussed. The voltage targets using NiCd and Mn batteries are 1.2 and 1.5 V respectively. The minimum voltage of a NiCd cell is 0.9 V. Also we consider the Voltage Down Converters (VDCs) which are used in memories and processors. No consideration is given to the detail of designing a complete memory chip because a single configuration requires an entire book.

6.1 STATIC RAM (SRAM)

Today, workstations, computers and super computers are demanding high-speed and high-density SRAMs, e.g., cache memories. These systems started to use 4-Mb fast SRAMs and will require, in the future, larger density memories with faster access time. Many 1-to-4-Mb BiCMOS SRAMs [1, 2, 3, 4, 5, 6] have achieved access times of 5 to 10 ns. In these SRAMs, the power dissipation is 275 to 1000 mW, which is not acceptable in many applications. On the other hand, high-density, low-power SRAMs are needed for applications such as hand-held terminals, laptops, notebooks and IC memory cards. Table 6.1 shows examples for high-density SRAMs with low-power characteristics. The standby current is in the order of 1 μA and sub-μA which is suitable for battery-backup operation.

Memory size (Ref.)	Power supply	CMOS technology	Access time	Power dissipation
1-Mb [7]	3.0 V	0.35-μm	7 ns	140 mW @ 100 MHz
4-Mb [8]	5.0 V	0.50-μm	23 ns	100 mW @ 10 MHz
4-Mb [9]	3.0 V	0.60-μm	68 ns	21 mW @ 10 MHz
16-Mb [10]	2.5 V	0.25-μm	15 ns	120 mW @ 20 MHz
16-Mb [11]	3.0 V	0.40-μm	15 ns	165 mW @ 30 MHz
16-Mb [12]	3.3 V	0.35-μm	9 ns	238 mW @ 30 MHz

Table 6.1 Low-power SRAMs performance comparison.

The power dissipation reduction in SRAMs is not only due to power supply voltage reduction, but also to low-power circuit techniques. In this section we review some of these circuit techniques for low-power applications.

SRAMs have several advantages over Dynamic RAMs (DRAMs) such as:

- No refresh operation of the memory cells are needed.
- The speed of an SRAM is higher because of the differential pair of bit-lines.
- The operational modes are simpler because the row and column address signals are simultaneously loaded.
- A low data retention current which is required by battery applications.

However, SRAMs have the great disadvantage of a large memory cell compared to DRAMs. For this reason, their capacities are smaller than that of DRAMs.

6.1.1 Basics of SRAMs

In order to treat the different circuit parts of an SRAM, it is important to understand some characteristics of these memories. In general the pins of a SRAM are :

1. Addresses (A_0 ... A_n); which define the memory location;

2. Write Enable ($\overline{WE}$); which selects between the read and write modes;
3. Chip Select ($\overline{CS}$); which selects one memory out of several within a system;
4. Output Enable ($\overline{OE}$); which is used to enable the output buffer; and
5. Input/Output data (I/O).
6. Power supply pins.

A timing diagram during read cycle is shown in Fig. 6.1(a). During this time the data stored in a specific SRAM location (defined by the address) is read out. For a read cycle, two times are shown in the figure; the read cycle time, t_{RC}, and the address access time, t_{AA}. Fig. 6.1(b) shows the write cycle which permits change to the data in an SRAM. Two times are indicated, the write cycle time, t_{WC}, and the write recovery time, t_{WR}. Some of this information is used in this chapter. For more detail on the timing, the reader can refer to any memory data book.

A typical SRAM architecture is shown in Fig. 6.2. The memory array contains the memory cells which are readable and writable. The row decoder (X-decoder) selects 1 out of $n = 2^k$ rows, while the column decoder (Y-decoder) selects $l = 2^i$ out of $m = 2^j$ columns. The address (row and column) are not multiplexed as in the case of a DRAM. Sense amplifiers detect small voltage variations on the memory complementary bit-line which reduces the reading time. The conditioning circuit permits the precharge of the bit-lines. The access time is determined by the critical path from the address input to the data output as shown in Fig. 6.3. This path contains address input buffer, row decoder, memory cell array, sense amplifier and output buffer circuits. The word-line decoding and bit-lines sensing delay times are critical delay components. To reduce the sensing time during a read operation, the swing on the bit-lines should be as small as possible.

For an asynchronous[1] SRAM, a special circuit called an Address Detection Transition (ATD) permits the generation of internal pulses. These pulses are of two types; activation and equalization. Activation pulses selectively activate particular circuits, while equalization pulses permit the reduction of the delay by restoring and equalizing differential nodes prior to being selected. In this section we treat only asynchronous SRAMs.

[1] Not clocked externally.

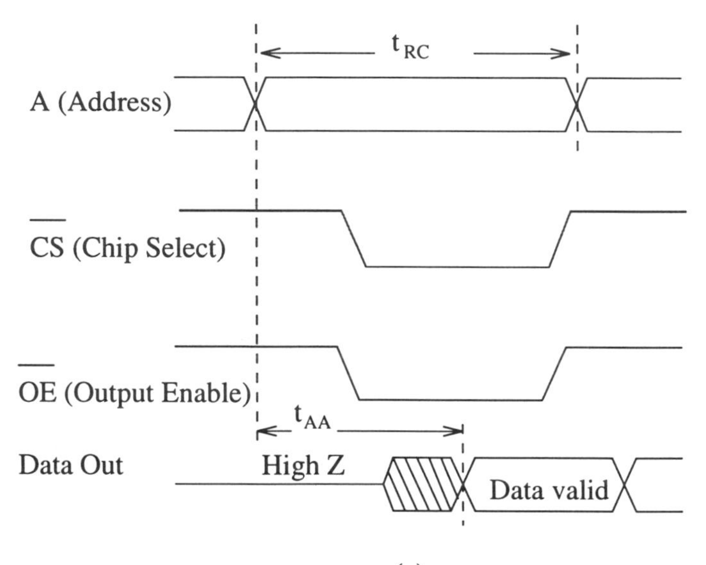

(a)

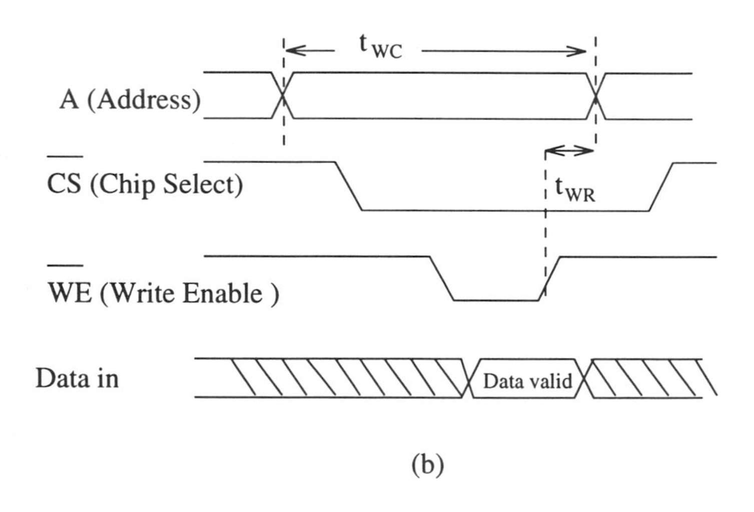

(b)

Figure 6.1 Typical timing of a SRAM: (a) read cycle; (b) write cycle.

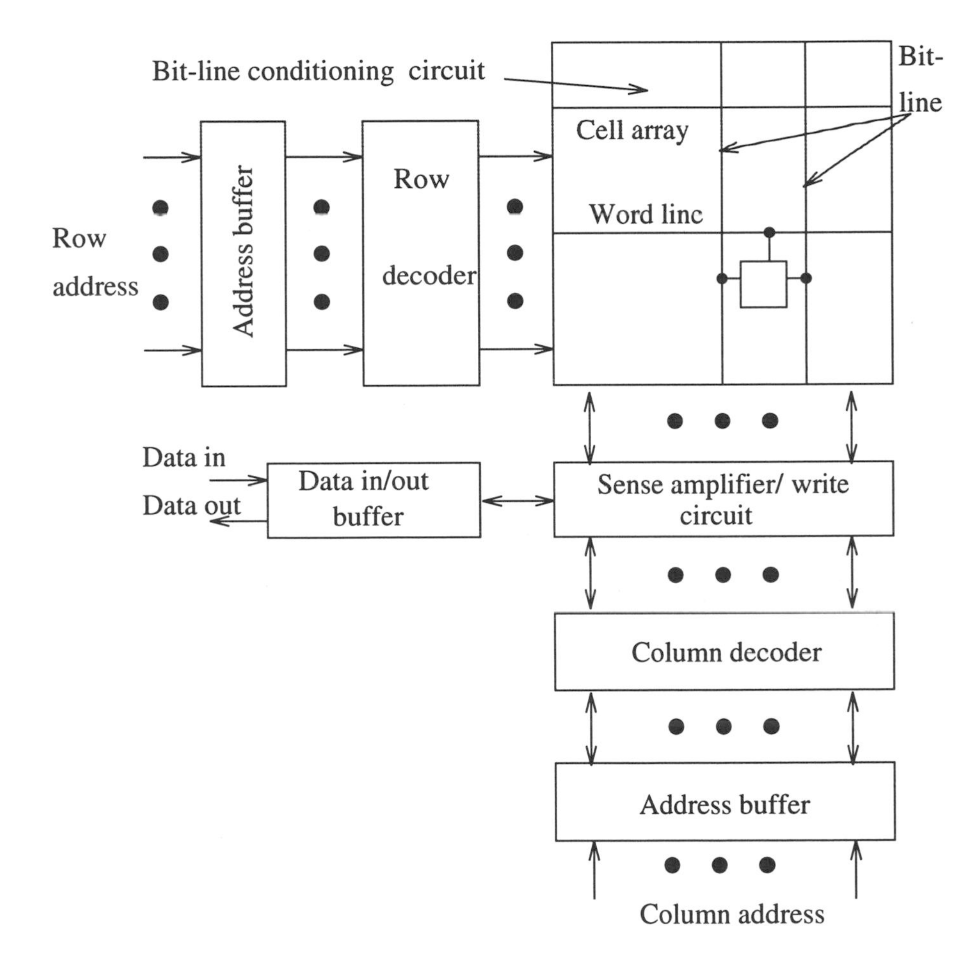

Figure 6.2 Typical SRAM architecture.

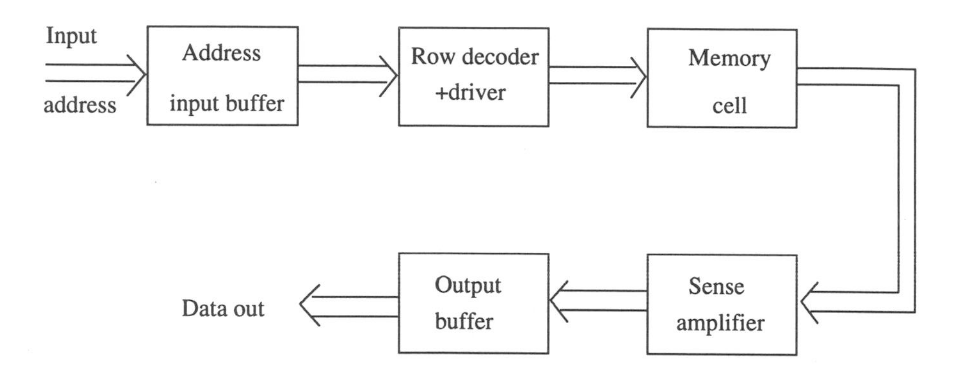

Figure 6.3 The critical path for read access in SRAM.

6.1.2 Static RAM Cells

The memory cell is an important circuit in the design of low-power and high-density SRAMs because the memory size is dominated by the cell area. There are various static memory cells. The cell of Fig. 6.4 has six transistors, in the form of two inverters, cross-coupled with two pass-transistors, connected to two complementary bit-lines BL and $\overline{BL}$. The pass-transistors are controlled by the signal WL (word-line).

During the read cycle, the bit-lines are held high (precharged). Assume that a "0" is stored at node A and a "1" is stored at node B. When the cell is selected; i.e., WL set to "1", $\overline{BL}$ is discharged through $N1$ and $N3$.

To write in the cell, one of the bit-lines is pulled low and the other high and then the cell is selected by WL. Assume that $\overline{BL}$ is set to "0" while initially a "1" is stored at node A ("0" at B). $N1$ and $P1$ should be sized such that node A is pulled down enough to turn $P2$ ON. This in turn causes node B to be pulled up. The cross-coupled inverter pair have a high gain to cause the nodes A and B to switch to opposite voltages. The data retention (standby) current of this cell can be as low as $10^{-15}A$. Although this full-CMOS cell has low retention current, the cell area is so large that it does not allow high-density SRAMs. A typical cell area using a 0.8 μm design rules is 75 μm^2.

The stability of the memory cell is its ability to hold a stable state. Fig. 6.5(a) shows the transfer curves of full CMOS SRAMs. The box between the two

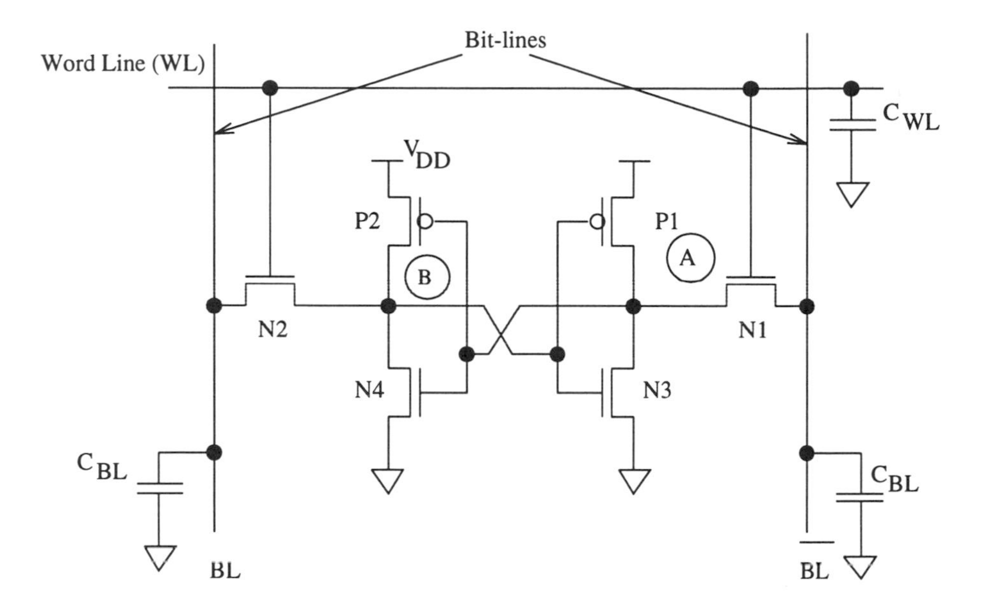

Figure 6.4 CMOS memory cell with PMOS load.

characteristics (I and II) defines the **Static Noise Margin** (SNM). Static noise is DC disturbance, such as offsets and mismatches, due to the processing and variations in process conditions. The SNM is defined as the maximum value of V_n (static noise source as shown in Fig. 6.5(b)) that can be tolerated by the cross-coupled inverters before altering state. An important parameter in SNM is the memory cell ratio, r, defined by

$$r = \frac{W_{N4}/L_{N4}}{W_{N2}/L_{N2}} \tag{6.1}$$

where transistors N_2 and N_4 are the access and driver NMOS transistors shown in Fig. 6.4. An analysis of SNM for memory cells is given in [13]. This static noise margin parameter increases with the ratio r. However, it is limited by the cell area constraint. The stability of the cell is maintained even if V_{DD} is scaled down.

Another memory cell configuration is shown in Fig. 6.6. This cell is similar to the full CMOS memory cell, except that the PMOS pull-up devices are replaced by high-resistance polysilicon loads. The memory cell area can be

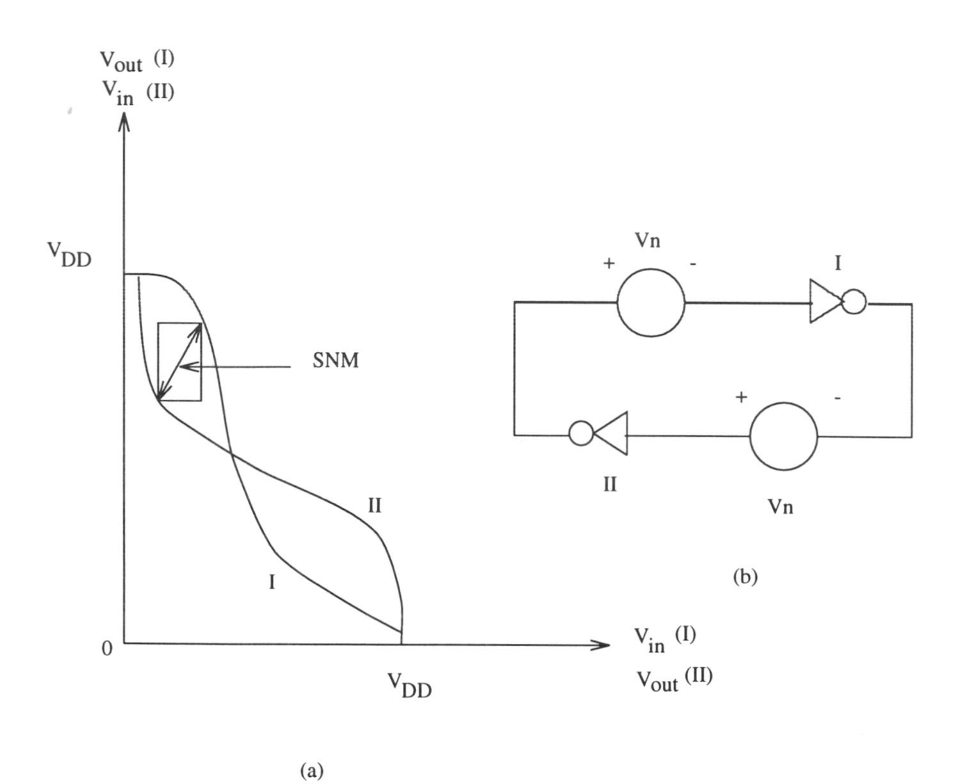

Figure 6.5 (a) Full CMOS memory cell transfer curves; (b) cross-coupled inverters with static noise.

about 30% to 40% smaller than the CMOS six-transistor memory cell, because the two polysilicon resistances can be formed on top of the two NMOS driver transistors. The High Resistive Load (HRL) memory cell has been used in several SRAM generations from 4-Kb. The high state storage node of Fig. 6.6 can be pulled down with time due to two kinds of leakage current; the leakage current of the drain junction and the subthreshold current. The voltage drop across the resistance R prevents regular cell operation, if the leakage current reaches the level of the poly-Si resistor current. In several SRAMs generations using HRL memory cell, the total standby current was set to 1 μA per chip at room temperature for battery-backup applications. Thus, for each memory generation with quadrupled density, the poly-Si resistance value is also quadrupled. For 4-Mb chip which has a total standby current less than 1 μA,

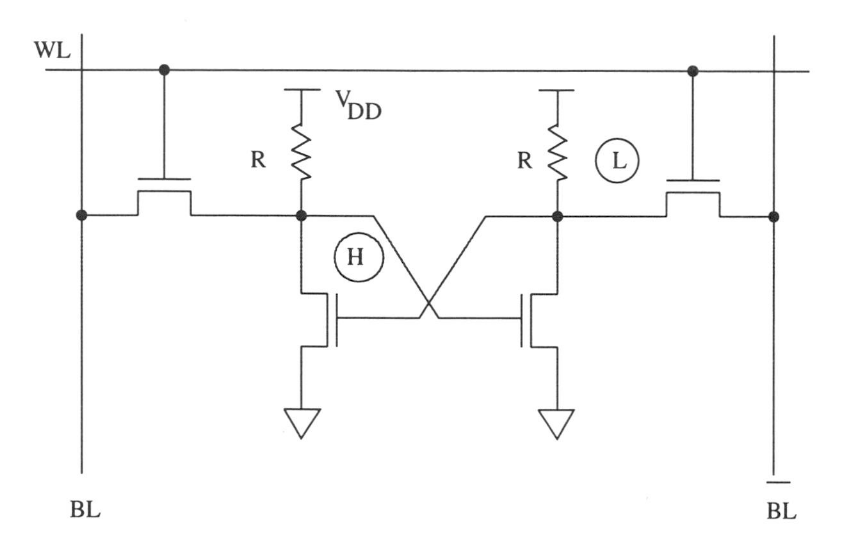

Figure 6.6 Static memory cell with high resistive load.

typical values of resistance are in the 5×10^{12} Ω range and the resistance current is limited to $10^{-13}A$. This current should be much larger than the total leakage current of the storage node of the cell to improve the data retention margin. The leakage current cannot be scaled because, first, the subthreshold current per channel width, tends to increase; particularly with the trend to decrease the threshold voltage for low-voltage. Second, the leakage current of the drain junction per area unit tends to increase with technology scaling. Moreover the junction area is shrunk with a rate lower than the SRAM density increase rate. In [14], it was determined that the maximum SRAM capacity for low-power applications, using an HRL memory cell is 4 Mb where the retention current is 1 μA.

Note that the high-level node voltages of all poly-Si load memory cells are $(V_{DD} - V_T)$ after write cycle, where V_T is the threshold voltage of the access transistor, subject to body effect. These nodes need a time of several ms to charge up to V_{DD}. The SNM of the poly-Si load memory cell is more sensitive to cell ratio r, than the full CMOS cell [13]. A typical value of r is 3. Also the cell stability is drastically degraded when V_{DD} is 3 V or less. The transfer curves in the read mode can be easily plotted for different V_{DD} to find out that the cell cannot store the data at a certain low-voltage.

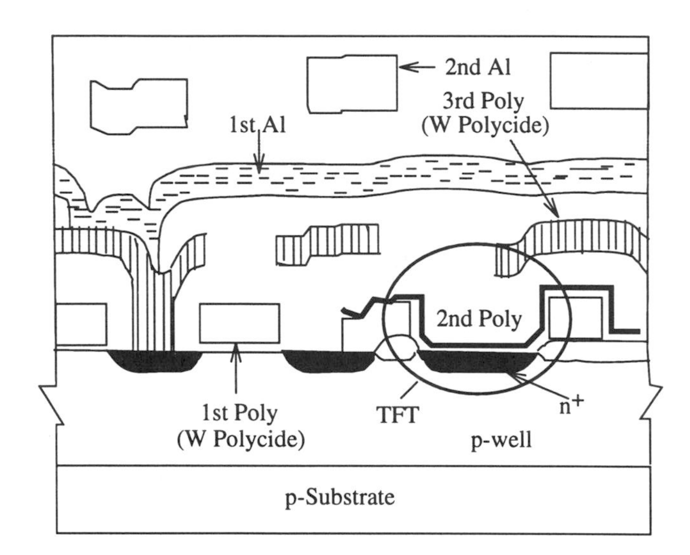

(a)

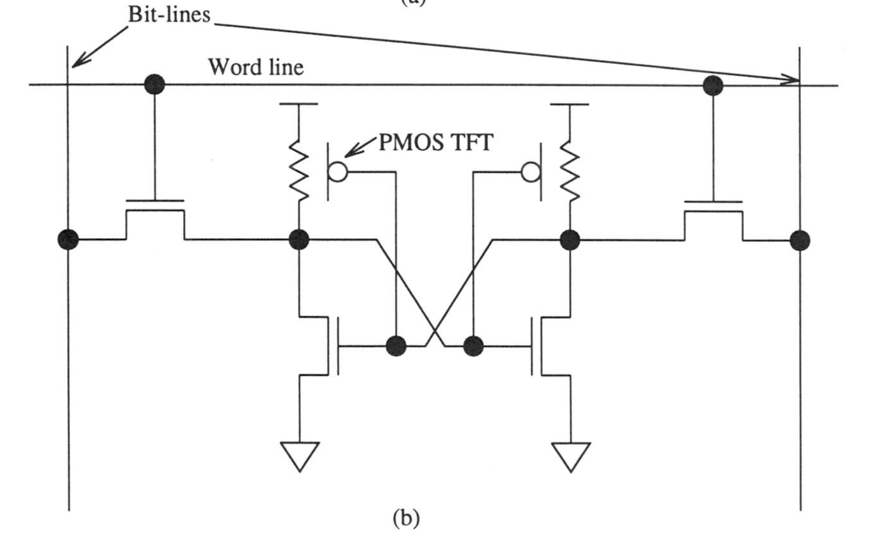

(b)

Figure 6.7 Example of thin-film PMOS transistor [8]: (a) Cross section; (b) PMOS TFT load memory cell.

For 4 Mb and higher density SRAMs, the polysilicon load cell starts to be replaced by a polysilicon PMOS load called PMOS Thin Film Transistor (TFT) for low-power applications [8, 9, 15]. Fig. 6.7 shows a cross section and circuit diagram of the poly-Si PMOS load memory cell [8]. The TFT device is fabricated from amorphous silicon ($\alpha - Si$). This material has a grain size of 2 μm while that of the conventional poly-Si material is 0.03 μm. The thickness of this $\alpha - Si$ is 100 nm and the gate oxide thickness of the TFT is 40 nm. This technology results in improved ON/OFF currents compared to the one using poly-Si. The N^+ drain area of the NMOS transistor is used as the gate electrode for the PMOS TFT. To obtain a small area, the polysilicon PMOS must be stacked on the NMOS driver. The second polysilicon layer forms the channel regions. The TFT memory cell area is more than 40% smaller than the full CMOS one.

Fig. 6.8 shows the drain current of a PMOS TFT used in a 4-Mb SRAM as a function of the gate voltage. An ON current more than $10^{-7}A$ is obtained at a supply voltage of 3 V, while an OFF current of $10^{-13}A$ is attained. The ON current is larger by more than six order of magnitude than memory cell leakage currents which is much better than the current of the HRL cell. Thus, it results in an excellent data retention characteristic. Moreover, the very low OFF current results in a standby current less than 1 μA for 4-Mb SRAM. This current is low enough for battery back-up operation. At 1.2 V power supply, the current flowing in the PMOS TFT is more than one-and-a-half order of magnitude larger than the OFF current. This demonstrates the ability of this technology for low-voltage operation.

After write cycle, the high-storage node voltage in the cell becomes $V_{DD} - V_T$. The time needed for charging up this node to V_{DD} is

$$t_{ch} = \frac{C_p V_T}{I_l} \tag{6.2}$$

where I_l is the current flowing in the load device and C_p is the total parasitic capacitance of the node. Using 4-Mb data for TFT memory cell, $V_T = 1\ V$, $C_p = 10\ fF$ and $I_l = 10\ pA$ the t_{ch} is around 1 ms. For poly-Si load this charge-up time is larger than 100 ms because I_l is as low as 0.1 pA. The average interval time between two word-line selections (for the same word-line) is given by

$$t_i = \frac{N t_{cycle}}{M} \tag{6.3}$$

where N is the number of memory cells per SRAM chip, M is the number of memory cells per word-line, and t_{cycle} (or noted t_{RC}) is the operating cycle time. For 4-Mb, a typical value of t_i is 4.5 ms when the cycle time is 70 ns and

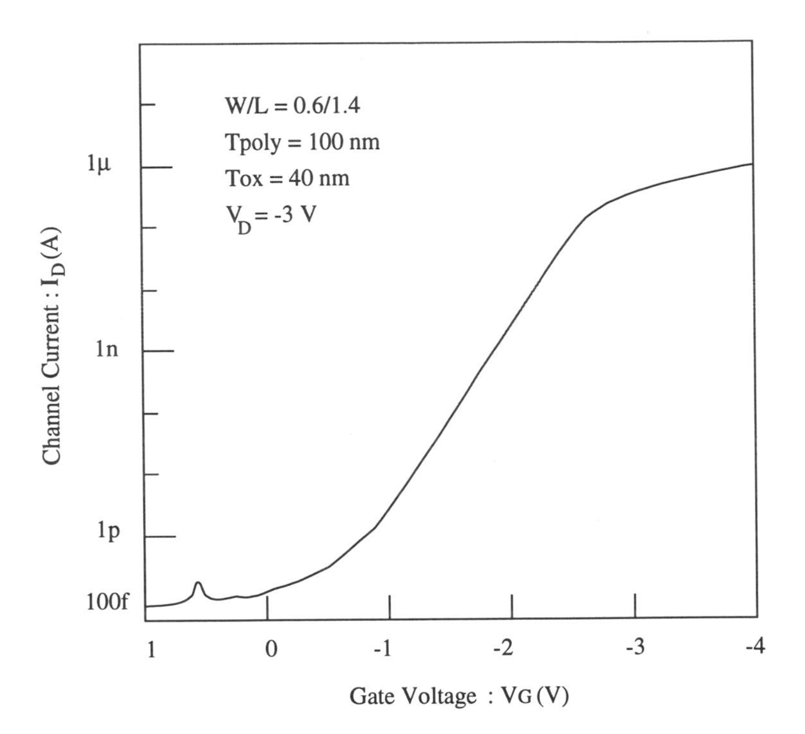

Figure 6.8 Electrical characteristic of TFT [8].

M equals 64-cell/word-line. Comparing t_i to t_{ch} for poly-Si load and PMOS TFT we have

$$t_{ch} < t_i \qquad For \quad PMOS \ \ TFT \tag{6.4}$$

$$t_{ch} \gg t_i \qquad For \quad poly - Si \ \ Load \tag{6.5}$$

Thus, the high-storage node, in the case of PMOS TFT cell, is charged-up quickly to V_{DD}. For this reason, the Soft Error Rate (SER) of the PMOS TFT cell is much lower than that of the poly-Si cell [16].

6.1.3 Read/Write Operation

Fig. 6.9 shows a simplified readout circuitry for an SRAM. The circuit has static bit-line loads composed of pull-up NMOS devices N_1 and N_2. The bit-lines are pulled-up to a voltage $(V_{DD} - V_{Tn})$, where V_{Tn} is the threshold voltage

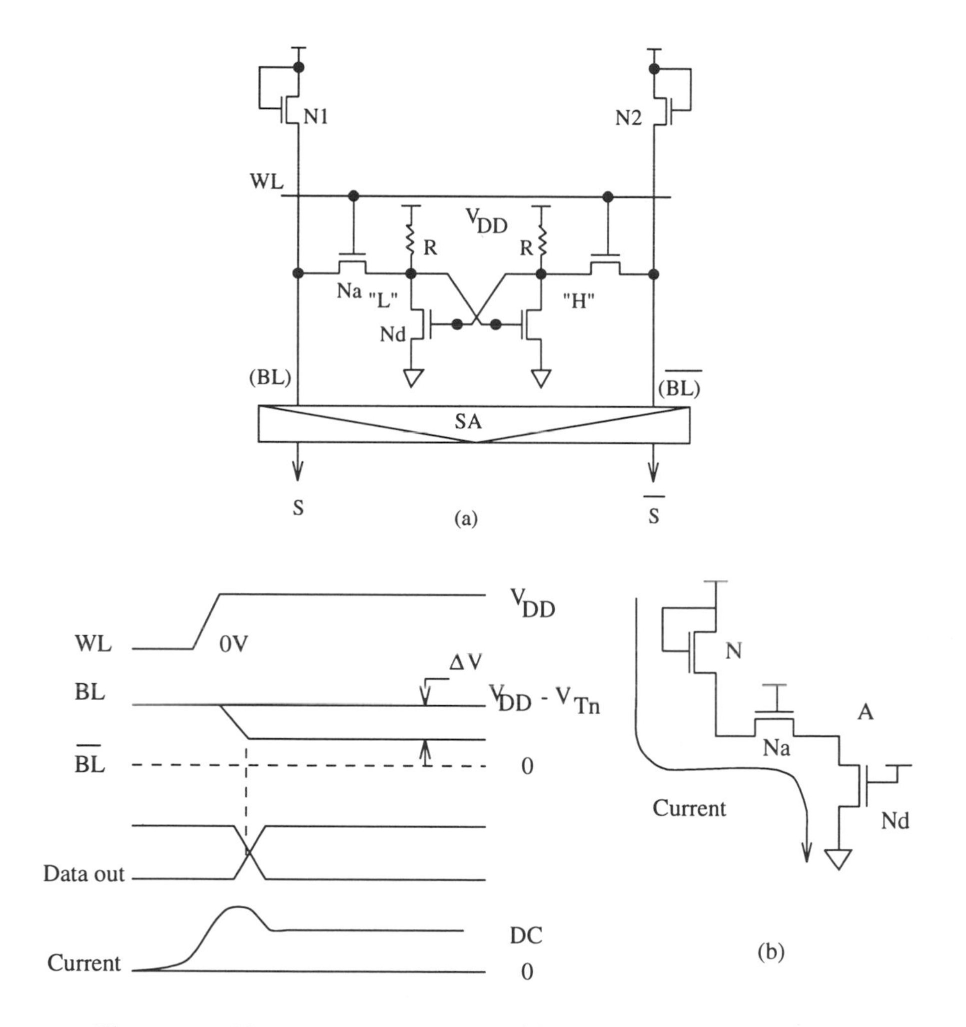

Figure 6.9 Memory cell read operation; (a) memory cell; (b) equivalent read circuit and signal waveforms.

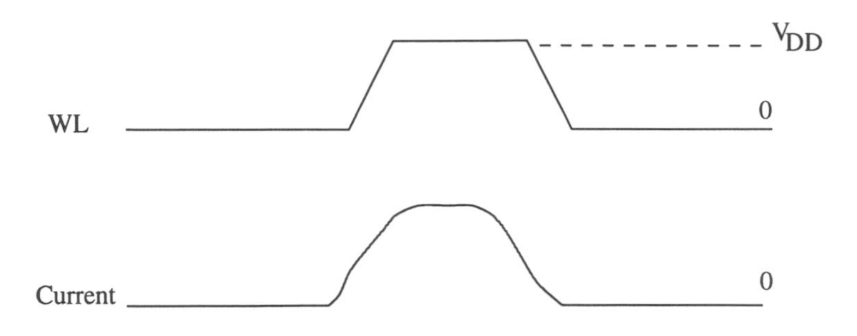

Figure 6.10 Power reduction by pulsing the word line.

subject to body effect. When the word-line WL is asserted, one word is selected. At this time, the bit-line BL is pulled down to a level determined by the pull-up NMOS N_1, the word-line transistor N_a, and the driver NMOS transistor Nd as shown in Fig. 6.9(b). The voltage at the node A should be low (near ground) to not alter the RAM content during this read operation. A small swing change on BL is desirable to achieve the high-speed readout, particularly if C_{BL} is high. The Sense Amplifier (SA) amplifies the small swing, ΔV on the bit-line. Typical values of ΔV and C_{BL} are 100 mV and 1-2 pF, respectively. It should be noted that this SA should provide a wide operating margin over all process, temperature, and voltage corners.

If the WL signal stays asserted, all selected columns consume a DC current flowing through the NMOS devices N_1, N_a and Nd. Thus, the shortening of read mode duration is necessary to reduce the power dissipation during this active mode. This is possible by pulsing WL with enough time to read the cell as shown in Fig. 6.10. The generation of pulsed WL signal is possible owing to the Address Transition Detection (ATD) technique as will be discussed in Section 6.1.5.

Fig. 6.11(a) shows a simplified circuit configuration for SRAM write operation. For a write operation the memory cell state should be flipped. When the write signal WE is asserted, the input data and its complement are placed on the bit-lines. If for example, a zero has to be stored in the node A initially at V_{DD}, the voltage at this node should be below the threshold voltage of the cell, as shown in equivalent circuit of Fig. 6.11(b). The bit-line in this case is pulled-down to almost 0 V. The design of write circuitry should provide a wide operating margin over all process, temperature, and voltage corners. Note that a DC current is consumed during a write mode, hence the WE signal should

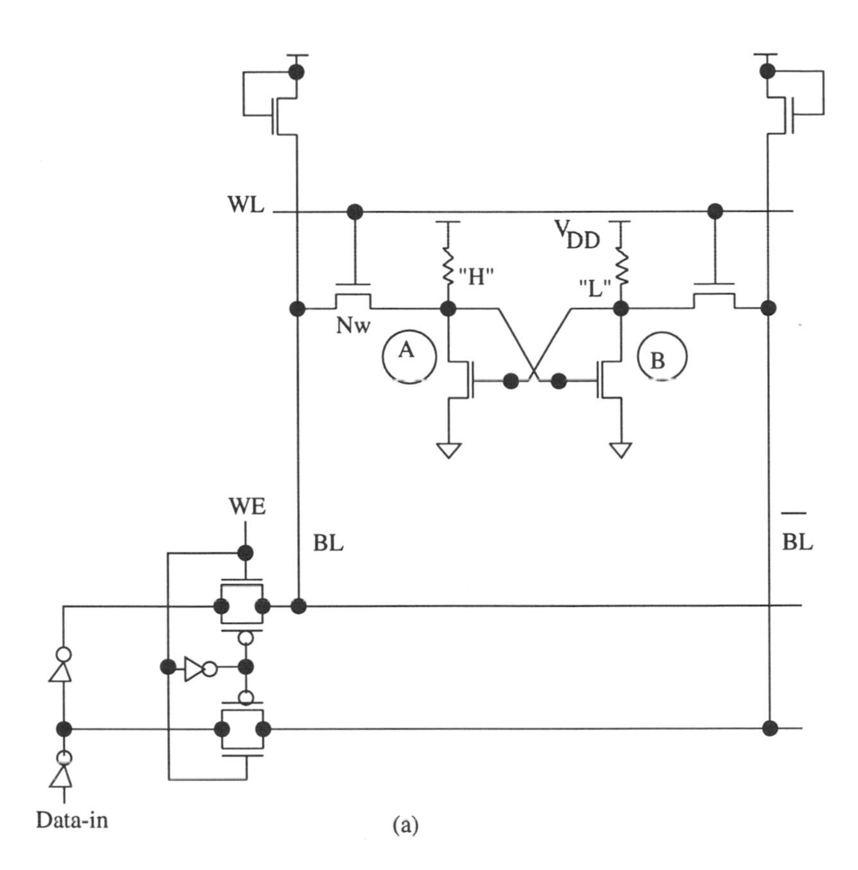

Figure 6.11 Memory cell write operation; (a) memory cell with write circuitry; (b) equivalent write circuit with signal waveforms.

also be short to cut this current at the end of the write operation. In high-speed SRAMs, write recovery time is an important component of the write cycle time. It is defined as the time necessary to recover from the write cycle to the read state after the *WE* signal is disabled. Note that the swing on bit-lines after write operation is large. Thus, an equalizer circuit is needed to reduce this swing, so that the read operation is performed quickly.

Fig. 6.12 illustrates a simplified schematic of an SRAM with read/write circuitry. At the end of the memory cycle a differential voltage existed on the bit-lines. A PMOS equalizing device is used to equalize the bit-lines after each read and write operation. The differential voltages on the bit-lines are restored

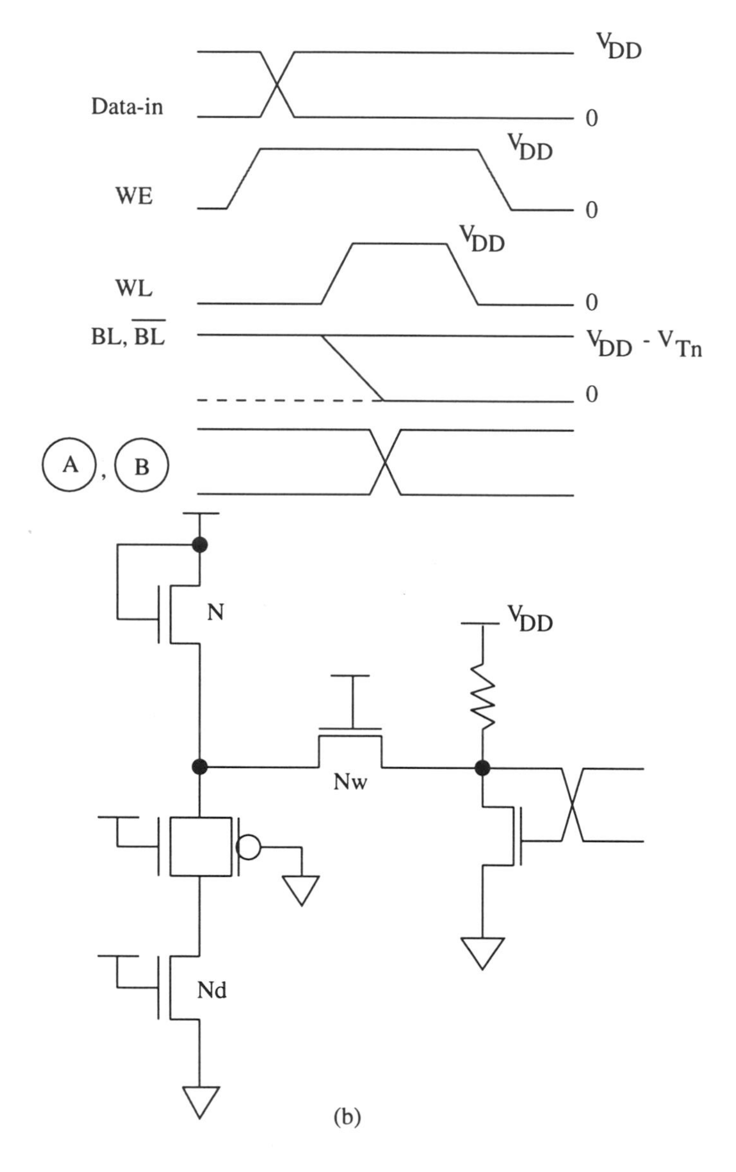

(b)

Figure 6.11 *(continued)*

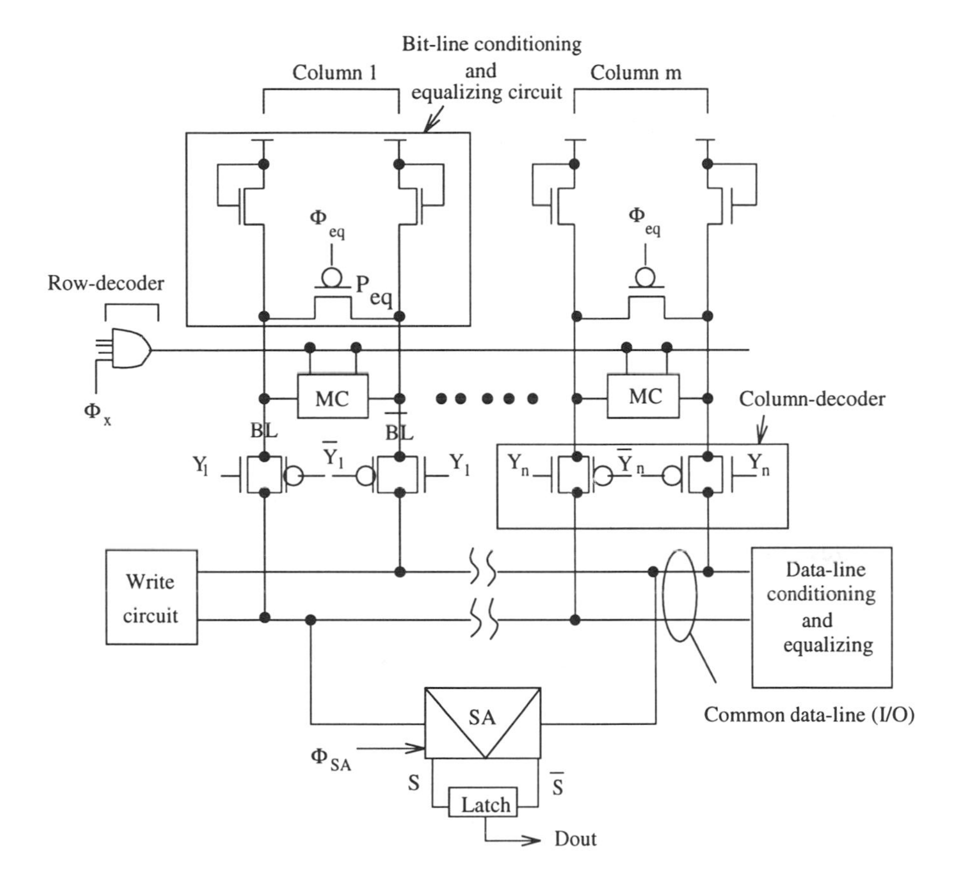

Figure 6.12 Simplified SRAM architecture.

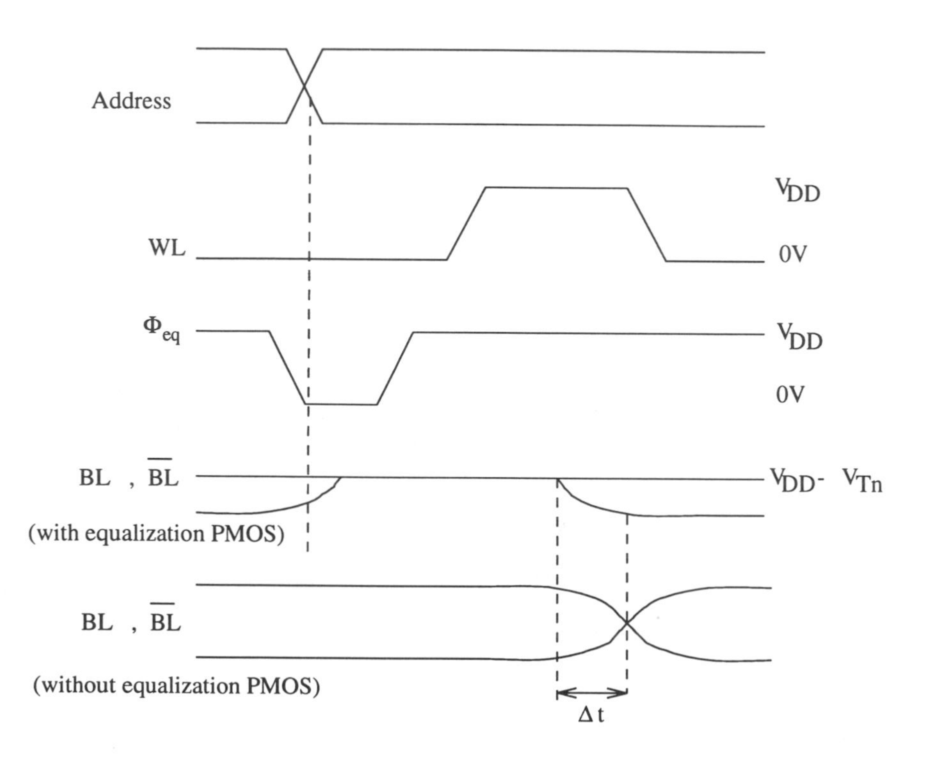

Figure 6.13 Bit-line equalizing and conditioning waveforms.

prior to being used. This results in an improved read time compared to the case without equalization, as illustrated by the waveforms of Fig. 6.13. When the address changes, immediately an equalizing falling pulse ϕ_{eq} is generated by an ATD circuit. Equalization, using PMOS, is very efficient if the bit-line load is PMOS. For the column decoder, transmission gates are used because they result in improved transfer time between the bit-lines and the common data-lines.

6.1.4 Low-Power Techniques

Different sources of power dissipation of an SRAM can be identified by using the typical architecture of Fig. 6.2. The total power dissipated is the addition of two components; the active power and the standby power. The active power is the sum of the power dissipated by the following components:

- The decoders (row and column);
- The memory array. If m memory cells are connected to the word-line, the active power of memory array (in read mode) is given by

  $$P_{mem-array} = mP_{act} + (n-1)mP_{leak} + mI_{DC}\Delta t f V_{DD} \quad (6.6)$$

  Where P_{act} is the power dissipated in active mode when selecting the m cells and P_{leak} is the data retention (standby) power of the unselected memory cells in the $m \times n$ array. The second term is negligible. The third term is due to the DC current, I_{DC}, during the read operation. Δt is the activation time of the DC consuming parts and f is the operating frequency ($f = 1/t_{RC}$). An example of such a current is the DC current flowing from the bit-line load to the ground through the memory cell;
- Sense amplifiers. They are dominated mainly by a DC current; and
- Remaining periphery such as input/output buffer, write circuitry etc.

Note that the power dissipated by the pads is not included. The power dissipation of the components, other than the memory array, depends on the total capacitances, the operating frequency and the internal voltage swing. It can include a DC component with a major contribution from the sense amplifier.

To reduce the active power consumption many techniques can be used and are summarized as follows :

- Reducing the capacitances of the word-line and the number of m cells connected to it. This is possible by using Hierarchical Word-Line (HWL) techniques.
- Reducing the DC current by using the pulse operation technique for the word-line and the periphery circuits (including sense amplifier).
- Use of multi-stage static CMOS decoding to reduce the AC current.
- Lowering the operating power supply voltage.

The standby power (or sometimes called retention current) of an SRAM has a major contribution from the memory cells in the array if the sense amplifiers are disabled in this mode. It is given by

$$P_{standby} = mnP_{leak} \quad (6.7)$$

One way to reduce the standby current is to reduce the operating voltage. However, note that the data-retention current will increase with memory capacity. Moreover, the leakage current, per cell, tends to increase because the threshold voltage is expected to be reduced for low-voltage operation.

In the following sections, many key circuits in an SRAM are reviewed. The circuit techniques and memory organization to reduce the active and data-retention currents are presented.

6.1.5 Address Transition Detector (ATD) Circuit

To generate the different timing signals for word-lines, equalization and sensing, an on-chip pulse generator, which detects the address change, is needed. It is based on address transition detection technique. The ATD is a key technique to reduce the active power of memories. Fig. 6.14(a) shows the schematic diagram of an ATD pulse generator. Short pulses are generated with XOR circuits when the address changes from "L" to "H" or "H" to "L"; then summed through an OR gate. The overall pulse width is controlled by the RC delay line shown in Fig. 6.14(b). The corresponding waveforms are shown in Fig. 6.14(c). The ϕ_{ATD} pulse is usually stretched out with a delay circuit to generate the different pulses needed in the SRAM. Note that the $\overline{CS}$ signal is also included as an input to the ATD generator.

6.1.6 Decoders

Usually the decoding in an SRAM is performed by using complementary CMOS. Two kinds of decoders are used ; the row and the column decoders. Fast static decoders are based on OR//NOR and AND/NAND gates. Fig. 6.15 shows an example of a two-bit input address row decoder. The input buffers have to drive the interconnect capacitance of the address lines and the input capacitance of the NAND gates. To match the pitch of the memory cell and to perform decoding for severals blocks, two-stages decoders are used. The first stage performs predecoding and the second one performs the final decoding function [Fig. 6.16]. The two-stages decoder circuit has other advantages over the one stage decoder such as to reduce the number of transistors and fanin. Also it reduces the loading on the address input buffers. This predecoding technique optimizes both speed and power. In the last stage an additional signal ϕ_x is included in the AND gate. This signal is generated from an ATD pulse generator to enable the decoder and ensure the pulse activated word-line. There

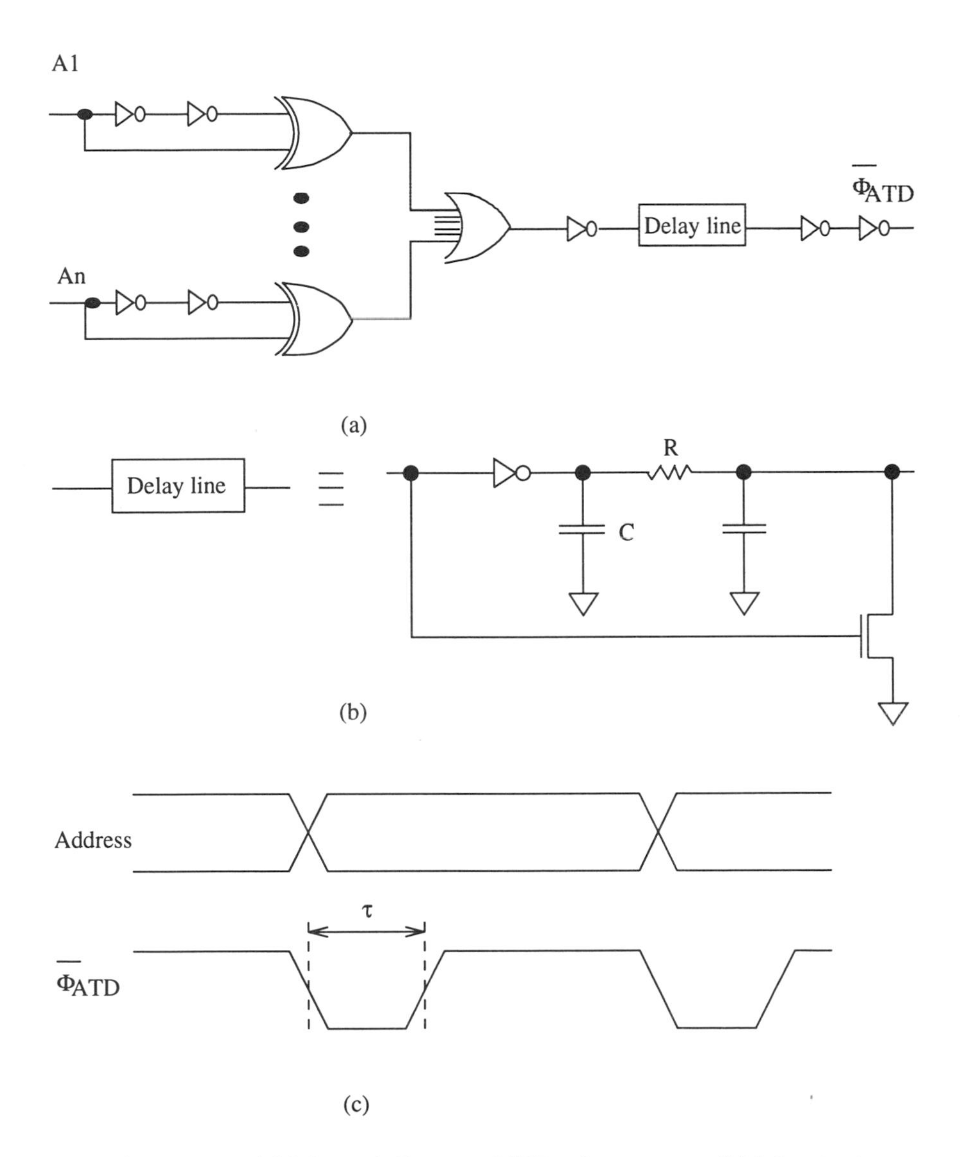

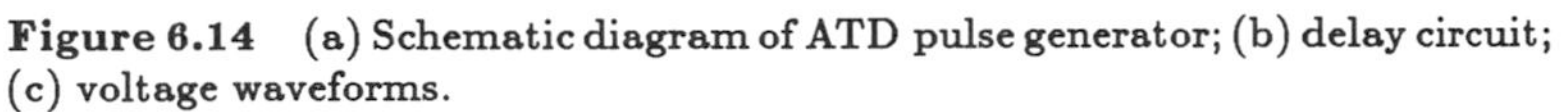
Figure 6.14 (a) Schematic diagram of ATD pulse generator; (b) delay circuit; (c) voltage waveforms.

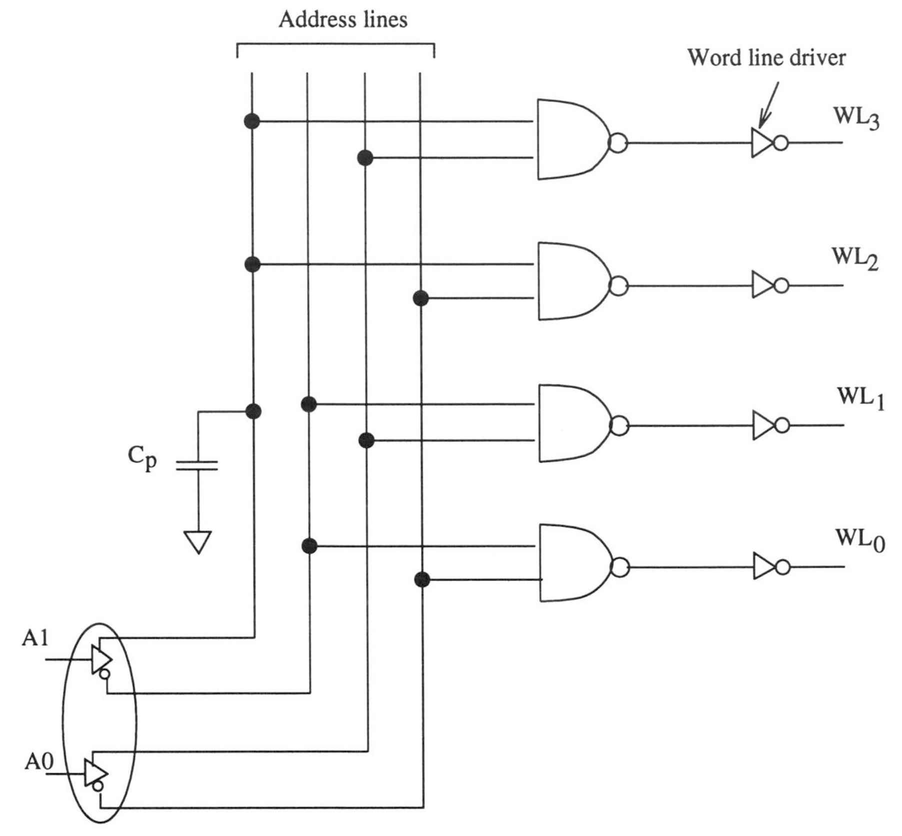

Figure 6.15 Row decoder example for two-bit address.

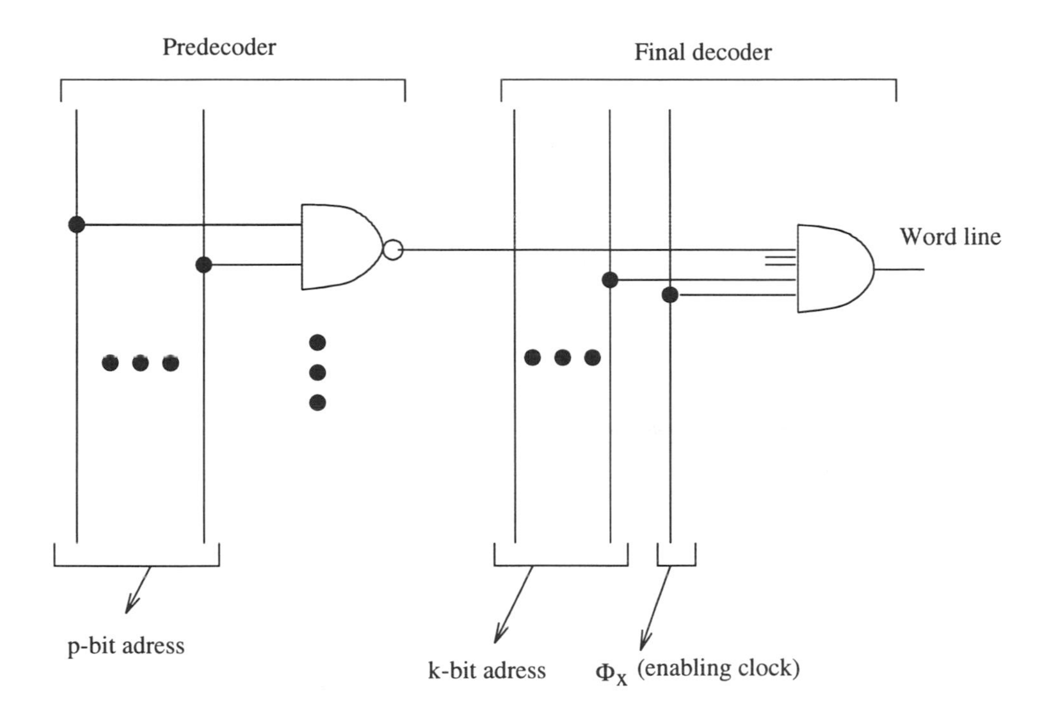

Figure 6.16 Example of two-stages row decoder circuit.

are several ways to build row-decoders and it depends on the RAM architecture division.

The column decoder permits the selection of 1 out of m bits of the accessed row. Fig. 6.17(a) shows the circuits involved for column selection using an example of 4 columns. The selected gate permits the transferring of the data from the bit-lines to the common data-lines I/O. The signals Y_i are controlled by the AND/NAND column decoder as shown in Fig. 6.17(b).

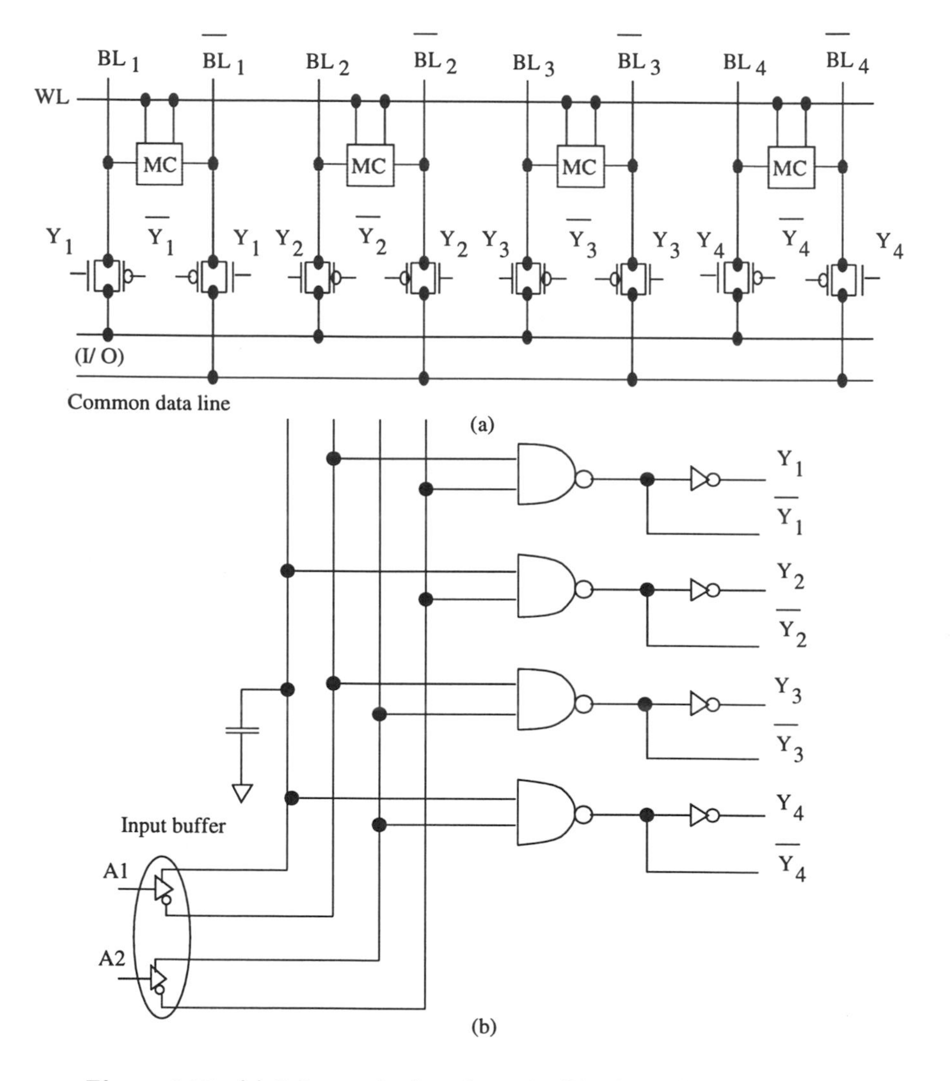

Figure 6.17 (a) Column selection schematic; (b) column decoder for two-bit column address.

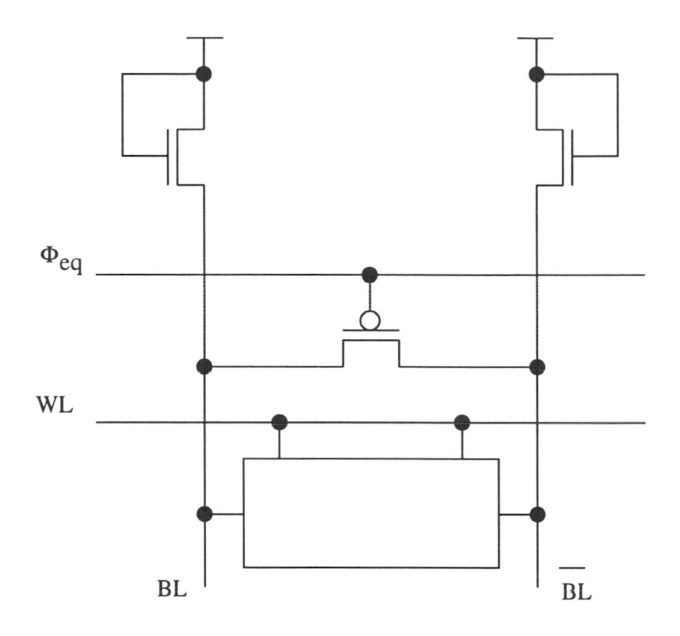

Figure 6.18 Bit-lines precharging circuit with NMOS loads.

6.1.7 Bit-line Conditioning Circuitry

The NMOS bit-lines' loads [Fig. 6.18] have been used in many SRAMs at 5 V power supply. They provide a precharge level on the bit-lines of $V_{DD} - V_T$. The threshold voltage of the load, V_T is subject to the body effect. A typical value of this precharge level for 5 V power supply is 3.5 V. This level is suitable for voltage-type sense amplifiers to provide large gain and fast sensing delay.

To reduce the DC current, during the write circuit, a variable bit-line load technique can be employed [Fig. 6.19]. It realizes fast sensing in the read cycle and a short write pulse width in the write cycle. For fast sensing, the voltage swing of the bit-line should be small. To achieve this, the load impedance should be low. On the other hand, to obtain a low current during write cycle, the load impedance of the bit-lines should be high. As shown in Fig. 6.19, during the read operation, all four NMOS transistors N_1, N_2, N_3, and N_4 are turned ON. The bit-lines are switched into a low-impedance state so that the voltage swing of the bit-lines is limited to a small value (e.g., 100 mV). During the write operation, the NMOS devices N_3 and N_4 are switched OFF and only the small size transistors N_1 and N_2 are turned ON.

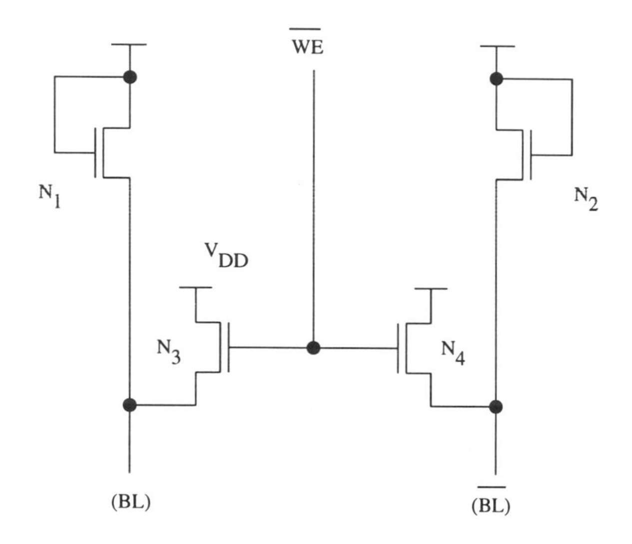

Figure 6.19 Variable load bit-lines.

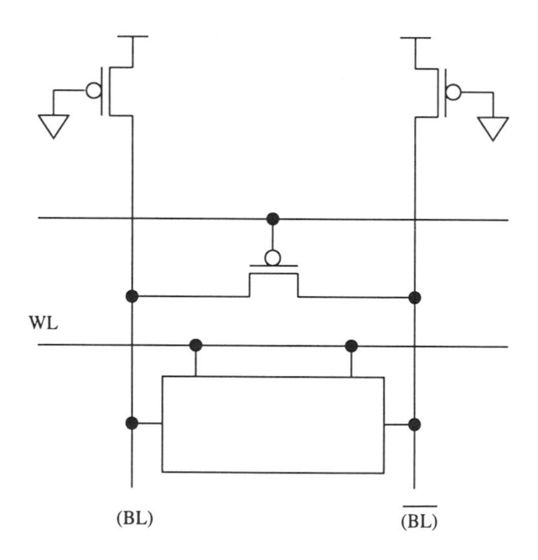

Figure 6.20 Bit-lines' precharge circuit with PMOS loads for low-voltage.

As the power supply voltage is scaled down to 3 V, the precharge level can be lower than 2 V. Thus, during read operation the high-level node of the memory cell can become equal to the bit-line voltage. Hence, the noise margin of the memory cell is drastically degraded and consequently the cell stability and soft error are degraded. Therefore, at 3 V power supply voltage, a PMOS transistor can be used as bit-lines' load [Fig. 6. 20]. The bit-lines precharge voltage is V_{DD}. For low-voltage bit-lines precharge voltage, special sense amplifiers should be used because conventional sensing circuits have poor voltage gain (less than 10). A variable impedance bit-line, using PMOS transistors, can also be implemented.

6.1.8 Sense Amplifier

When reading a memory cell, the bit-lines are initially precharged, then one of the two bit-lines goes down, while the other stays high. The operation of pulling down the bit-line is very slow because the discharging MOS device, in the memory cell, is small and the bit-line capacitance is high. This results in very slow memory read time. Sense amplifiers are used to detect the small variation on the bit-lines and amplify it to get at the end full-swing signal. A simple unbalanced inverter with a high logic threshold voltage can be used. Since its input is single and has very small noise margin, it is very sensitive to noise on the bit-line. Thus, sense amplification, for the data-lines, is a key to achieve fast access time and low-power dissipation. In general, the delay of a sense amplifier (from the time of word-line activation) represents 30 to 40 % of the whole read access time.

Various kinds of sense amplifiers have been devised for fast sensing operation and low-power dissipation. Fig. 6.21(a) shows a single-end sense amplifier with an active current-mirror. This structure forms the basis for many SRAMs' sense amplifier circuits. It has two differential inputs, DL and $\overline{DL}$. The noise equally affects both the two inputs and only the difference is detected. The transistor N_s acts as a current source. Before the signal ϕ_{SA} is asserted, the data-lines DL and $\overline{D}L$ are high. All the nodes, A, B and C, are high. The signal ϕ_{SA} is asserted when DL starts, for example, to drop slowly. In this case, the NMOS transistor N_s is ON. The output voltage (node C) drops suddenly to a certain voltage. Thus, the input signal is amplified by the gain of this differential amplifier.

Fig. 6.21(b) shows the voltage waveforms of the single-end sense amplifier using SPICE simulation. The signal ϕ_{SA} is generated with an ATD pulse. It is

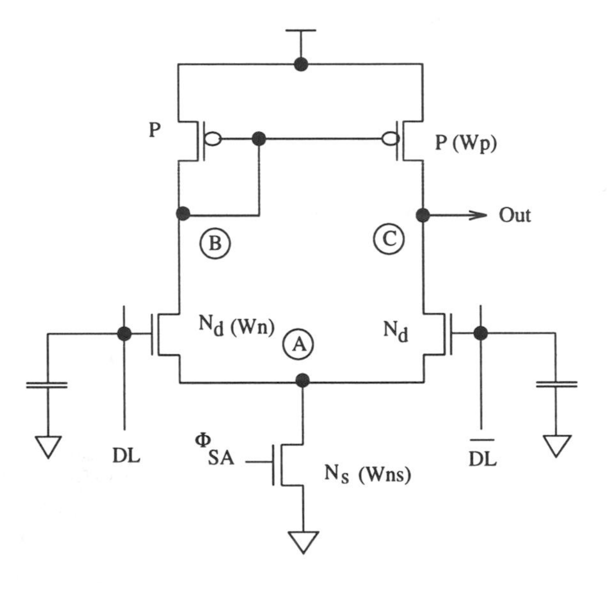

(a)

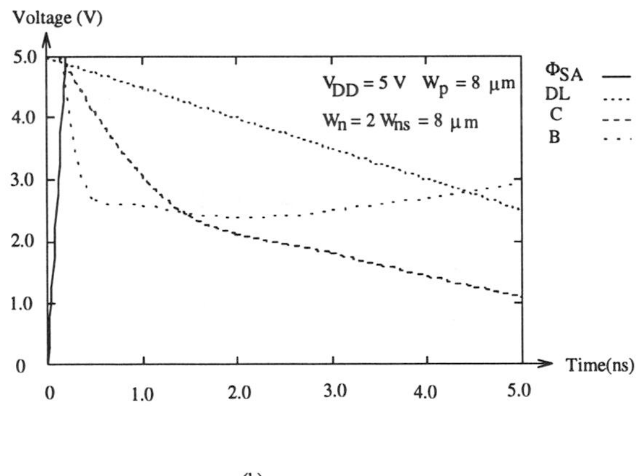

(b)

Figure 6.21 (a) Single-ended sense amplifier; (b) simulated voltage waveforms.

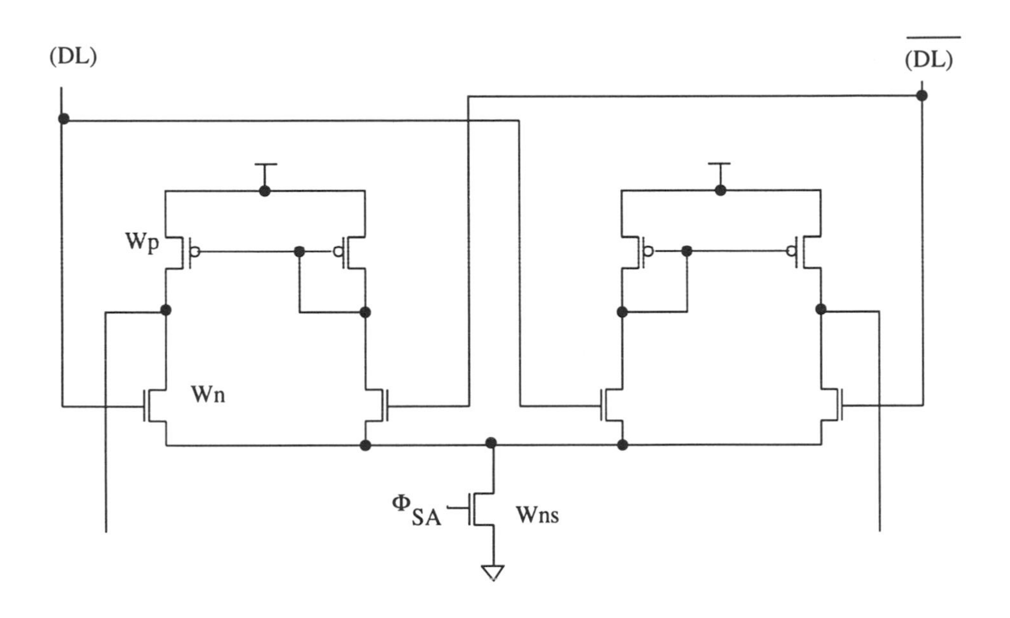

Figure 6.22 Double-ended sense amplifier.

asserted for a time, enough to amplify the small variation (few hundreds of mV) on data-lines[2], then it is disactivated. In this scheme the DC current consumed by the sense amplifier is cut off. Usually the sense amplifier is common to many columns through the common data-lines. The small signal gain of this amplifier is given by

$$A = \frac{g_{mn}}{g_o} \tag{6.8}$$

where g_{mn} is the transconductance of the driver NMOS N_d and g_o is the combined output conductance of the PMOS load and the NMOS driver.

In many SRAMs multi-stage sense amplifiers are needed to attain large voltage gain. In this case, the double-end sense amplifier is used as shown in Fig. 6.22. This circuit has often been used in many SRAMs. To attain high-speed data sense, a two and three-stage sense amplifier technique can be adopted. Fig. 6.23 shows a two-stage amplifier structure. An equalization technique is used for the data-lines, using the equalization pulse ϕ_{eq}, which is generated with an ATD pulse. It is indispensable, not only to attain faster data transfer

[2]The output of the sense amplifier is then latched.

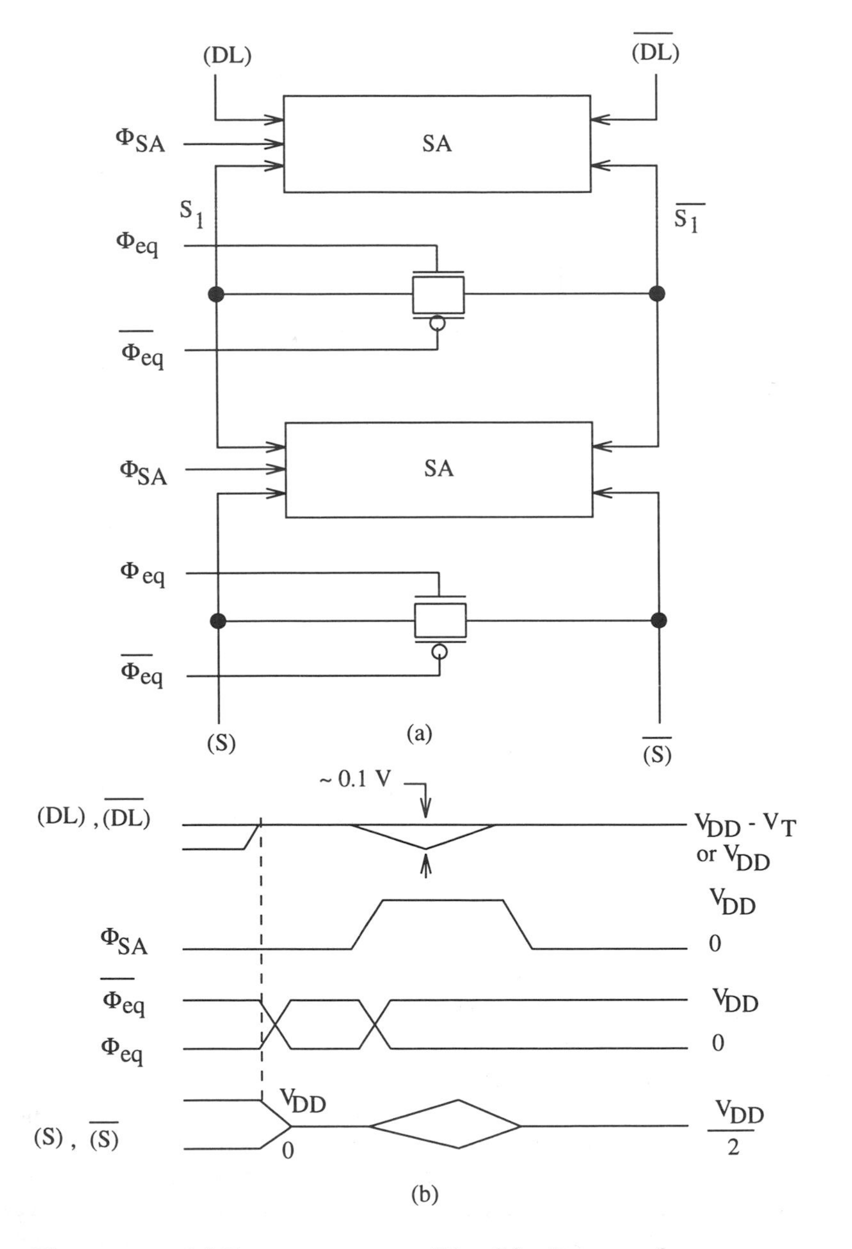

Figure 6.23 (a) Two-stage sense amplifier; (b) voltage waveforms.

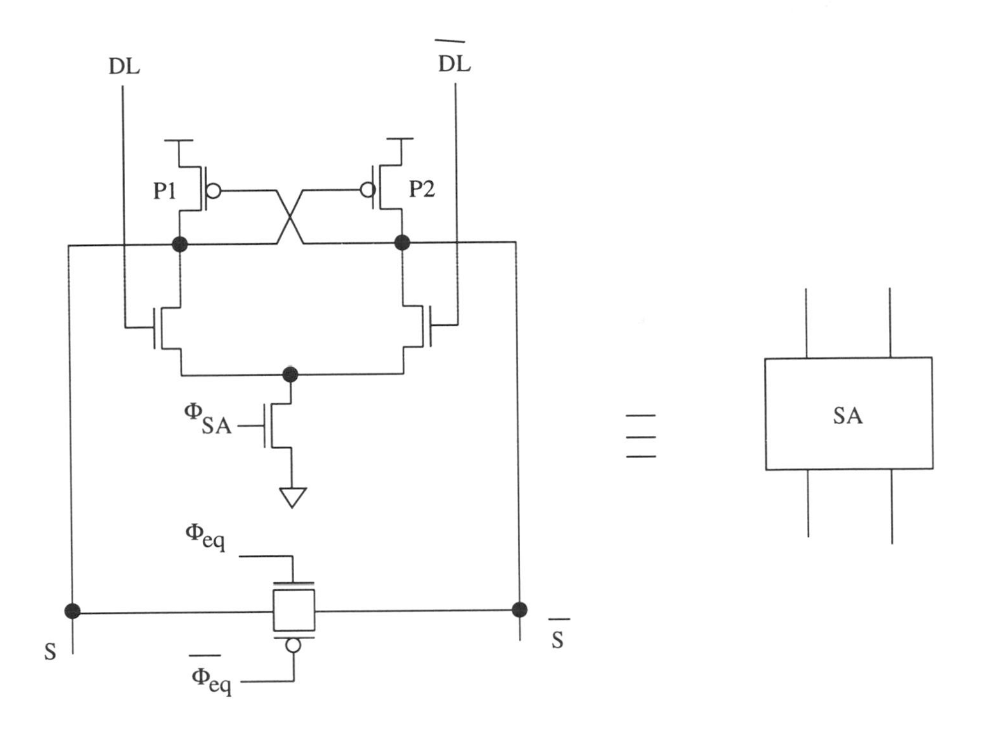

Figure 6.24 PMOS cross-coupled sense amplifier.

during read operation, but also to suppress incorrect data before the correct data appears in the sense amplifier [17]. For low-power applications and also due to the plastic packaging limitations of static memories, this type of sense amplifier can result in high power dissipation for high-density memories even if the current source is pulsed.

Many circuits have been proposed to reduce the power of the sense amplifier while improving their sensing delay time. One of them is the PMOS cross-coupled amplifier [18] shown in Fig. 6.24. The PMOS loads, P_1 and P_2, are cross-coupled and the differential outputs S ans $\bar{S}$ are connected to their gates. The positive feedback in this latch amplifier permits much faster sense speed than the conventional one. In this circuit the equalization technique is used for the reasons discussed above. Fig. 6.25 shows the sense delays of both the PMOS cross-coupled amplifier and the double-end current-mirror amplifier as a function of the average current of the amplifier. The input voltages simulate

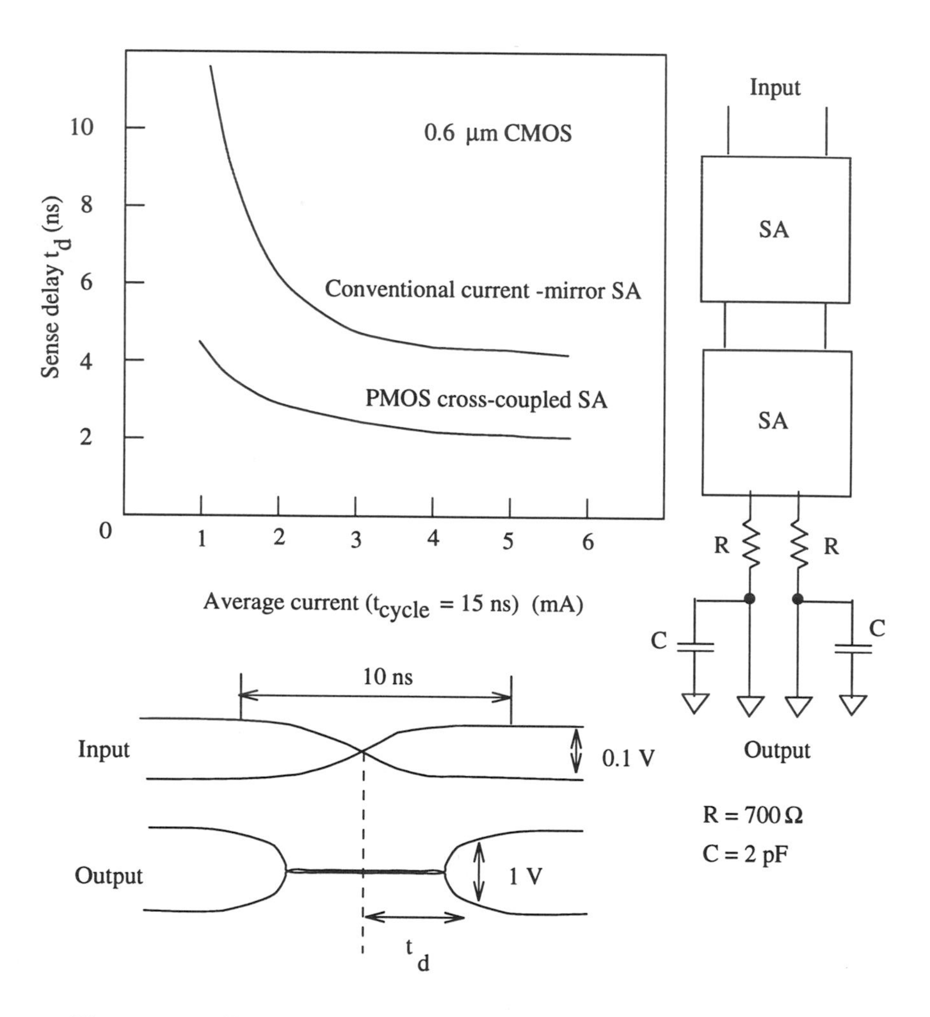

Figure 6.25 Simulated sense delays of the PMOS cross-coupled and current-mirror SAs [18].

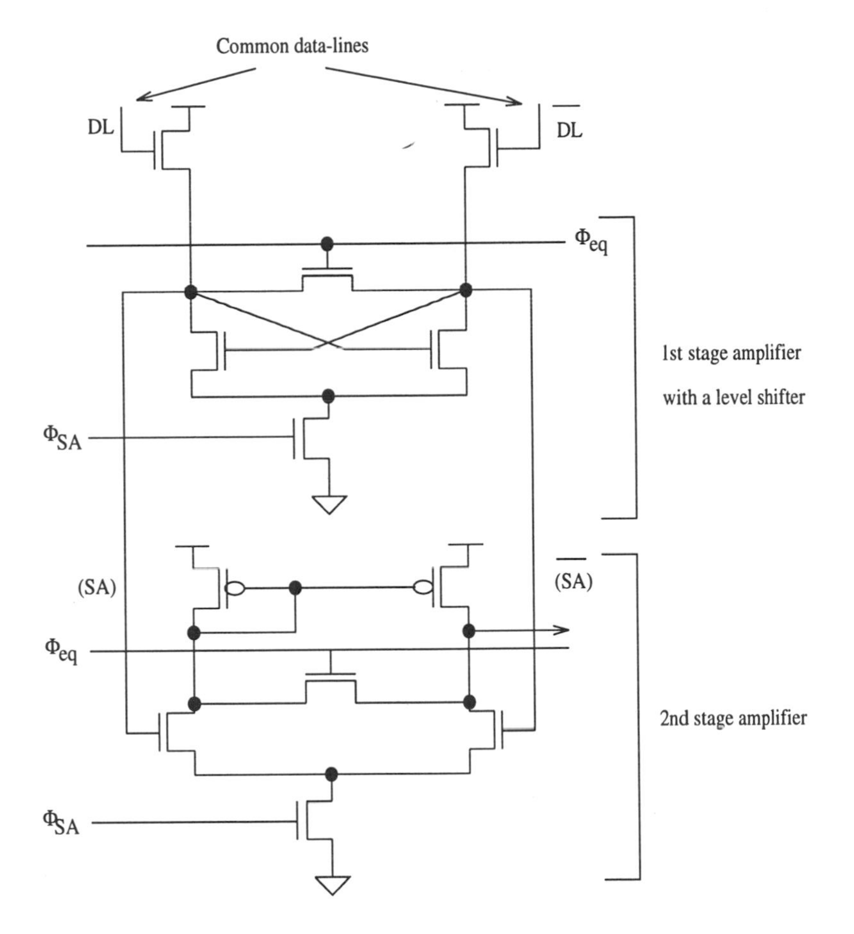

Figure 6.26 Sense-amplifier circuit with level shifting.

the common data-lines' voltages and the sense delay t_d is defined as the delay time from the crossover point of the input voltages to the point when the output reaches 1 V difference. The PMOS cross-coupled amplifier has less than half the delay of the conventional current-mirror sense amplifier. Moreover, this latch amplifier consumes less than one-fifth of the power of a current-mirror amplifier. The PMOS cross-coupled latch amplifier requires much more accurate timing for ϕ_{eq} to optimize the sensing delay [18]. This circuit also has low-power property compared to the current-mirror amplifier since it has nearly full-swing outputs with positive feedback.

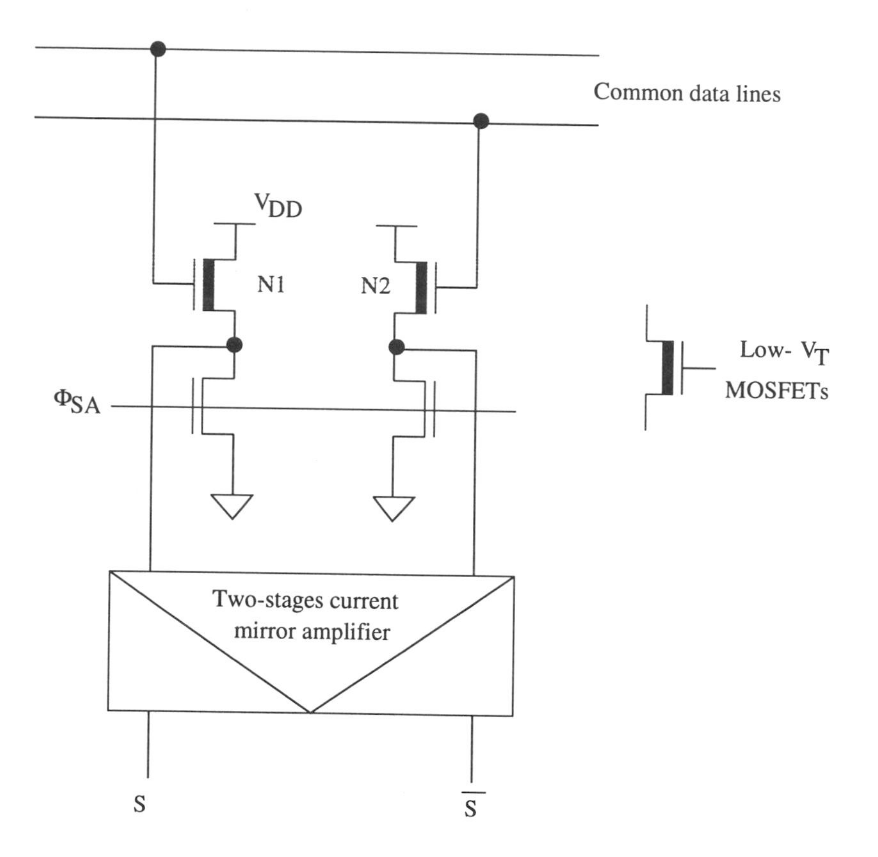

Figure 6.27 low-voltage sense amplifier [19].

When the voltage is scaled to 3 V power supply, the data-line voltage is near V_{DD}, then a level shifting can be performed. Fig. 6.26 shows a two stage sense amplifier used for 3.3 V supply. The first stage is a cross-coupled NMOS amplifier which also performs level shifting of the common data-line voltage. In the second stage, a conventional sense amplifier is used which operates at the maximum gain point since the level on SA and $\overline{SA}$ are medium levels.

Fig. 6.27 shows another sense amplifier developed for low-voltage power supply [19]. This circuit is used when the bit-lines are close to V_{DD}, where the gain of a conventional current-mirror amplifier is poor. The circuit is composed of a level-shift circuit and a conventional current-mirror amplifier. The level-shifter shifts the bit-line voltage to a medium voltage; 0.6 to 0.7 V, (@ 1 V power supply voltage) where the gain is maximum. Low-V_T NMOS devices N_1 and N_2 are used to provide these medium levels. These devices are subject to the body effect.

Recently current sense-amplifiers have been proposed to overcome the gain reduction of voltage amplifiers at low power supply [7, 12]. Also they reduce the power dissipation of the sensing operation compared to voltage sense amplifiers at the same delay. These circuits require very careful design.

6.1.9 Output Latch

In low-power SRAM, the pulse technique for word-line and sense amplifier is indispensable in order to reduce the DC current. In such a pulse mode, a data-latch circuit is required to store the amplified data by the sense amplifier from the memory cell for the data output circuitry. Fig. 6.28 shows an example of an output latch placed after the sense amplifier. The requirements of such an output latch are the following :

- The latch circuit must not delay the read access time. Such a requirement is attained by connecting the latch with data-bus lines in parallel. One input transmission gate, controlled by ϕ_I, is used to enter the data to the latch. Another transmission gate, controlled by ϕ_O, is used to put the data back into the data-bus.

- The latched data must not be destroyed by the noise entering the SRAM. A noise in an SRAM is generated and propagated by the following mechanism. On the system board, a ground noise can enter the SRAM. When the peak level of the ground noise becomes large enough for the first gate of the address buffer to change the logic value of the address input, an ATD pulse noise is generated. This noise pulse could turn on the word-line and the sense amplifier for a short time resulting in an expected signal on the data-bus. Therefore, the latched data could be destroyed if the input signal is ON. To avoid such a problem, two circuit techniques are included in the circuit of Fig. 6.28. The first one is the generation of ϕ_I only when the pulse width of the ATD is large enough, compared to that of the noise. The other circuit technique is to place latch-protecting inverters [Fig. 6.28] in the front of the output gates. The inverters prevent noise from entering the output gates.

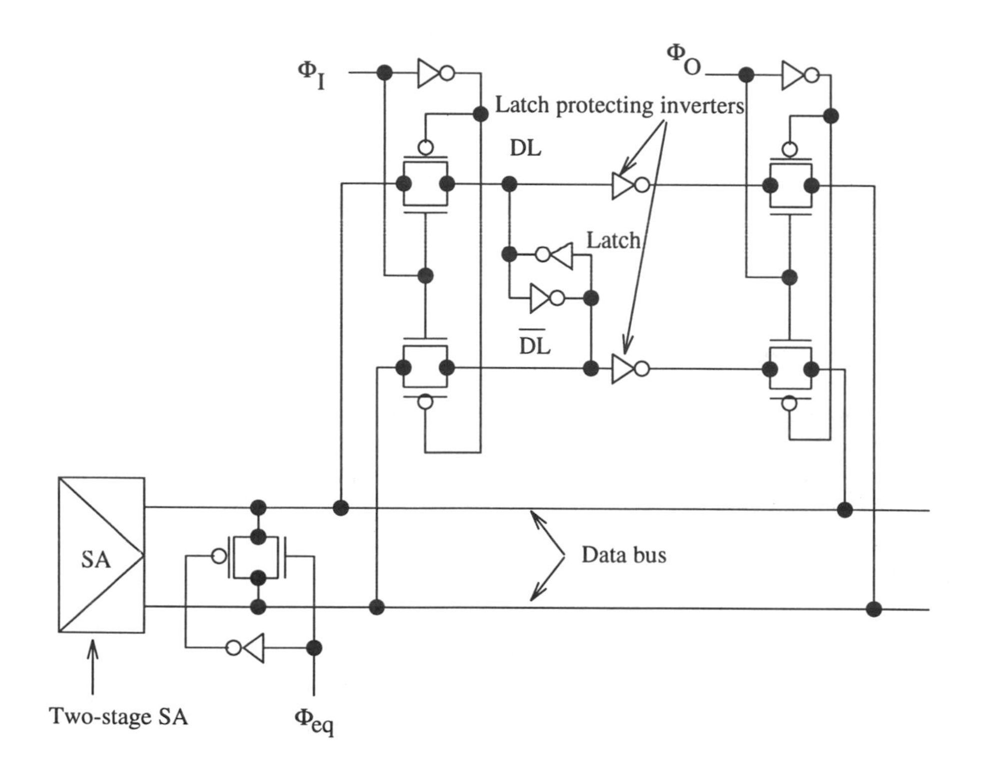

Figure 6.28 Noise-immune data latch circuit [15].

- The new data must be quickly latched into the data-latch. The circuit of Fig. 6.28 can be optimized for fast operation.

6.1.10 Hierarchical Word-Line for Low-Power Memory

With the increased memory size, the word-line delay and the column power increase. To solve this problem, a Divided Word-Line (DWL) structure was proposed [20]. The concept of DWL is shown in Fig. 6.29. The cell array and the word-line are divided into n_B blocks (sub-arrays). If the SRAM has n_C columns, each block has n_C/n_B columns. The divided word-line of each block is activated by the main word-line and the corresponding block select signal. Consequently, only the memory cells connected to one divided word-line within a selected block are accessed in a cycle. Hence, the column current

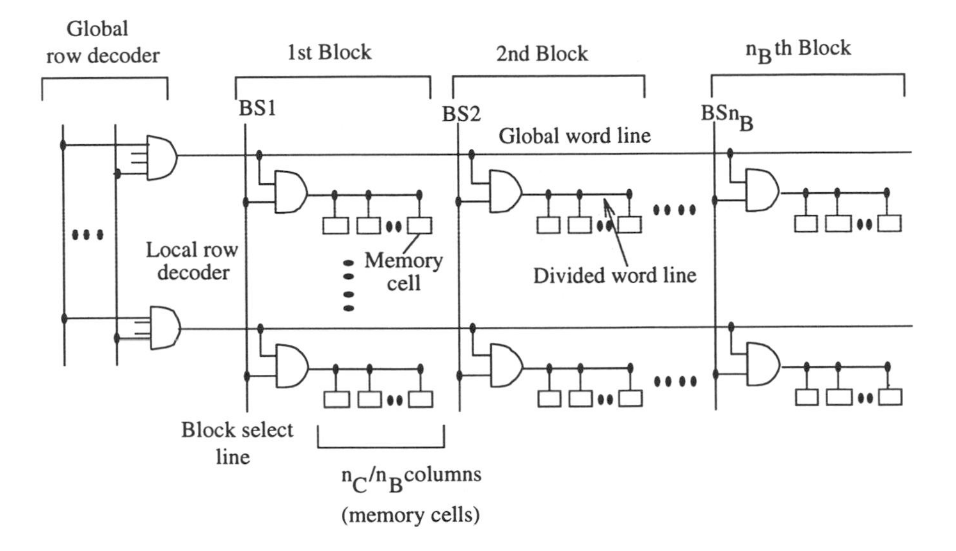

Figure 6.29 Divided Word-Line (DWL) concept [20].

is reduced, since only the selected columns switch. Moreover, the word-line selection delay, which is the delay time from the address input to the divided word-line, is reduced. This delay is composed of the main word-line select delay and the divided word-line select delay. The main word-line selection delay is reduced compared to the conventional one, because the total capacitance of connected transistors is reduced. In a conventional SRAM, the word-line has all the row memory cells' gates of a row connected to it. The main word-line delay increases as the number of blocks increase because the number of block select gates increases. On the other hand, the divided word-line delay decreases as the number of connected cells is reduced with the increasing number of blocks. Consequently, the word-line selection delay has a minimum for a certain number of blocks.

Fig. 6.30 shows the effect of the number of blocks in DWL structure on the word-line select delay and the column power for 64-Kb SRAM [20]. In this example, a number of blocks of eight can be chosen. The area penalty for this case is only 5%, compared to the conventional memory. As an example, for 1-Mb SRAM, the cell array is divided into 16 blocks and each block consists of 512 rows by 128 columns. 9-bit address ($A_1 ... A_9$) is used to select a row within

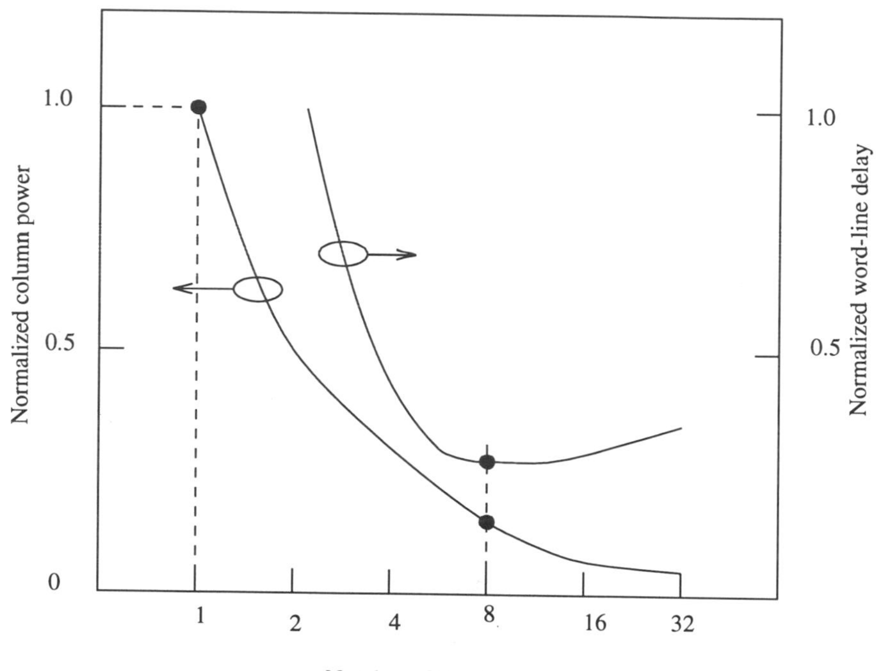

Figure 6.30 Normalized column power and word-line delay of a 64-Kb SRAM using DWL structure [20].

a block using two-stage row decoder. Global block selection is done using 5-bit address.

The DWL structure has been widely used in high-density SRAMs for its low-power, high-speed characteristics. However, in high-density SRAMs, with a capacity more than 4-Mb, the number of blocks in the DWL structure will have to increase. Therefore, the capacitance of the global word-line increases causing the delay and power increase. To solve this problem, the concept of Hierarchical Word Decoding (HWD) was proposed in [21] as shown in Fig. 6.31. The word select line is divided into more than two levels. The number of levels (hierarchy) is determined by the total load capacitance of the word select line to efficiently distribute it. Hence, the delay and the power are reduced. For 4-Mb, three levels of hierarchy have been used with 32 blocks; each block having 128 columns by 1024 rows. Fig. 6.32 shows the delay time and the total

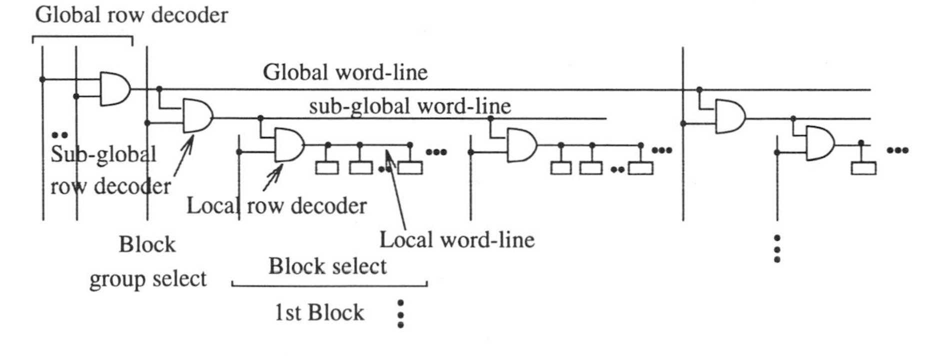

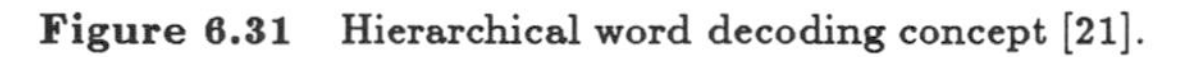
Figure 6.31 Hierarchical word decoding concept [21].

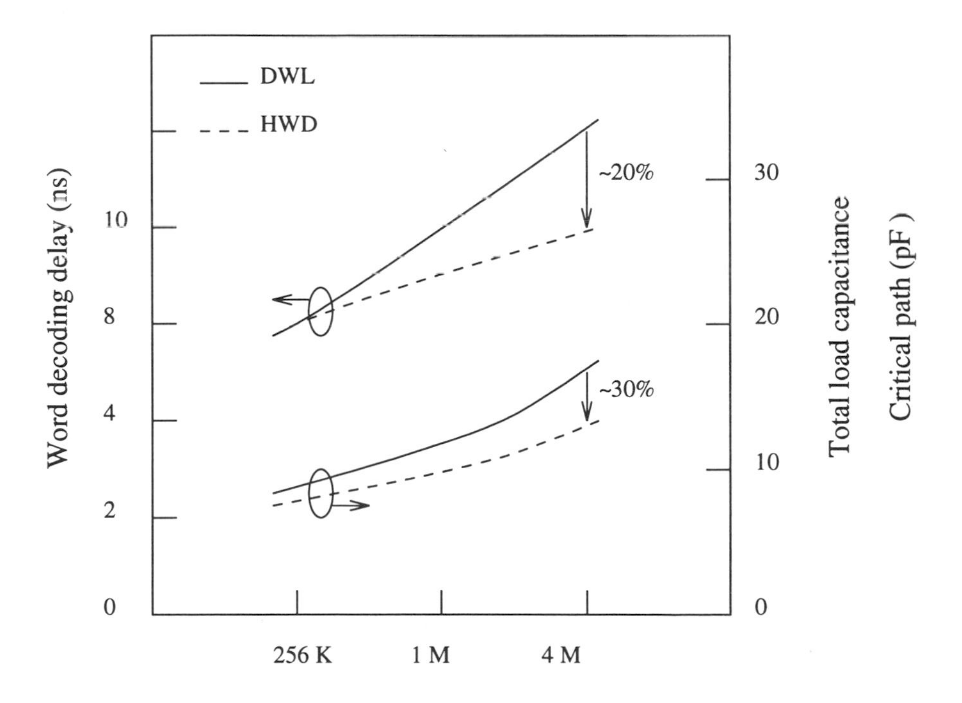

Figure 6.32 Comparison of DWL and HWD [21].

capacitance of the word decoding path comparison for the optimized DWL and HWD structures of 256-Kb, 1-Mb, and 4-Mb SRAMs. For 256-Kb SRAM there is no significant advantage of HWD over DWL. However, for high-density SRAMs the performance, of HWD in terms of power and delay, becomes clear. The three-levels scheme can be used efficiently for 16-Mb SRAMs.

6.1.11 Low-Voltage SRAM Operation and Circuitry

There are several applications which need a 1.2 V battery power supply. For such a application 1 V SRAMs are needed. At 1 V power supply, a stable operation is targeted and it is very important that the noise is reduced. Moreover, the active and standby powers should be reduced to meet the requirement of battery operation.

For 1 V power supply, a full CMOS memory cell has a lower power dissipation in standby mode and greater immunity to transient noise and voltage variation than other cells. It can also operate at the lowest supply voltages. Although a full CMOS cell operates well at ultralow-voltage, its area is almost double of that of PMOS TFT. Hence it is not suitable for high-density memories (size > 4-Mb).

When the full CMOS memory cell is operated at 1 V power supply, a typical cell ratio is 3 for stable operation. The SNM of this cell, at 1 V, can be almost the same as for a poly-Si load memory cell at 5 V. When using the full CMOS cell, no boosting of the word-line is needed to write a high voltage level in the cell. However, the PMOS TFT cell requires a boosted voltage ($V_{ch} > V_{DD}$) on the word-line during the write cycle [19]. If the voltage of the word-line is raised only to V_{DD} in the write cycle, the high node B of Fig 6.33 is initially at $V_{DD} - V_T$, where V_T is the threshold voltage of the access device subject to the body effect. This low-level ($V_{DD} - V_T$) of the node B can not charge up to V_{DD} because of the poor drivability of the PMOS TFT device.

When the boosted word-line technique is applied to the PMOS TFT cell during a write cycle, a problem can arise. The unselected cells connected to the boosted common word-line suffer from an instability problem because a large current flows through the low node of the cell. This large current is due to the high voltage on the access transistor. Consequently, this technique is not suitable for 1 V operation.

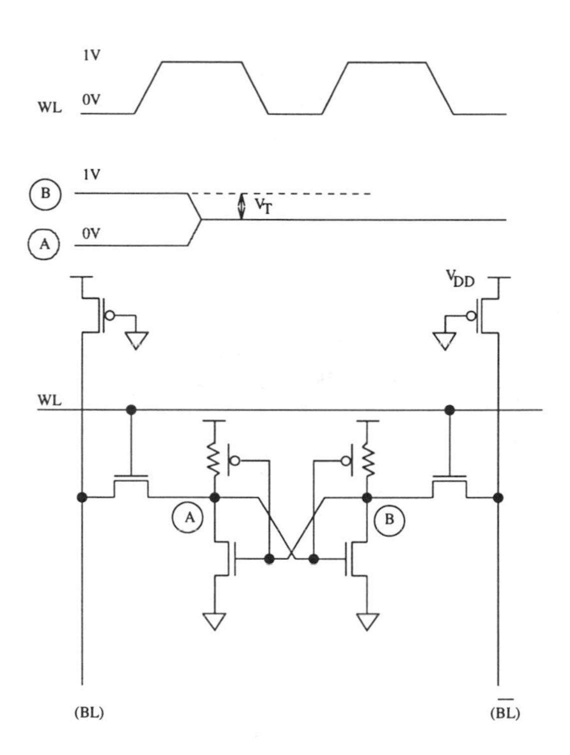

Figure 6.33 PMOS TFT memory cell operated at 1 V power supply voltage.

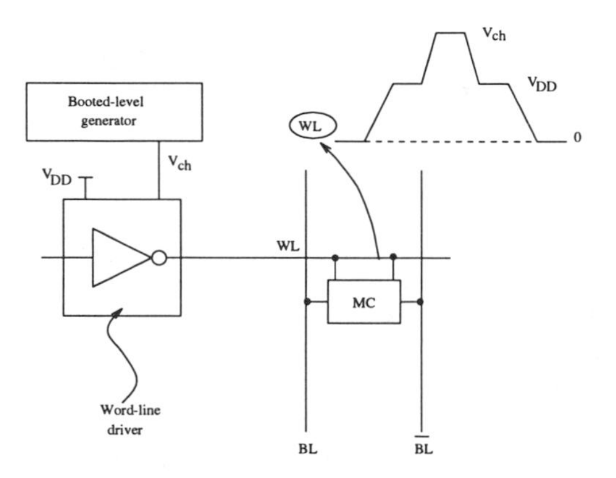

Figure 6.34 Two-step technique for 1 V operation [19].

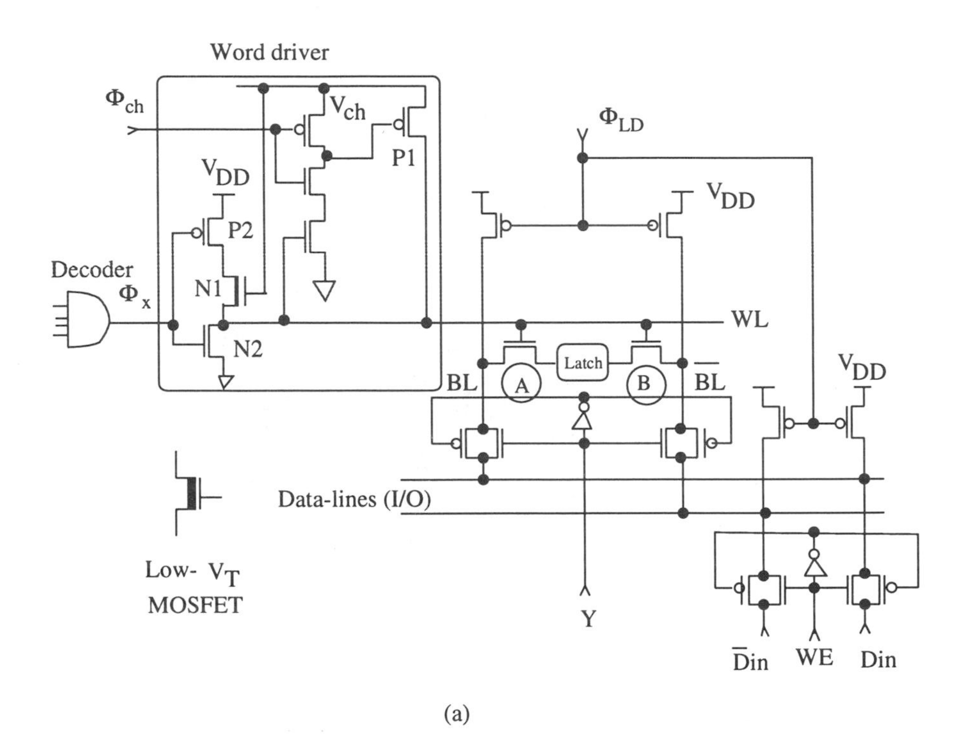

Figure 6.35 (a) TSW read/write circuitry [19].

A Two-Step Word (TSW) voltage technique has been proposed by Ishibashi et al. [19] to solve the cited problem. Fig. 6.34 shows the block diagram of the proposed memory. The boosted-level generator[3] generates a voltage $V_{ch} = 1.5V$ for $V_{DD} = 1V$. The word-line voltage has two-steps, one is V_{DD} and the other is V_{ch}. The circuitry for the TSW method is shown in Fig. 6.35(a). When ϕ_x goes to zero, the signal WL is raised to $V_{w1} = V_{DD}$. Then when ϕ_{ch} is asserted with a high level, equal to V_{ch}, the transistor P_1 turns ON and then the WL level is increased to $V_{w2} = V_{ch}$. In this case, the low threshold voltage device N_1 turns OFF and the inverter formed by the transistors P_2 and N_2 is isolated to reduce any leakage current.

Fig. 6.35(b) shows the voltage waveforms for the TSW circuitry in read/write modes. During the write cycle, the high node A is first charged to a low voltage,

[3]The boosted level generator is presented in Section 6.2.11.

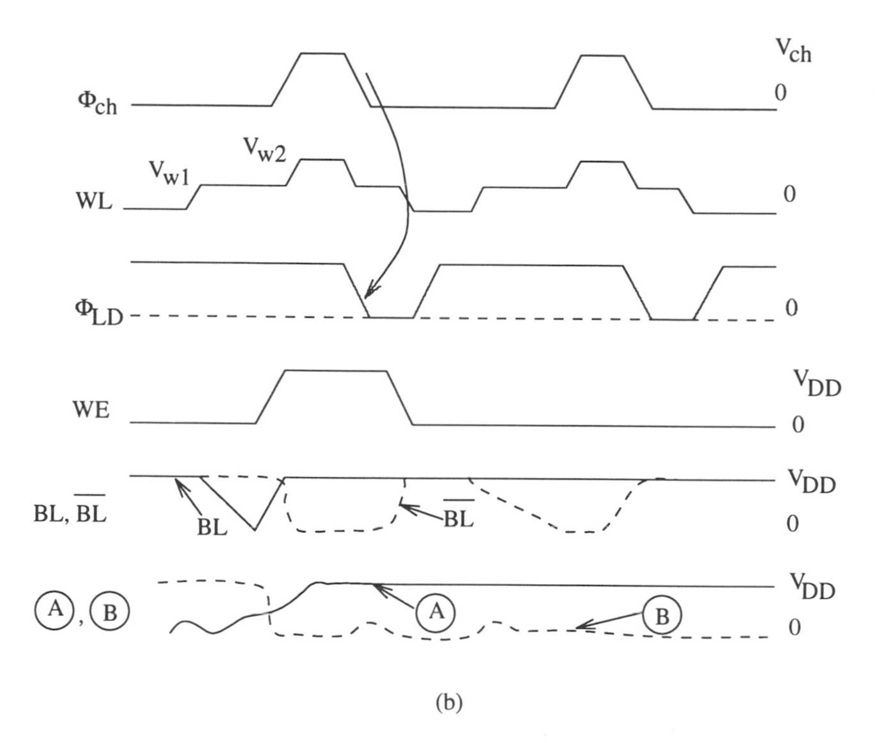

Figure 6.35 *(continued)* (b) Voltage waveforms [19].

then raised to V_{w2}. The bit-lines are initially floating, then precharged at the end of write cycle. In the next read cycle, the bit-lines are floating. Before the word-line voltages rise to V_{w2}, the cell discharges $\overline{BL}$ through the low node B. Thus, when the word-line has risen to V_{w2}, current does not flow in the cell and the node B stays at low level voltage. Note that this technique requires multi-V_T CMOS devices and causes delay in writing because the bit-lines are discharged before writing.

However, the low-voltage SRAMs discussed above require a relatively high threshold voltage $V_T \geq 0.5V$. Thus, their speed is quite slow. As an example, a 256-Kb SRAM with full CMOS memory cells attained 3 μs access time at 1 V power supply using 0.8 μm CMOS technology [22]. The active power at 0.1 MHz is 0.2 mW and the standby power is 5 nW. Another example is a 1-Mb SRAM with full CMOS memory cells which achieves 200 ns access time at 1 V power supply using 0.5 μm CMOS technology [23]. The active current at 1 MHz is 0.1 mW and the standby current is 10 nW. Note that if the threshold voltage is too low for ultra-low voltage applications, all the circuits composing the SRAM will suffer from the subthreshold current leakage. Thus, the retention current increases drastically causing a serious problem for low-power applications. Moreover, the temperature effect and the threshold voltage variation enhance this current. So far, no practical solution has been proposed.

6.2 DYNAMIC RAM

The first dynamic RAM (DRAM) was introduced in 1970 with a capacity of 1-Kb. Since then, the density has quadrupled every three years (one generation). Recently, some experimental 256-Mb DRAMs were reported [24, 25, 26]. At present, low-voltage 16-Mb DRAMs run in high-volume production. The development of these higher densities have made DRAMs the cheapest per bit compared with other types of memories. They are widely used as the main memory of mainframes, PCs, and workstations. The access time has been decreased from few hundreds of ns for 4-Kb DRAMs to less than 50 ns for 256-Mb. Also the power dissipation has been reduced by an order of magnitude from 4-Kb capacity to 256-Mb capacity reaching 50 mW at 1.5 V power supply. The area of the memory cell has been reduced from more than 100 μm^2 for 64-Kb DRAM to 1.28 μm^2 for 64-Mb DRAM.

In addition to the trend for higher-density standard DRAMs, there are two other trends: Low-Power (LP) DRAMs, and high-speed DRAMs. The high-speed DRAMs sacrifice the retention current as well as density for faster access time. Low-voltage low-power DRAMs are becoming important particularly for battery operation. LP DRAMs extend the time of the battery operation as well as battery back-up operation. The active current of LP DRAMs has been lowered. The data-retention current has also been reduced but still it is about one order of magnitude higher than those of SRAMs[4]. The 5 V power supply standard has been used for many DRAM generations from 64-Kb to 16-Mb externally. This was followed by 64-Mb DRAM powered with external 3.3 V not only to reduce the power dissipation, but also to ensure reliability. The gate oxide reliability limits the maximum voltage which is related to the boosted voltage inside the chip. Regarding the internal voltage, the 5 V can be used to a maximum DRAM capacity of 4-Mb. At 16-Mb generation, the internal voltage is 3.3 V while maintaining external 5 V with on chip voltage

[4] This comparison is made for 4-Mb memories.

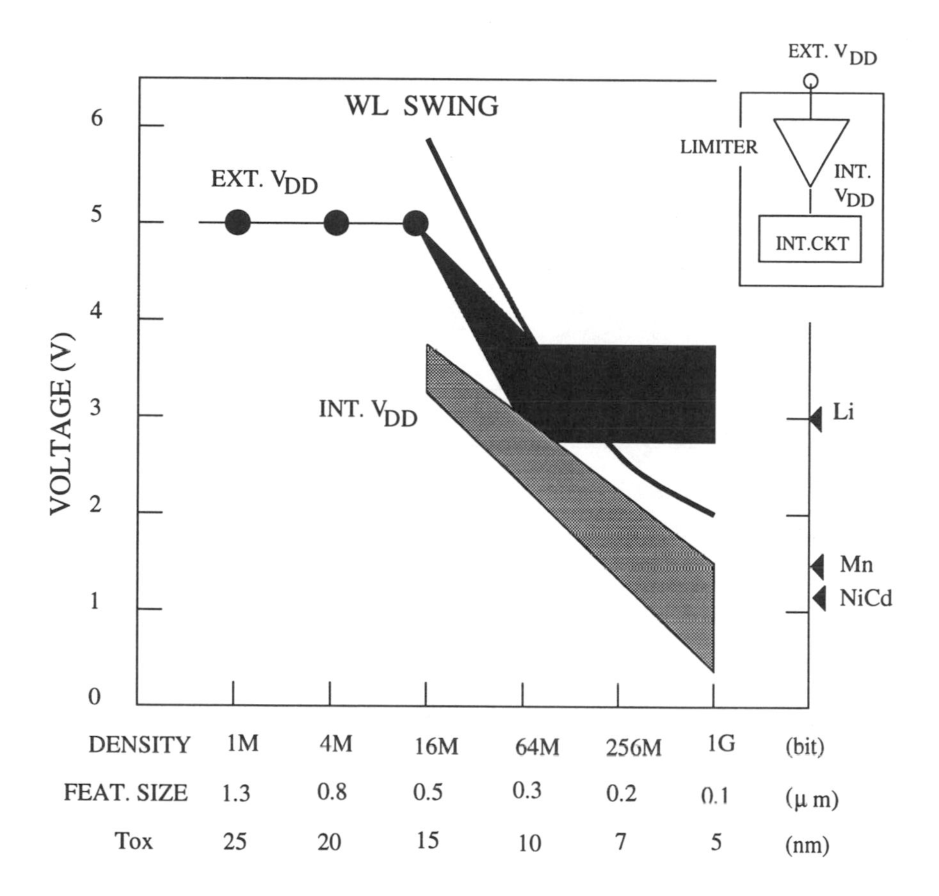

Figure 6.36 Trends of DRAM supply [28].

down converter [see Section 6.3]. However the 3.3 V external power supply will dominate.

Recently, activities to realize 1.5 V battery-operated DRAMs are accelerating the trend in low-voltage operation [27, 28, 29]. Fig. 6.36 shows the trend of DRAM supply [28]. In battery operation, the chip must be operated on a variety of batteries with various supply voltages for a long-term and under supply fluctuations.

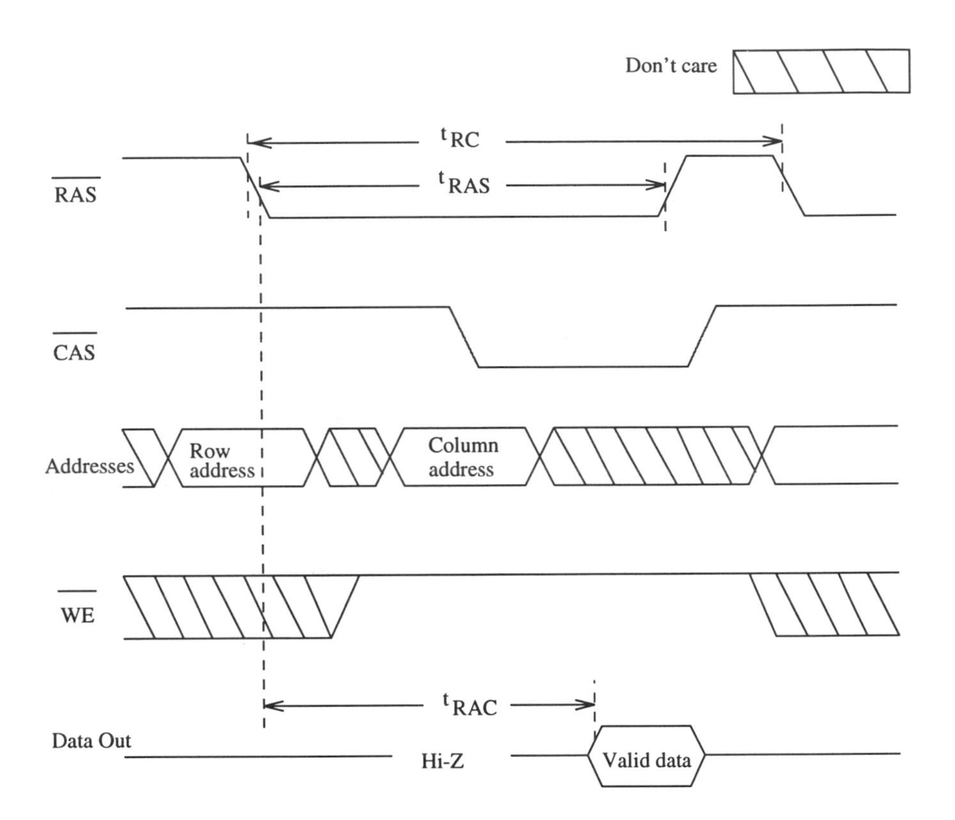

Figure 6.37 Example of address multiplexed DRAM timing during read cycle.

6.2.1 Basics of a DRAM

In general the pins of a DRAM are :

- Address; which is separated in time with two separate fields. These fields are the row and column address.
- Row Address Strobe ($\overline{RAS}$). The row address is clocked by this signal.
- Column Address Strobe ($\overline{CAS}$). The column address on the multiplexed pins is clocked by this signal.
- Write Enable ($\overline{WE}$).

- Input/output data pins.
- External power supply pins.

It is clear that the multiplexed address penalizes the access delay so for fast DRAMs separate address input pins can be used. The multiplexing permits the reduction of the pin count and the cost of packaging. An example of DRAM timing, using the address multiplexing during read mode, is shown in Fig. 6.37. Some important times are shown, such as the access time from $\overline{RAS}$ low, t_{RAS}, the row address strobe cycle time (or cycle time), t_{RC}, and the row address strobe low-state time, t_{RAS}.

Fig. 6.38 shows a generic 4-Mb DRAM architecture. It uses almost the same circuit techniques as SRAM except for memory array. Some additional circuits are needed such as a Back Bias Generator (BBG), a Half-Voltage Generator (HVG), an optional Voltage-Down Converter (VDC), a Reference Voltage Generator (RVG), and a boosted voltage generator circuit. The substrate back-bias voltage is indispensable for stable operation of the DRAM array. The half-voltage generator permits generation of the precharge level for the bit-lines to half-V_{DD} as it is explained in the following sections. The reference voltage generator is needed for the VDC. The boosted voltage generator uses a charge-pump circuit and permits overdriving of the word-line WL to a voltage higher than V_{DD}. More details on these circuits, composing the DRAM, are given in the following sections.

6.2.2 DRAM Memory Cell

CMOS DRAMs, with three-transistor and four-transistor cells, were used in 1- and 4-kb generations. One-transistor (1T) cell offers smaller chip size and low cost. These justify the process complexity to fabricate the 1T cell, particularly its capacitor.

A schematic of a 1T DRAM cell is illustrated in Fig. 6.39(a). The charge is stored in capacitor C_s. To prevent loss of the stored information, the capacitor must be refreshed within a specific time with special circuitry. The bit line has a capacity C_{BL} including the parasitic load of the connected circuits. Typical values for the storage and the bit-line capacitors are 30 fF And 250 fF, respectively. The ratio $R = C_{BL}/C_s$ is very important for the sensing operation.

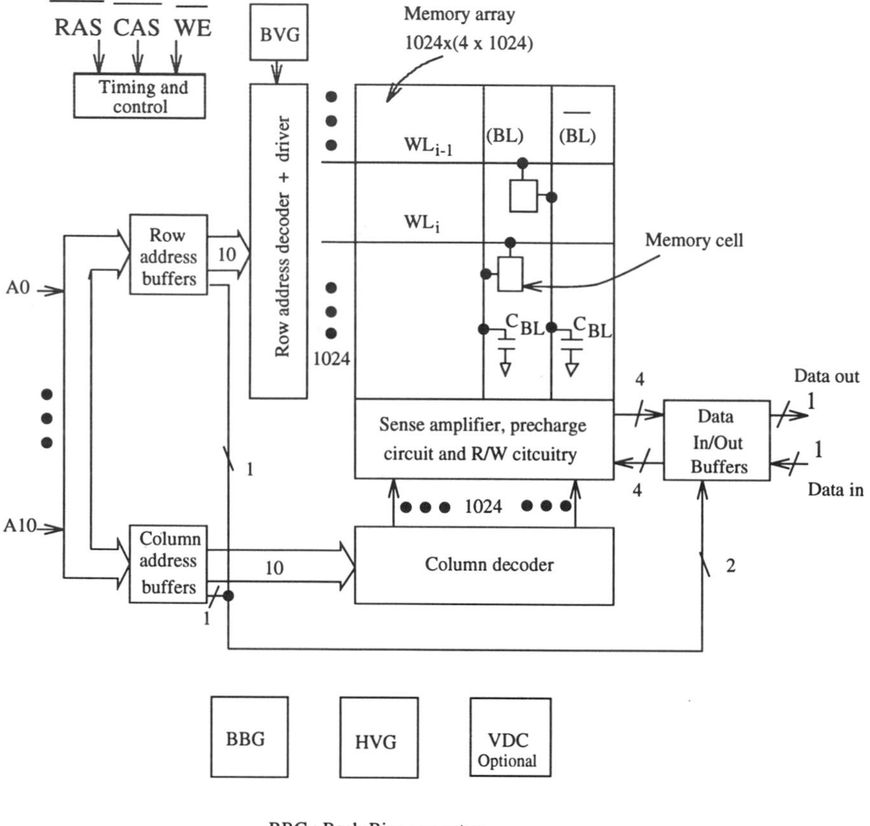

BBG : Back-Bias generator
BVG : Boosted voltage generator
HVG : Half-voltage generator
VDC : Voltage down converter

Figure 6.38 Generic 4-Mb DRAM architecture.

During the read operation (WL is selected) the bit-line voltage changes by

$$V_s = \Delta V_{BL} = (V_{MC} - V_{BL})\frac{C_s}{C_{BL} + C_s} \tag{6.9}$$

where $(V_{MC} - V_{BL})$ is the difference between the memory cell voltage and the bit-line voltage before the selection of the cell. A typical value of the difference is $V_{DD}/2$. Hence, we have for the bit-line sense signal

$$V_s = \frac{V_{DD}}{2(1+R)} \tag{6.10}$$

For 3.3 V supply voltage, and using a ratio $R = 8$ for 16-Mb DRAM, the sense signal $V_s = 180\ mV$. This small voltage change, of the bit-line, requires sensing circuits. For low-voltage operation, V_s decreases, thus a low ratio R is required. This is possible by reducing C_{BL} and increasing C_s.

C_s was implemented using a simple planar-type capacitor as shown in the structure of Fig. 6.39(b). This structure was used in DRAMs with capacity up to 1-Mb. With the increased density, many three-dimensional approaches were used for DRAMs with capacity higher than 1-Mb. One approach is to stack the capacitor over the access transistor (STC cell). Another approach is to use a trench capacitor. For more details on advanced cell structure the reader can consult [30, 31].

The signal charge ($Q_{sig} = C_s \Delta V_s$) transferred to the bit-line during a read operation should have enough margin against noise. The sources of noise are the following :

- bit-line noise; which is caused by capacitive couplings and other sources;
- leakage charge; which is mainly due to the leakage in the junction of the NMOS transistor of a 1T memory cell; and
- α-particle-induced soft errors.

In the early DRAM, the plate of the capacitor was grounded to reduce the noise injection from the V_{DD} power supply. However, for multi-Mb DRAMs, a $V_{DD}/2$ bias for the cell plate was used. This scheme has several advantages such as, the reduction of the stress on the thinner oxide of the storage capacitor, and the reduction of supply voltage noise. Many 1-Mb DRAMs have used this cell biasing scheme.

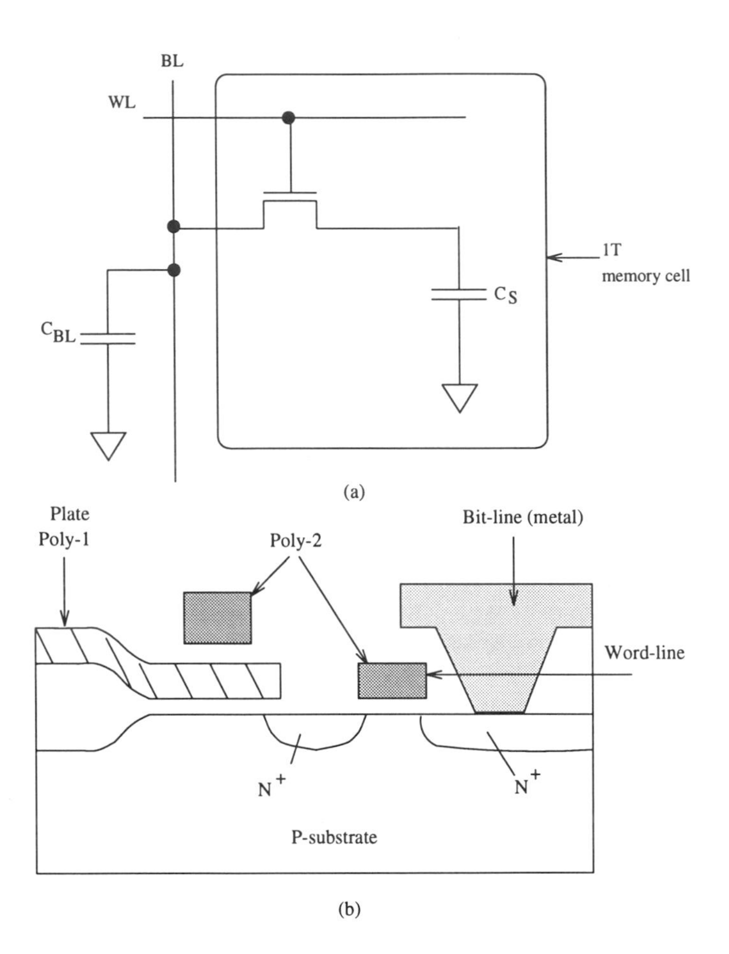

Figure 6.39 (a) One-transistor (1T) dynamic RAM cell; (b) planar-type capacitor using two polysilicon layers.

For Gb DRAM cell design with reduced V_{DD}, the ratio R should be reduced. This is possible by reducing the bit-line capacitance, C_{BL} and increasing the storage capacitance C_s. On the other hand, the area occupied by C_s should be reduced to increase the chip capacity. One solution for C_s reduction is the use of a capacitor insulator with extremely high permittivity ϵ such as Ferroelectric materials such as $BaSrTiO_3$ film. Consequently a simple planar-type capacitor can be used in that case.

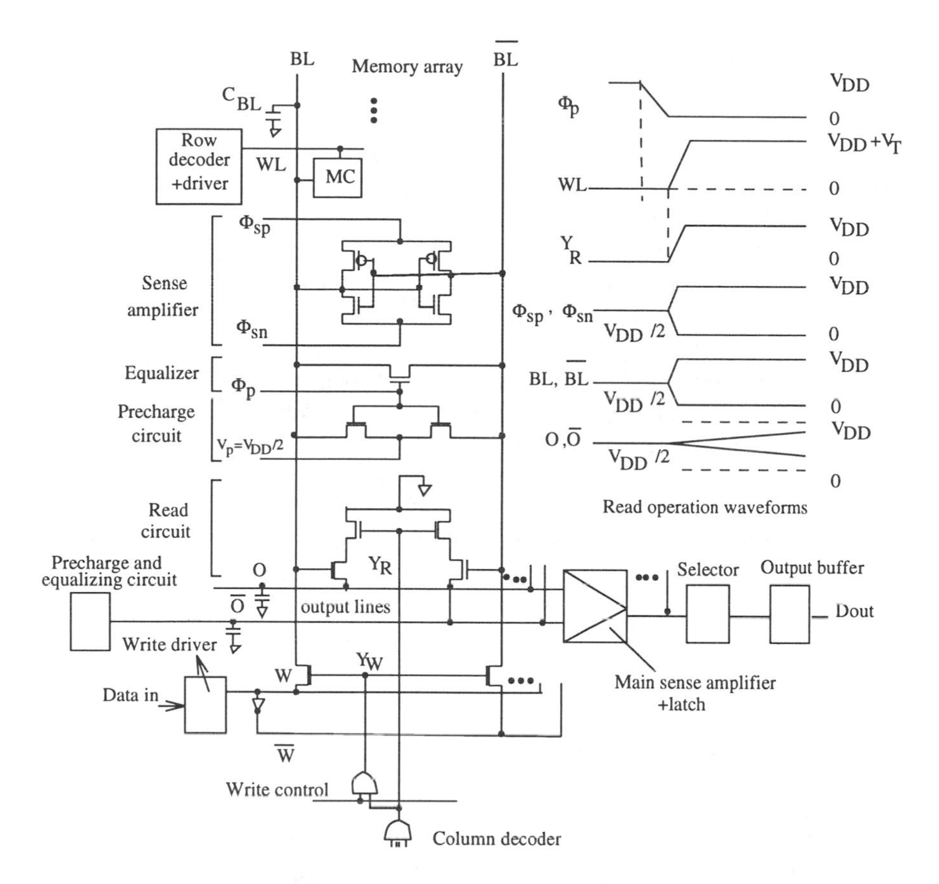

Figure 6.40 Read/write operation in the 1T dynamic memory cell.

6.2.3 Read/Write Circuitry

Fig. 6.40 illustrates the different circuits for read, write precharge, and equalization functions. The read operation is performed as follows. Initially both the bit-lines (BL and $\overline{BL}$) are precharged to V_p which is equal to $V_{DD}/2$ and equalized before the data reading operation. This half-V_{DD} precharge technique permits the reduction of the active power dissipation as discussed in Section 6.2.9. The signal WL is selected by the row decoder. The high level of the word-line voltage has to be greater than V_{DD} to increase the stored charge in the memory cell. The selected memory cell is connected to one bit-line. Then ΔV_{BL} (100 to 200 mV) appears between the bit-lines, immediately after the word-line rises. Then it is amplified by the latch-type CMOS sense amplifier which is connected to both bit-lines. After the sensing and the restoring operations, the voltage levels of the bit-lines have a full-swing condition. The bit-line differential voltage signal is transferred to the differential output-lines (O and $\bar{O}$), through a read circuit. The signal YR is selected almost at the same time with WL. The parasitic capacitance of the output-lines is large (a typical value $2\,pF$ for 4-Mb DRAM), and the readout circuit would need a long time to amplify the output-line signal. A main sense amplifier is used to read the output-lines, then the data is selected among several main SAs connected to different sub-arrays. Finally it is transferred to the output buffer.

The DRAM cell readout mechanism is destructive, and hence the same data must be written to the cell on every read access. Consequently, on each bit-line pair, a CMOS amplifier is needed to amplify and restore the level. This mechanism is not needed in SRAMs since the read operation is non-destructive.

In the write mode, the YW signal is selected by a column decoder as shown in Fig. 6.40. In this case, the write control signal is activated. The selected bit-lines are connected to a pair of write-lines W and $\bar{W}$ and the data are transferred to the memory cell when WL goes HIGH.

6.2.4 Low-Power Techniques

Fig. 6.38 can be used to identify the different sources of power dissipation in a DRAM. For simplicity we assume that the internal supply voltage is the same compared to the external one. The total power dissipated is the addition of two components; the active power and the data-retention power. The active power is the sum of the power dissipated by the following components;

- The decoders (row and column);
- The memory array. This is the dominant one. If m memory cells are connected to the word-line, the active power of memory array is given by
$$P_{mem\_array} = m \times P_{actm} \tag{6.11}$$
Where P_{actm} is the power dissipated in active mode when selecting the m cells. It is given by
$$P_{actm} = C_{BL} \Delta V_{BL} V_{DD} f \tag{6.12}$$
- The sense amplifier;

- Other circuits such as refresh circuit, substrate back-bias generator, boosted level generator, a voltage reference circuit, and a half-V_{DD} generator. These circuits also dissipate a DC current;
- The rest of periphery such as main sense amplifier, input/output buffers, write circuitry etc.

Note that the power dissipated by the pads is not included.

To reduce this active power, many techniques can be used and are summarized as follows :

- Reducing all capacitances; particularly the bit-lines and word-lines capacitances. As seen from Equations (6.11) and (6.12) $m \times C_{BL}$ should be reduced. Techniques which permit this are partial activation multi-divided bit-line and shared I/O [see Section 6.2.7]. Also to reduce the word-line capacitance, a technique such as partial activation of multi-divided word-line can be used [see Section 6.2.8];
- Lowering the internal V_{DD}. This includes the generation of half-V_{DD} for precharging the bit-lines and reducing the external supply voltage; and
- Reducing the DC power required by periphery circuits. This is possible by using static CMOS decoders and pulse operation technique using an ATD circuit (as in SRAMs).

The data retention power in a DRAM is mainly due to refresh operation and the DC power (I_{DC}) due to peripheral circuits such as BBG, BVG, VRG, HVG. The refresh process is performed by reading the m cells connected on each word-line and restoring them. Thus, n refresh cycles are needed for $n \times m$ DRAM. It can be estimated by

$$P_{ret} = \frac{P_{dyn}}{f} \frac{n}{t_{refresh}} + P_o \tag{6.13}$$

where $\frac{P_{dyn}}{f}$ is the total dynamic energy (f is the operating frequency) and $n/t_{refresh}$ is the refresh time of m cells. To reduce the power dissipation due to the refresh mode, one obvious technique is to increase $t_{refresh}$ and decrease n. P_o is the AC and DC power dissipated by the other circuits such as VDC, BBG, RVG, HVG, and boosted level generator. To reduce this power many

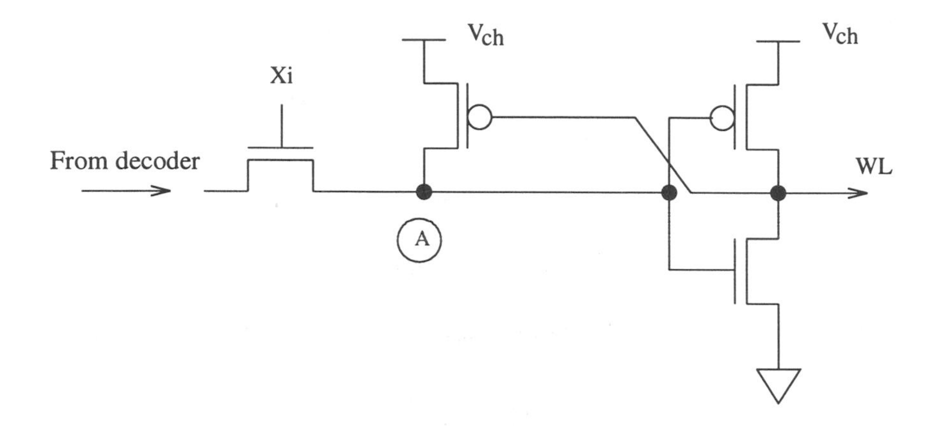

Figure 6.41 Static CMOS word-line driver.

techniques can be used. One of them is to reduce the frequency of operation of circuits which have high-power during active mode when operating in data retention mode. Another one is to reduce the DC current of these circuits using, for example, dynamic concept.

In the following sections, the circuit techniques to reduce the active and data-retention power dissipation are presented. Also, different circuits constituting a DRAM are described and low-power issues of these circuits are discussed.

6.2.5 Decoder

In a DRAM, the static CMOS NAND decoders are used. The power is reduced by using the predecoding technique. This topic is discussed more in Section 6.1.6 for SRAMs. Fig. 6.41 shows a static CMOS word-line driver. The boosted level, V_{ch}, generated by an internal charge pump circuit, is used in the output stage. When node A is high at $(V_{DD} - V_T)$, the output inverter leaks a high DC current because this is lower than V_{ch} by at least two threshold voltages, subject to body effect. Therefore, a small size PMOS transistor P_f is used to restore the level of the node A to V_{ch} level. Also this transistor permits the latching of the low output level (ground). The X_i signal, when selected, is normally at V_{DD}. The unselected X_i is discharged to ground in the selected block before the row decoder selection.

6.2.6 Sense Amplifier

The main sense amplifier is the main source of DC current during the active mode. It employs the same sense amplifier discussed in Section 6.1.8 for SRAMs. The DC current can be shut down using the ATD technique.

6.2.7 Bit-Line Capacitance Reduction

Reducing the bit-line capacitance not only reduces the power dissipation but also improves the signal-to-noise ratio of the memory cell. This is possible by two approaches :

1. Reducing the number of memory cells n per bit-line. In this case, multi-divided bit-line technique is used.
2. Reducing the junction capacitances of connected transistors such as access devices. One possible solution is the back-bias of the substrate containing these devices. A negative voltage on the substrate permits to reduce the junction capacitance. In addition, the use of the trench isolation technique for CMOS devices rather than the LOCOS isolation results in almost 50% reduction in capacitance.

Fig. 6.42 shows the principle of multi-divided bit-line architecture for the memory array. The $m \times n$ array is now divided into m columns by k sub-arrays. Each sub-array contains n/k word-lines. In this scheme the bit-line capacitance C_{BL} is reduced by dividing it into k sections. Also the signal-to-noise of the cell is improved. Fig. 6.43 illustrates an example of 1-Mb DRAM [32]. The memory is divided into two parts; upper and lower. One part is divided into $N = 16$ sub-arrays and the total number of sub-arrays is $k = 32$. Two sub-bit-lines share one amplifier which are selected by isolation signals, ISO and $\overline{ISO}$. Thus, a partial activation is performed by selecting only one SA along the bit-line. The switch SW is controlled by the Y signal from the shared column decoder. This signal runs in parallel to the bit-lines and uses metal-2. Thus, the I/O is shared by two sub-bit-lines. This principle results in reduced power dissipation and chip-size. It has been used for many DRAM generations up to 16-Mb.

6.2.8 Multi-Divided Word-Line

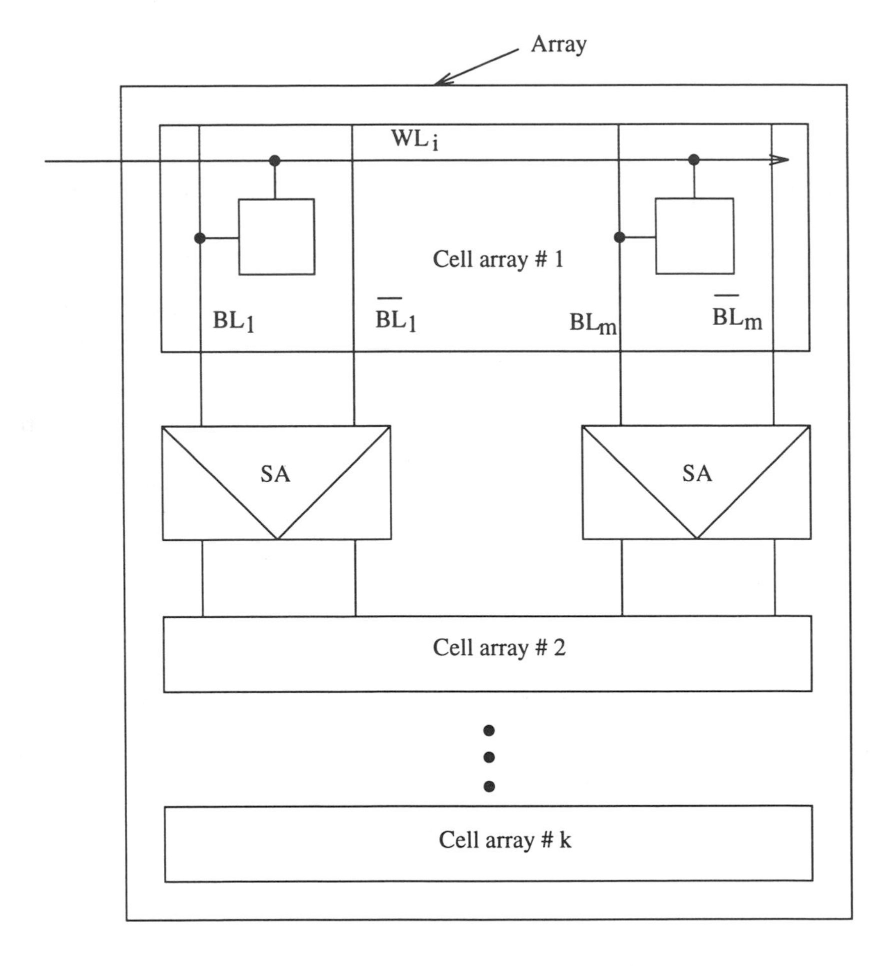

Figure 6.42 Principle of multi-divided array.

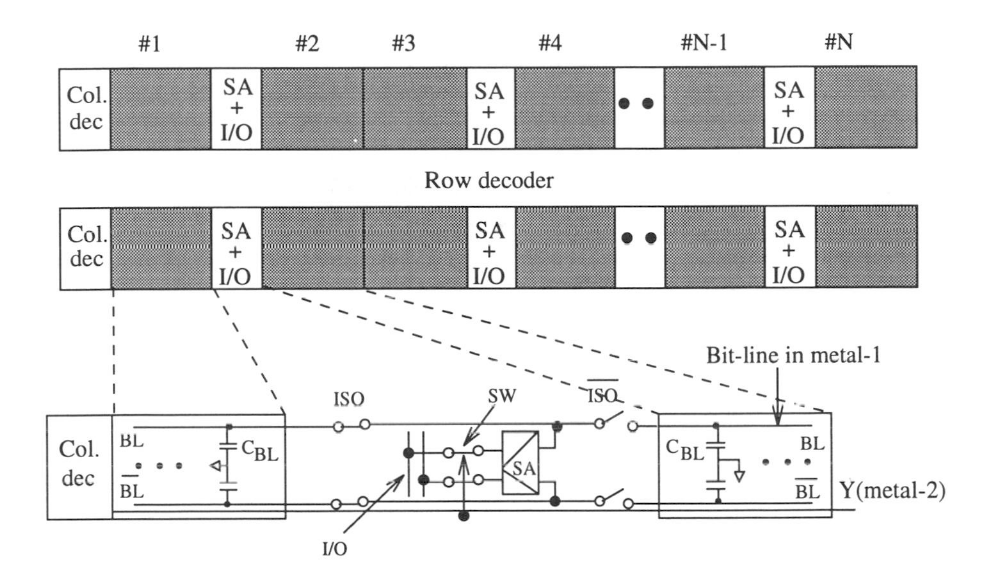

Figure 6.43 Multi-divided bit-line architecture with shared SA, I/O and column decoder [32].

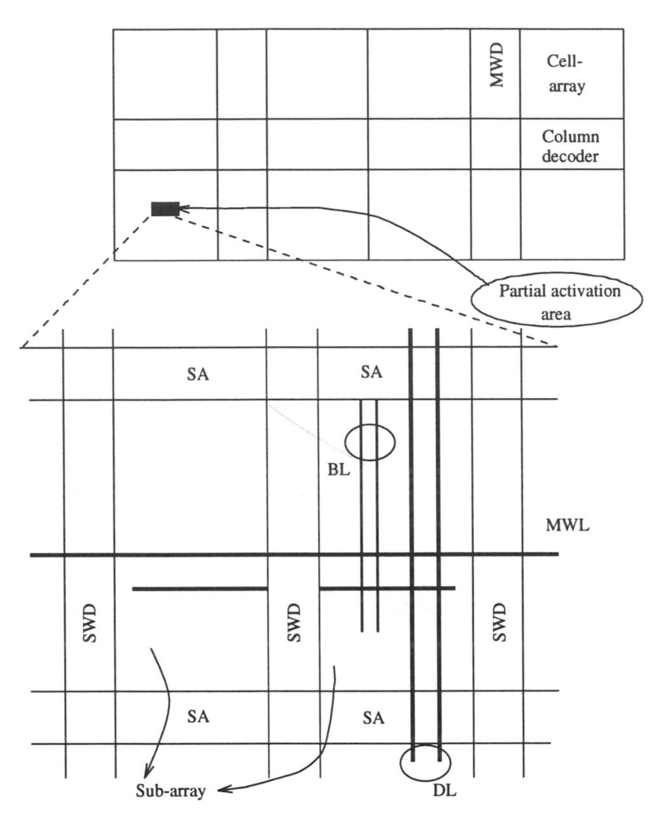

Figure 6.44 Multi-divided array structure with dual-word line.

Fig. 6.44 shows the hierarchical word-line structure proposed for a 256-Mb DRAM [26]. This scheme resembles the one used in the SRAM. The DRAM cell array is divided into several blocks and each one itself is divided into sub-arrays. The Sub-Word-Line (SWL) circuitry is embedded in the sub-array. Only one SWL is activated by the Main-Word-Line (MWL) and the row select signal. It is common to two sub-arrays as shown in Fig. 6.44. Thus, only two cell sub-arrays are activated which represents a very small portion of the total cell arrays. In the case of the 256-Mb, the active cell array size is 1/1024 of the total number. This structure results in reduced active current and ground bounce.

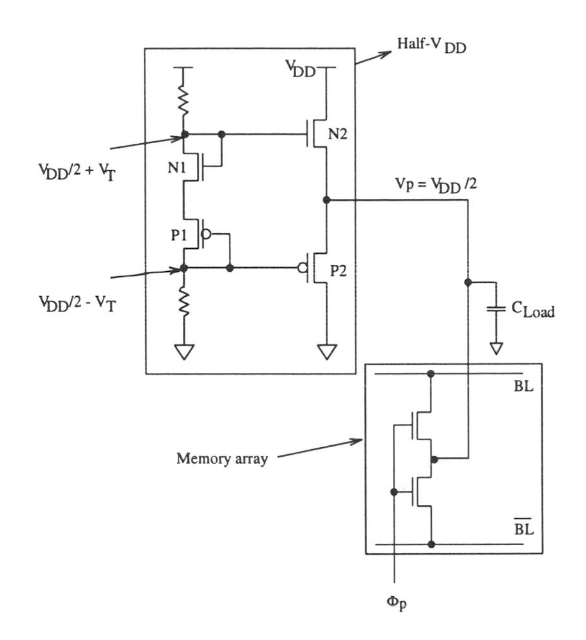

Figure 6.45 Conventional half-V_{DD} voltage generator.

6.2.9 Half-voltage Generator

One efficient technique to reduce the memory array operating current is half-V_{DD} bit-line precharge [33, 34]. During the sensing operation, one bit-line switches from $V_{DD}/2$ to V_{DD} and the other switches to zero. This results in a power saving of almost half, compared to the full-V_{DD} precharge case, as well as peak current. Note that the reduction in peak current leads to suppression of noise. In addition, the precharge time is reduced and the cycle time is shortened. This precharging technique has been used starting from 1-Mb DRAM generation.

A simple circuit which permits the generation of this half-V_{DD} is shown in Fig. 6.45. The HVG circuit is composed of two stages. One stage is a bias generator which generates two voltage levels; $(V_{DD}/2 + V_T)$ and $(V_{DD}/2 - V_T)$. The second one is the push-pull output stage which generates the level $V_{DD}/2$ distributed to the memory array. The load capacitance, seen by the push-pull output stage, is huge. A typical value is a few tens of nF. A typical response time when the circuit is powered-up is few tens of μs at 3.3 V power supply voltage for 16-Mb DRAM. This HVG circuit has many disadvantages such as

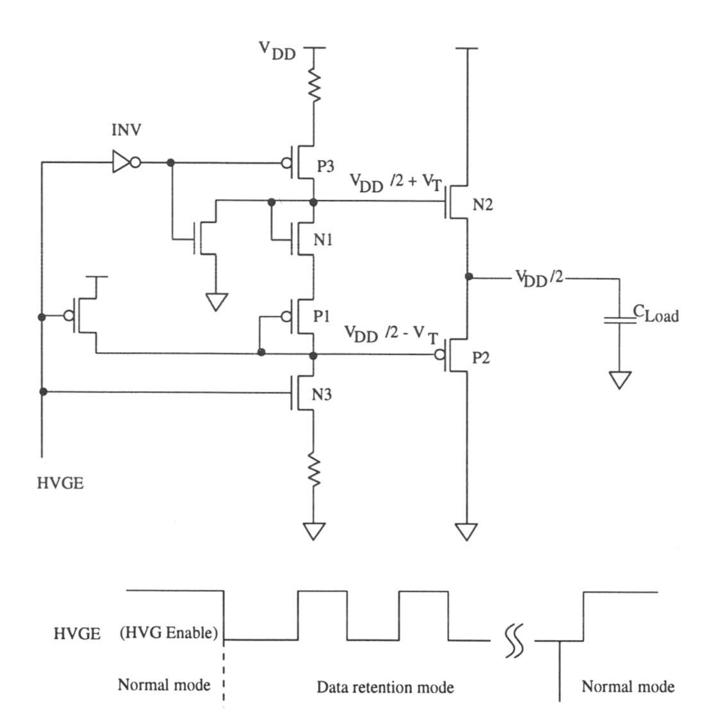

Figure 6.46 HVG circuit with clocked mode in data retention mode.

sensitivity to V_T variation due to N_1 and P_1 devices. Another problem is the poor drivability of the two MOS devices N_2 and P_2, because their $V_{GS} - V_T$ is almost 0. Moreover, this circuit consumes a DC current which is not acceptable for battery-operation applications.

To reduce the DC current of the HVG circuit of Fig. 6.45 to sub-μA range, the concept of dynamic HVG can be used as shown in Fig. 6.46. In the normal mode, the HVG Enable ($HVGE$) signal is ON. Then P_3 and N_3 are ON and the HVG circuit operates exactly like the one discussed in Fig. 6.45. The load capacitor C_{load} permits the holding of the V_p level. In the data retention mode, N_3 and P_3 are clocked with low-frequency signal generated by $HVGE$. When $HVGE$ is low, N_3 and P_3 are OFF and no DC current is consumed. The output transistors N_1 and P_1, are OFF and the HVG circuit is in the hold mode. When $HVGE$ is high, the circuit consumes a DC current for a short time and the output is sampled. The average current consumed is related to the duty ratio of the *HVGE* signal in the data-retention mode. To solve the other problems cited an HVG circuit was proposed in [28] but this circuit dissipates a DC current.

6.2.10 Back-Bias Generator

The back-bias voltage V_{BB} is utilized in a DRAM to reduce the subthreshold current and the junction capacitances, to improve device isolation, to enhance latch-up immunity, and protect the circuit against voltage undershoots of the input signals. Also this voltage can be used to compensate for some device parameter variations.

For NMOS devices with P-well (substrate) a negative V_{BB} is generated by pumping electrons out of the ground node and into the substrate. A typical V_{BB} generator configuration is shown in Fig. 6.47. This circuit is known as charge pump. The node A oscillates between V_T and $(V_T - V_{DD})$. During the high side of the cycle, the node A must be at least at V_T to pump the charge from the ground. On the low side of the cycle, the node A must be a V_T drop below V_{BB}. The output node V_{BB} stablizes at a voltage level equal to $(2V_T - V_{DD})$, since the load capacitance is huge. The clock (clk) is generated by a ring oscillator with N (N is an odd number) stage. The frequency f of oscillation, is approximately $1/(2Nt_d)$, where t_d is the delay of one inverter. The buffer is needed to drive the huge C_{pump} capacitance. The average current pumped out of the substrate is approximated by

$$I_{pump} = (V_{BB} - V_{BBmin})C_{pump}f \tag{6.14}$$

where V_{BBmin} is the back-bias voltage when no current is pumped and is equal to $(2V_T - V_{DD})$ (optimum value). During the start-up a large current is pumped; equal to $(-V_{BBmin}C_{pump}f)$.

Another PMOS version, of the charge-pump circuit, is shown in Fig. 6.48. Since the gate voltage of P_1 only reaches $-V_{DD}$, V_{BB} is pumped to a limit of $(V_T - V_{DD})$. For $V_{DD} = 5V$, the NMOS and PMOS charge pump circuits generates typical voltages of -3 and -4 V, respectively. However, for 3.3 V power supply, the PMOS version can generate a low negative voltage of -2.5 V which is lower than the one generated by the NMOS version at this power supply.

Fig. 6.49 shows a pumping circuit which avoids the V_T losses and hence is suitable for low-voltage operation [35]. When the clock (clk) is low, the voltage of the node A reaches $(|V_{Tp}| - V_{DD})$, and the PMOS transistor P_1 clamps

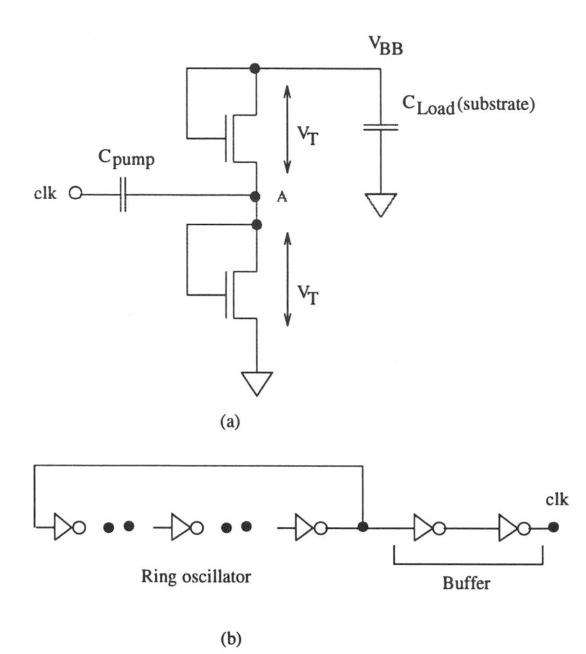

Figure 6.47 (a) Conventional charge pump circuit; (b) buffered clock generator.

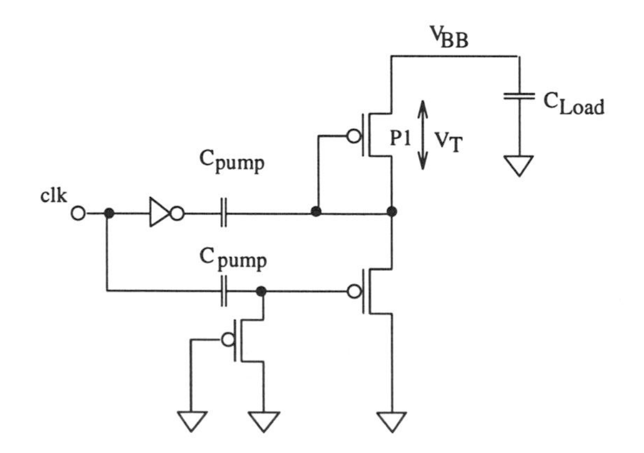

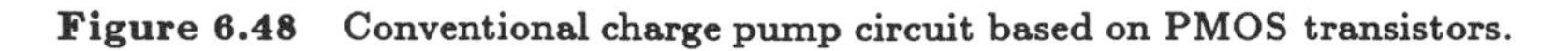

Figure 6.48 Conventional charge pump circuit based on PMOS transistors.

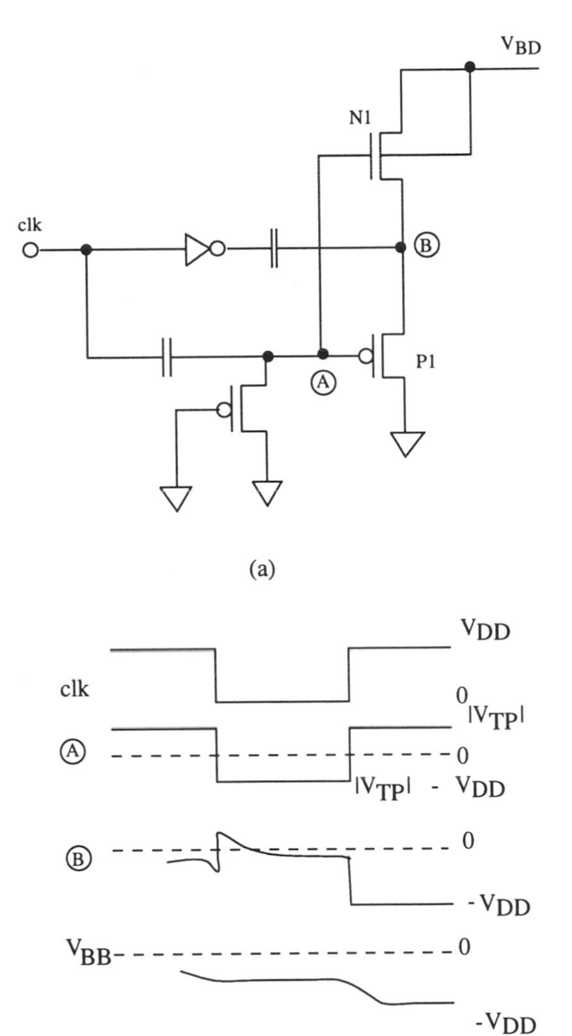

Figure 6.49 (a) Low-voltage pumping circuit; (b) voltage waveforms.

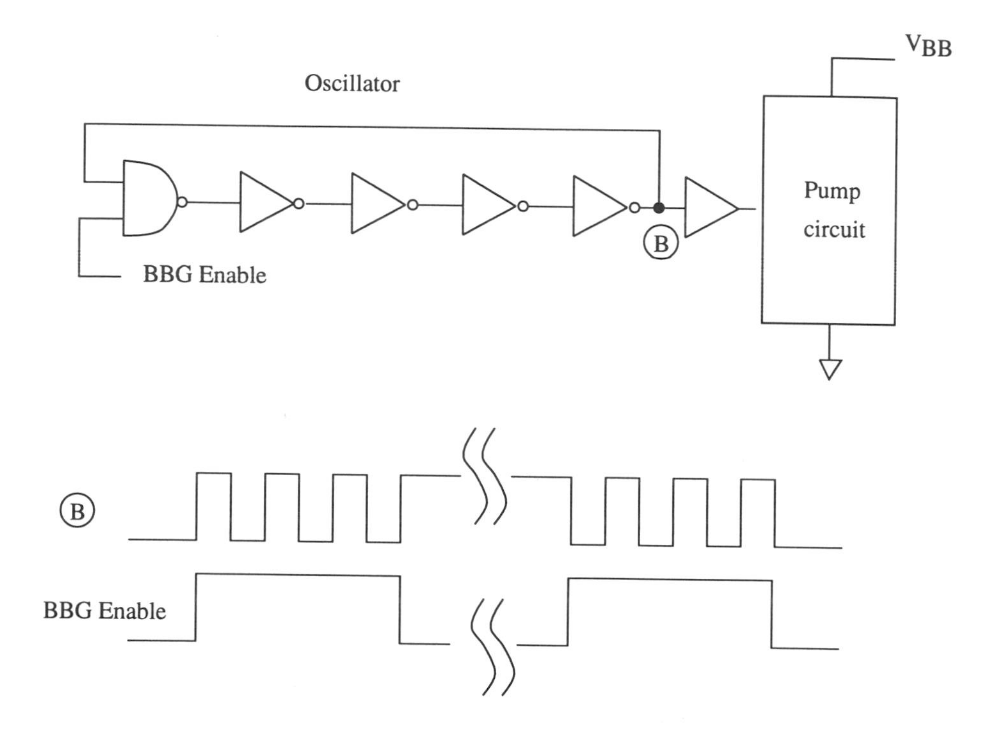

Figure 6.50 BBG for battery-backup mode.

the voltage of node B to the ground level. The V_{BB} level is in that case, $(|V_{Tp}| - V_{DD} - V_{Tn})$. When *clk* goes to a high level, the voltage of A rises to V_{Tp} and the voltage of B, by capacitive coupling, becomes $-V_{DD}$, causing V_{BB} to be equal to $-V_{DD}$. Therefore the V_{BB} will be

$$V_{BB} = max\{-V_{DD}, |V_{Tp}| - V_{DD} - V_{Tn}\} \tag{6.15}$$

This circuit needs a special triple-well structure to avoid minority carrier injection of the NMOS transistor N_1 as discussed in [35].

To reduce the power dissipation of the BBG circuit, while the DRAM is not in an active mode, the BBG can be operated at low frequency. Fig. 6.50 shows a simplified circuit diagram of the BBG circuits for low-power operation [36]. In the normal mode, the ring oscillator works all the time to retain the V_{BB} level. In the data retention mode, the BBG Enable ($BBGE$) signal is clocked with a low duty ratio. Then the ring oscillator is operating with low-frequency to refresh the pumping circuit.

6.2.11 Boosted Voltage Generator

A Boosted level circuit is needed to generate a voltage level above V_{DD} by at least V_T. The word-line driver is powered with this voltage V_{ch}. A simple boosted voltage generator is shown in Fig. 6.51. It uses the charge pump circuit technique discussed in Section 6.2.10. The output of this circuit is switching between $(V_{DD} - V_T)$ and $(2V_{DD} - V_T)$. The clock ϕ is generated by a simple ring oscillator. Another circuit which switches between V_{DD} and $2V_{DD}$ is shown in Fig. 6.51(b). It uses two non-overlapping clock phases. This second circuit configuration uses feedback NMOS devices, N_1 and N_2, to eliminate the threshold voltage loss and boost the voltage at higher voltage. This circuit is not sensitive to power supply voltage reduction.

The boosted level can not be directly used to drive the load. Thus a pass transistor is needed to isolate the switching boosted level from the load as shown in the example of the circuit of Fig. 6.52(a) [28]. The charge pump circuit $CP1$ generates at the node A, a boosted signal switching between V_{DD} and $2V_{DD}$. To control the pass transistor N, two pump circuits $CP2$ and $CP3$, and an inverter INV are needed. The pump circuit CP generates, at node B, a signal switching between $2V_{DD}$ and $3V_{DD}$ and uses the boosted voltage V_{ch}. The other pump circuit $CP3$, controls the inverter INV. The output of this inverter (node D) switches between V_{DD} and $3V_{DD}$. The output of this HVG circuit is $V_{ch} = 2V_{DD}$ and it is stable since C_{load} is large. The voltage waveforms are shown in Figure 6.52(b). This circuit is insensitive to V_{DD} reduction and can work down to sub-1 V power supply.

6.2.12 Self-Refresh Technique

Standard DRAMs require an external DRAM controller[5] to control the refresh process of memory cells. The stored charge in the memory cell decreases due to the leakage current with high rate at high temperature. The refresh time (period) $t_{refresh}$ is determined from the time needed for the stored charge in the memory cell to keep enough margin against leakage at high temperature. This indicates that $t_{refresh}$ can be lower than what is expected at room tem-

[5] Some DRAMs have an internal controller.

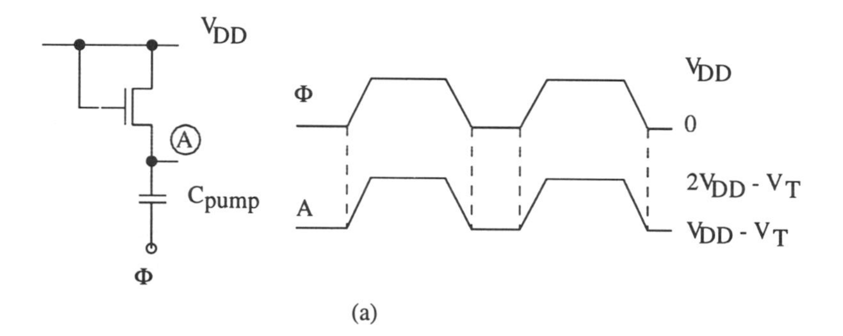

(a)

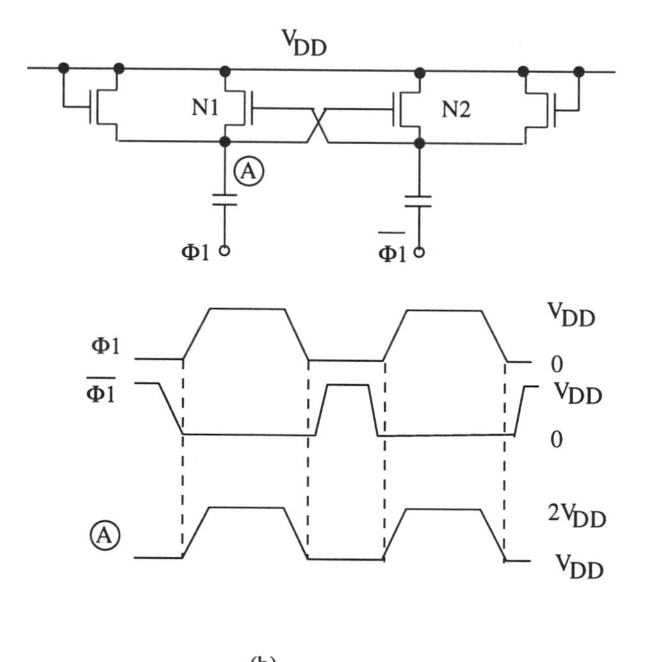

(b)

Figure 6.51 Charge pump circuits for boosted voltage.

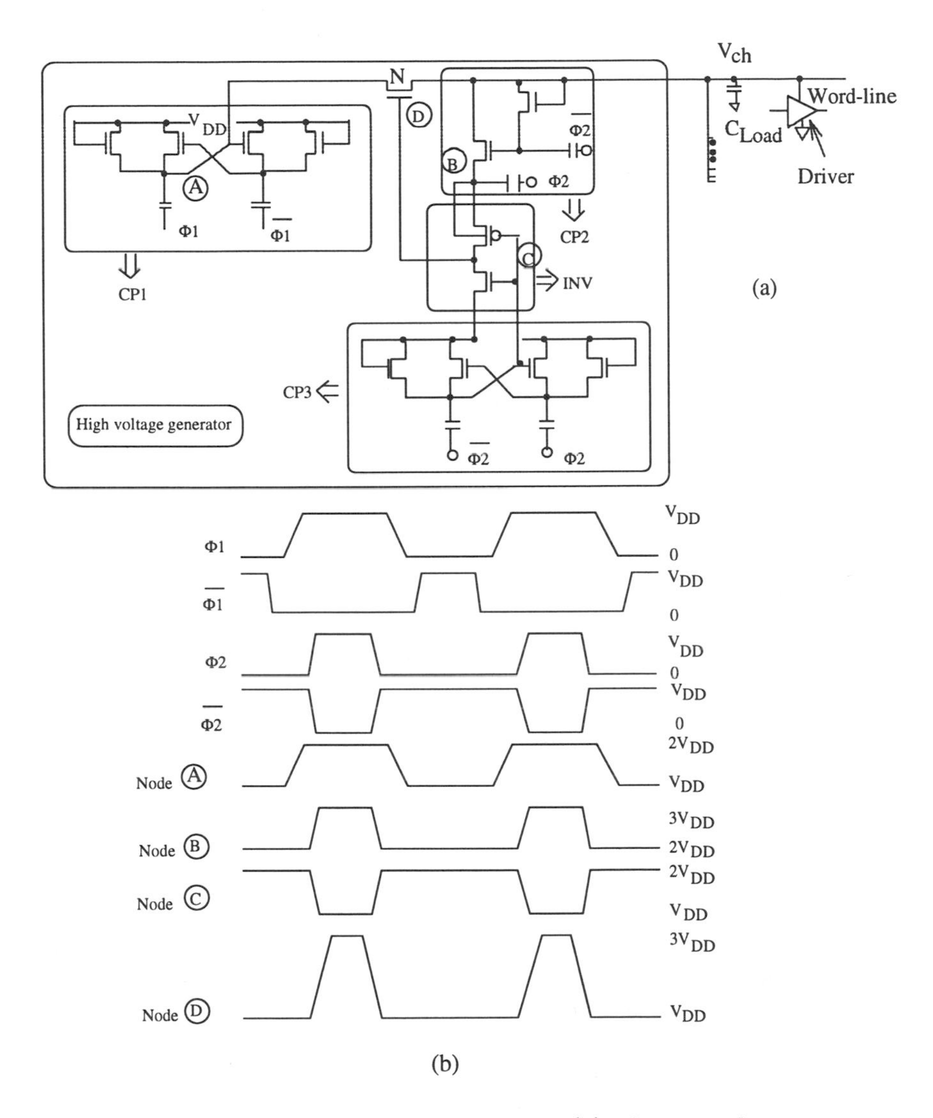

Figure 6.52 (a) High-voltage generator circuit; (b) voltage waveforms.

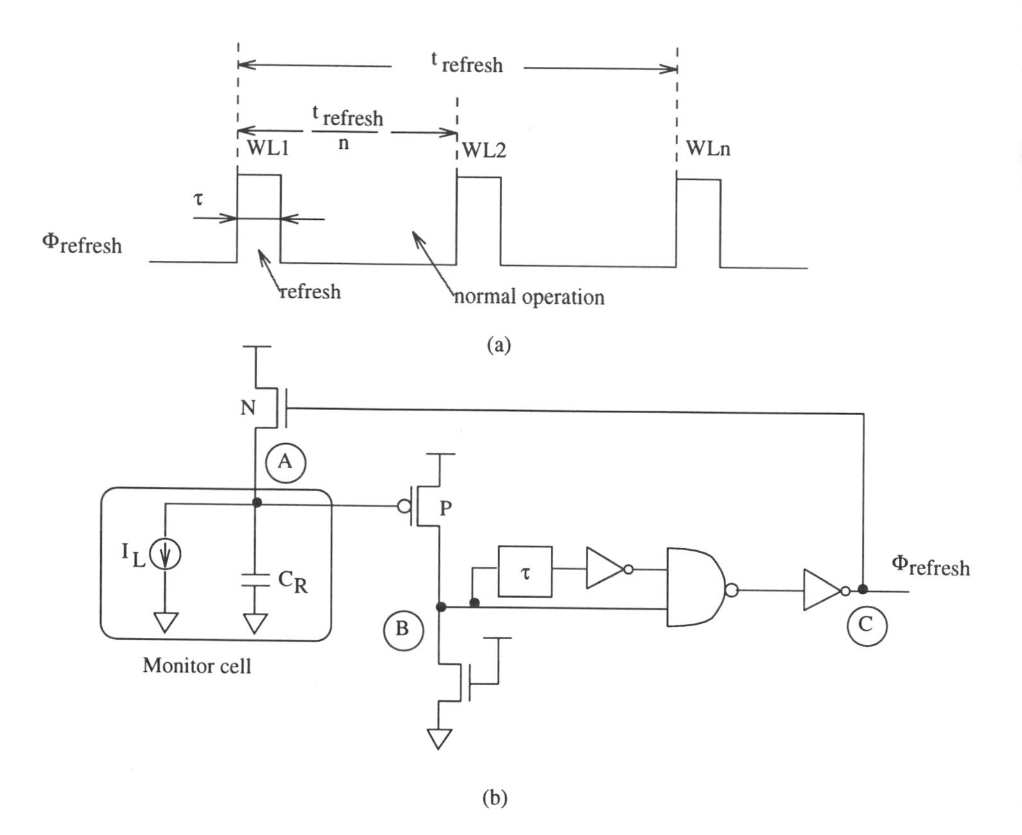

Figure 6.53 Self-refreshing technique.

perature. One way to increase this time, and hence reduce the data retention power dissipation, is to control the refresh period function of the chip temperature. Fig. 6.53 shows an on-chip self-refresh control circuit with a memory-cell leakage monitoring scheme. A refresh clock $\phi_{refresh}$ is generated automatically with a period of $t_{refresh}$. The monitor cell, which has a leakage current I_L, controls the refresh period. Initially node A is high, the NMOS transistor N is OFF, and node B is low. When the charge on node A is decreased to the point that the PMOS transistor P turns ON, node B rises up. Then, during time τ a high pulse is generated at the node C, which in turn charges up node A to high level.

6.2.13 Low-Voltage DRAM Operation and Circuitry

Low-voltage operation is required to reduce the power dissipation and to assure the reliability of deep-submicrometer MOS devices in future DRAMs. The power supply voltage can be as low as 1 V to meet the requirement of battery operation for portable applications. To get high performance in a high-density DRAM, at low supply voltage, the threshold voltage of MOS devices should be reduced. This results in an increased subthreshold current and hence circuit techniques are neeeded to reduce the standby current. In this section, circuit techniques to reduce the subthreshold current for the DRAM array (equalizer, precharge and sense amplifier) circuits, memory-cell access, and word-line driver are described.

6.2.13.1 DRAM Array Circuits

Fig. 6.54 shows the conventional DRAM array circuit with the half-V_{DD} bit-lines precharging technique. This circuit has already been discussed in Section 6.2.3. When V_{DD} is scaled down, this half-V_{DD} scheme causes several problems with respect to the CMOS latch-type SA and the equalizer. For example, for the NMOS transistor, N_{SA}, of the N-type SA (N-SA) the following problem can exist. When the signal ϕ_{sn} is pulled-down during the readout operation, the sensing operation starts when the voltage V_{GS1} [See Fig. 6.54] becomes larger than the V_T of the NMOS transistor of the SA. However, if $V_{DD}/2$ is low enough, approaching the value of V_T, then the sensing operation is very slow due to the low value of V_{GS1}. Note that V_T is subject to the body effect when the common source of the N-SA is falling to ground.

Another problem arises during the equalization period. The equalization is carried out by the NMOS device, N_{EQ}, when the signal ϕ_p is activated. In the final stage of equalization, the drive current of the NMOS equalizer decreases drastically, particularly when $V_{DD}/2$ is not higher than V_T. Note that the threshold voltage of the equalizer is also subject to the body effect.

One solution to these problems is the use of low-V_T devices in the DRAM array for the CMOS SA, precharge and equalizing circuits. However, this leads to a drastic increase in the leakage current during the active period. The leakage current paths are shown in Fig. 6.55. To significantly reduce this leakage current the concept of Well-Synchronized Sensing and Equalizing (WSSE) concept was proposed [37]. It is based on the following two concepts:

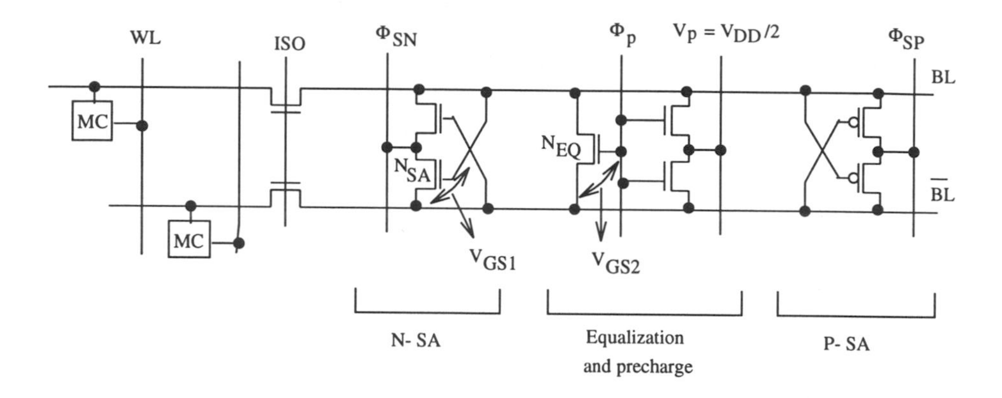

Figure 6.54 DRAM array circuits.

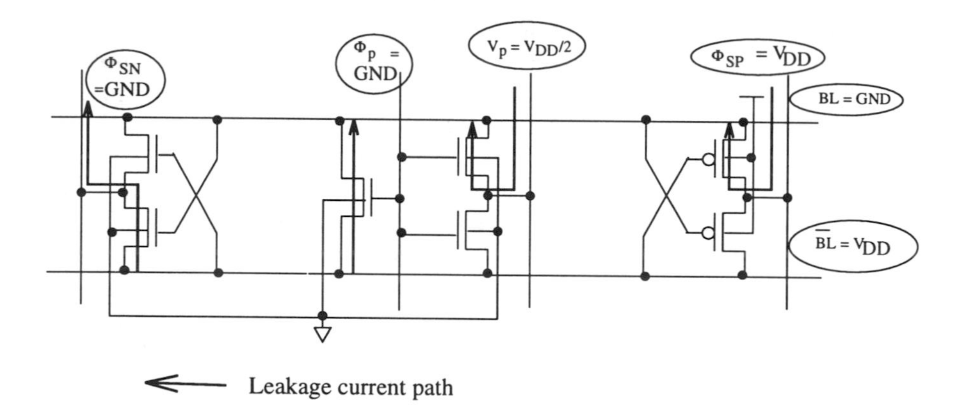

Figure 6.55 Leakage current paths in DRAM array circuits.

- The voltage levels of the transistor sources and the well are equaled during the sensing, the restoring, and the equalizing period. This eliminates the body effect.

- A negative (positive) bias, V_{BB} (V_{ch}) is applied to P-well (N-well), respectively, during the active period. Thus, the leakage current is reduced because V_T increases due to the body effect.

Fig. 6.56(a) shows the WSSE circuits using a triple-well structure. The N-well and the P-well control voltages, V_{Wn} and V_{Wp}, respectively, are controlled by a special logic. Fig. 6.56(b) illustrates the voltage waveforms. Before the word-line is activated, the bit-lines and ϕ_{sn} and ϕ_{sp}, are equalized to half-V_{DD}. The P-well and N-well levels are precharged to $(1/2V_{DD} - V_{Tn})$ and $(1/2V_{DD} - V_{Tp})$, respectively. These voltage levels permit to avoid any drain-well voltage forward-biasing during the initial time, after WL activation. During this initial time, one bit-line is different than $V_{DD}/2$. In the sensing and restoring period, the signals ϕ_{sn} and V_{Wp} are pulled-down while the signals ϕ_{sp} and V_{Wn} are pull-up; each pair is synchronized. After this period, the bit-lines BL and $\overline{BL}$ are in full-swing condition. Then, the level V_{Wp} is pulled below GND to V_{BB} and isolated from ϕ_{sn}, while the level V_{Wn} is pulled above V_{DD} to V_{ch} and isolated from ϕ_{sp}.

6.2.13.2 Memory Cell

First, let's discuss the requirements for the memory cell, particularly at low-voltage. Fig. 6.57 shows the memory cell in the restoring operation. To restore the high-level, V_{DD}, from the bit-line to the storage capacitor, the word-line must be boosted to a level V_{ch}. This level has the following requirement

$$V_{ch} > V_{DD} + V_T(V_{DD}) + \alpha \tag{6.16}$$

where α is the voltage margin and $V_T(V_{DD})$ is the threshold voltage of the access NMOS transistor when its source is at V_{DD}. Note that the NMOS device has $(V_{DD}+|V_{BB}|)$ as an effective back-bias voltage. For transistor reliability, V_{ch} should be as small as possible. This means that $V_T(V_{DD})$ is required to be small. This threshold voltage is given by

$$V_T(V_{DD}) \;=\; V_{TO} \;+\; \gamma\sqrt{V_{DD} + |V_{BB}| + 2\phi_f} \tag{6.17}$$

where V_{TO} is threshold at zero source and substrate bias, γ is the body effect coefficient and ϕ_f is the Fermi potential.

Fig. 6.58 shows the unselected memory cell in long cycle operation. The bit-line has completed the sensing operation and is at ground level (GND). In this situation, the memory cell is exposed to worst case leakage condition. The charge stored in the cell leaks rapidly due to the subthreshold current. This situation sets the lower limit of the threshold voltage. Note that the access transistor of the memory cell has $|V_{BB}|$ as back-bias voltage. The threshold voltage in this mode is given by

$$V_T(0) = V_{TO} + \gamma\sqrt{|V_{BB}| + 2\phi_f} \tag{6.18}$$

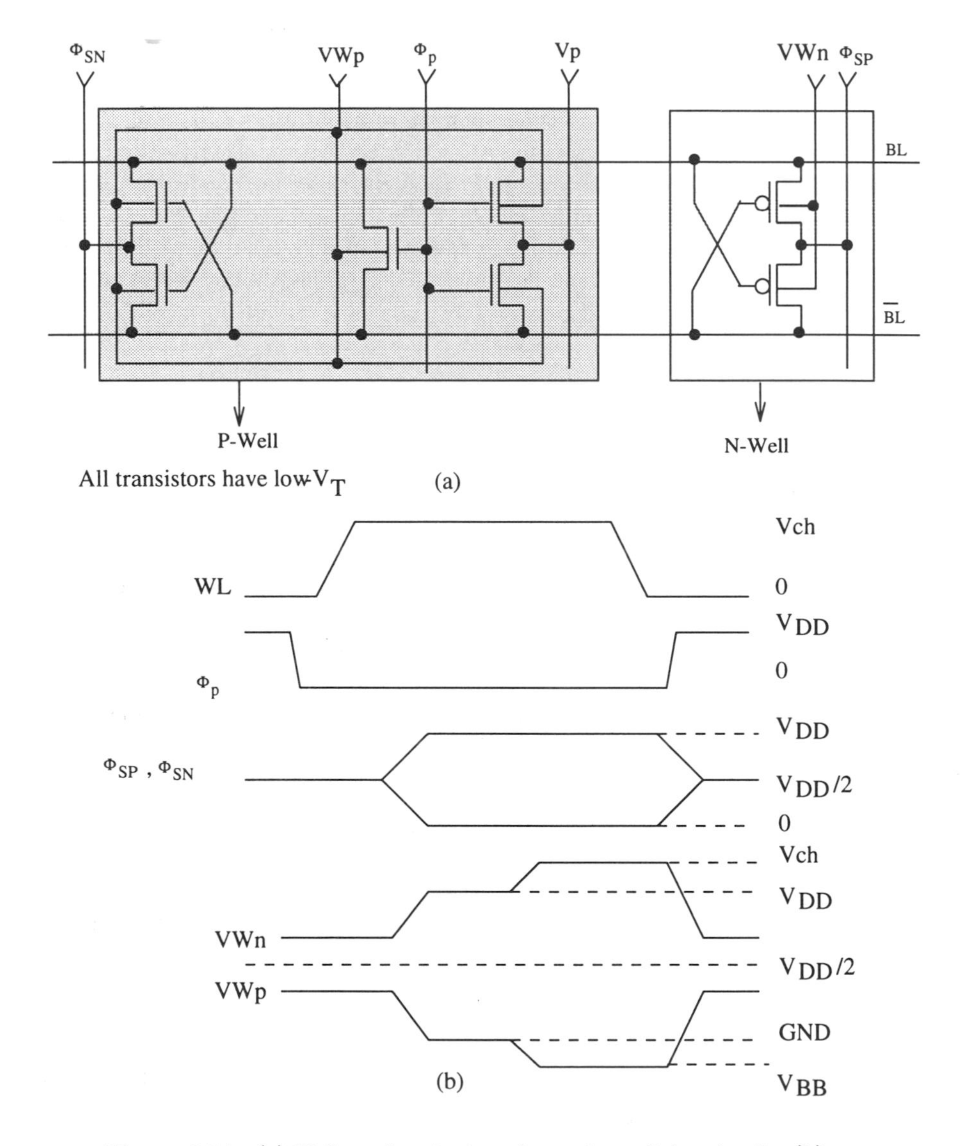

Figure 6.56 (a) Well-synchronized sensing and equalizing circuits; (b) operation waveforms.

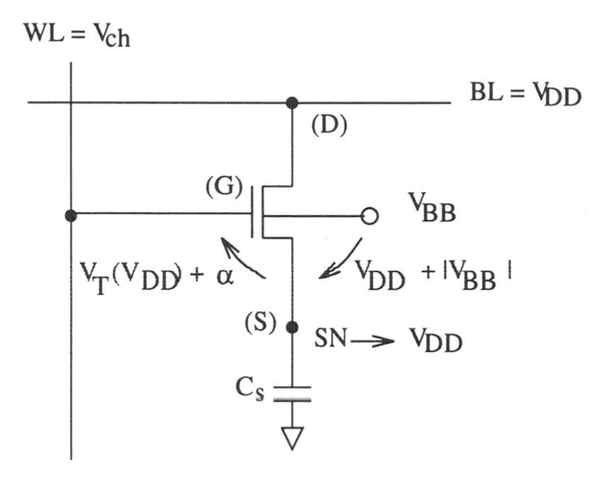

Figure 6.57 Memory cell: restoring operation.

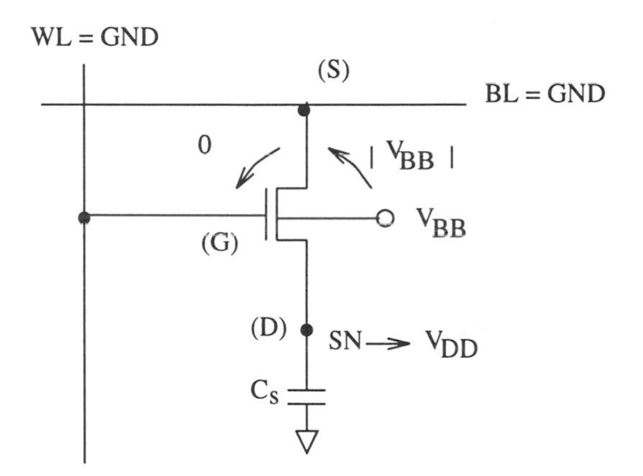

Figure 6.58 Memory cell: long cycle operation.

To meet these two requirements of the threshold voltage, the substrate voltage should have a sufficient back-bias voltage to suppress the body effect.

For example when the internal supply voltage is $V_{DD} = 1.5$ V, the $|V_{BB}|$ is set to -1. The $V_T(1.5\ V)$[6] is 1 V and the $V_T(0)$ is 0.75 V and S = 90 mV/decade.

[6] Extrapolated threshold voltage.

Therefore, the leakage current of a transistor with $W = 1\ \mu m$, is $\sim 10\ fF$. In this case, V_{ch} must be larger than $(V_{DD} + V_T(V_{DD}))$ which is 3 V.

When the V_T of the memory cell is reduced, the leakage current increases drastically. The concept of Boosted Sense Ground (BSG) [38] was proposed to shut down the subthreshold current in the memory cell access transistor. This is achieved by slightly boosting the low-level voltage of the bit-line. This level is called BSG level, and is set at 0.5 V. During a long cycle operation, the gate-source of an unselected cell is negative (-0.5 V), then the subthreshold current is reduced by 6 orders of magnitude (for S = 80 mV/decade). Fig. 6.59 shows the BSG circuit applied to a memory cell. The BSG line is common to all N-channel sense amplifiers. The BSG level is generated by a circuit similar to the VDC circuit [see Section 6.3]. In active mode, the differential amplifier and N_1 are activated and the voltage of the sense ground becomes V_{ref}. The N_2 transistor has a large width and is activated by the signal SE at the beginning of the sensing period to suppress an unnecessary rise in the BSG level by the sensing current. In the standby mode, the differential amplifier is made inactive to reduce the standby current and also N_1 and N_2. The BSG level is clamped to the threshold voltage of N_3. Note that the boosted level, V_{ch}, is reduced compared to the conventional scheme because V_T is reduced.

6.2.13.3 Word-Line Driver

Scaling the threshold voltage down increases the subthreshold current of a DRAM, particularly for iterative circuits such as word-line drivers or decoders. If the DRAM is divided into k blocks, each block has n drivers, then the total of word drivers is $k.n$. Fig. 6.60 shows an example of DRAM drivers. During the active mode, one driver out of $k.n$ drivers is selected by the row decoder and the word-line is at the boosted level V_{ch}, generated by the internal charge pump circuit.

When the threshold voltage is low, the subthreshold current of each driver is important. Then for a DRAM the total subthreshold current of the drivers is

$$I_{subdr} = k.n.I_{sub} \tag{6.19}$$

where I_{sub} is the subthreshold current of NMOS and PMOS transistors (assumed the same). For a high-capacity DRAM, the current I_{subdr} would be huge. For example, a multi-Giga-bit DRAM has a 1 million drivers, and each driver has a subthreshold current of $\sim 10\ nA$ at room temperature, then the total subthreshold current would be 10 mA. At 75 C, this current can be hundreds of mA. This high DC current destroys the V_{ch} level because the charge

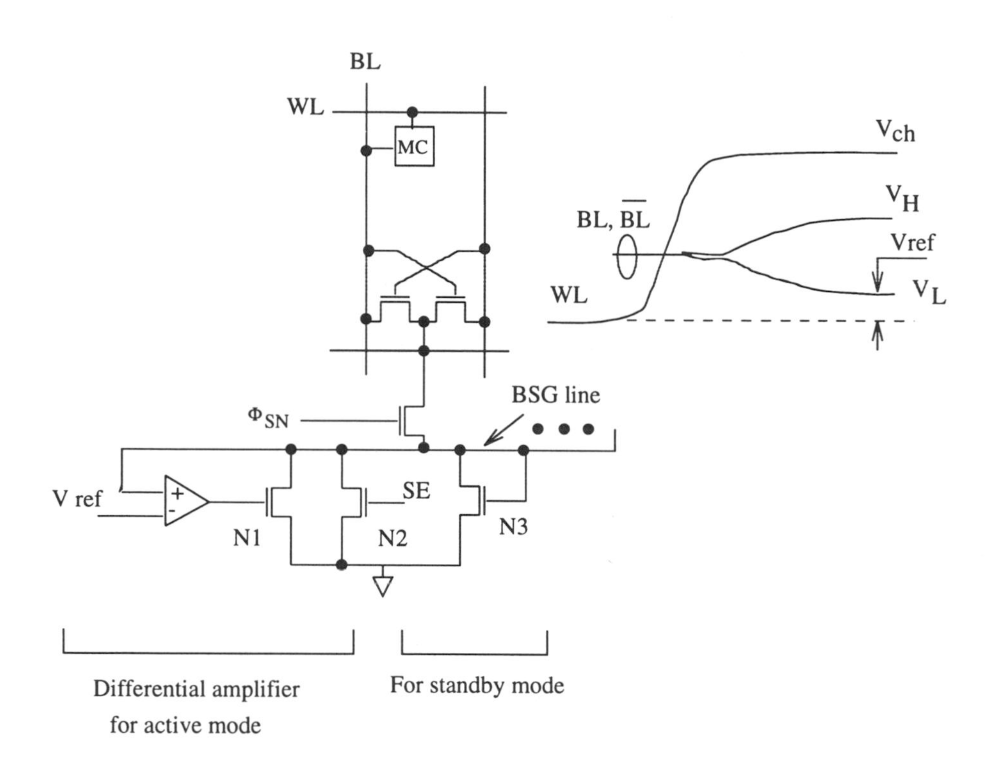

Figure 6.59 Boosted Sense Ground (BSG) circuit.

pump circuit cannot handle such a DC current. Note that this current should always be evaluated in the worst case; maximum temperature, and the lowest value of V_T. In the standby mode, all the drivers are turned OFF. The current I_{subdr} is still the same.

To solve this problem, the concept of Self-Reverse-Biasing (SRB) scheme can be used [24]. This concept has already been discussed in Section 4.10 [Chapter 4]. Fig. 6.61 shows the application of the SRB scheme to word-line drivers. During the active mode, the control signal $\bar{\phi}$ is low and the node SL is equal to V_{ch}. Only one word-line is selected. When $\bar{\phi}$ goes to high (standby mode), the PMOS device P_S limits the subthreshold current. In this mode, all drivers are OFF, even the selected one. Fig. 6.62 shows the technique to turn off the

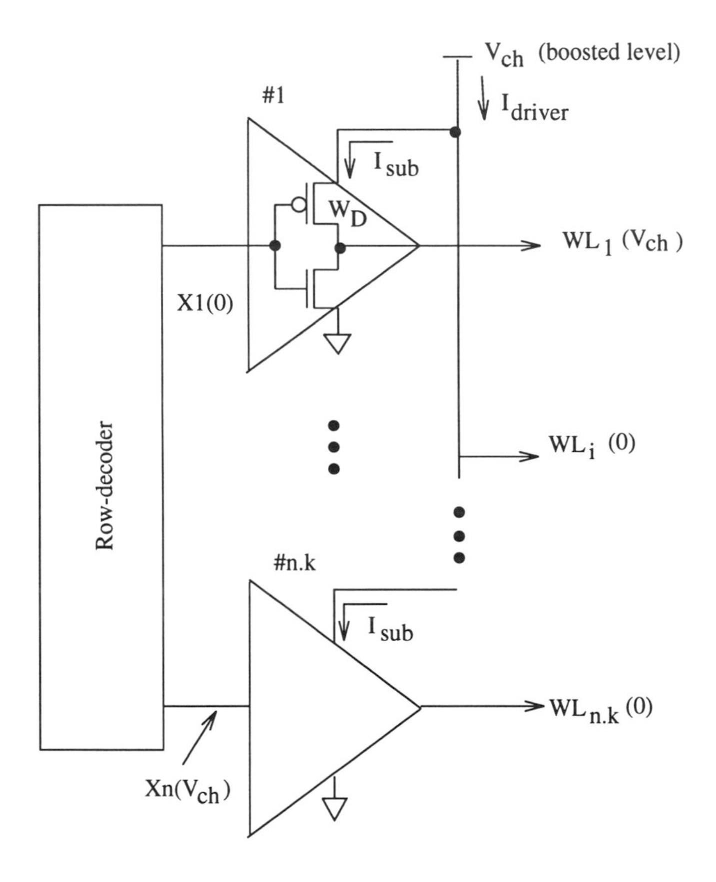

Figure 6.60 Word-line drivers.

selected driver in standby mode. When ϕ is low, node A_i is high, then the selected word driver is low.

One problem associated with the SRB scheme is that during the active mode, after one selected word-line driver is activated, all the other drivers are leaking thereby substantially contributing to the active current. This problem is solved by the partial activation of hierarchical power-line scheme [39]. Fig. 6.63 shows the principle of the 2-D selection scheme. In this scheme, the array of k blocks by n drivers is divided into k sub-blocks in columns and l sub-blocks in rows. The total of sub-blocks, each containing a set of drivers, is $k \times l$. During the active mode, only one sub-block is activated. Thus the subthreshold current in the active mode is drastically reduced.

6.3 ON-CHIP VOLTAGE DOWN CONVERTER

Chip makers prefer to scale down V_{DD} to enhance the device reliability, while the users prefer it the same power supply voltage and dislike the frequent changes. The reduction of V_{DD} is also important to achieve low-power characteristic. The strategy to meet these contradictory requirements is to use an on-chip Voltage Down Converter (VDC). A VDC can be used to convert the old power supply voltage standard of 5 V to 3.3 V to power CMOS circuits using 0.5 μm and sub-0.5 μm technology. For the state-of-the-art 0.25 μm CMOS technology, the power supply voltage must be 2.5 V. However, the new standard is becoming 3.3 V and is likely to stay that way for many years. Thus a 3.3/2.5-V VDC is required.

On-chip VDCs are used for DRAMs as well as SRAMs, ASICs and digital processors. They are employed in commercial 16-Mb DRAMs to reduce the external 5 V to an internal voltage of 3.3 V. For SRAMs, they have not been commonly used as in DRAMs, particularly in commercial ones. The SRAMs can operate over a wide range of power supply. Moreover, they already have low data retention current, enough for battery-operated applications. In this section, we discuss the VDC circuit techniques for DRAMs which are basically the same as for SRAMs and other circuits.

Numerous papers have reported designs of the VDC circuit for a DRAM [32, 40, 41, 42, 43, 44, 45] and for an SRAM [46]. Fig. 6.64 shows one approach using a VDC to reduce the internal voltage for a DRAM. Memory cell array and the periphery circuits are powered from the internal supply voltage, while the I/O

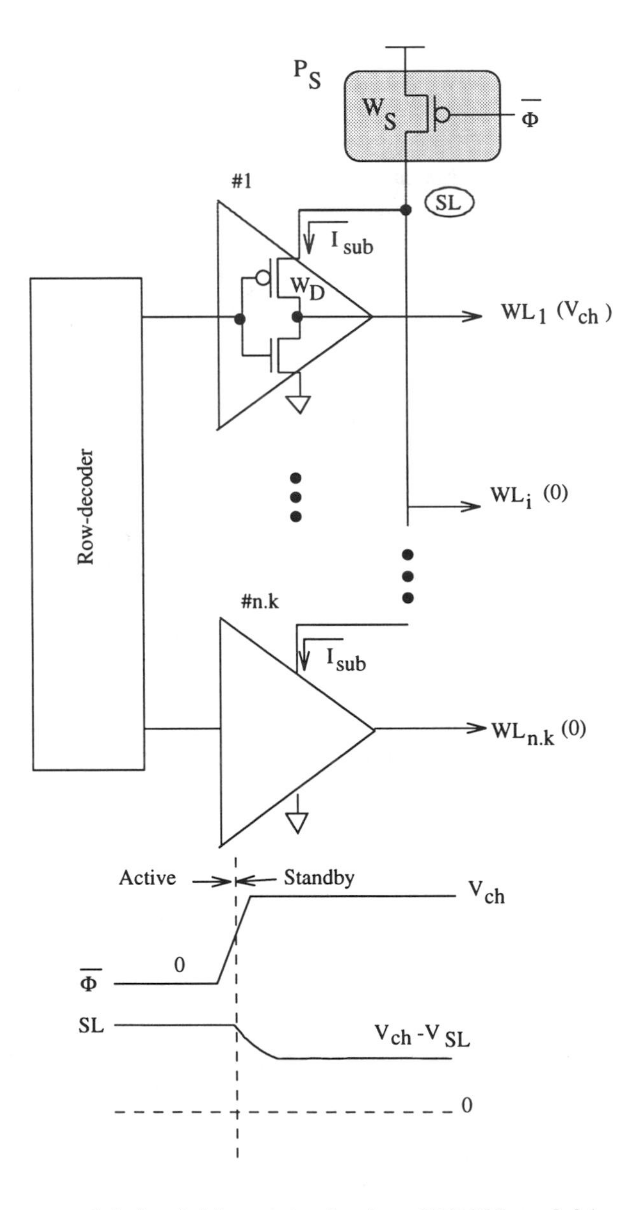

Figure 6.61 Subthreshold current reduction of DRAM word-drivers.

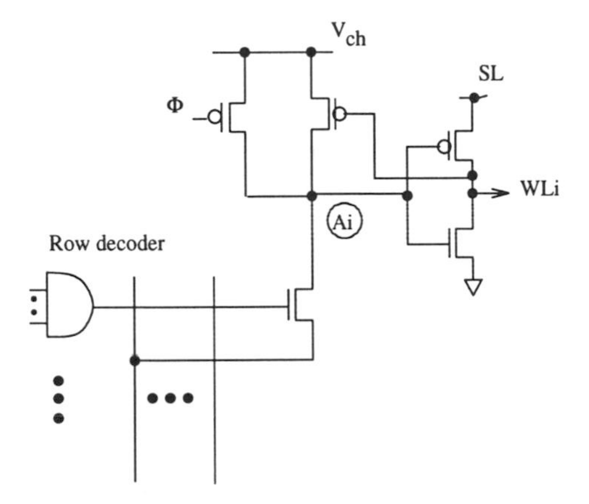

Figure 6.62 Detail of word-driver with voltage shifter.

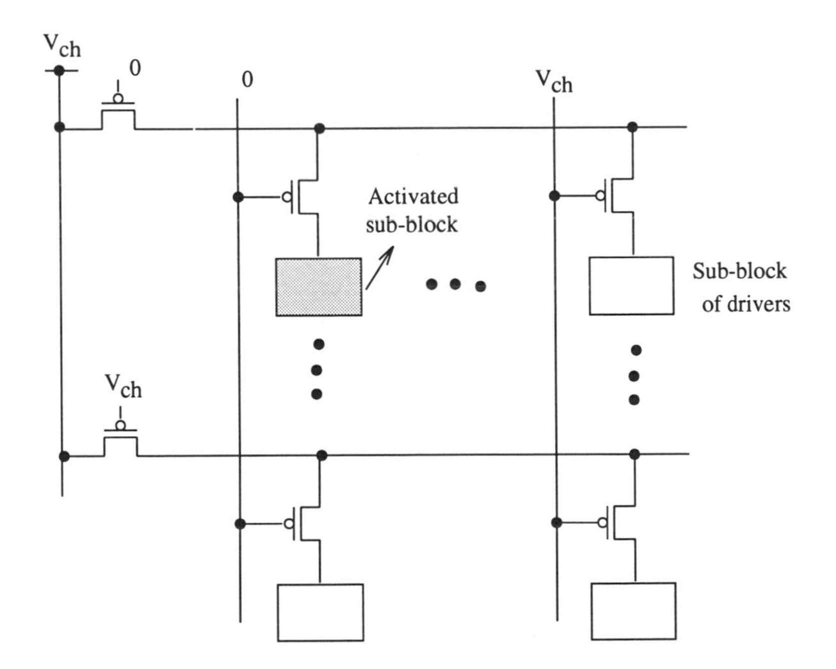

Figure 6.63 Partial activation of 2-D hierarchical power-line scheme.

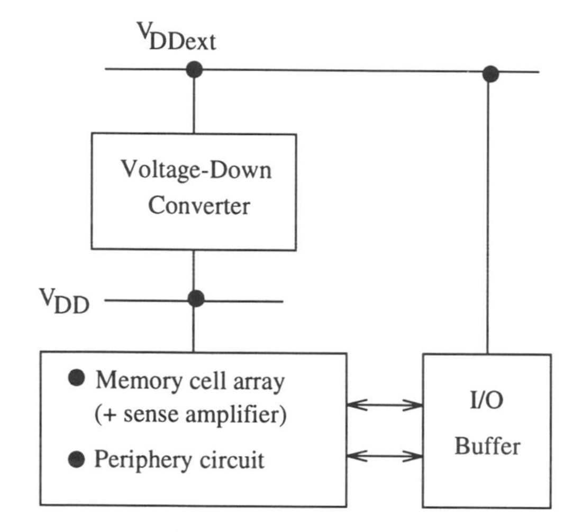

Figure 6.64 Voltage down converter for low-internal power supply.

buffers are powered with the external voltage to maintain the compatibility. However, the VDC, in this situation, should be stable when supplying a large current to periphery and memory array. When the VDC is used for battery-operated applications, the standby current should be less than 1 μA over a wide range of temperatures (0-70 C).

Fig. 6.65 shows a schematic of the VDC structure for a DRAM, used to convert a 5 V to 3.3 V. It is composed of a Reference Voltage (V_{ref}) Generator (RVG), a driver circuit and a time-dependent load. The buffer circuit consists of a differential amplifier [Fig. 6.66] and common-source drive PMOS transistor P_b. The current load has a peak, for the memory array, of more than 100 mA in 10-30 ns time and more than 100 mA in few ns for the periphery circuit. To deliver such a large current, the width of the PMOS P_b of the output stage should be large. Moreover, when the output current changes rapidly, the output voltage V_{DD} decreases by ΔV_{DD}. To minimize ΔV_{DD}, the gate control voltage, V_G, has to change quickly. This is possible by increasing the differential amplifier tail current, I_s. The current source, I_o, is needed to clamp the output voltage V_{DD} when the load current becomes almost zero.

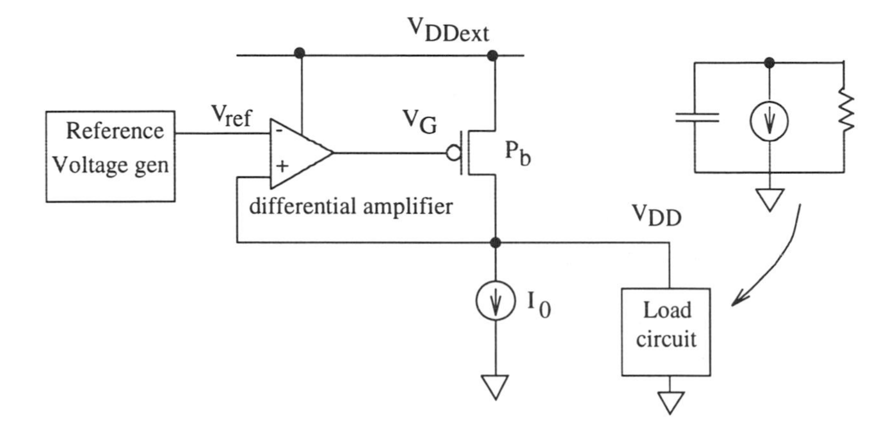

Figure 6.65 Schematic of the principle of a VDC.

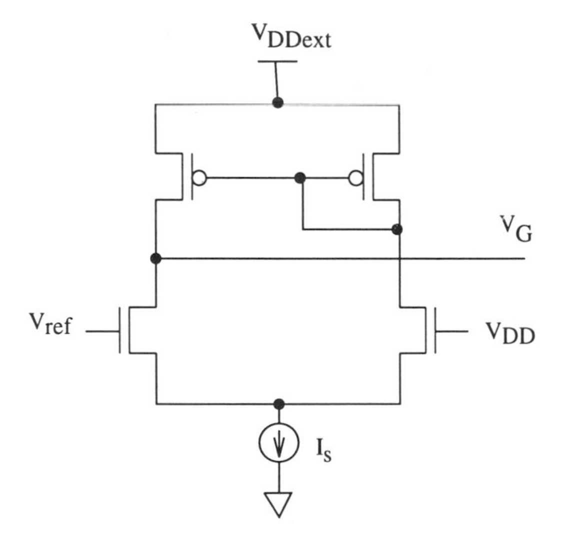

Figure 6.66 Schematic of the differential amplifier.

A VDC circuit is one of the keys for achieving a DRAM with data-retention current that can be used in battery based applications. The requirements for low-power are the following :

- The standby current must be less than 1 μA over a wide range of temperature, process and power supply voltage variations; and
- The output impedance of the VDC should be low.

6.3.1 Driver Design Issues

The internal voltage generated by the VDC can have many sources of fluctuations which are as follows. DC changes in the reference voltage due to process and temperature variations. Transient variations caused by the noise in the external power supply and by the load current. The variation of the internal voltage with respect to the reference voltage should be less than 3%. The variation with respect to the load have to be less than 10% and with respect to the power supply less than 1%.

The stability of this circuit is essential for the operation of the VDC. To study the stability, ac small-signal analysis is carried out. Fig. 6.67 shows the simplified equivalent circuit using the MOS small-signal techniques [47]. The gate capacitance of the output PMOS C_G, is huge and is taken into account. g_{m1} and g_{m2} are the transconductances of the differential amplifier and the output stage, respectively. r_1 and r_2 are their respective equivalent output resistance. C_L is the output load capacitance composed of the wire capacitance C_w[7], and the switched capacitance of the memory core C_m[8].

The frequency response of this circuit is expressed by

$$A = \frac{v_o}{v_i} = -\frac{g_{m1}r_1}{1+sC_Gr_1} \cdot \frac{g_{m2}r_2}{1+sC_Lr_2} \tag{6.20}$$

The circuit has two poles: $p_1 = 1/C_Gr_1$, for the differential amplifier and $p_2 = 1/C_Lr_2$ for the output stage. The two poles must be sufficiently separated from each other to ensure a good phase margin [48]. For a DRAM application, the pole p_2 varies drastically, because of the load variation. Thus, the circuit can fail to ensure a sufficient phase margin and hence it can generate ringing or oscillation. Therefore, phase compensation has to be applied. One

[7] A typical value of C_w is 100 pF.
[8] A typical value of C_m is 1200 pF.

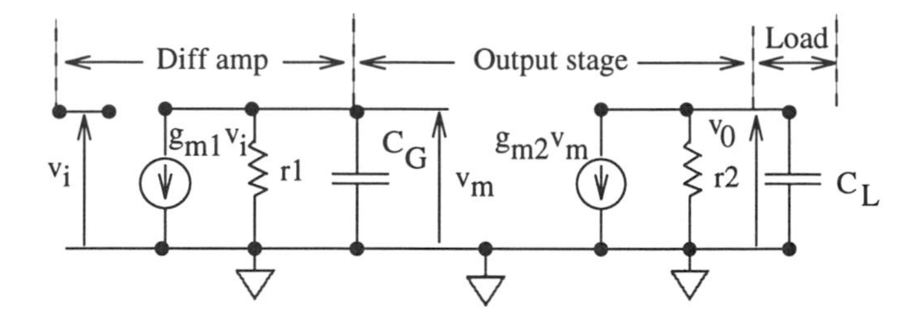

Figure 6.67 Simplified small signal equivalent circuit of the VDC with a load capacitance.

possible compensation technique is shown in Fig. 6.68(a) and it is called *Miller* compensation technique. The compensation capacitor C_c is connected between the input and the output of the second stage. It shifts the pole p_1 towards lower frequency p_1', as shown in Fig. 6.68(b). Thus, the phase margin is improved.

The condition of the stablization is defined at the point of 0 dB loop gain where the phase margin is larger than 45 degrees. Using the small-signal analysis with the compensation capacitor C_c the condition can be extracted. This capacitor is a function of g_{m2}, g_{m1}, C_L and C_G. To determine it, g_{m2} has to be known, using large-signal analysis. The PMOS driver P_b has to be sized to satisfy the condition on $\Delta V_{DD}/V_{DD}$ (less than 10%), due to the transient load current variation. Hence g_{m2} can be determined from the size of P_b. For a 16-Mb DRAM, the width of the output PMOS P_b can be as high as 30,000 μm and C_c equals to 200 pF. This is for 3.3 V internal power supply generation from 5 V.

The current tail of the differential amplifier can be high (few mA) in active mode. The driver can be disactivated in standby mode to consume only a very small current by Chip Select (CS) signal. In this case, the internal voltage can be supplied by a low-power voltage follower [46]. The voltage follower has the same configuration as the driver but the tail current is in the sub-μA range.

6.3.2 Reference Voltage Generator

The Reference Voltage Generator (RVG) must provide a high accuracy over a wide variation of V_{DD}, process, and temperature. So far, the RVGs have been based on the band-gap reference and on the threshold voltage generator.

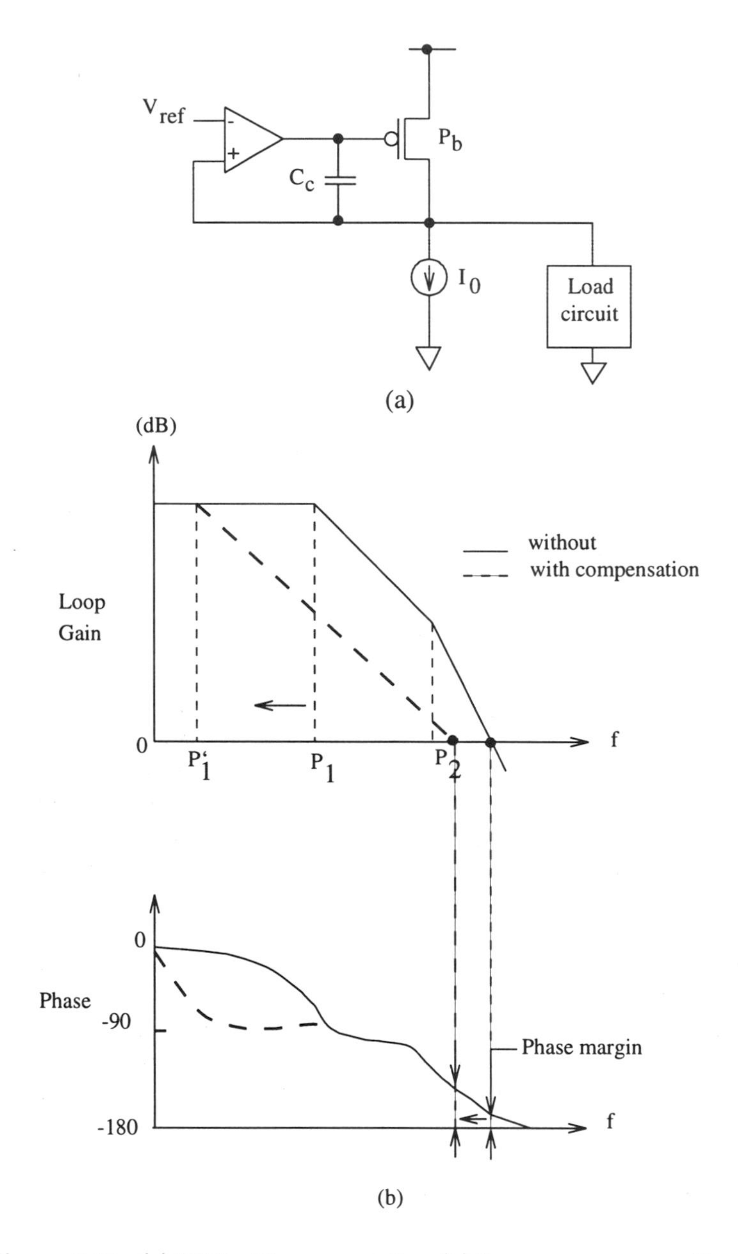

Figure 6.68 (a) VDC with compensation; (b) frequency response with and without compensation.

The former consumes a DC current which is not low enough for low-power applications. The latter is more suitable for a CMOS technology.

Fig. 6.69(a) shows a PMOS-V_T difference generator with an output voltage $\Delta V_T = |V_{Tp1}| - |V_{Tp2}|$ ($V_{Tp1} < V_{Tp2} < 0$). The equivalent circuit is shown in Fig. 6.69(b). This circuit needs a PMOS device with high threshold voltage. A typical value for the threshold voltage difference is 1.1 V. The PMOS transistors are chosen as threshold voltage difference generator because they are in N-wells and therefore the difference is independent of back-bias (V_{BB}). The circuit of Fig. 6.69(a) does not suffer much from V_{DDext} bounce. The temperature dependency of the V_T difference is expressed by [49]

$$\frac{d(\Delta V_T)}{dT} = \frac{k}{q} \ln \frac{N_{si1}}{N_{si2}} \tag{6.21}$$

where N_{si1} and N_{si2} are the surface impurity concentrations of P_1 and P_2, respectively. For a stable-temperature design, the concentration ratio N_{si1}/N_{si2} and, therefore the threshold voltage difference, should not be excessively large. A typical value of temperature dependency is 0.4 mV/C, which is small for the VDC circuit.

Since the ΔV_T is around 1 V, the circuit of Fig. 6.70 is used to convert this difference to the required internal supply voltage. The voltage-up converter amplifies ΔV_T to:

$$V_{ref} = \Delta V_T \left(1 + \frac{R_1}{R_2}\right) \tag{6.22}$$

The mismatch between the two PMOS devices P_1 and P_2 of Fig. 6.69 can be minimized by using large channel widths and lengths. But still the deviation on V_T, due to the fabrication process, has to be eliminated. This can be done by using fuse trimming technique to control the ratio of the resistors R_1 and R_2. The total current consumed by this RVG circuit is

$$I_{RVG} = \frac{V_{ref}}{R_1 + R_2} + 3I + I_s \tag{6.23}$$

where $3I$ is the current consumed by the voltage regulator [see Fig. 6.69(a)] and I_s is the current of the differential amplifier. $I_L = V_{ref}/(R_1 + R_2)$ is the current of the output stage. I can be made $\ll 1\mu A$, however I_s and I_L can not be made smaller, particularly I_L. The resistor is implemented, for example, by using doped polysilicon. Typical values of the resistances are of the order of 100 $K\Omega$. They can not be increased excessively, otherwise the area of the RVG can be significantly high. Moreover, the substrate noise can affect the reference

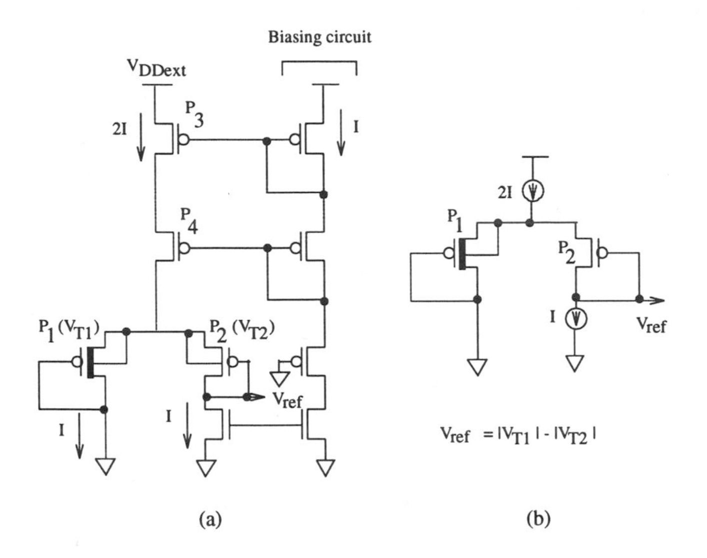

Figure 6.69 (a) Voltage regulator based on PMOS-V_T difference with biasing circuit; (b) equivalent circuit.

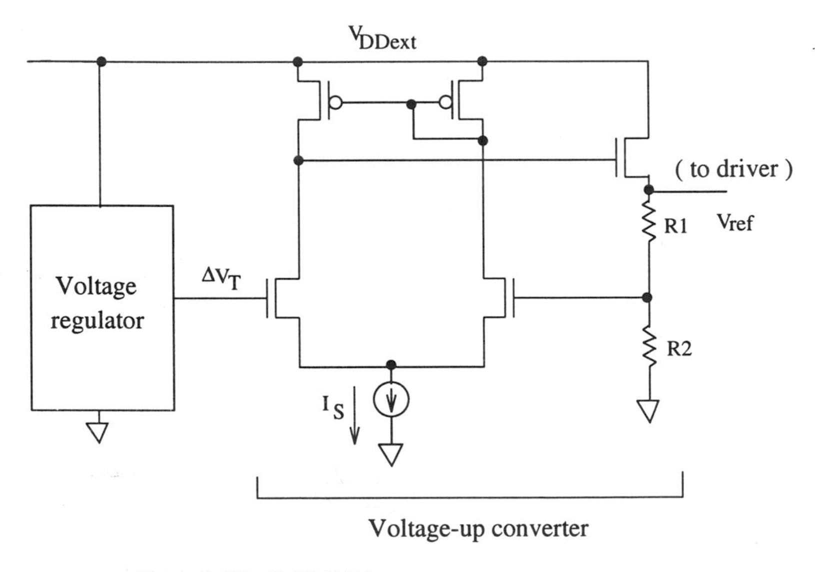

Figure 6.70 Reference voltage generator by voltage-up converter.

voltage through the coupling capacitances of the resistors. The total current of this type of RVG is in the order of few a tens of μA.

To reduce the current of the RVG to sub-μA range for battery-operated DRAMs, the concept of dynamic RVG can be used [50] as shown in Fig. 6.71. A PMOS transistor P_1 with low $|V_T|$ is used. During the sampling period (ϕ_1 is high), all switches $S_1 - S_4$ are closed. The threshold voltage difference, ΔV_T, between the two PMOS devices, P_1 and P_2, appears across the resistor R_R. If the transistor dimensions of the pairs P_1 and P_2, and N_1 and N_2 are identical, the reference voltage is given by

$$I_R = \frac{\Delta V_T}{R_R} \tag{6.24}$$

This current is mirrored to the output node. If the dimension of P is identical to that of P_1, the output voltage V_{ref} is given by

$$V_{ref} = \Delta V_T \frac{R_L}{R_R} \tag{6.25}$$

This shows that the reference voltage can be adjusted to any voltage. Moreover, with trimming technique V_{ref} can be adjusted against process variation effect (ΔV_T variation). The output voltage is sampled on the hold capacitor C_H. When ϕ_1 is low, the circuit is in hold mode. Clock ϕ_2 is delayed to clock ϕ_1 to minimize fluctuation of the output voltage. These clocks are generated from the self-refresh clock circuit in a DRAM. The circuit consumes a DC current only when ϕ_1 is applied. The average current consumed by this circuit is

$$I_{avr} = 3I_R\gamma_\phi = 3(\Delta V_T/R_R)\gamma_\phi \tag{6.26}$$

where γ_ϕ is the duty ratio of ϕ_1. The current of this circuit can be reduced to a low-level in sub-μA range by controlling the duty ratio. For example to generate a reference voltage of 2.4 V from an external power supply voltage of 3.3 V, R_R and R_L are 9 $k\Omega$ and 72 $k\Omega$, respectively. ΔV_T has a typical value of 0.3 V. The total DC is 100 μA. So with a duty ratio lower than 1/100, the average current can be reduced below 1 μA. It can be easily shown that this circuit has a low sensitivity to power supply voltage and temperature variations.

6.4 CHAPTER SUMMARY

Low-power architectures/circuits techniques for SRAMs, DRAMs and VDCs were reviewed. The obvious technique to reduce the power dissipation is the

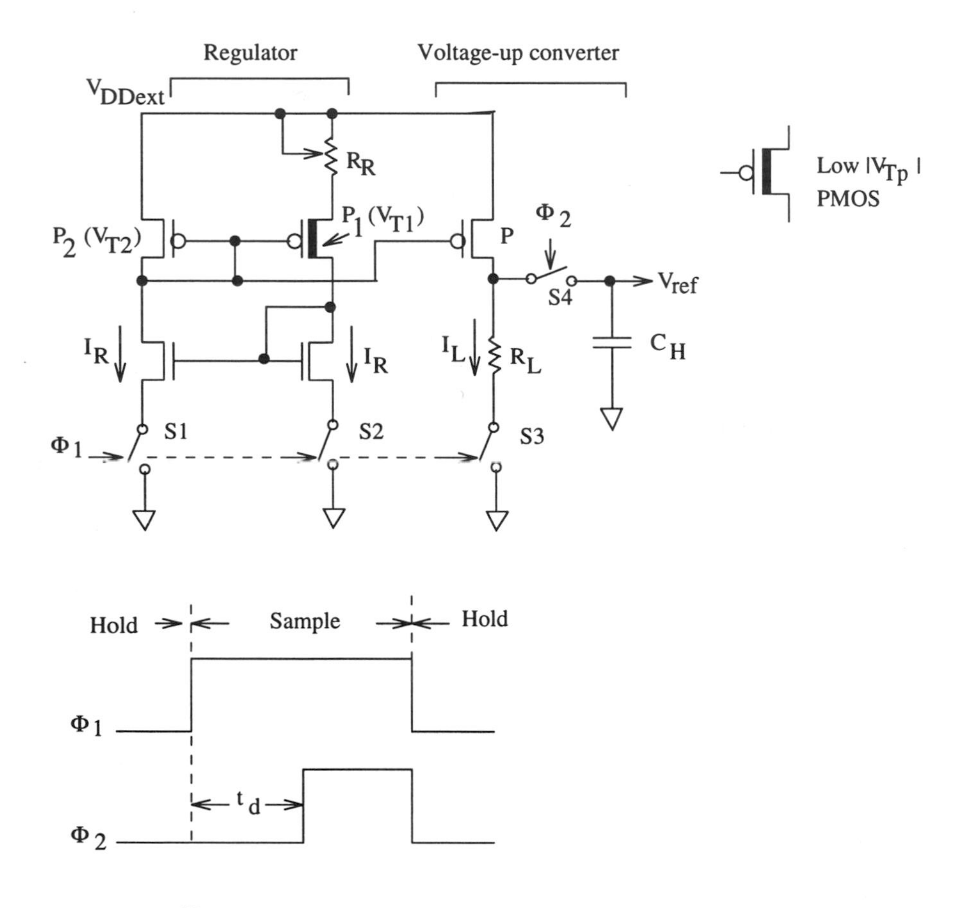

Figure 6.71 Low-power dynamic reference voltage generator.

voltage scaling. The reduction of power supply voltage to 1- and sub-1 V range requires new circuit innovations and breakthroughs, particularly when low threshold voltage devices are used. It was shown that not only the power supply voltage scaling contributes to the power consumption reduction but also the reduction of capacitances and DC currents using sophisticated techniques. Many of the techniques presented for memories can be useful to other applications such as : ASICs, DSPs, etc. Design issues for stable operation of a VDC and low-standby current techniques were investigated.

REFERENCES

[1] H. Tran et al., "An 8-ns 1-Mb ECL BiCMOS SRAM with a Configurable Memory Array Size," International Solid-State Circuits Conf. Tech. Dig., pp. 36-37, February 1989.

[2] M. Matsui et al., "An 8-ns 1-Mb ECL BiCMOS SRAM," International Solid-State Circuits Conf. Tech. Dig., pp. 38-39, February 1989.

[3] Y. Maki et al., "A 6.5-ns 1 Mb BiCMOS ECL SRAM," International Solid-State Circuits Conf. Tech. Dig., pp. 136-137, February 1990.

[4] M. Takada et al., "A 5-ns 1-Mb ECL BiCMOS SRAM," IEEE Journal of Solid State Circuits, vol. 25, no. 5, pp. 1057-1062, October 1990.

[5] A. Ohba et al., "A 7-ns 1-Mb BiCMOS ECL SRAM with Program-Free Redundancy," in Symp. VLSI Circuits Conf. Tech. Dig., pp. 41-42, May 1990.

[6] Y. Okajima et al., "A 7-ns 4-Mb BiCMOS SRAM with a Parallel Testing Circuit," International Solid-State Circuits Conf. Tech. Dig., pp. 54-55, February 1991.

[7] K. Sasaki et al., "A 7-ns 140-mW 1-Mb CMOS SRAM with Current Sense Amplifier," IEEE Journal of Solid-State Circuits, vol. 27, no. 11, pp. 1511-1518, November 1992.

[8] T. Ootani et al., "A 4-Mb CMOS SRAM with a PMOS Thin-Film Transistor Load Cell," IEEE Journal of Solid-State Circuits, vol. 25, no. 5, pp. 1082-1092, October 1990.

[9] S. Murakami et al., "A 21-mW 4-Mb CMOS SRAM for Battery Operation," IEEE Journal of Solid-State Circuits, vol. 26, no. 11, pp. 1563-1570, November 1991.

[10] K. Sasaki et al., "16-Mb CMOS SRAM with a 2.3-μm^2 Single-Bit-Line Memory Cell," IEEE Journal of Solid-State Circuits, vol. 28, no. 11, pp. 1125-1130, November 1993.

[11] M. Matsumiya et al., "A 15-ns 16-Mb CMOS SRAM with Interdigitated Bit-Line Architecture," IEEE Journal of Solid-State Circuits, vol. 27, no. 11, pp. 1497-1503, November 1992.

[12] K. Seno et al., "A 9-ns 16-Mb CMOS SRAM with Offset-Compensated Current Sense Amplifier," IEEE Journal of Solid-State Circuits, vol. 28, no. 11, pp. 1119-1124, November 1993.

[13] E. Seevinck, F. J. List, and J. Lohstroh, " Static-Noise Margin Analysis of MOS SRAM Cells," IEEE Journal of Solid-State Circuits, vol. SC-22, no. 5, pp. 748-754, October 1987.

[14] H. Kato et al., "Consideration of Poly-Si Loaded Cell Capacity Limits for Low-Power and High-Speed," IEEE Journal of Solid-State Circuits, vol. 27, no. 4, pp. 683-685, April 1992.

[15] K. Sasaki et al., "A 23-ns 4-Mb CMOS SRAM with 0.2-μA Standby Current," IEEE Journal of Solid-State Circuits, vol. 25, no. 5, pp. 1075-1081, October 1990.

[16] K. Ishibashi, T. Yamanaka, and K. Shimohigashi, "An α-Immune, 2-V Supply Voltage SRAM using a Polysilicon PMOS Load Cell," IEEE Journal of Solid-State Circuits, vol. 25, no. 1, pp. 55-60, February 1990.

[17] K. Sasaki et al., "A 15-ns 1-Mbit CMOS SRAM," IEEE Journal of Solid-State Circuits, vol. 23, no. 5, pp. 1067-1072, October 1988.

[18] K. Sasaki et al., "A 9-ns 1-Mbit CMOS SRAM," IEEE Journal of Solid-State Circuits, vol. 24, no. 5, pp. 1219-1225, October 1989.

[19] K. Ishibashi, K. Takasugi, T. Yamanaka, T. Hashimoto, K. Sasaki, " A 1-V TFT-Load SRAM using a Two-Step Word-Voltage Method," IEEE Journal of Solid-State Circuits, vol. 27, no. 11, pp. 1519-1524, May 1992.

[20] M. Yoshimito, K. Anami, H. Shinohara, T. Yoshihara, H. Takagi, S. Nagao, S. Kayano, and T. Nakano, "A Divided Word-Line Structure in the Static RAM and its Application to a 64K Full CMOS RAM," IEEE Journal of Solid-State Circuits, vol. SC-18, no. 5, pp. 479-485, October 1983.

[21] T. Hirose, H. Kuriyama, S. Murakami, K. Yuzuriha, T. Mukai, K. Tsutsumi, Y. Nishimura, Y. Kohno, and K. Anami, "A 20-ns 4-Mb CMOS SRAM with Hierarchical Word Decoding Architecture," IEEE Journal of Solid-State Circuits, vol. 25, no. 5, pp. 1068-1074, October 1990.

[22] A. Sekiyama, T. Seki, S. Nagai, A. Iwase, N. Suziki, and M. Hayasaka, "A 1-V Operating 256-Kb Full-CMOS SRAM," IEEE Journal of Solid-State Circuits, vol. 27, no. 5, pp. 776-782, May 1992.

[23] T. Yabe, et al., "High-Speed and Low-Standby-Power Circuit Design of 1 to 5 V Operating 1 Mb Full CMOS SRAM," Symposium on VLSI Circuits Tech. Dig., pp. 107-108, May 1993.

[24] G. Kitsukawa, et al., "256-Mb DRAM Circuit Technologies for File Applications," IEEE Journal of Solid-State Circuits, vol. 28, no. 11, pp. 1105-1113, November 1993.

[25] T. Hasegawa, et al., "An Experimental DRAM with a NAND-Structured Cell," IEEE Journal of Solid-State Circuits, vol. 28, no. 11, pp. 1099-1104, November 1993.

[26] T. Sugibayashi, et al., "A 30-ns 256-Mb DRAM with a Multidivided Array Structure," IEEE Journal of Solid-State Circuits, vol. 28, no. 11, pp. 1092-1099, November 1993.

[27] M. Aoki, J. Etoh, K. Itoh, S-I. Kimura, and Y. Kawamoto, "A 1.5-V DRAM for Battery-Based Applications," IEEE Journal of Solid-State Circuits, vol. 24, no. 5, pp. 1206-1212, October 1989.

[28] Y. Nakagome, et al., "An Experimental 1.5-V 64-Mb DRAM," IEEE Journal of Solid-State Circuits, vol. 26, no. 4, pp. 465-471, April 1991.

[29] H. Yamauchi, et al., "A Circuit Technology for High-Speed Battery-Operated 16-Mb CMOS DRAMs," IEEE Journal of Solid-State Circuits, vol. 28, no. 11, pp. 1084-1091, November 1993.

[30] N. C. C. Lu, " Advanced Cell Structures for Dynamic RAMs," IEEE Circuits and Devices Magazine, no. 1, pp. 27-36, January 1989.

[31] M. Takada, "DRAM Technology for Giga-bit Age," International Conf. Solid State Devices and Materials, Tech. Dig., pp. 874-876, 1993.

[32] L. Itoh, et al., "An Experimental 1-Mb DRAM with on Chip Voltage Limiter," in International Solid-State Circuits Conf., Tech. Dig., pp. 282-283, 1984.

[33] N. C-C. Lu, and H. H. Chao, " Half-V_{DD} Bit-Line Sensing Scheme in CMOS DRAMs," IEEE Journal of Solid-State Circuits, vol. SC-19, no. 5, pp. 451-454, August 1984.

[34] H. Kawamoto, T. Shinoda, Y. Yamaguchi, S. Shimizu, K. Ohishi, N. Tanimura, T. Yasui, "A 288K CMOS Pseudostatic RAM," IEEE Journal of Solid-State Circuits, vol. SC-19, no. 5, pp. 619-623, October 1984.

[35] Y. Tsikikawa et al., "An Efficient Back-Bias Generator with Hybrid Pumping Circuit for 1.5 V DRAMs," in Symposium of VLSI Circuits, Tech. Dig., pp. 85-86, May 1993.

[36] Y. Konishi, et al., "A 38-ns 4-Mb DRAM with a Battery-Backup (BBU) Mode," IEEE Journal of Solid-State Circuits, vol. 25, no. 5, pp. 1112-1117, October 1990.

[37] T. Ooishi, et al., "A Well-Synchronized Sensing/Equalizing Method for Sub-1 V Operating Advanced DRAMs," in Symposium on VLSI Circuits, Tech. Dig., pp. 81-82, May 1993.

[38] M. Asakura, et al., "An Experimental 256-Mb DRAM with Boosted Sense-Ground Scheme," IEEE Journal of Solid-State Circuits, vol. 29, no. 11, pp. 1303-1309, November 1994.

[39] T. Sakata et al., "Subthreshold-Current Reduction Circuits for Multi-Gigabit DRAMs," in Symposium on VLSI Circuits, Tech. Dig., pp. 45-46, May 1993.

[40] T. Furuyama, et al., "A New On-Chip Voltage Converter for Submicrometer High-Density DRAMs," IEEE Journal of Solid-State Circuits, vol. 22, no. 3, pp. 437-441, June 1987.

[41] M. Takada, et al., "A 4-Mb DRAM with Half Internal Voltage Bit-Line Precharge," IEEE Journal of Solid-State Circuits, vol. 21, no. 5, pp. 612-617, October 1986.

[42] M. Hiroguchi, et al., "Dual-Operation-Voltage Scheme for a Single 5-V, 16-Mb DRAM," IEEE Journal of Solid-State Circuits, vol. 23, no. 5, pp. 1128-1132, October 1988.

[43] G. Kitsukawa, et al., "A 1-Mb BiCMOS DRAM using Temperature-Compensation Circuit Techniques," IEEE Journal of Solid-State Circuits, vol. 24, no. 3, pp. 597-602, June 1989.

[44] M. Horiguchi, et al., "A Tunable CMOS-DRAM Voltage Limiter with Stabilized Feedback Amplifier," IEEE Journal of Solid-State Circuits, vol. 25, no. 5, pp. 1129-1135, October 1990.

[45] M. Horiguchi, et al., "Dual-Regulator Dual-Decoding-Trimmer DRAM Voltage Limiter for Burn-in Test," IEEE Journal of Solid-State Circuits, vol. 26, no. 11, pp. 1544-1549, November 1991.

[46] K. Ishibashi, K. Sasaki, and H. Toyoshima, " A Voltage Down Converter with Submicroampere Standby Current for Low-Power Static RAMs," IEEE Journal of Solid-State Circuits, vol. 27, no. 6, pp. 920-926, June 1992.

[47] P. E. Allen, and D. R. Holberg, "CMOS Analog Circuit Design," Holt, Rinehart and Winston Publisher, 1987.

[48] P. R. Gray, and R. G. Meyer, "Analysis and Design of Analog Integrated Circuit," 2nd Edition Wiley Publisher, 1984.

[49] R. A. Blauschild et al., "A New NMOS Temperature Stable Voltage Reference," IEEE Journal of Solid-State Circuits, vol. SC-13, pp. 767-774, December 1978.

[50] H. Tanaka, Y. Nakagome, J. Etoh, E. Yamasaki, M. Aoki, and K. Miyazawa, "Sub-1-μm Dynamic Reference Voltage Generator for Battery-Operated DRAMs," in Symp. VLSI Circuits, Tech. Dig., pp. 87-88, May 1993.

7

VLSI CMOS SUBSYSTEM DESIGN

In this chapter, we study the application of the circuit techniques developed through Chapter 4 in the implementation of CMOS building blocks such as adders, multipliers, ALUs, data-path, and regular structures, etc. The power dissipation constraint is also included through the several options presented for each circuit. The use of Phase locked Loop (PLL) in high-speed CMOS systems for deskewing the internal clock is also examined. Low-power issues of the circuits presented are also discussed.

7.1 PARALLEL ADDERS

Parallel adders are the most important elements used in arithmetic operations of microprocessors, DSPs, etc. As in any logic design they are constrained by parameters such as speed, area, and power dissipation. The adder cell is also an element of multipliers, dividers, multiplier-accumulators (MACs), etc. Among the various adder's implementations used in many designs, we can cite the following classes:

- Ripple Carry Adders (RCA);
- Carry Look-Ahead Adders (CLA);
- Carry Select Adders (CS); and
- Conditional Sum Adders (CSA).

This section is devoted to describing all these adder classes.

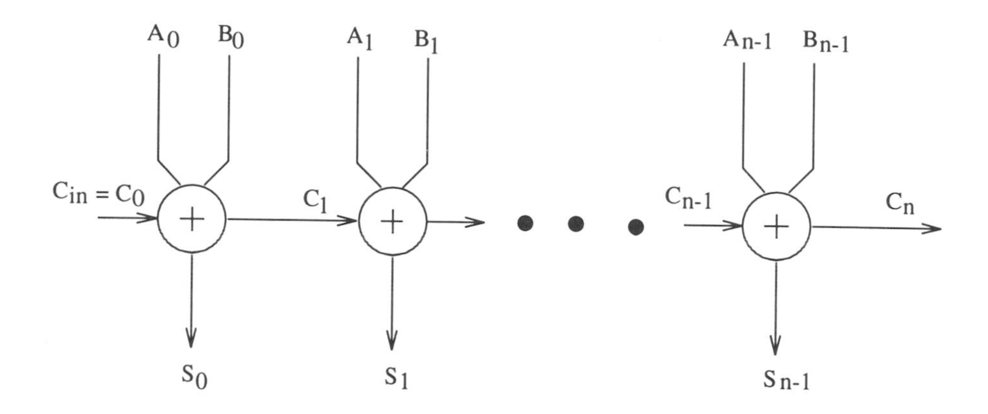

Figure 7.1 Ripple carry adder.

7.1.1 Ripple Carry Adders

In Chapter 4, a description of the functionality of an adder cell was presented. In an n-bit adder, a propagation of the carry always occurs. This propagation limits the speed of the adder. The simplest way to construct an n-bit adder is to cascade n 1-bit adders as shown in Fig. 7.1. This adder is called *Ripple Carry Adder* (RCA). Because the carry ripples through the n-stages, the sum of the nth bit cannot be performed until the carry C_{n-1} is evaluated. The delay of n-bit addition is given by

$$t_{rca} = (n-1)t_c + t_s \tag{7.1}$$

where t_c is the carry delay and t_s is the sum delay. Since the carry propagation path is a critical stage for the delay, the full-adder cell should be optimized. The sum and carry out are given by

$$S = A \oplus B \oplus C \tag{7.2}$$

$$C_{out} = A \cdot B + (A + B).C_{in} \tag{7.3}$$

The schematic of Fig. 7.2 can be generated to efficiently implement the adder cell. Compared to the conventional CMOS full-adder implementation, there is no inverter stage. Therefore, the carry delay is reduced. To optimize the cell, the transistors in the carry path W_p and W_n, can be sized up [see Fig. 7.2]. The other devices can be kept small to reduce the load on the carry and the power dissipation. The transistors, driven by the carry in C_{in}, are placed close to the output. This will reduce the body effect, since the carry signal is the

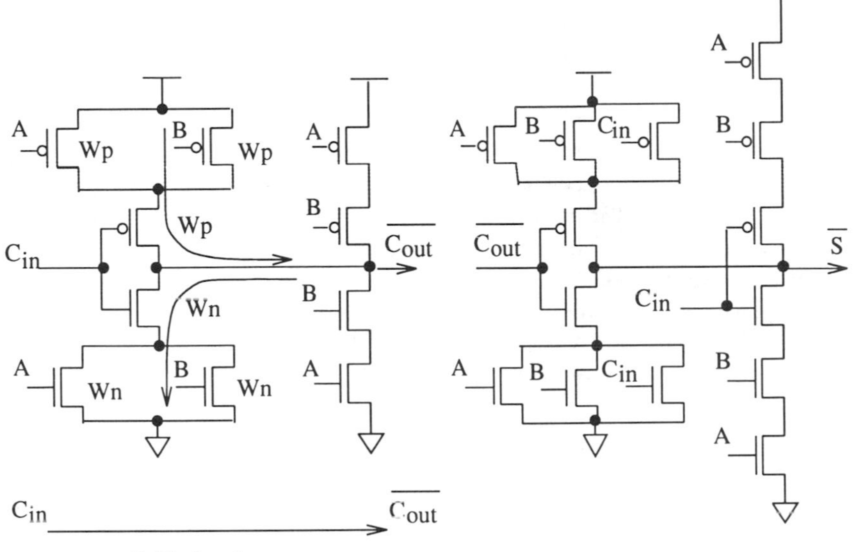

Figure 7.2 Symmetrical optimized full-adder cell.

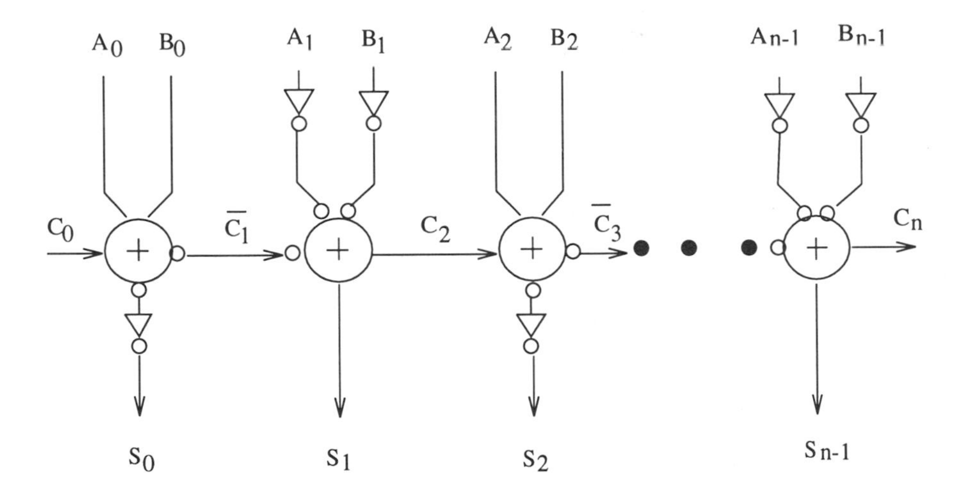

Figure 7.3 Parallel ripple carry adder for the optimized cell.

latest one in an adder chain. The schematic of Fig. 7.2 is symmetrical and leads to better layout and small area. Since the outputs are complemented, and in order to implement an RCA circuit, the configuration of Fig. 7.3 can be used. In this case, many cells use inverted inputs.

Note that an n-bit RCA circuit is subject to the glitching problem. Fig. 7.4 shows a static simulation of a 4-bit adder, with the inputs A_i set to zero (0), and the inputs B_i and C_{in} rising from 0 to 1. The outputs S_i should stay at 0, however, due to the delay of the carry signal, through the chain of full-adders, the outputs exhibit spurious transitions (glitching). These dynamic transitions dissipate extra power and can represent an important portion of the total power. With careful design this glitching problem can be minimized. One advantage of the RCA is its low-power characteristic. However, its speed is very limited, particularly when the adder is wide.

Another efficient full-adder cell is based on Transmission Gates (TGs). Fig. 7.5 shows an optimized version of the full-adder cell using TGs already discussed in Chapter 4. The carry signal propagates only through one TG. Hence, an n-bit RCA would be faster and more compact than the conventional one[1]. Fig. 7.6 shows the construction of an n-bit adder. Practically, an inverter is added every four stages to reduce the degradation of the carry signal due to the distributed RC effect. When the carry signal is inverted after 4 1-bit stages, complementary carry path adders are used for the next 4-bit stages. This adder structure is sometimes called Manchester adder. This circuit is faster than the RCA and may have lower power dissipation.

7.1.2 Carry Look-Ahead Adders

To avoid the linear growth of the carry delay, we use a Carry Lookahead Adder (CLA) in which the carries can be generated in parallel. The carry of each bit is generated from the propagate and the generate signals (P_i, G_i) as well as the input carry (C_0). The propagate and the generate signals (P_i, G_i) are derived from the operands A_i and B_i by

$$G_i = A_i \cdot B_i \quad (7.4)$$

$$P_i = A_i + B_i \quad (7.5)$$

[1] The conventional adder is an RCA using a conventional static full-adder

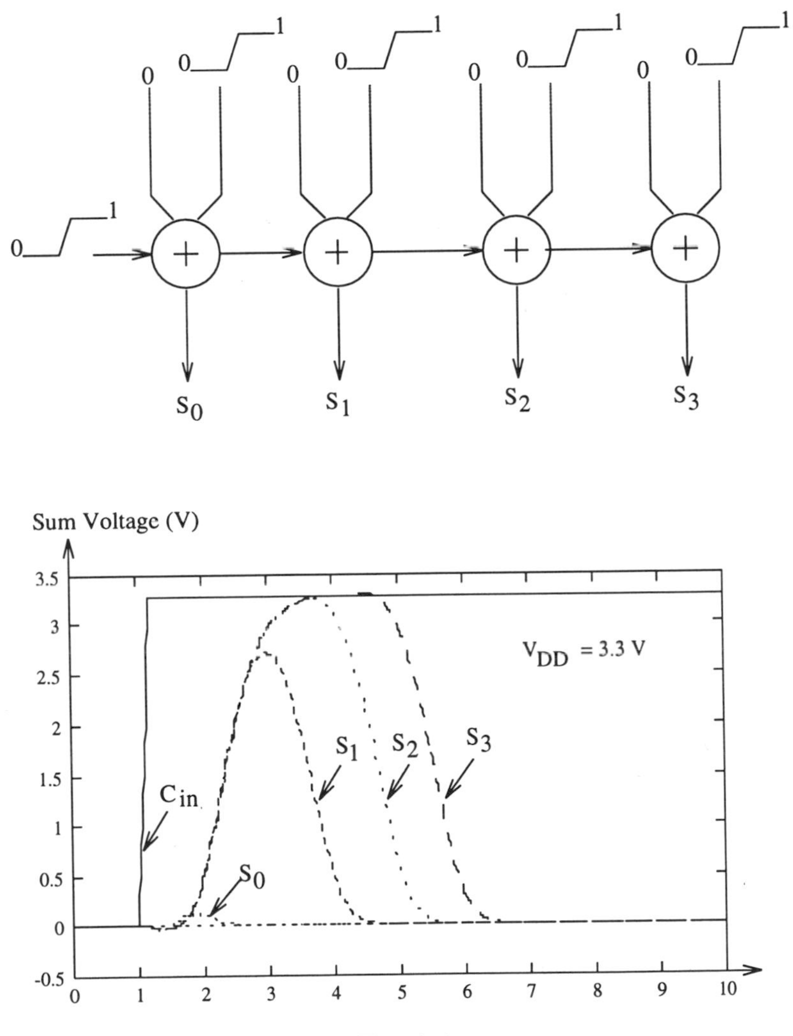

Figure 7.4 Sum voltage waveforms with glitching phenomenon.

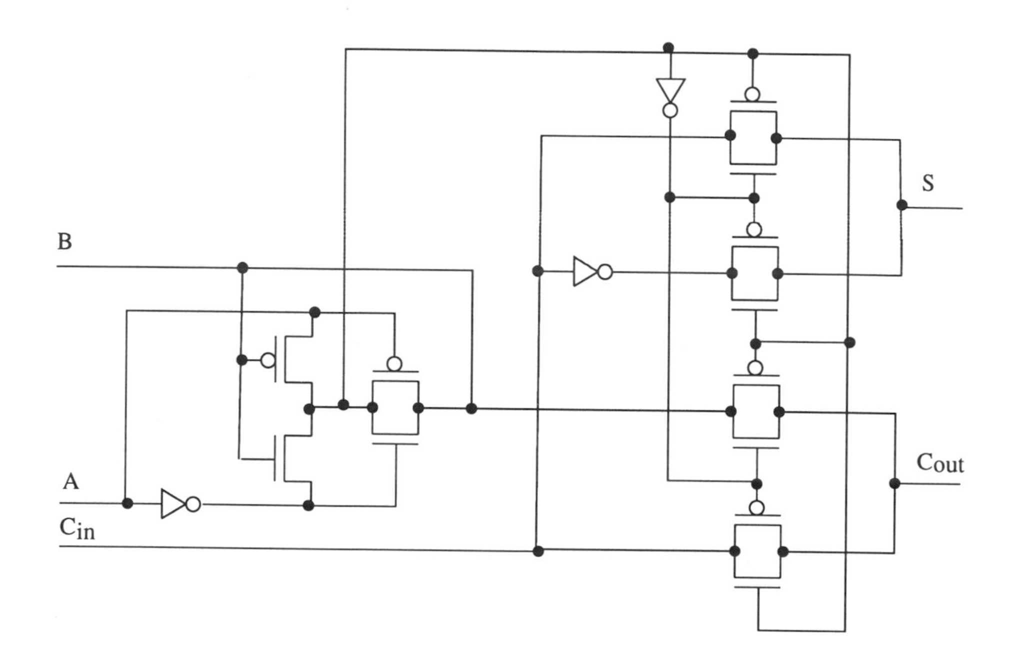

Figure 7.5 Optimized TG full-adder cell.

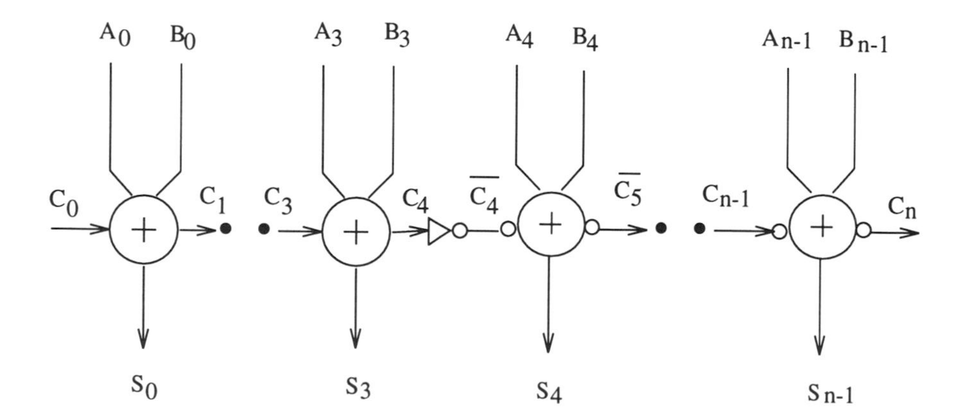

Figure 7.6 Ripple carry adder for the optimized TG cell.

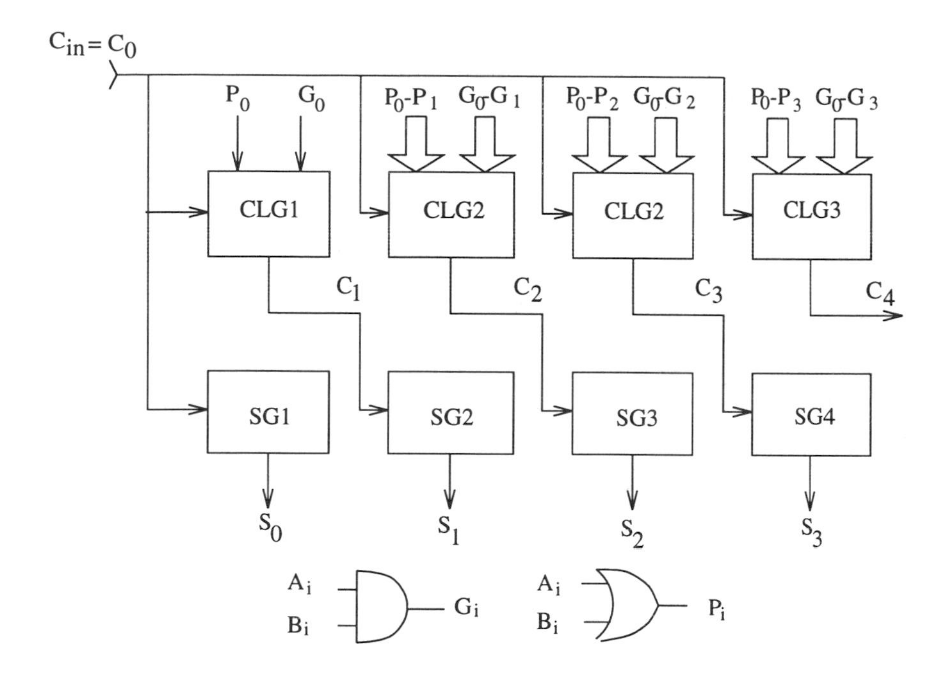

Figure 7.7 Block diagram of 4-bit carry lookahead adder (CLA).

The carries of the four stages are given by

$$C_1 = G_0 + P_0C_0 \tag{7.6}$$

$$C_2 = G_1 + P_1G_0 + P_1P_0C_0 \tag{7.7}$$

$$C_3 = G_2 + P_2G_1 + P_2P_1G_0 + P_2P_1P_0C_0 \tag{7.8}$$

$$C_4 = G_3 + P_3G_2 + P_3P_2G_1 + P_3P_2P_1G_0 + P_3P_2P_1P_0C_0 \tag{7.9}$$

Fig. 7.7 shows the block diagram of a 4-bit CLA adder. The carry generator blocks ($CLG1$ to $CLG4$) generate the carries C_1 to C_4, in parallel, from the carry in signal C_0. The different P_i and G_i signals are implemented following the expressions given by Equations (7.4) and (7.5). The sum generator blocks ($SG1$ to $SG4$) generate the sums. The sum, S_i, is generated by

$$S_i = C_{i-1} \oplus A_i \oplus B_i \tag{7.10}$$

or

$$S_i = C_{i-1} \oplus P_i \tag{7.11}$$

if the propagate signal is given by

$$P_i = A_i \oplus B_i \tag{7.12}$$

In general, an n-bit CLA adder can be implemented efficiently using 4-bit blocks.

Fig. 7.8(a) and 7.8(b) show the first and the fourth CMOS carry lookahead generator circuits, respectively. The generate and propagate signals are generated in parallel and are fed to all carry generators with the input carry signal C_0. The carry signals are generated simultaneously. However, because the number of stacked MOS transistors increases, the delay of the fourth carry is greater than that of the first and limits the adder speed. The sum generator of the CMOS adder of Fig. 7.2 can be used in this case. The same circuit is used for all four bits. This implementation is slow because of the large numbers of stacked MOS transistors which represent a high equivalent resistance in the pull-up and pull-down paths.

Another CLA circuit implementation in static CMOS design which improves the critical carry path delay is shown in Fig. 7.9(a). In this circuit, the number of stacked devices is reduced. The same cell of Fig. 7.9(a) can be used to generate each carry within a 4-bit block. P and G are the global propagate and generate signals, respectively. The inverter of the circuit of Fig. 7.9(a) is used to reduce the load on the fourth carry, $\bar{C}_{i+4}$, when it is used to drive the next fourth CLG circuit. The output of this inverter, I, drives many blocks such as the next first-bit, the next second-bit, the next third-bit CLGs, and the next sum blocks. For the fourth bit stage, P and G are given by

$$P = P_{i+3}P_{i+2}P_{i+1}P_i \tag{7.13}$$

$$G = G_{i+3} + P_{i+3}G_{i+2} + P_{i+3}P_{i+2}G_{i+1} + P_{i+3}P_{i+2}P_{i+1}G_i \tag{7.14}$$

The circuits of Fig. 7.9(b) and Fig. 7.9(c) show the implementations of the global functions P and G. Similarly, the P and G signals for the third, second and first bit stages can be constructed. For an n-bit adder, all the P and G signals are computed in parallel. Hence, the critical path is the carry path $C_i \rightarrow C_{i+4}$, except for the first 4-bit adder block, where the critical path can be from one of the inputs (A_0 or B_0) to the carry out C_4.

The sum generator is implemented using the propagate signals, P_i and $\bar{P}_i$. Fig. 7.10(a) illustrates one possible circuit using a static CMOS implementation.

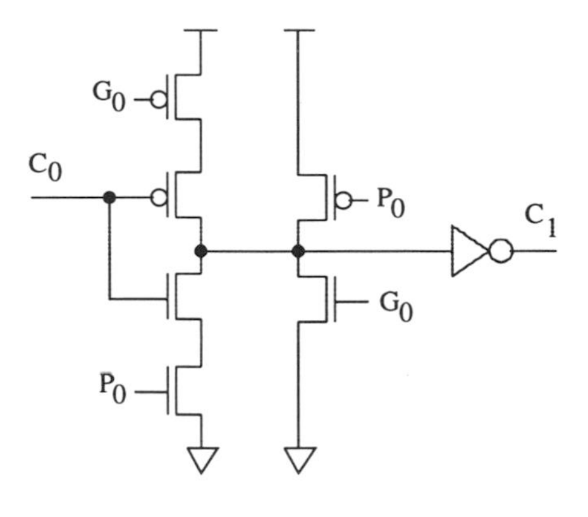

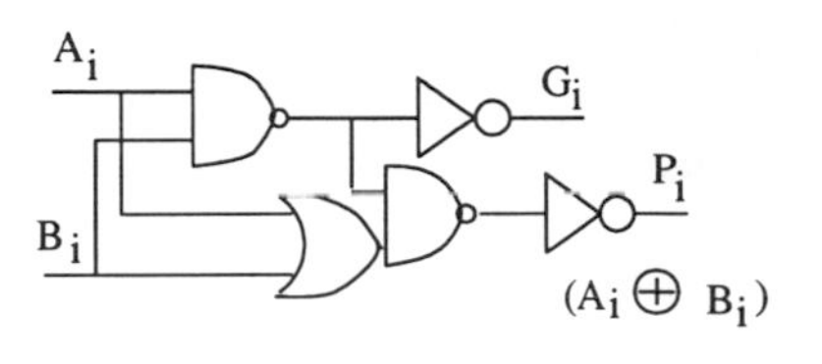

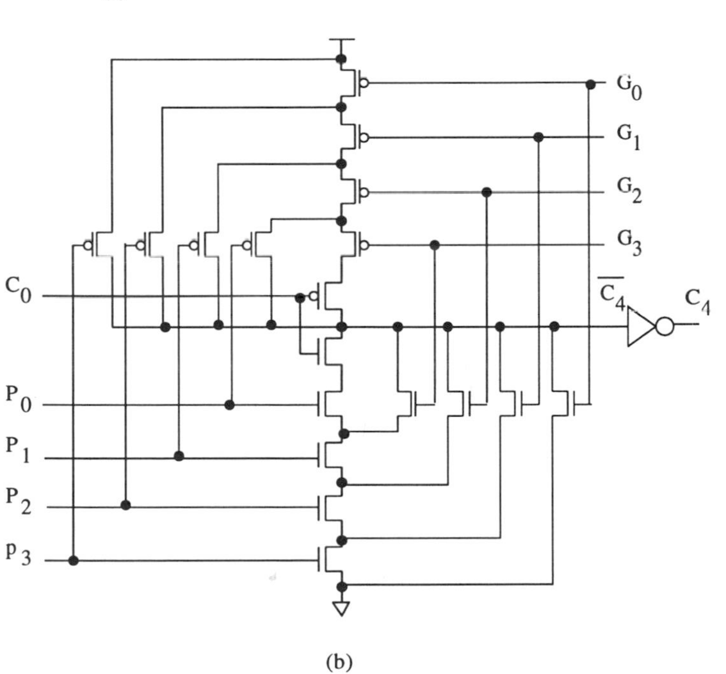

Figure 7.8 CMOS carry lookahead generators: (a) first bit; (b) fourth bit; (c) propagate and generate signals circuit.

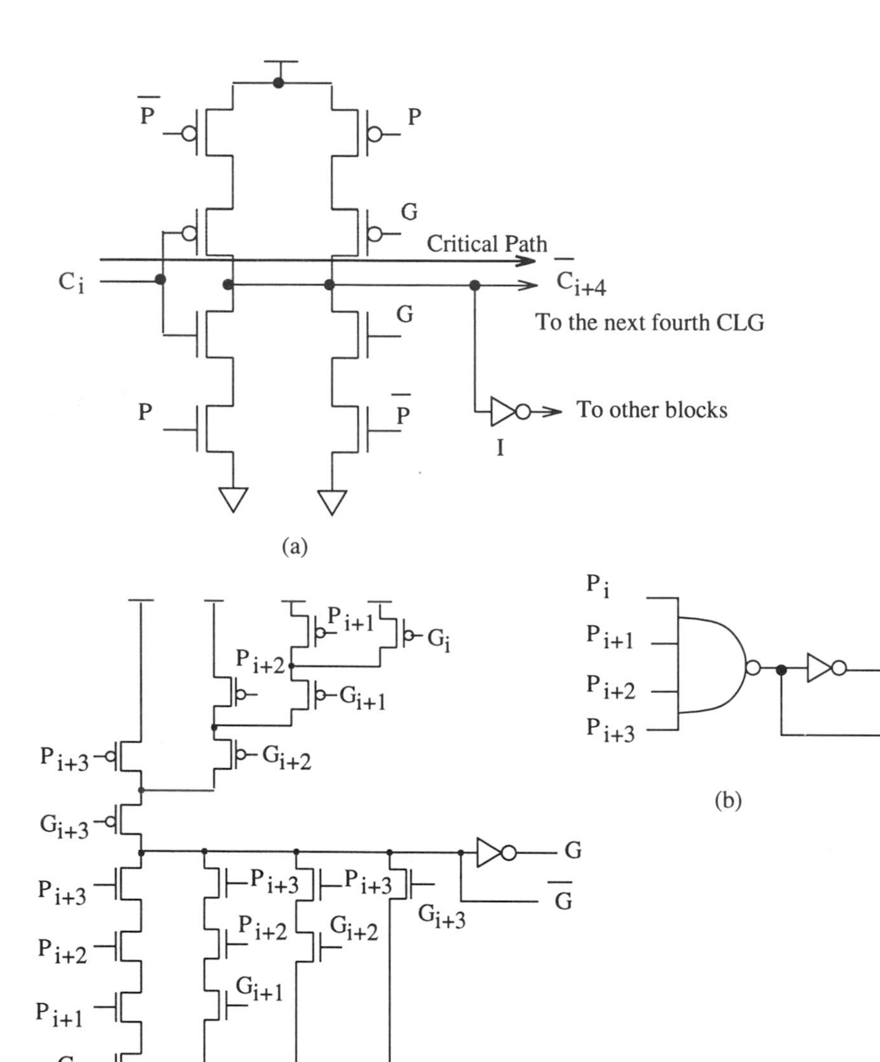

Figure 7.9 (a) The fourth carry lookahead generator; (b) the P signal generator; (c) the G signal generator.

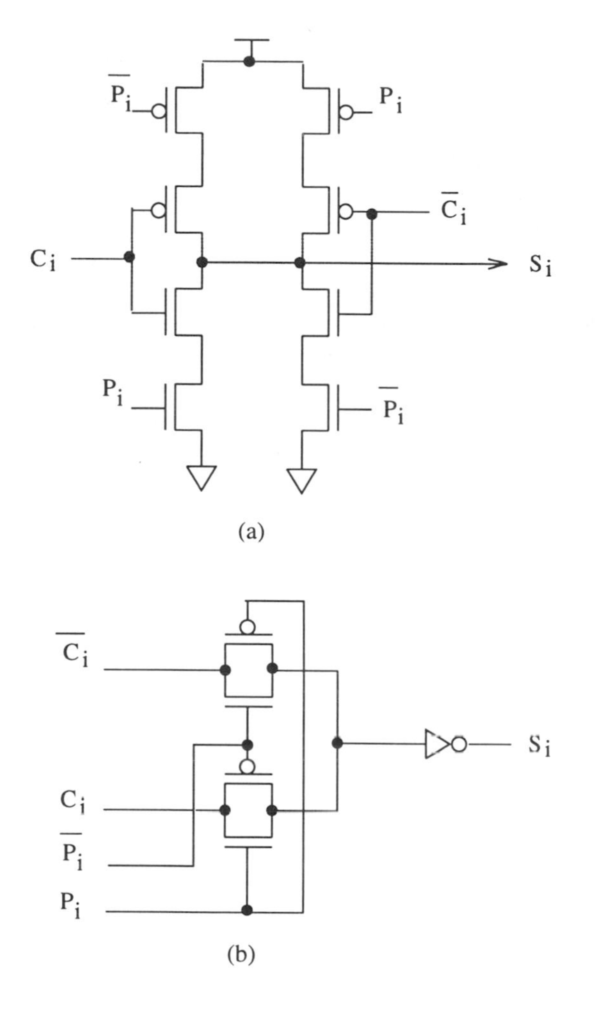

Figure 7.10 Sum generator circuits: (a) static CMOS; (b) transmission gate version.

Another circuit more compact and faster is shown in Fig. 7.10(b). It uses transmission gates and needs only 6 transistors.

Many circuit techniques for high-speed carry lookahead adders have been proposed. One of them uses the pseudo-NMOS like style [1]. The adder was used in a multiplier and achieved a high-speed static operation. However, it consumes a DC current and it is not suitable for low-power applications.

Other CLA implementations, to improve the carry path delay, are based on the transmission gates and CPL families. In this section we present the one based on CPL. The TG version is left to the reader to design. Fig. 7.11 shows the block digram of a 32-bit PMOS latch CPL carry lookahead adder using 4-bit blocks. The carry generators (CLGs) of each 4-bit block generate the carries C_{i+1} through C_{i+4} in parallel from the carry in, C_i. The different P_i and G_i signals, required by each 4-bit block, are not shown for clarity reasons. When the carry C_{i+4} is fed to the next 4-bit block it uses a buffer to distribute this carry to other CLGs and SGs. Therefore, the carry path is not significantly loaded. This results in a fast operation. Fig. 7.12 shows the CPL implementation of the CLG of the fourth bit. This circuit is located in the critical path of the carry signal. It is compact and uses only NMOS pass transistors. P and G are the global propagate and generate signals, respectively. The fourth carry is generated from the carry in or G signals through only one NMOS device. The P signal block is implemented using AND/NAND CPL style. After each 4 CLG blocks of the critical path, the carry is buffered and restored using PMOS latch buffers. The PMOS latch restores the reduced high level to full-swing to avoid any DC leakage current as shown in Fig. 7.11. Fig. 7.13 shows the G signal block for the fourth-bit CLG as an example. The same circuit style can be used to generate this G signal for the third-bit, the second-bit, and the first-bit CLGs. In addition the output inverter uses a PMOS latch to restore the swing. The PMOS latch circuit is incorporated only when dual rail signals are available. However, for a single-ended signal, a feed-back PMOS transistor is added to restore the full rail high-level as in the case of the sum generator of Fig. 7.14.

7.1.3 Carry-Select Adder

Another adder implementation which improves the speed of the RCA is the Carry Select adder (CS). It provides a regular layout, as in the case of an RCA. A CS adder basically consists of blocks; each executing two additions. One assumes that the carry in is "1"; the other assumes the carry in is "0". The real carry in is computed from the previous block and selects one of the two sum outputs with a simple TG multiplexer. Fig. 7.15 shows an example of an 8-bit carry select adder implementation with 4-4 staging. The carry signal, C_4, selects the next four sums and the carry C_8. The 4-bit adder blocks usually use RCA with transmission gate implementation. For a 32-b adder, the use of the normal staging 4-4-4-4-4-4-4-4, does not lead to an optimum delay. This is due to the multiplexing delay of the next carry. Optimal staging depends on the technology. For example, for the 0.8 μm CMOS device parameters presented

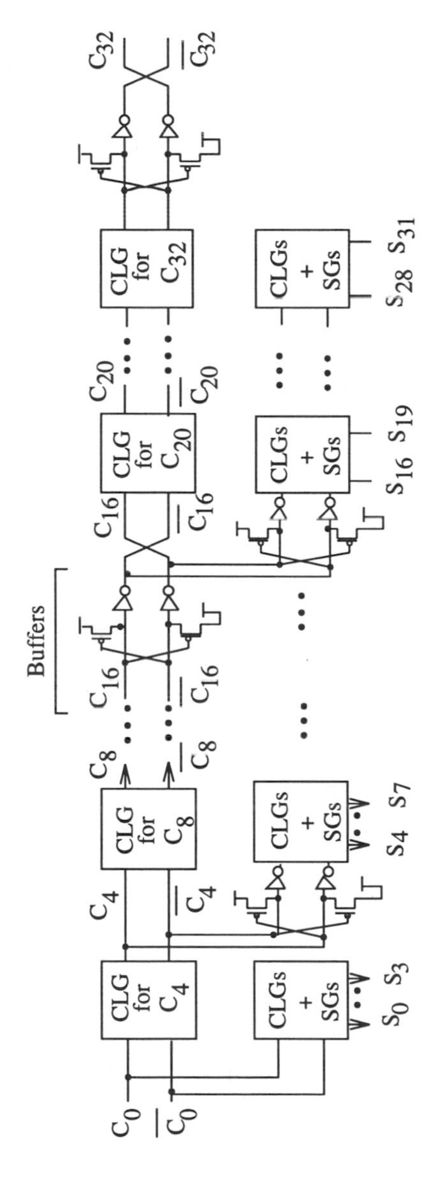

Figure 7.11 Block diagram of the CPL implementation.

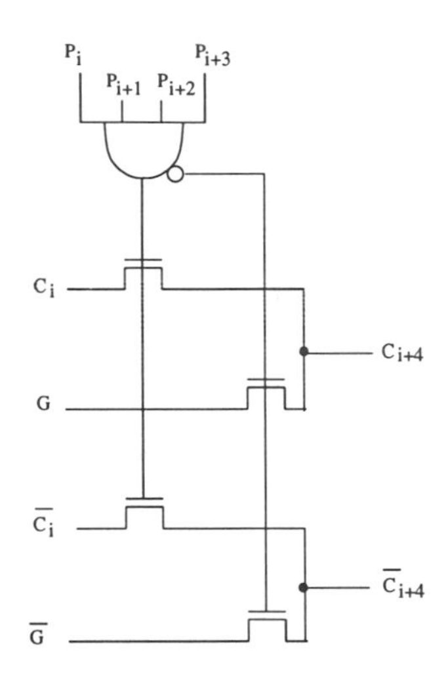

Figure 7.12 CPL carry lookahead generator.

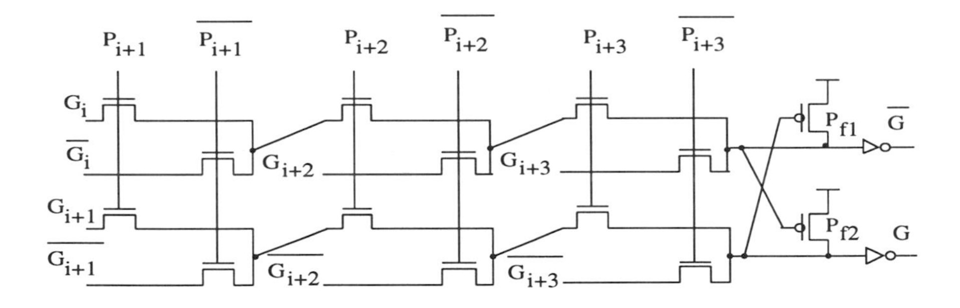

Figure 7.13 G block in CPL logic.

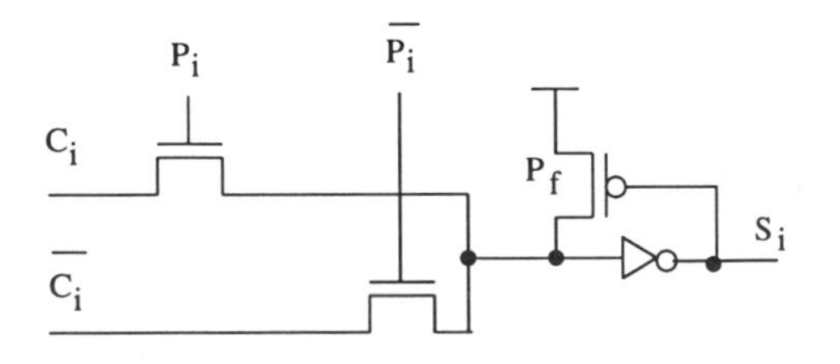

Figure 7.14 CPL sum generator circuit.

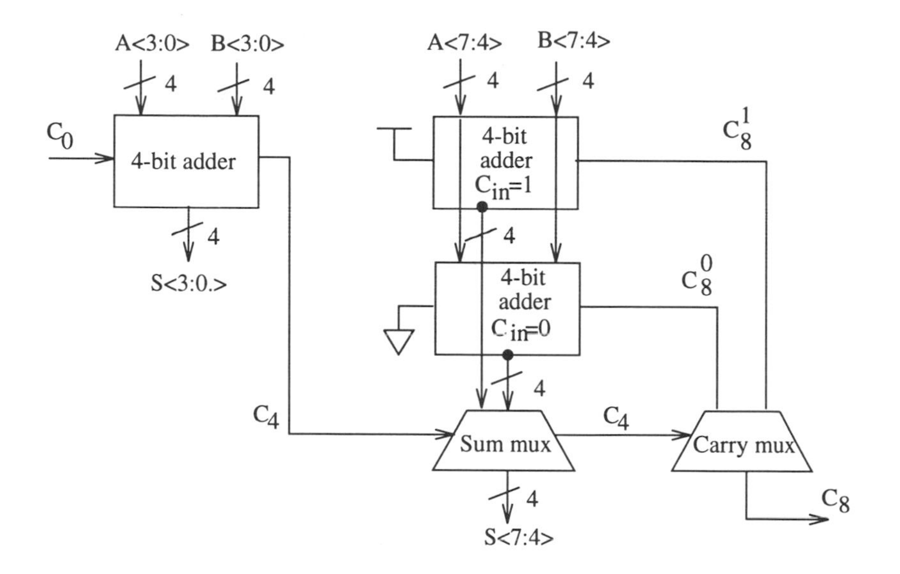

Figure 7.15 An 8 bit architecture of carry select adder.

in Chapter 3, simulations show that the optimal staging of a 32-bit CS adder using TGs is 4-4-7-9-8 at 3.3 V power supply voltage. This implementation is regular and easy to layout, however it has a higher occupied area than the RCA.

7.1.4 Conditional Sum Adders

In 1960 Sklansky considered the Conditional Sum Adder (CSA) as the fastest one, from a theoretical point of view [2, 3]. The concept behind this architecture is explained using the basic circuit of Fig. 7.16. This example is for a 4-bit conditional sum adder. It uses two types of cells: i) the conditional cell, and ii) the multiplexer. For each bit there is one conditional cell circuit. It computes two sums and two carries: S^0 and C^0 are calculated for a carry in zero, and S^1 and C^1 are calculated for a carry in one. The selection of the true sum is done with the first carry in and the previous carries. The true final carry out (C_4 in Fig. 7.16) is also selected.

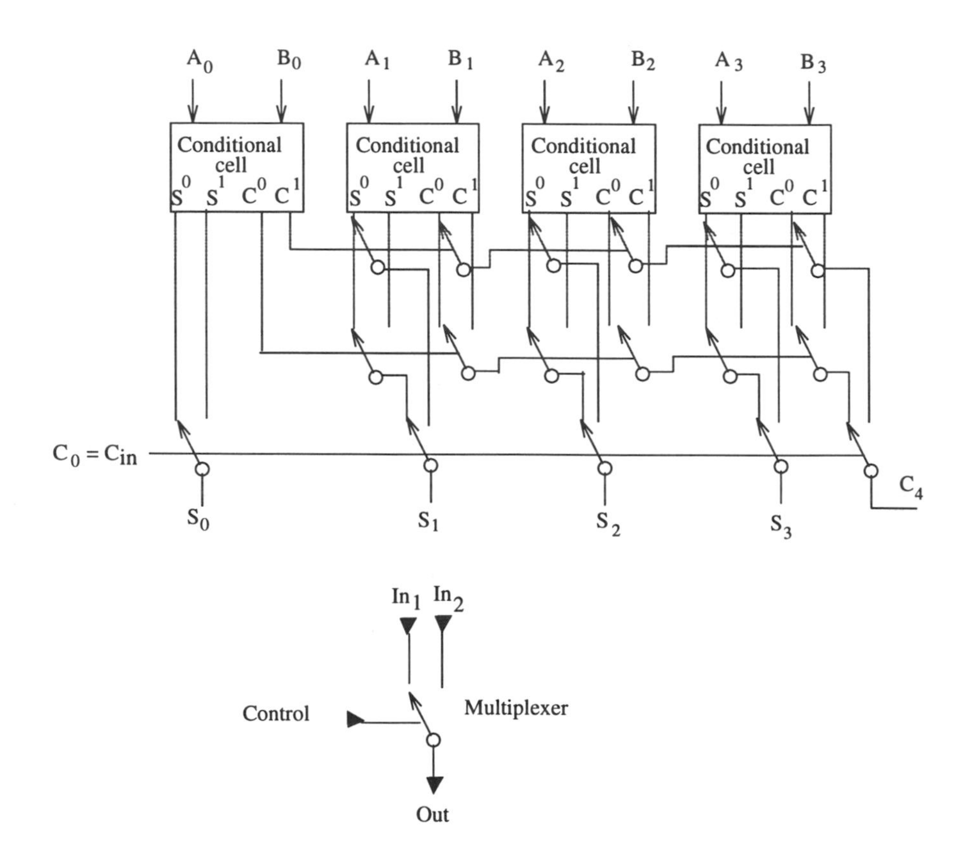

Figure 7.16 A simplified schematic of a 4 bit conditional sum adder.

A possible implementation of the conditional sum adder is shown in Fig. 7.17 for the case of a 4-bit adder [4]. The conditional cell can be implemented with the compact logic elements of Fig. 7.17(b). The different signals of the conditional cell are constructed using the following relations

$$\bar{S}_i^0 = A_i.B_i + \bar{A}_i.\bar{B}_i \tag{7.15}$$

$$\bar{C}_i^0 = \overline{A_i.B_i} \tag{7.16}$$

$$\bar{S}_i^1 = \bar{A}_i.B_i + A_i.\bar{B}_i \tag{7.17}$$

$$\bar{C}_i^1 = \overline{A_i + B_i} \tag{7.18}$$

The adder uses mainly for the multiplexers transmission gates as shown in Fig. 7.17(c). Note that the architecture uses the signals and their complements (dual-rail architecture) to avoid the use of inverters for the multiplexers. Otherwise the delay of the carry path will be penalized by the addition of inverters.

To design an n-bit (e.g., 32-bit) adder, one possible technique for fast operation is to use staged blocks of constant width or variable width. In this case, all the conditional sum blocks compute their respective double sums and double output carries in parallel. The true sum and carry out signals of each block are then selected by the carry in generated by the previous stage. The architecture at the block level uses a carry-select like technique where the carry in of each block is the true carry out of the previous block. The optimal staging can be determined from circuit simulation. The architecture has two critical delay paths within a block. One from the carry in to the carry out which is affected by the layout routing since the carry in of a block is distributed to all the final multiplexers. The other critical delay path is the one from the LSB-input of a block to the carry out.

To reduce the power dissipation and the delay of the CSA adder, a CPL-like circuit style can be used. Fig. 7.18 shows the different circuit cells needed to implement such an adder. In Fig. 7.18(a), the conditional cell schematic is shown. The output signals have a high level voltage equal to $V_{DD} - V_T$. Fig. 7.18(b) shows the compact multiplexer using NMOS pass-transistors. The control signals of the multiplexers should have full-rail swing. When using these reduced swing circuits in the adder, whenever a full-rail swing is needed it can be generated with the double-rail swing restored circuit of Fig. 7.18(c). The output inverter of the sum signal is shown in Fig. 7.18(d). The feedback PMOS transistor is needed to restore the high level when only a single-rail exists. The layout of such an adder is regular. Only three cells of the first, second and third bits have to be drawn. Fig. 7.19 illustrates the layout of a 4-bit block in 0.8 μm design rules.

7.1.5 Adder's Architectures Comparison

The ripple adder has the smallest area compared to the other classes and the lowest power in many cases. So it should be limited to applications where the area and/or the power must minimized, while the speed is not important. For fast adders, usually the CLA circuit is used, however its power dissipation can be relatively high. The carry select adders are widely used as the optimum compromise between high-speed operation of the CLAs and the small area of

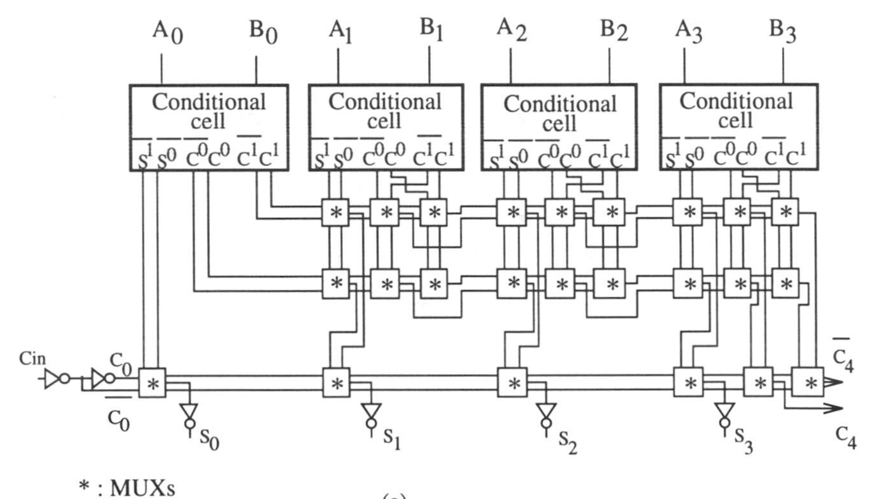

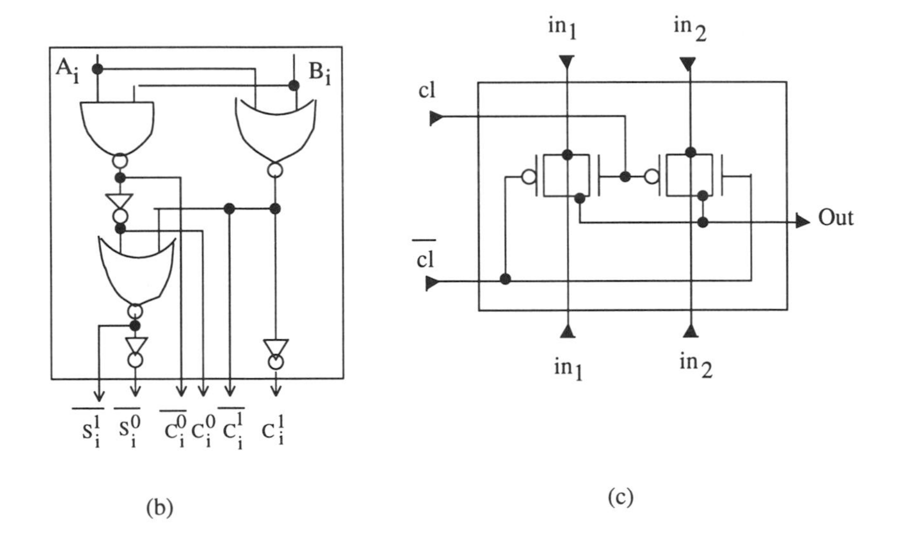

Figure 7.17 Conditional sum adder architecture: (a) 4-bit block; (b) conditional sum circuit; (c) multiplexer.

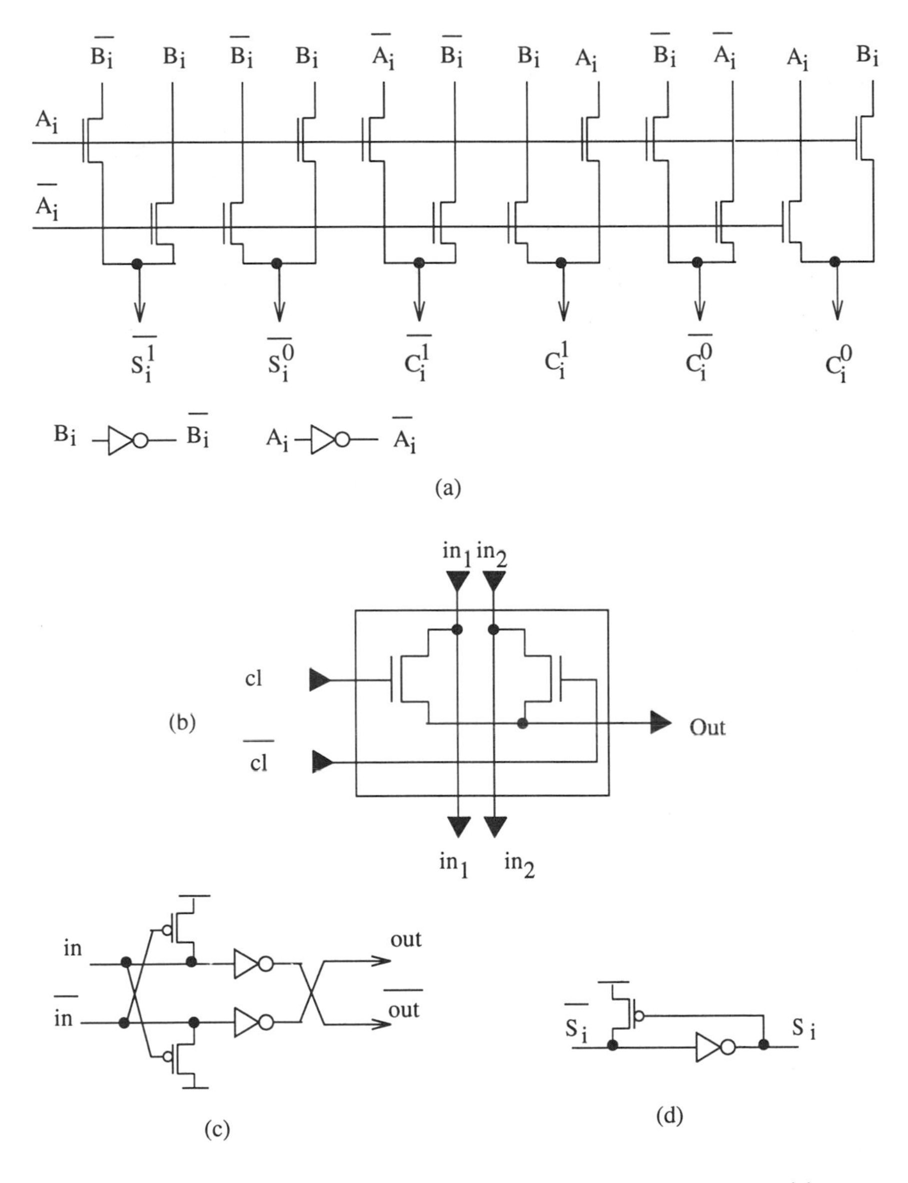

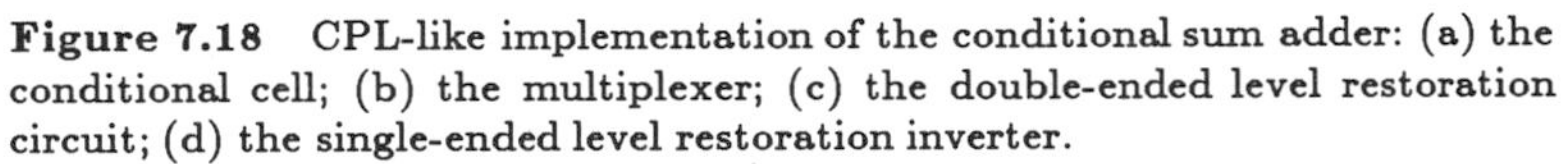

Figure 7.18 CPL-like implementation of the conditional sum adder: (a) the conditional cell; (b) the multiplexer; (c) the double-ended level restoration circuit; (d) the single-ended level restoration inverter.

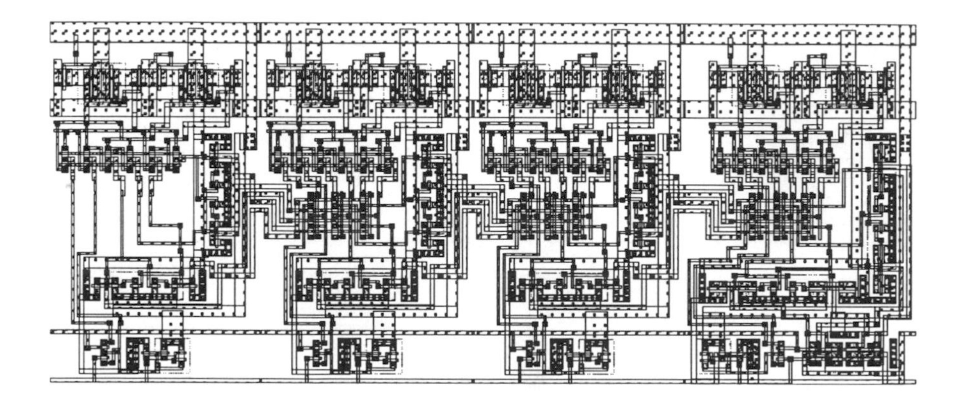

Figure 7.19 4 bit conditional sum adder layout.

RCAs. The conditional sum adder, with variable block staging, combined with carry select like style can result in the fastest adder if well optimized. The power dissipation of this adder can be comparable or maybe less than that of the RCA because it uses a reduced internal swing and a relatively small transistor count if the CPL-like style is used. When considering all the criteria such as the power, the area and the speed, a tool can be developed to select the adder class which satisfies the specified requirements.

For wide adders, having operand's size more than 32-bit, the different architectures can still be utilized. However, to optimize the speed and power of such a wide adder, several additional algorithms can be combined. Examples of wide adders can be found in [5, 6].

7.2 PARALLEL MULTIPLIERS

High-speed parallel multipliers are becoming one of the keys in RISCs (Reduced Instruction Set Computers), DSPs (Digital Signal Processors), graphics accelerators and so on. Parallel multipliers are used in data processors as well as digital signal processors. For example, for multi-media applications 16×16 fast multipliers are needed. For floating-point unit using double-precision multiplication (IEEE-754 standard), the mantissa data has 52-bit. Then 54×54 are required for such an operation. The two added bits are the sign bit and the guard bit. In this section we discuss several parallel multiplier algorithms which have been used in VLSI. The reader can consult references [7, 8] for more details on array multiplication algorithms.

7.2.1 Braun Multiplier

Consider two unsigned numbers $X = X_{n-1}...X_1X_0$ and $Y = Y_{n-1}...Y_1Y_0$

$$X = \sum_{i=0}^{i=n-1} X_i 2^i \tag{7.19}$$

$$Y = \sum_{i=0}^{i=n-1} Y_i 2^i \tag{7.20}$$

The product $P = P_{2n-1}...P_1P_0$, which results from multiplying the multiplicand X by the multiplier Y, can be written in the following form

$$P = \sum_{i=0}^{i=n-1} \sum_{j=0}^{j=n-1} (X_iY_j)2^{i+j} \tag{7.21}$$

Each of the partial product terms $P_k = X_iY_j$ is called summand. Fig. 7.20(a) shows an example of 4×4 multiplication. The summands are generated in parallel with AND gates. Fig. 7.20(b) shows the Braun's array multiplier [7]. Such a multiplier of $n \times n$ requires $n(n-1)$ adders and n^2 AND gates. The adder can be implemented efficiently by arranging the array for a regular layout. Fig. 7.21 shows a regular 4×4 array implementation of the multiplier of Fig. 7.20 using three different cells. The first cell contains an AND gate [Fig. 7.21(b)]. The second cell shown in Fig. 721(c) contains a full-adder and an AND gate. The routing lines are also illustrated in these cells. The last cell represents a full-adder composing the final carry propagate adder. The multiplier array is using what is called carry-save adders.

The delay of such a multiplier is dependent on the delay of the full-adder cell and the final adder in the last row. In the multiplier array, an adder with balanced carry and sum delays is desirable because sum and carry signals are both on the critical path. This is different than the case of a parallel adder where the carry path should be optimized and speed up compared to the sum path. For large arrays, the speed and power of the full-adder are very important. CPL-like styles discussed in Chapter 4 can result in reduced power dissipation and high-speed of operation. The final adder in the last row can use the techniques presented in Section 7.1.

$$
\begin{array}{cccccccccl}
 & & & & X_3 & X_2 & X_1 & X_0 & & = X \\
 & & & & Y_3 & Y_2 & Y_1 & Y_0 & & = Y \\
\hline
 & & & & X_3Y_0 & X_2Y_0 & X_1Y_0 & X_0Y_0 & & \\
 & & & X_3Y_1 & X_2Y_1 & X_1Y_1 & X_0Y_1 & & & \\
 & & X_3Y_2 & X_2Y_2 & X_1Y_2 & X_0Y_2 & & & & \\
 & X_3Y_3 & X_2Y_3 & X_1Y_3 & X_0Y_3 & & & & & \\
\hline
P_7 & P_6 & P_5 & P_4 & P_3 & P_2 & P_1 & P_0 & & = P = X*Y
\end{array}
$$

(a)

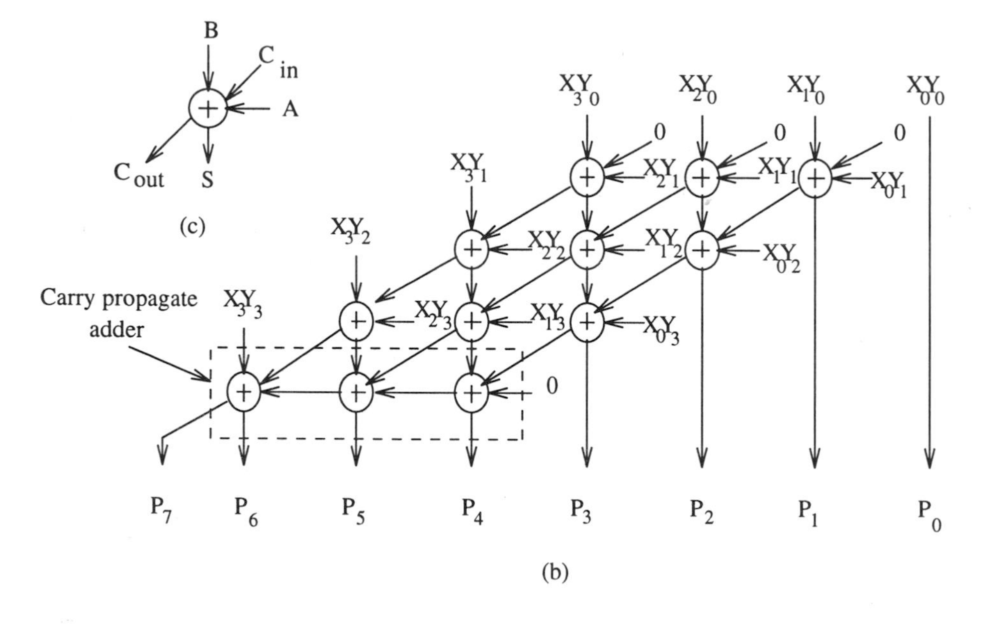

Figure 7.20 (a) Partial products of a 4×4 unsigned integer multiplication; (b) multiplier array; (c) full-adder schematic.

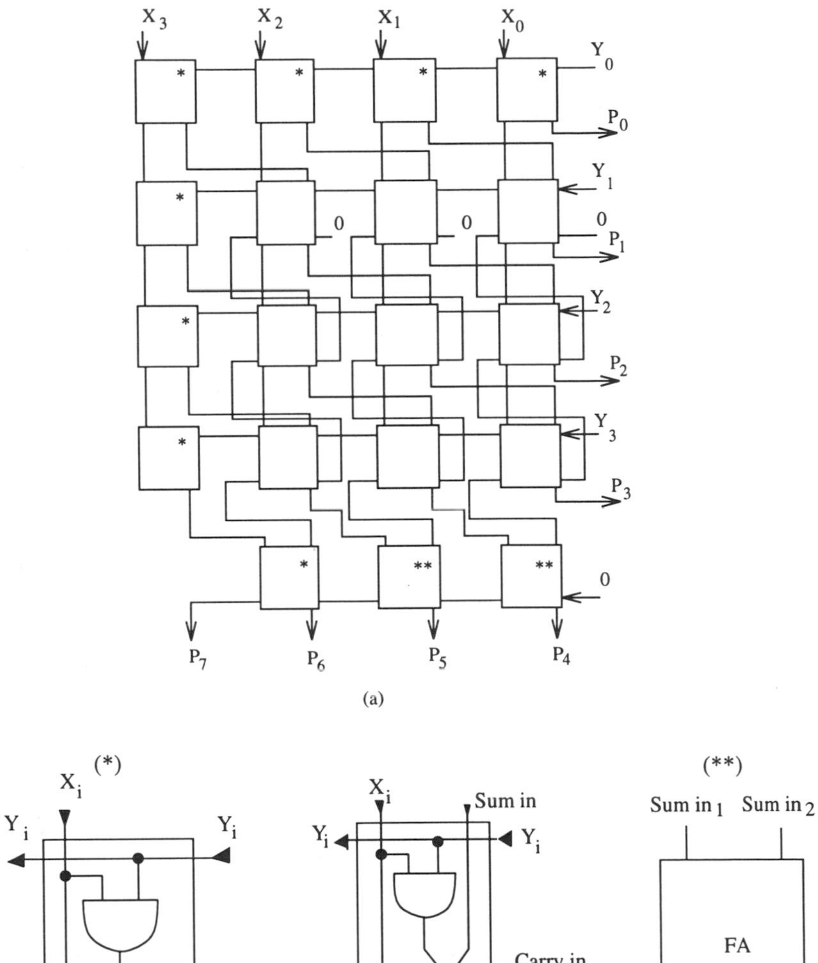

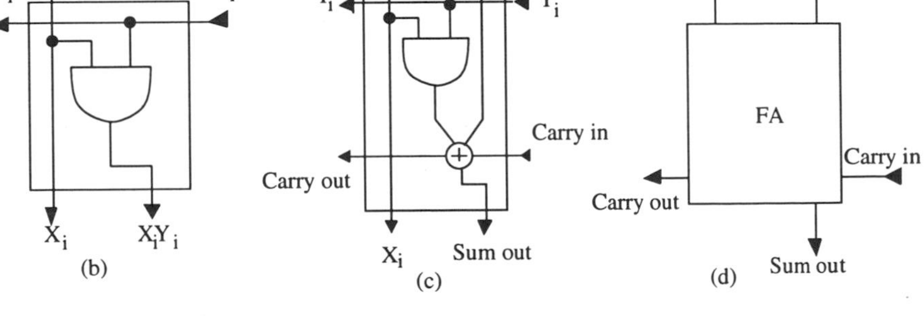

Figure 7.21 (a) Regular array of the 4×4 multiplier; (b) AND cell; (c) AND with full-adder cell; (d) full-adder cell.

7.2.2 Baugh-Wooley Multiplier

It was noted that Braun multiplier performs multiplication of unsigned numbers. The Baugh-Wooley technique [7] was developed to design regular direct multipliers for two's complement numbers. This direct approach does not need any two's complementing operations prior to multiplication. Let us consider two-numbers X and Y with the following form

$$X = -X_{n-1}2^{n-1} + \sum_{i=0}^{i=n-2} X_i 2^i \tag{7.22}$$

$$Y = -Y_{n-1}2^{n-1} + \sum_{i=0}^{i=n-2} Y_i 2^i \tag{7.23}$$

The product $P = XY$ is given by the following equation

$$\begin{aligned} P = XY \quad = \quad & X_{n-1}Y_{n-1}2^{2n-2} + \sum_{i=0}^{i=n-2}\sum_{j=0}^{j=n-2} X_iY_j2^{i+j} \\ & - X_{n-1}\sum_{i=0}^{i=n-2} Y_i2^{n+i-1} - Y_{n-1}\sum_{i=0}^{i=n-2} X_i2^{n+i-1} \end{aligned} \tag{7.24}$$

In order to avoid the use of subtractor cells and use only adders, the negative terms should be transformed. So

$$-X_{n-1}\sum_{i=0}^{i=n-2} Y_i2^{n+i-1} = X_{n-1}\left(-2^{2n-2} + 2^{n-1} + \sum_{i=0}^{i=n-2} \bar{Y}_i2^{n+i-1}\right) \tag{7.25}$$

Using this property in Equation (7.23), the product P becomes

$$\begin{aligned} P = XY \quad = \quad & -2^{2n-1} + (\bar{X}_{n-1} + \bar{Y}_{n-1} + X_{n-1}Y_{n-1}) \cdot 2^{2n-2} \\ & + \sum_{i=0}^{i=n-2}\sum_{j=0}^{j=n-2} X_iY_j2^{i+j} + (X_{n-1} + Y_{n-1}) \cdot 2^{n-1} \\ & + X_{n-1}\sum_{i=0}^{i=n-2} \bar{Y}_i2^{n+i-1} + Y_{n-1}\sum_{i=0}^{i=n-2} \bar{X}_i2^{n+i-1} \end{aligned} \tag{7.26}$$

Using the above relation an $n \times n$ multiplier, using only adders, can be implemented. The schematic circuit diagram of a 4×4 two's complement multiplier based on Baugh-Wooley's algorithm is shown in Fig. 7.22. The different cells composing the array are also shown. In this scheme $n(n-1)+3$ full-adders are

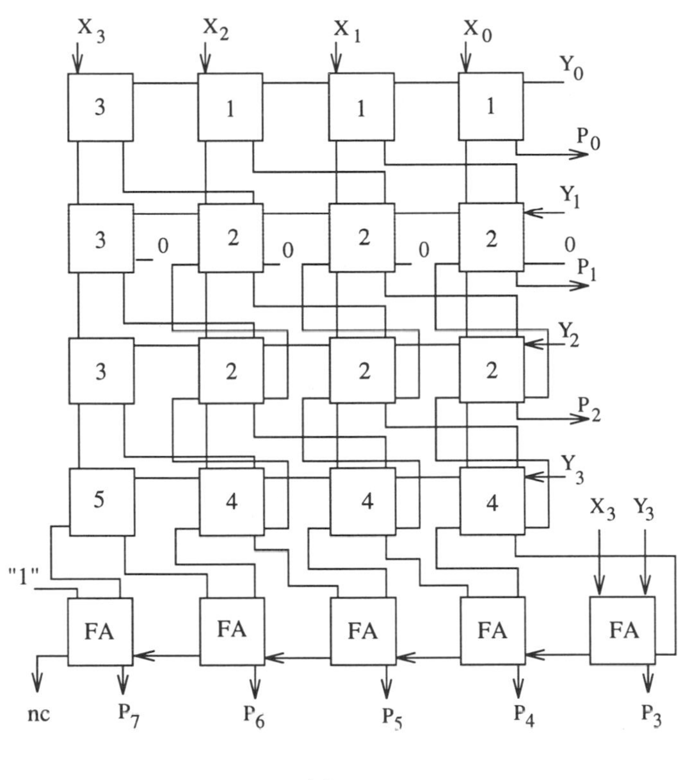

Figure 7.22 (a) 4 × 4 Baugh-Wooley two's complement regular array (FA : Full-Adder).

required. So for the case of $n = 4$ the array needs 15 adders. When n is relatively large, the final adder stage in the multiplier array can be implemented with the techniques discussed in Section 7.1.

This type of multiplier is suitable for applications where operands with less than 16 bits are to be processed. Applications for such a multiplier are, for example, for digital filters where small operands are used (e.g., 6, 8 and 12). For low-power and high-speed of operation, the array uses a CPL-like adder as mentioned previously in Section 7.2.1, while a CSA scheme, combined with carry select, can be utilized in the final adder. For operands equal or greater than 16-bit, the Baugh-Wooley scheme becomes too area-consuming and slow.

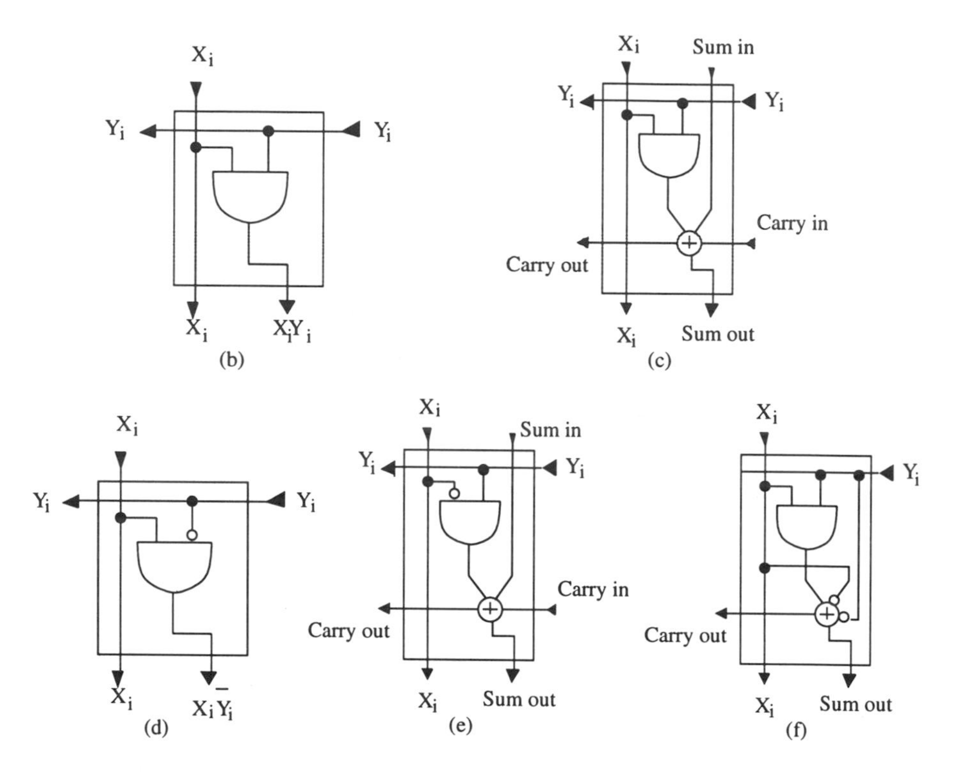

Figure 7.22 *(continued)*
(b) cell 1; (c) cell 2; (d) cell 3; (e) cell 4; (f) cell 5.

Hence, techniques to reduce the size of the array, while maintaining the regularity are required.

7.2.3 The Modified Booth Multiplier

For operands equal or greater than 16-bits, the modified Booth algorithm [8] have been used in almost all the designed multipliers. It is based on recoding the two's complement operand (i.e., multiplier) in order to reduce the number of partial products to be added. This makes the multiplier faster and uses less hardware (area). For example, the modified Radix-2 algorithm is based on partitioning the multiplier into overlapping groups of 3-bits, and each group is decoded to generate the correct partial product.

Let us write the multiplier, Y, in two's complement

$$Y = -Y_{n-1}2^{n-1} + \sum_{i=0}^{i=n-2} Y_i 2^i \qquad (7.27)$$

It can be rewritten as follows

$$Y = \sum_{i=0}^{i=n/2-1} (Y_{2i-1} + Y_{2i} - 2Y_{2i+1}) \cdot 2^{2i} \quad with \ Y_{-1} = 0 \qquad (7.28)$$

In this equation, the terms in brackets have values in the set$\{-2, -1, 0, 1, +2\}$. The recoding of Y, using the modified Booth algorithm, generates another number with the following five signed digits, -2, -1, 0, +1, +2. Each recoded digit in the multiplier performs a certain operation on the multiplicand, X, as illustrated in Table 7.1

Table 7.1 Partial product selection.

Y_{2i+1}	Y_{2i}	Y_{2i-1}	Recoded digit	Operation on X
0	0	0	0	$0 \times X$
0	0	1	+1	$+1 \times X$
0	1	0	+1	$+1 \times X$
0	1	1	+2	$+2 \times X$
1	0	0	-2	$-2 \times X$
1	0	1	-1	$-1 \times X$
1	1	0	-1	$-1 \times X$
1	1	1	0	$0 \times X$

So the bits of the multiplier are partitioned into groups of overlapped 3-bits, each group permits generation of a certain partial product. The five possible multiples of the multiplicand are relatively easy to generate following the explanation given in Table 7.2

The generated partial product is related to the multiplicand for each recoded digit by the relationships presented in Table 7.3. PP_i is the partial product and PP_n is the sign bit of the partial product with $P_n = P_{n-1}$ when no shifting of the partial product is performed. Note that the partial product is represented on $n + 1$ bits.

Table 7.2 Partial product generation process.

Recoded Digit	Operation on X
0	Add 0 to the partial product
+1	Add X to the partial product
+2	Shift left X one position and add it to the partial product
-1	Add two's complement of X to the partial product
-2	Take two's complement of X and shift left one position

Table 7.3 Partial product generation relations.

Recoded Digit	Operation on X		Added to LSB
0	$PP_i = 0$	for $i = 0, \cdots n$	0
+1	$PP_i = X_i$	for $i = 0, \cdots n$	0
+2	$PP_i = X_{i-1}$	for $i = 0, \cdots n$	0
-1	$PP_i = \bar{X}_i$	for $i = 0, \cdots n$	1
-2	$PP_i = \bar{X}_{i-1}$	for $i = 0, \cdots n$	1

To clarify this algorithm, an example is presented in Fig. 7.23. Let $X = 10010101$ and $Y = 01101001$. The recoded digits of Y are

$$01101001 \rightarrow +2 \; -1 \; -2 \; +1$$

The bits are grouped into 3-bit groups overlapped by one bit and a bit with a value of zero is added on the right side of Y as Y_{-1}. So the multiplication of two 8-bit numbers generates only 4 partial products. The number is then reduced by half. The partial product in this example is represented on 9 bits. For a correct partial product's addition, the signs are extended as shown in Fig. 7.23. The shape of the multiplier is then trapezoidal due to the sign extension.

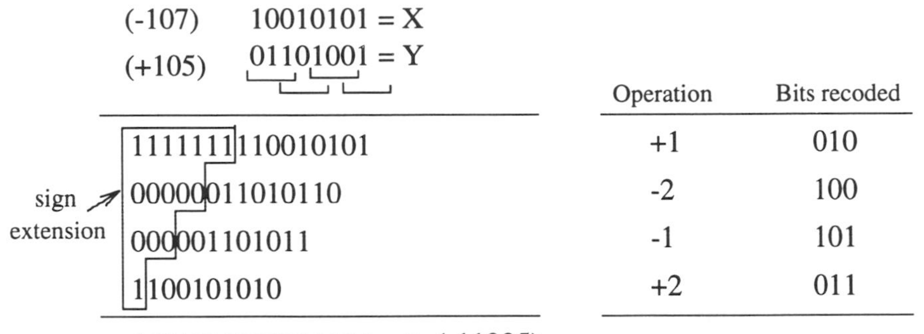

Figure 7.23 8 bit Booth multiplication example using two complement's number.

In order to make the array rectangular, and then more regular for VLSI implementation, the problem of sign extension must be addressed. This problem is more crucial when the operand lengths are wide, where each partial product must be sign-extended to the length of the product. In this section we will not deal with the techniques to solve the problem of the sign extension. But we will discuss one technique which is shown in Fig. 7.24 for the example of Fig. 7.23. The basic idea is to use two extra bits in the partial product. For the first partial product, the two additional bits, PP_{n+1} and PP_{n+2} are equal to the sign bit of the partial product

$$PP_{n+2} = PP_{n+1} = PP_n \tag{7.29}$$

For the second partial product, if the first partial product was positive, then the two additional bits for this second partial product are given by the expression above, otherwise we have two cases

$$PP_{n+2} = PP_{n+1} = 1 \quad if \;\; PP_n = 0 \tag{7.30}$$

and

$$PP_{n+2} = \overline{PP}_{n+1} = 1 \quad if \;\; PP_n = 1 \tag{7.31}$$

So it is more interesting to use a third bit, F, as a flag to indicate whether there is, from the previous partial, a negative sign bit to be propagated. F_1 is the flag generated by the first partial product to the next one. For the example of Fig. 7.24, $F_0 = 0$ (no PP before the first one), and $F_1 = F_2 = F_3 = 1$. So for the first partial product there is a sign propagation to all the others. This

			Operation	Bits recoded
(-107)	10010101 = X			
(+105)	01101001 = Y			
	11110010101		+1	010
	11011010110		-2	100
	11001101011		-1	101
	0100101010		+2	011
	1101010000011101 = P (-11235)			

[dashed box] Additional bits to be generated [sign extension]

[box] Additional bits generated from the previous sign and the present sign

Figure 7.24 The previous example of Figure 7.23 with simplified sign extensions.

flag is expressed by the following Boolean equation

$$F_{j+1} = F_j + PP_{n,j} \tag{7.32}$$

where $PP_{n,j}$ is the sign bit of the jth partial product.

Let us now see the implementation of the $n \times n$ modified Booth multiplier. Fig. 7.25 shows the block diagram of the multiplier. Also it gives an idea about the floorplan of this subsystem. It is composed of the following blocks:

- The multiplier array containing partial product's generators and 1-bit adders;
- The Booth encoder and the sign extension bits (PP_{n+2}, PP_{n+1}, F). The Booth encoder generates the five signals $(0, +1\times, +2\times, -1\times$, and $-2\times)$ for each group of 3-bit of Y; and
- The final stage adder performs $2n$ bits addition.

For the sake of simplicity, we treat the case of a 6×6 multiplier. All the cells described in this example are the basic cells of any multiplier size. Fig. 7.26

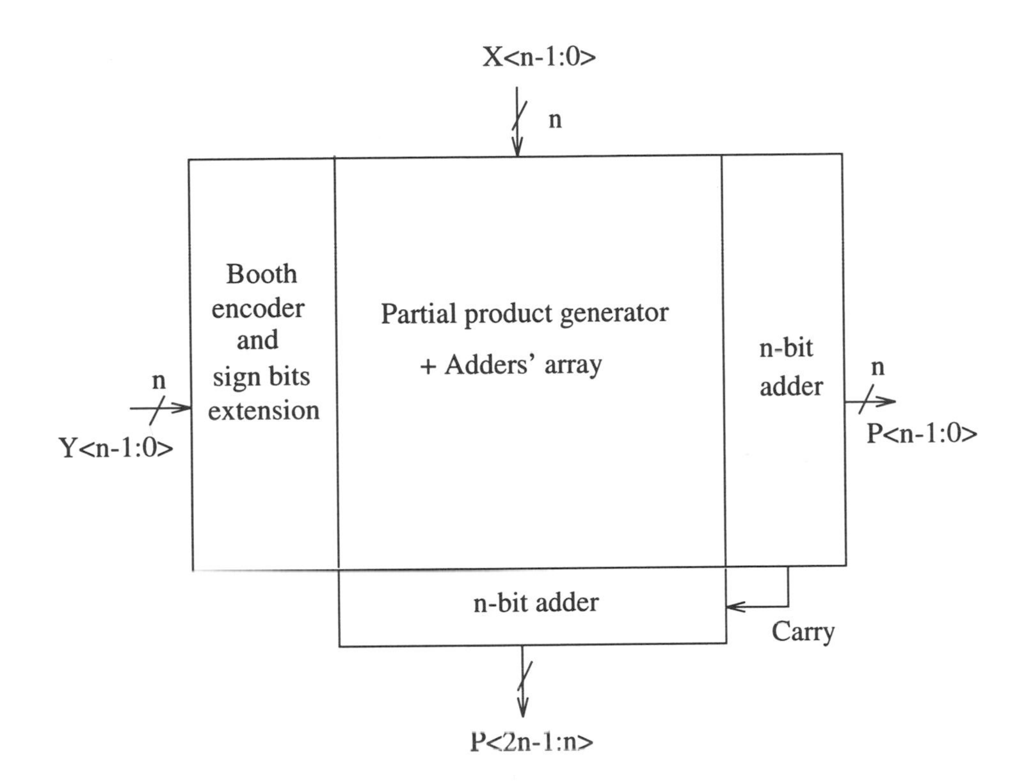

Figure 7.25 Block diagram of the $n \times n$ multiplier using modified Booth algorithm.

shows the implementation of such a multiplier. Four types of cells are used plus the final adder. These cells are:

- The ADD cell which generates 0 or 1 [see Table 7.3]. The schematic circuit of this cell is shown in Fig. 7.27(a). Two implementations are possible: one using pass-transistors controlled by the five signals defining the recoded digit code, and the other one is an AND2 gate of the two signals $-1\times$ and $-2\times$.
- The partial product MUX (PP-MUX) which generates the partial product. Fig. 7.27(b) shows the schematic of PP-MUX using CPL type logic. The feedback PMOS, P_f in this figure or in the one of Fig.

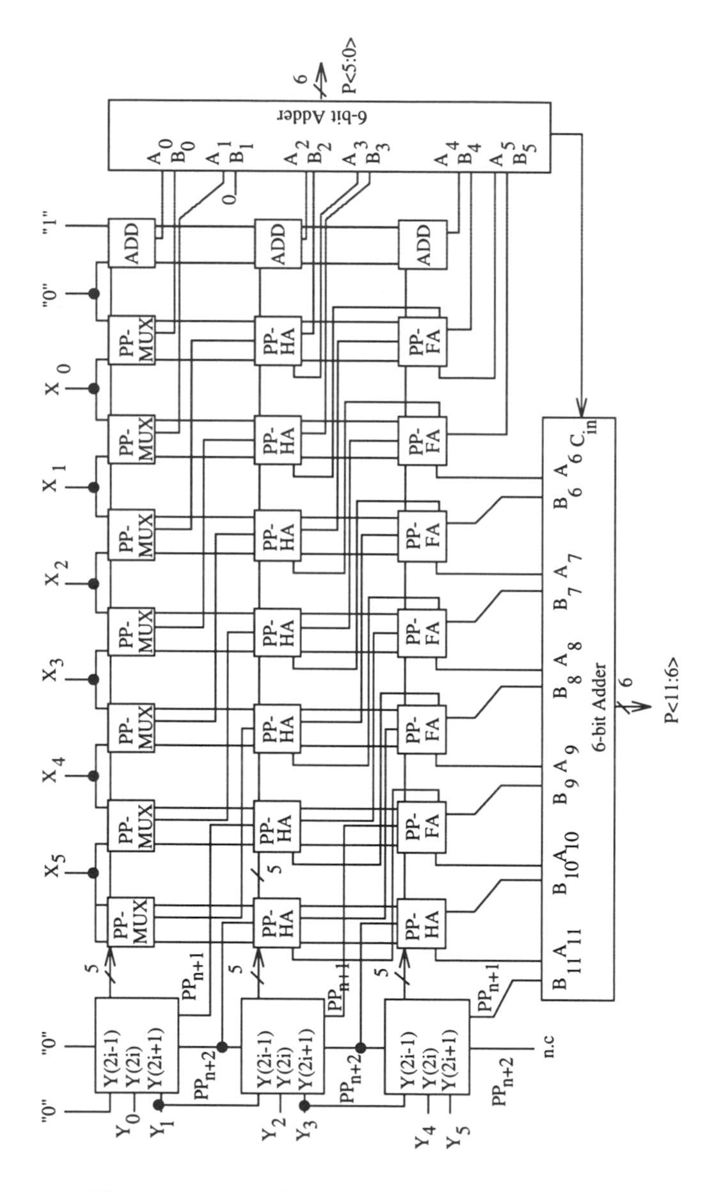

Figure 7.26 A schematic of a 6-bit Booth multiplier.

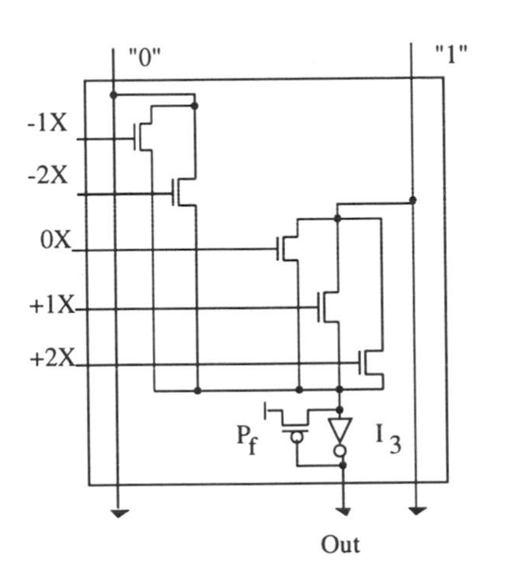

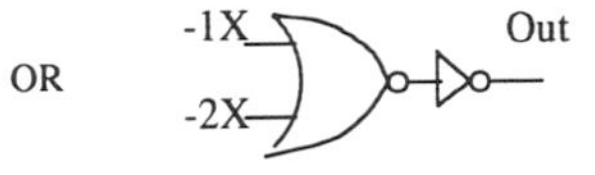

(a)

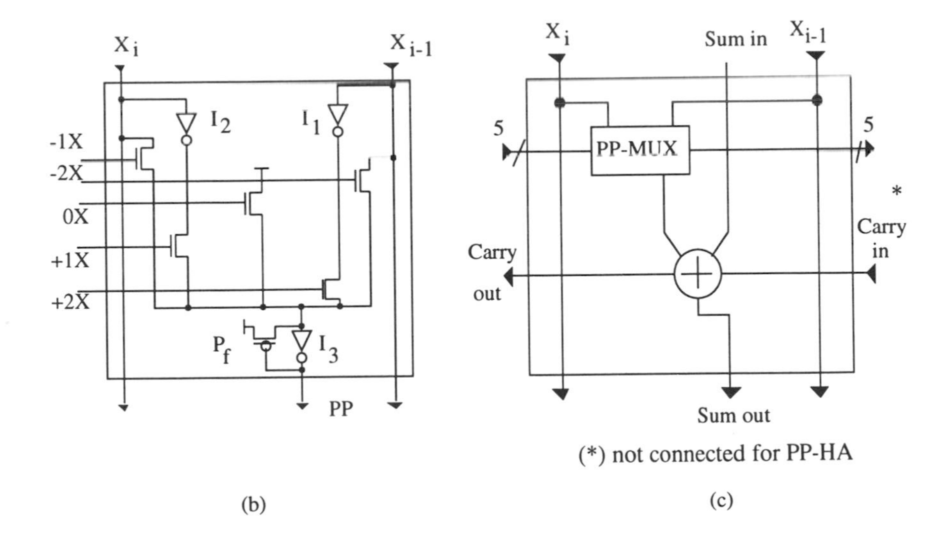

(b) (c)

Figure 7.27 Booth multiplier cells: (a) ADD; (b) PP-MUX; (c) PP-FA (or PP-HA).

7.27(a) are used to restore the high level to eliminate any DC current. This implementation permits fast operation and low-power operation.

- The PP-FA (PP-HA) cells. They merge the PP-MUX circuit and a full-adder (half-adder), respectively. CPL-like adder can be utilized for fast operation and low-power.
- The Booth Encoder (BE). It generates the five control signals $0\times$, $+1\times$, $+2\times$, $-1\times$, and $-2\times$ from a group of three bits of the multiplier Y. Fig. 7.28 shows the schematic of the different circuits involved in the BE block. The additional circuits of the two bits $PP_{n+1,j}$ and $PP_{n+2,j}$ of the jth PP are also illustrated. F_j and F_{j+1} are the previous and the next flags, respectively. $PP_{n,j}$ is the sign bit of the jth PP. Note that F_0 is 0.

The Booth multiplier exhibits a lot of unnecessary glitches. The main reason for glitches is due to the race condition between the multiplicand and the multiplier due to the Booth encoder. The power dissipation associated with the glitches can be an important portion of the total power and hence it needs to be reduced by some techniques of signal synchronization.

7.2.4 Wallace Tree

By applying the Booth algorithm, the number of partial products is halfed. However for large multipliers, 32-bit and over, the number of the partial products is over 16-bit. In this case, the performance of the modified Booth algorithm is limited. One technique, to improve the performance of these multipliers, is to adopt the Wallace tree using 4-2 compressors. A 4-2 compressor accepts 4 numbers and a carry in, and sums them to produce 2 numbers and carry out (really it is a 5-3 compressor). Fig. 7.29(a) shows an example of such a tree on partial products of an unsigned 8×8 multiplier. Eight partial products are produced. Using 4-2 compressors, two levels of additions (stages) are needed. The final two summands are added using a fast 16-bit adder. Some zeros are added to the array. This example shows that the bits which are not used in the 1st stage (level) jump to the next one to be combined with the ones produced by the compressors. Fig. 7.29(b) shows the architecture of the 8×8 multiplier. For the first stage of the tree, two blocks, A and B, are required. The block A (B) of compressors group the first (last) four partial products, respectively.

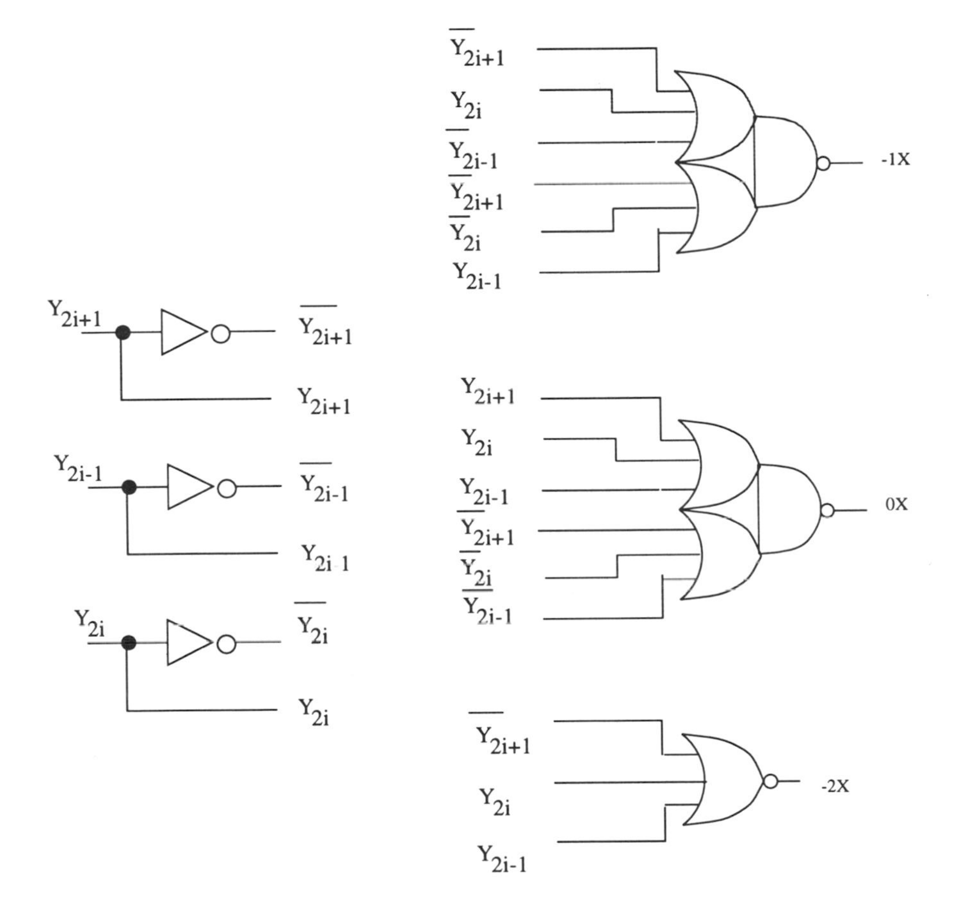

Figure 7.28 Logic schematic of the Booth encoder including the sign extension logic.

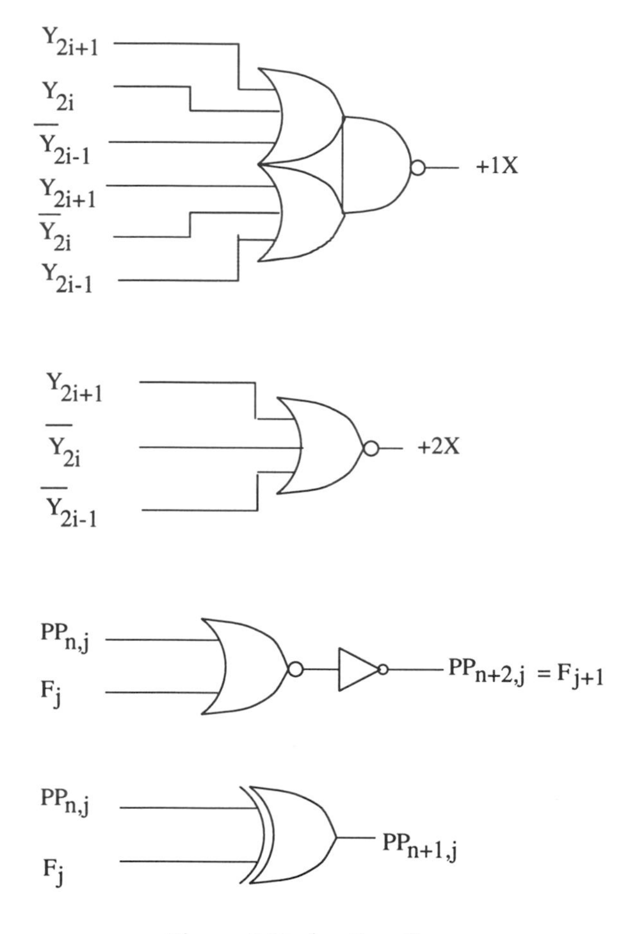

Figure 7.28 *(continued)*

Fig. 7.30 shows how the 4-2 compressor can be implemented by 2 full-adders or by custom static CMOS logic [9]. 4-bit $I_1, \cdots, I_4$, are added to produce 2 sums S and C. Hence, 4-bit of the partial product are compressed to produce two new partial products. The compressor is implemented, using carry-save adder construction, by two cascaded full-adders as shown in Fig. 7.30(b). Notice that carry-out2 is never generated by carry-in. Fig. 7.31 shows the 4-2 compressor circuit using a compact structure of multiplexers [10]. This structure is faster than the static complementary version. Fig. 7.32 shows the interconnection of the 4-2 compressors for block A of the example of Fig. 7.29. C_o is connected

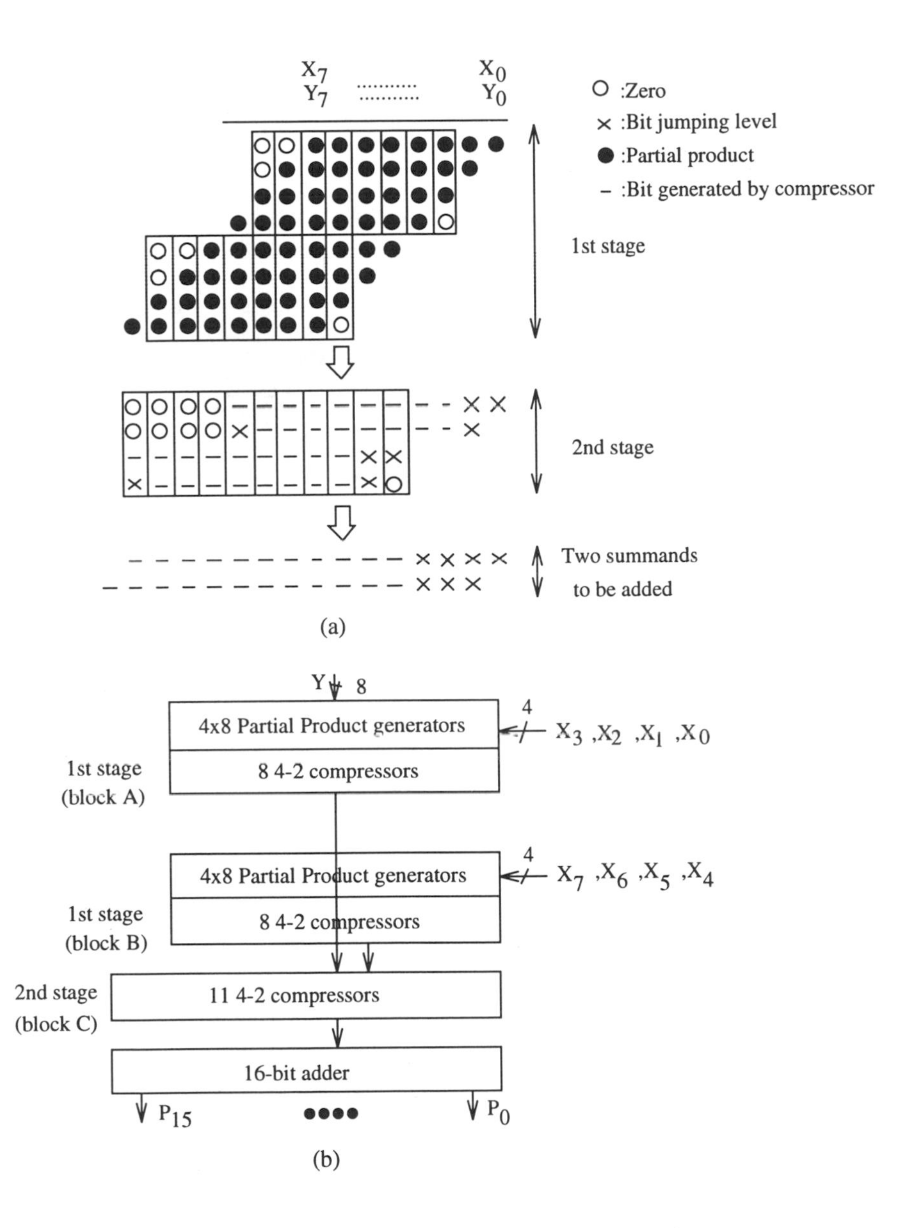

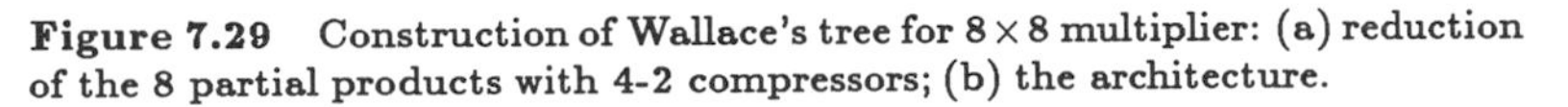
Figure 7.29 Construction of Wallace's tree for 8×8 multiplier: (a) reduction of the 8 partial products with 4-2 compressors; (b) the architecture.

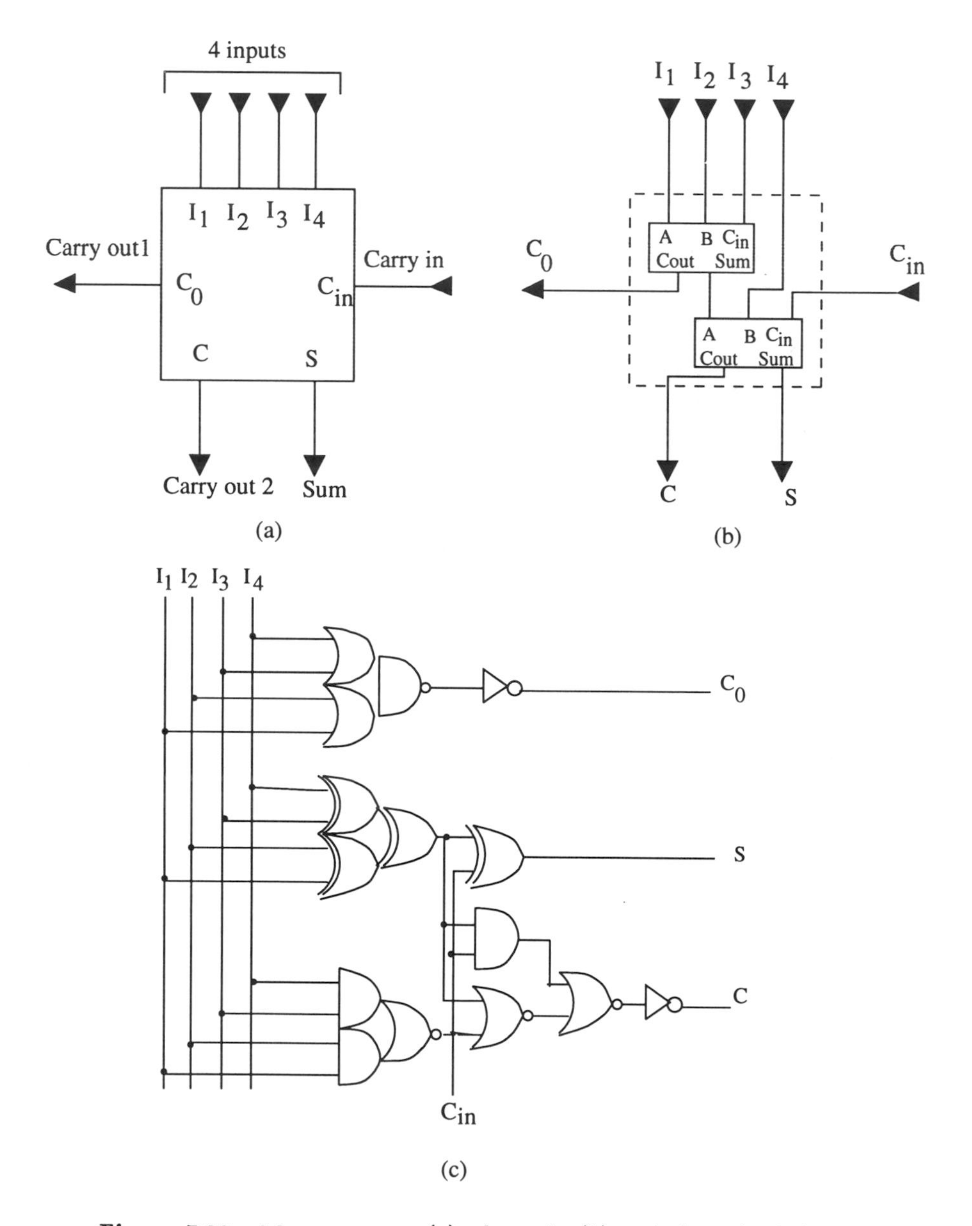

Figure 7.30 4-2 compressor: (a) schematic; (b) equivalent circuit by two-full-adders; (c) static CMOS logic implementation.

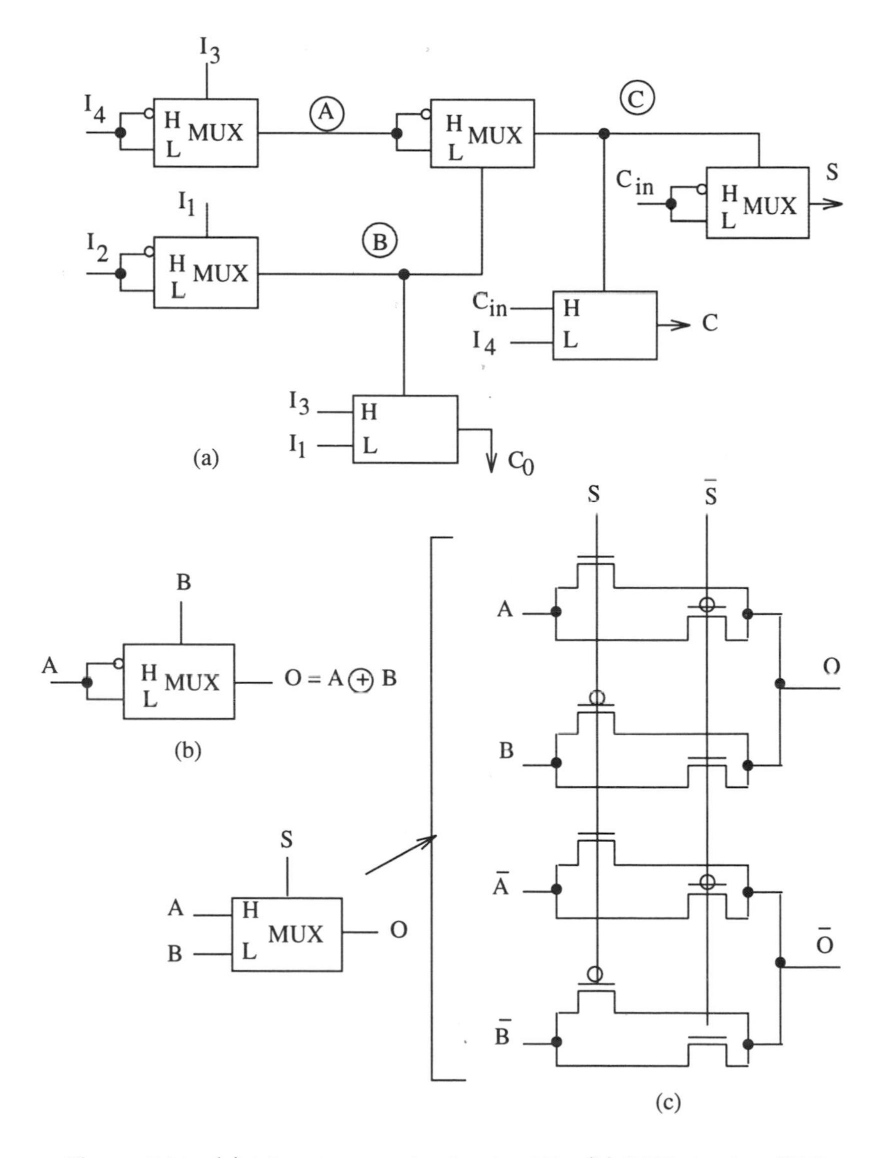

Figure 7.31 (a) 4-2 compressor circuit using TGs; (b) MUX circuit as XOR; (c) MUX implementation.

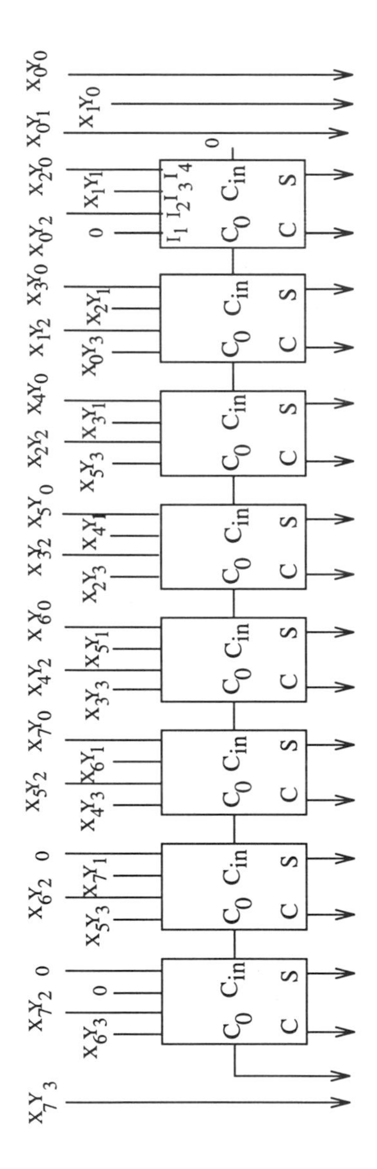

Figure 7.32 Interconnections of 4-2 compressors in block *A* of the 1st-stage in the 8 × 8 multiplier.

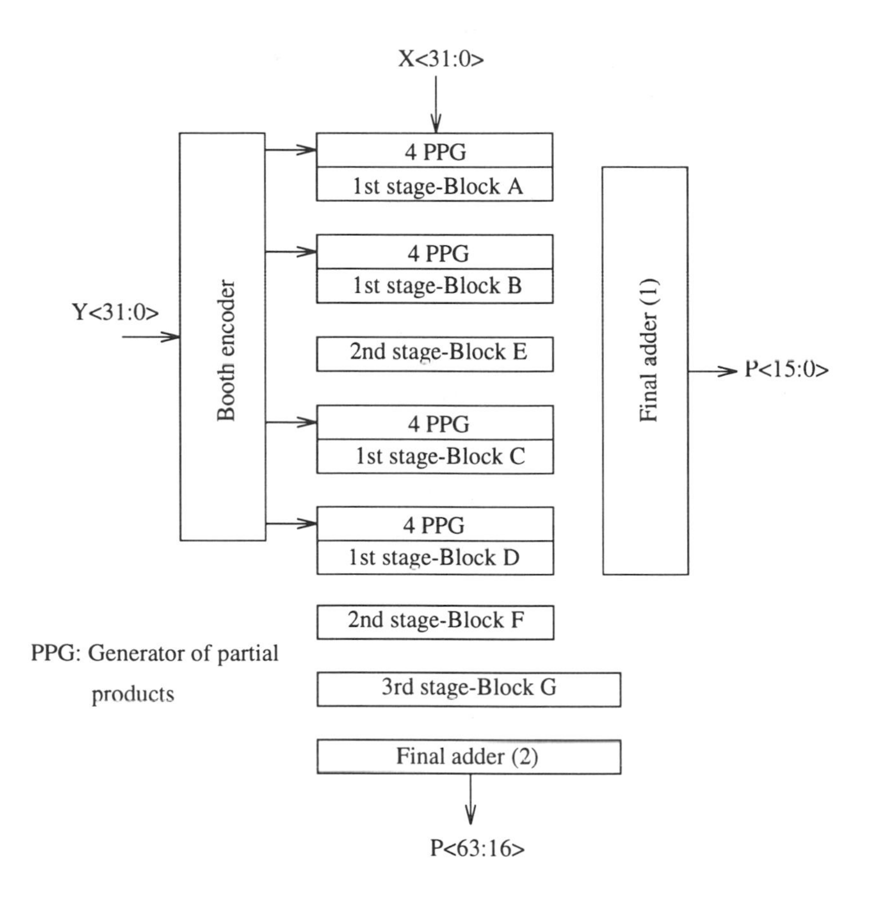

Figure 7.33 The architecture of 32 × 32 modified Booth multiplier with Wallace tree.

to the next carry-in C_{in}. Since these signals are independent, the carry is not propagated through the row.

To further enhance the Wallace tree multiplier, the modified Booth algorithm can be used to reduce the number of partial products by half in a carry-save adder array. One example of such combined construction is the architecture of the 32 × 32 multiplier shown in Fig. 7.33. It consists of four functions: the Booth encoder, the partial product's generator, the compressor blocks, and the final 64-bit adder. The Wallace tree is constructed with 3 stages (levels). The first stage has 4 blocks (A to D), with each block summing up 4 partial products among 16. The second stage sums up the 8 new generated partial products from the first stage. Hence, two blocks are needed, E and F. Finally, block G of the third stage of the tree generates two other new partial products to the final adder. This architecture exhibits some irregularities in the layout since it has a complicated interconnection scheme. Hence, the interconnection wires affect the speed and power dissipation of the adder.

7.2.5 Multiplier's Comparison

The basic array multipliers, like Baugh-Wooley scheme, consume low-power and have relatively good performance. However, their use can be limited to process operands with less than 16-bit (e.g., 8-bit). For operands of 16-bit and over, the modified Booth algorithm reduces the partial product's numbers by half. Therefore, the speed of the multiplier is reduced. Its power dissipation is comparable to the Baugh-Wooley multiplier due to the circuitry overhead in the Booth algorithm. However, circuit techniques can cause this multiplier to have low-power characteristics. The fastest multipliers adopt the Wallace tree with modified Booth encoding. A Wallace tree would lead, in general, to larger power dissipation and area, due to the interconnect wires. Hence, it is not recommended for low-power consumption applications. Dynamic multipliers are not discussed in this section since they introduce problems of control and timing. Hence extra area and power dissipation are added to the design.

7.3 DATA PATH

A VLSI chip can be partitioned in two parts; the data path (or execution unit) and the control unit. Data paths are often used in digital signal processors, microprocessors and application specific ICs (ASICs). The data path consists of a combination of an Arithmetic Logic Unit (ALU), a shifter, a file register, I/O ports, a multiplier, an adder, a magnitude comparator, and data busses, etc. It performs many operations on the data in the register file, to which the results are sent back. The data busses permit communication between the different units of the data path. The data busses are the communication means for the data transfer between the ALU, shifter, and file register, etc. These busses have a heavy load (few pF). In CMOS design, dynamic techniques are used to allow fast operation. One way to reduce the power dissipation, due to the precharging transistors, is to use static busses [11].

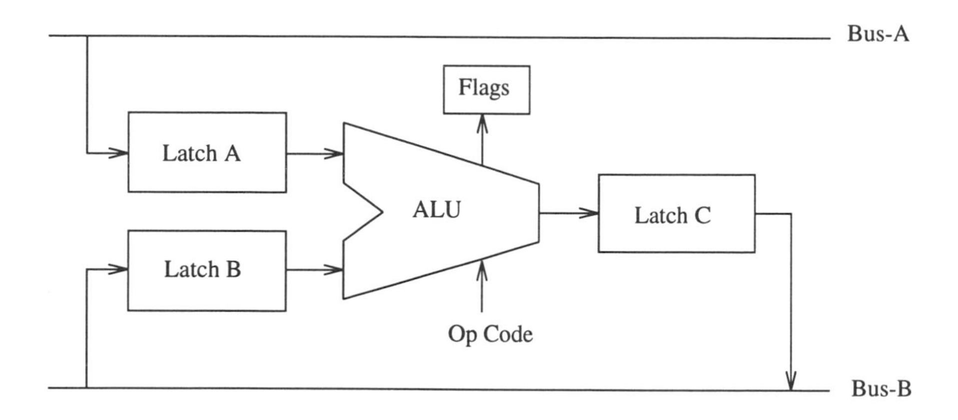

Figure 7.34 Arithmetic Logic Unit (ALU).

The control unit delivers the instructions to the data path. These instructions determine the operations that the data path has to perform. The control unit can be implemented using random logic, micro-ROM (Read Only Memory), PLA (Programmable Logic Array) or a combination of these three implementations. Other macrocells, such as TLB (Translation Lookaside Buffer), cache memory, etc., can be added to the data path and the control unit. In this section, several blocks of a data path are discussed.

7.3.1 Arithmetic Logic Unit

ALU is an important part of a data path. It is a macrocell which executes arithmetic operations such as multiplication, addition, subtraction, negation, and logic operations such as AND, OR, XOR, comparison, etc. It performs the operation on two operands stored in latch A and latch B and puts the result in latch C as shown in Fig. 7.34. The operation code (op code) selects the operation of the ALU to be executed. The flags indicate the status of the ALU, such as overflow, zero-result, and carry generation, etc. The input latches A and B are, in general, connected to two parallel data busses. Sometimes, the input latches are merged with MUXs to select many input sources to the ALU. The result latch is connected to one of the busses or, to a third one. The ALU described in this section is static for low-power applications.

The maximum clock frequency of a VLSI circuit may be limited by the ALU operations; especially the arithmetic ones. The critical delay of an arithmetic operation is due mainly to the carry propagation along the width of the ALU. There are many types of ALU, depending on the number of operations to be performed. Fig. 7.35 shows the block diagram of a 1-bit slice of an ALU. It has exactly the same structure as the adder, except that the P and G blocks are programmable. Fig. 7.35(a) shows the P block with 4 control signals ($OP_1 \cdots OP_4$). The feedback PMOS transistor, P_f, permits restoration of the high-level from $V_{DD} - V_{Tn}$ to V_{DD}. Hence the DC current of the first inverter, due to the reduced high-level, is eliminated. Fig. 7.35(b) shows the G block with 4 op code signals ($OP_5 ... OP_8$). The P and G blocks use the pass-transistor style. The techniques discussed in Section 7.1 can be applied to achieve low-power and fast operation. The carry and result (sum) blocks are shown in Fig. 7.35(c) and (d), respectively. Table 7.4 summarizes some of the functions that can be implemented with these blocks. Several other operations can be realized with this ALU.

Table 7.4 Examples of ALU operations (w. means with).

Operation	LSB-C_{in}	P function	G function	Op code ($OP_1 ... OP_8$)
Add w. carry	0	$P = A\ xor\ B$	$G = A\ or\ B$	10011101
Subtraction	1	$P = A\ xor\ \bar{B}$	$G = A\ or\ B$	10011101
Bit-wise AND	0	$P = A\ and\ B$	$G = 0$	01110000
Bit-wise OR	0	$P = A\ or\ B$	$G = 0$	00010000
Not A	0	$P = \bar{A}$	$G = 0$	10100000

Table 7.4 *(continued)*

Operation	Result
Add w. carry	$A + B$
Subtraction	$A + \bar{B} + 1$
Bit-wise AND	$A\ and\ B$
Bit-wise OR	$A\ or\ B$
Not A	$\bar{A}$

To implement an n-bit ALU, all the techniques discussed for carry speed-up in adders can be applied. Drivers are needed to distribute the op code signals for

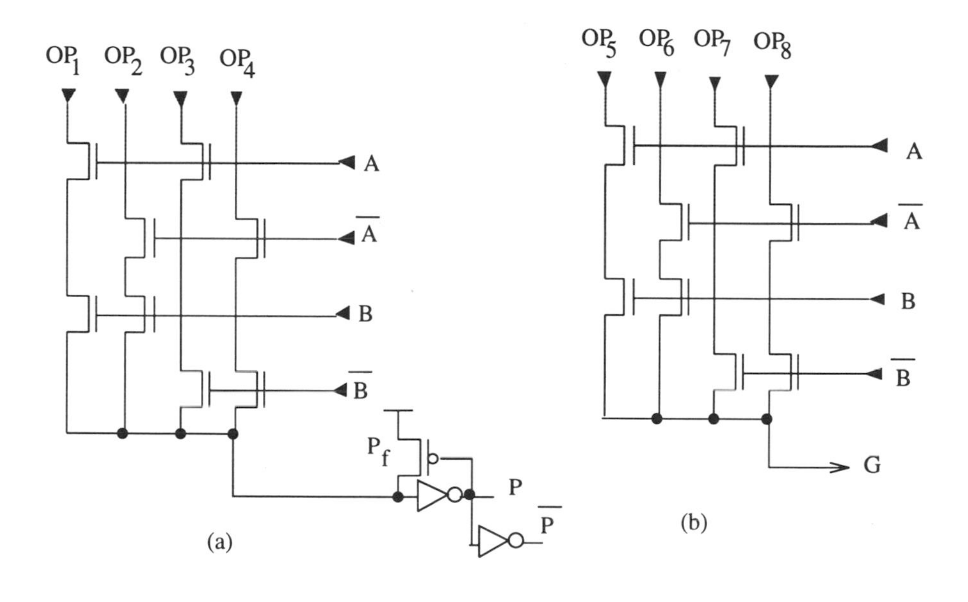

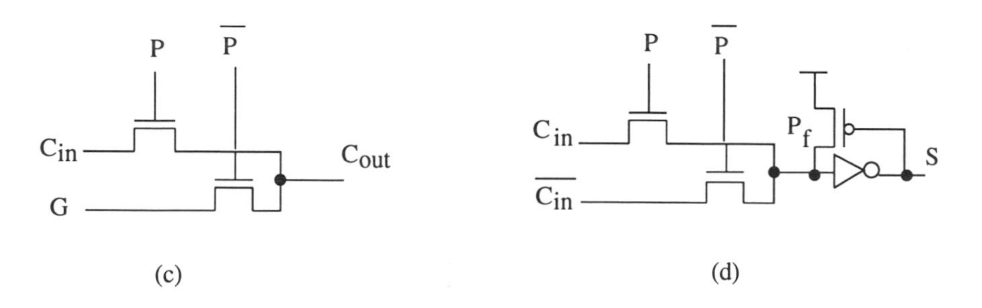

Figure 7.35 1-bit ALU blocks: (a) P block; (b) G block; (c) carry block; (d) result block.

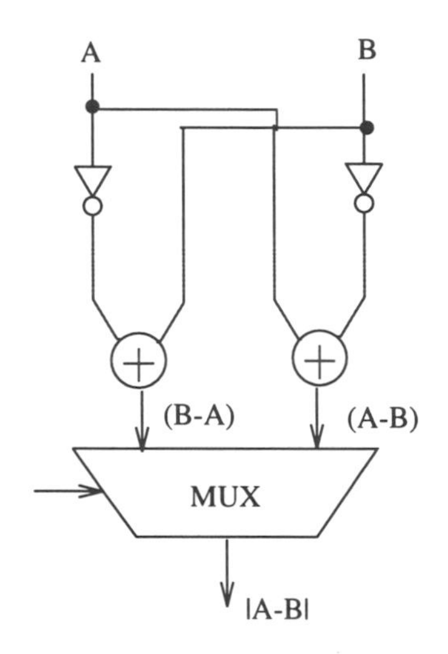

Figure 7.36 Absolute value calculator.

an n-bit ALU. For low-power design, the busses which communicate with the ALU are in general not precharged as in the case of many data paths.

7.3.2 Absolute Value Calculator

The Absolute Value Calculator (AVC) is, in general, used in data paths of video processors to compare the data of two pictures. Fig. 7.36 shows the architecture of the AVC. This parallel circuits performs two subtractions simultaneously, $A - B$ and $B - A$. Using the most significant bit of these two operations, the MUX circuit selects the positive one. Then the output gives the absolute value $|A - B|$.

To reduce the power dissipation and the area of an n-bit AVC, the logic of two n adders required can be reduced by the merging of the common functions for both operations. Also the techniques described in Section 7.1, for n-bit addition, should be used.

7.3.3 Comparator

A magnitude comparator is used in many DSP applications. It permits comparison of the magnitudes of two numbers A and B by providing if $A < B$, or $A = B$, or $A > B$. Fig. 7.37(a) shows an example of a two-bit comparator which requires two types of cells $C1$ and $C2$. The cell, $C1$, is constructed by the circuit of Fig. 7.37(b). Table 7.5 shows the truth table for this cell.

Table 7.5 Truth table for cell $C1$.

A	B	C	D	Note
0	1	0	0	$A < B$
0	0	0	1	$A = B$
1	1			
1	0	1	0	$A > B$

Let us explain how a 2-bit comparator works. When $A_1 < B_1$, then $C_1 = D_1 = 0$, and $A_1 A_0 < B_1 B_0$ regardless of the magnitudes of the lower bits. Similarly, for $A_1 > B_1$, then $C_1 = 1$, $D_1 = 0$, and $A_1 A_0 > B_1 B_0$ regardless of the magnitudes of the lower bits. When $A_1 = B_1 = 0$, the magnitudes of the two 2-b numbers depends on A_0 and B_0. In this situation, there are three different cases:

1. $A_1 A_0 < B_1 B_0$ for $A_0 < B_0$ (i.e., $C_0 = D_0 = 0$). Then we can set $E_0 = F_0 = 0$.

2. $A_1 A_0 = B_1 B_0$ for $A_0 = B_0$ (i.e., $C_0 = 0$, $D_0 = 1$). Then we can set $E_0 = 0$ and $F_0 = 1$.

3. $A_1 A_0 > B_1 B_0$ for $A_0 > B_0$ (i.e., $C_0 = 1$, $D_0 = 0$). Then we can set $E_0 = 1$ and $F_0 = 0$.

These relations can easily be used to implement the second cell, $C2$, of the comparator as shown in Fig. 7.37(c).

This technique, for the two-bit comparator, can be extended for an n-bit comparator. It can be constructed by using a parallel tree of the cells $C1$ and $C2$. A 4-bit comparator could, for example, be constructed with two 2-bit comparators connected in parallel and at the output the 4 E and F generated signals

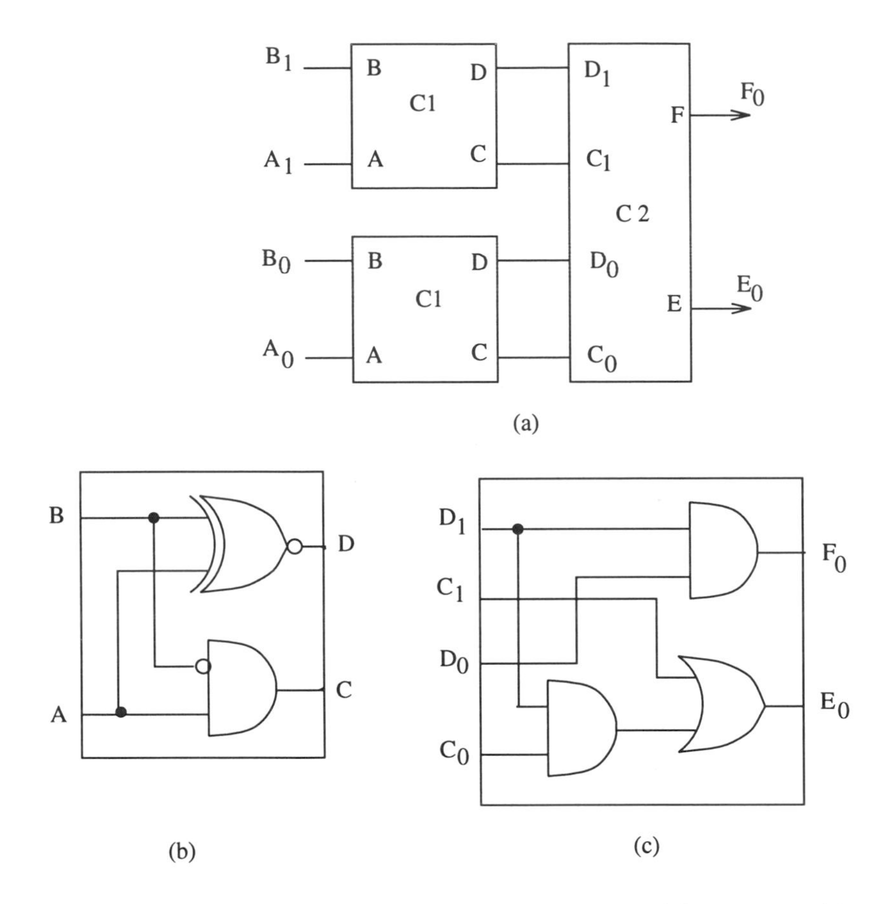

Figure 7.37 A two-bit comparator: (a) block diagram; (b) $C1$ circuit; (c) $C2$ circuit.

are fed to an added $C2$ cell. In this architecture, the glitching is reduced by equalizing the delay paths of each cell.

7.3.4 Shifter

Another macrocell of the data path is the shifter. It performs shift or rotate operations on the data. If the number of bits to be shifted is arbitrary, then a barrel shifter is used [12, 13]. Fig. 7.38 shows the CMOS implementation

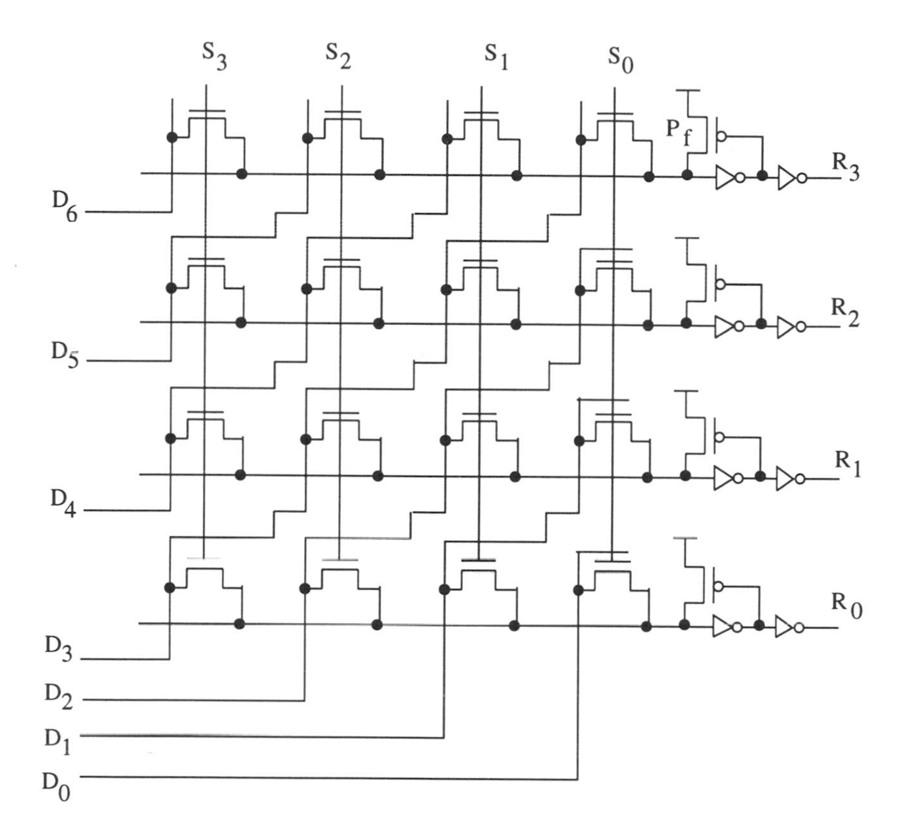

Figure 7.38 Barrel shifter circuit.

of a 4-bit barrel shifter. NMOS transistors are used as switches in the array. The input bus ($D_0 - D_6$) can be connected to the output bus ($R_0 - R_3$) via the pass transistors. The control signal S_0-S_3 selects the pass transistors to be switched. These signals determine the amount of shift and they are generated by a 2-bit decoder. Since the outputs have a high level of $V_{DD} - V_T$, due to the pass transistor, then the output buffer uses a feedback PMOS device, P_f, to restore the high level to V_{DD}. This eliminates any DC current in the first inverter of the buffer.

Table 7.6 shows the values of the output bus function of the input data. Depending on the values of $D < 6 : 0 >$, several shift operations can be performed. For example if $D < 6 : 4 >=$ "0", and $D < 3 : 0 >$ is the 4-bit input data, then a logical shift is realized. However, if $D < 6:4 >=$ "1" and $D < 3:0 >$ is the input data, then an arithmetic shift operation is performed.

Table 7.6 Output bus function of the shifting amount.

$S_3S_2S_1S_0$	$R_3R_2R_1R_0$
0001	$D_3D_2D_1D_0$
0010	$D_4D_3D_2D_1$
0100	$D_5D_4D_3D_2$
1000	$D_6D_5D_4D_3$

The barrel shifter is not a critical unit for the delay. A low-power operation is performed by using a static implementation. This shifter can be implemented with transmission gates and the feedback PMOS are not required. However for low-power, the use of NMOS array is more efficient. The feedback PMOS should be sized to minimum.

7.3.5 Register File

A register file is a set of registers which store data. It consists of a small array of static memory cells. Register files are used by microprocessors and DSPs and they permit multiple read and write ports [14, 15, 16, 17]. A typical array is 32 registers of 32-bit. For example an ALU needs two pieces of data from the register file. The array has dual-read single-write architecture.

Fig. 7.39 shows the schematic of the single-ended memory cell with 2 read ports and 1 write port (2R-1W). The read ports are the read bit-lines BL_R1 and BL_R2. The memory cell, composed of two cross-coupled inverters I_1 and I_2 is addressed by two read word-line signals, WL_R1 and WL_R2. The NMOS transistor N_1 is controlled by the Write Enable (WE) signal. N_1 is connected serially to the write access transistor N_2. The transistor N_2 is controlled by the write word-line (WL_W) signal. The transistor N_1 isolates the stored data from the write bit-line (BL_W). To write the data in the storage node A from the write bit-line, the inverters I_1 and I_2 should be sized carefully. The β[2] ratio of the inverter I_1 should be larger than 1 (e.g., 5) to set the threshold voltage of I_1 to a low-level. This is due to the fact that N_1 and N_2 weakly transfers a high level (only $V_{DD} - V_{Tn}$). Moreover, to ensure a correct write operation, the

[2] The definition of β is given in Chapter 4.

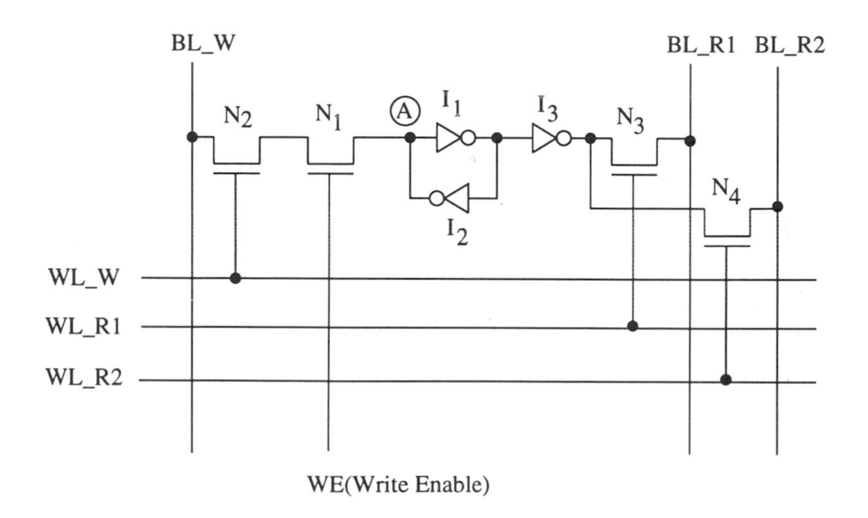

Figure 7.39 (2R-1W) register file cell.

feedback inverter I_2 should be weak so the access transistors N_2 and N_1 can change the state of node A. For example the NMOS and PMOS of I_2 should be minimum size except that the length of the NMOS is twice the minimum. Also the access transistors should have higher β compared to the transistors of I_2. For a given technology, the sizes should be determined by circuit simulation for a correct write operation. The inverter I_3 is a buffer for the storage node.

A pair of three-port memory cells is shown in Fig. 7.40. This structure has shared access transistor N_2 and write bit-line, BL_W. To read and write the memory cell, the simplified schematic of Fig. 7.41 is used. This schematic uses the column multiplexing scheme. For low-power, the register file uses static design and avoids the use of the conventional sense amplifier for bit-line's sensing. The sense amplifier consumes DC power. For a three port register file, two read and one write row decoders are required. Also, Write Enable (WE) and column addresses are needed to produce the column write enable for writing the data to the specified storage node. For fast operation AND gates can be used with a maximum of of 5-bit inputs.

During the read operation, if for example N_3 is asserted, then the data is put on the bit-line, BL_R1. The bit-line is selected through the pass-transistor N_s. The data is then sensed by the inverter I_1 in Fig. 7.41. During this period, the

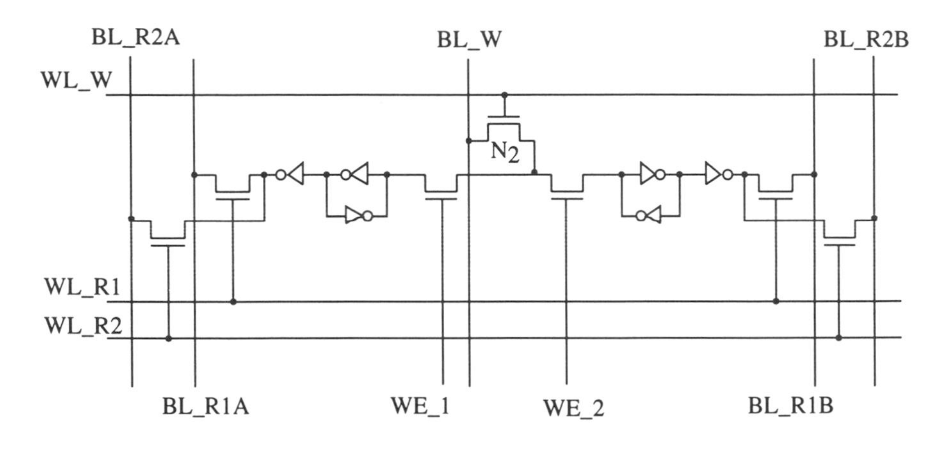

Figure 7.40 A pair of three-port memory cells (2R-1W).

read enable signal, RE, is asserted, N_l is OFF and only the feedback PMOS P_f is activated when a one ($V_{DD} - V_{Tn}$) is on the data-line. In this situation, the feedback PMOS charges up the data-line to V_{DD}. Also the DC current, which can be generated due to the reduced high level on the data-line, is completely eliminated. The β ratio of the inverter I_1 should be higher than one (e.g., 5) to achieve a symmetrical read access time for a zero and a one. When $RE = 0$, then the data-lines are isolated from the bit-lines and the NMOS transistor N_l is ON. Therefore, the latch formed by the pair of inverters I_1 and I_2 latches the old data.

The operation of such a register file is fully static and does not dissipate any static power at any mode of operation. Furthermore, the read and write operations are asynchronous. This type of register file is suitable for low-power applications.

7.4 REGULAR STRUCTURES

In this section we examine the design of large regular structures such as Programmable Logic Arrays (PLAs), Read Only Memories (ROMs) and Content Addressable Memories (CAMs). The ROMs and PLAs are not only used to implement controllers in a regular manner but they also can be applied to signal processing. RAMs are treated separately in Chapter 6. These large structures

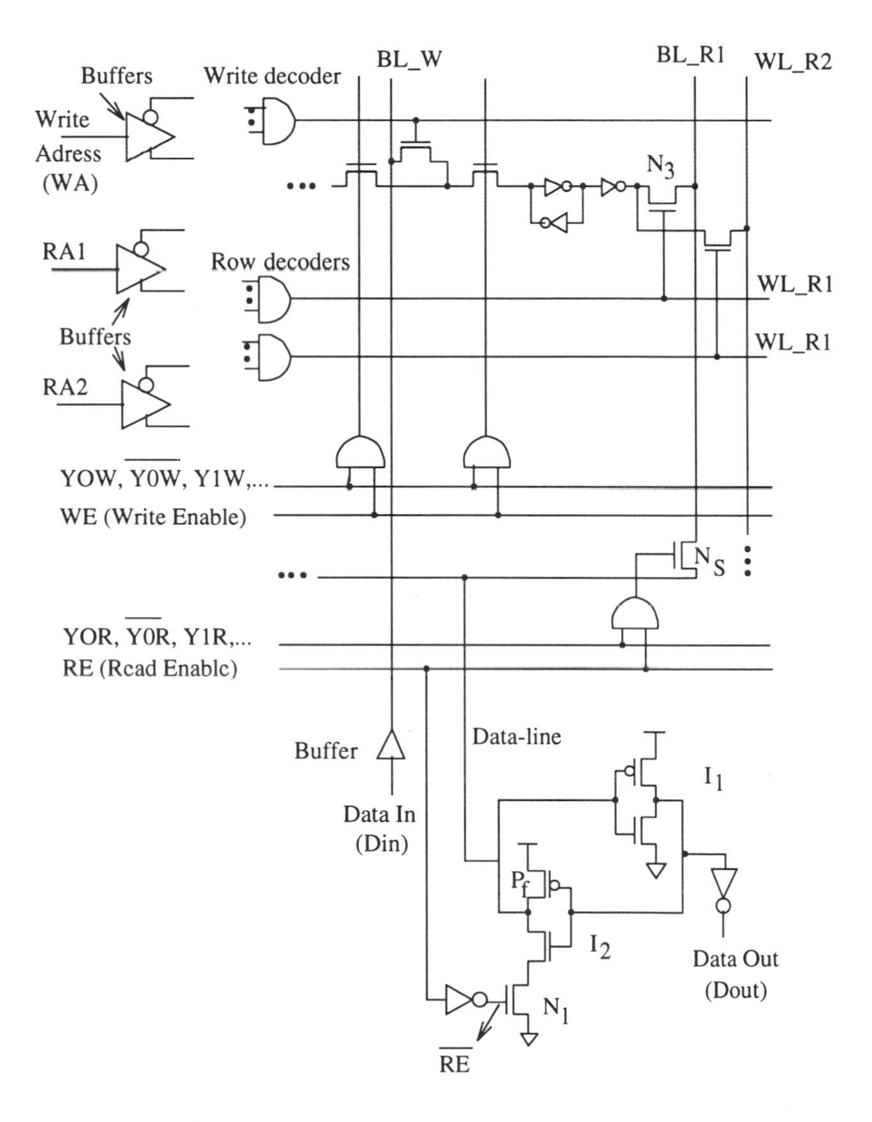

Figure 7.41 Simplified register file schematic.

are usually dynamic circuits for fast operation. These dynamic circuits can be shut down with a power management unit for power savings. If for example the clock is turned OFF, all dynamic circuits go into a precharge mode with all PMOS precharge devices are ON.

7.4.1 Programmable Logic Array

Logic functions such as those used in the control units of VLSI processors, or in finite-state machines, are hard to implement in random logic. One way of implementing these functions, in a regular structure, is the use of Programmable Logic Array (PLA) [18, 19].

PLAs have regular architecture divided mainly in two planes as shown in Fig. 7.42. These planes perform a specific function such as OR and AND. CMOS PLAs can be implemented in both static and dynamic styles. The style is chosen depending on the timing strategy in the chip. Other factors such as speed, power dissipation, and the allowed area, play an important role in the PLA design style. A CMOS PLA example, using pseudo-NMOS like style, is shown in Fig. 7.43. The output OR functions are realized with NOR gates. From Fig. 7.43(a), we have

$$P_1 = \overline{\bar{A} + B + \bar{C}} = A.\bar{B}.C \tag{7.33}$$

$$P_2 = \overline{A + C} = \bar{A}.\bar{C} \tag{7.34}$$

$$P_3 = \overline{\bar{B} + C} = B.\bar{C} \tag{7.35}$$

$$P_4 = \overline{A + \bar{C}} = \bar{A}.C \tag{7.36}$$

The buffers are used when the load on the bit-line is large. They consist in general of two inverter's stages. The OR plane is in principle similar to the AND plane [Fig. 7.43(b)]. From Fig. 7.43(b), we have

$$X = P_1 + P_2 + P_3 \tag{7.37}$$

$$Y = P_1 + P_4 \tag{7.38}$$

For this pseudo-NMOS PLA, NOR-NOR logic gate style is used. This example shows that the PLA organization is useful for implementing Sum Of Products (SOP) functions. Hence any SOP function can be realized by programming the array with the AND and OR cells. Any type of latch or register can be used at the input and output. This design style of PLAs has a small size area and

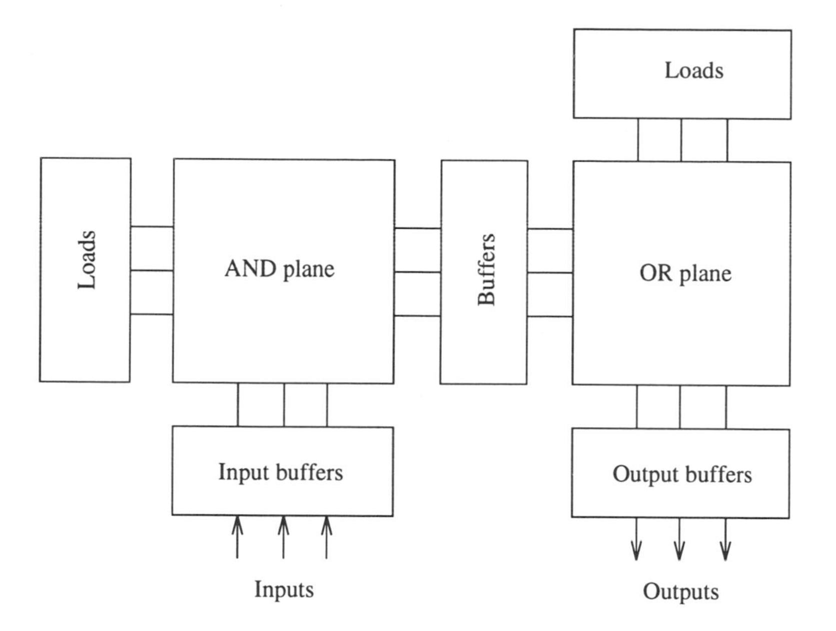

Figure 7.42 AND-OR PLA architecture.

it is simple to implement. However, it is not suitable for low-power application due to the high DC power dissipation, particularly when the PLA is large. Moreover, it has a speed problem.

In dynamic CMOS style, the circuit shown in Fig. 7.44 can be used. It is a self-timed PLA, where the AND and OR planes are both realized using precharged NOR configuration. In this structure, only a single clock phase is needed. When the clock, *clk*, is high the bit-lines are precharged in both planes. The NMOS transistors N_A and N_O are OFF, guaranteeing that there is no path to ground. Tracking lines in both planes are used to generate a delayed clock to the OR plane. When the clock is low, the precharge PMOS transistors, in the AND plane, turn OFF, N_A turns ON and the products are evaluated. The tracking lines ensure that N_O turns ON only when the inputs to the OR planes are stable. Otherwise the outputs can be spuriously discharged. This PLA is fast, but it has a lot of wasted dynamic power. The wasted power has several sources such as:

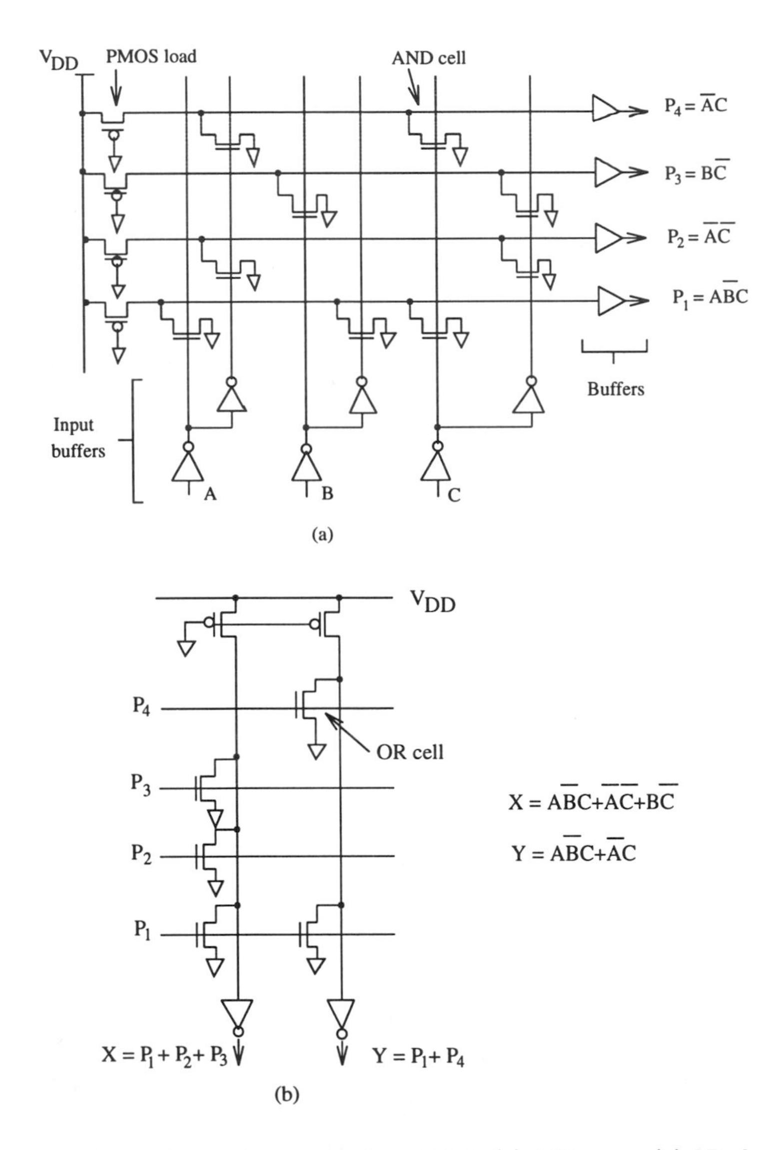

Figure 7.43 Pseudo-NMOS like CMOS PLA: (a) AND plane; (b) OR plane.

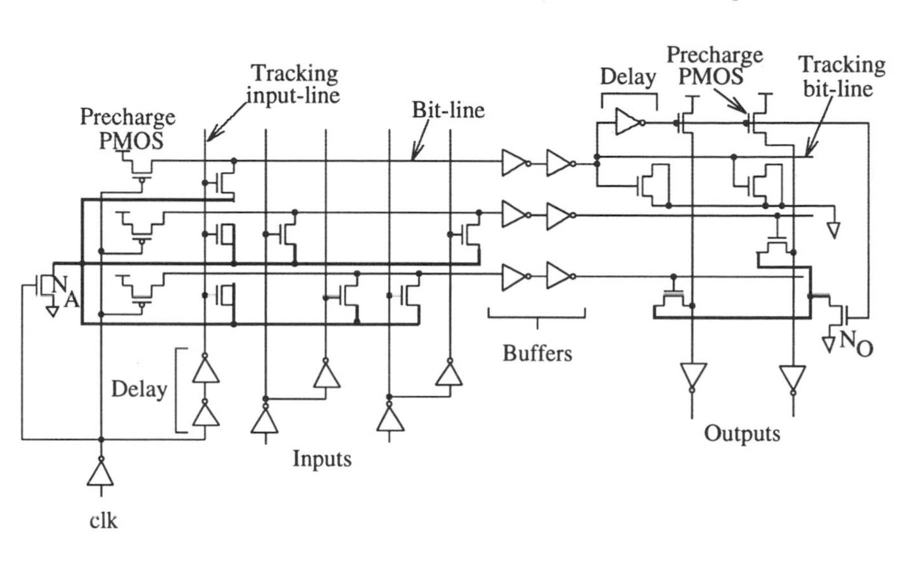

Figure 7.44 Self-timed dynamic PLA using NOR-NOR style.

- The virtual ground lines are charged and discharged every cycle. The total capacitance of the virtual ground is important, particularly for large PLAs because for the purpose of layout compactness the ground lines are in diffusion. This capacitance can be reduced using metal level in multi metal's technology;
- The number of inverters forming the buffers are important. Then, during the evaluation, several of them switch; and
- The switching activity of dynamic NOR implementation is high [see Chapter 4].

Consider now the PLA shown in Fig. 7.45 with AND-NOR structure. The OR plane is still the same compared to the PLA of Fig. 7.44. However, the AND plane is considerably simplified because:

- The virtual ground lines disappear; and

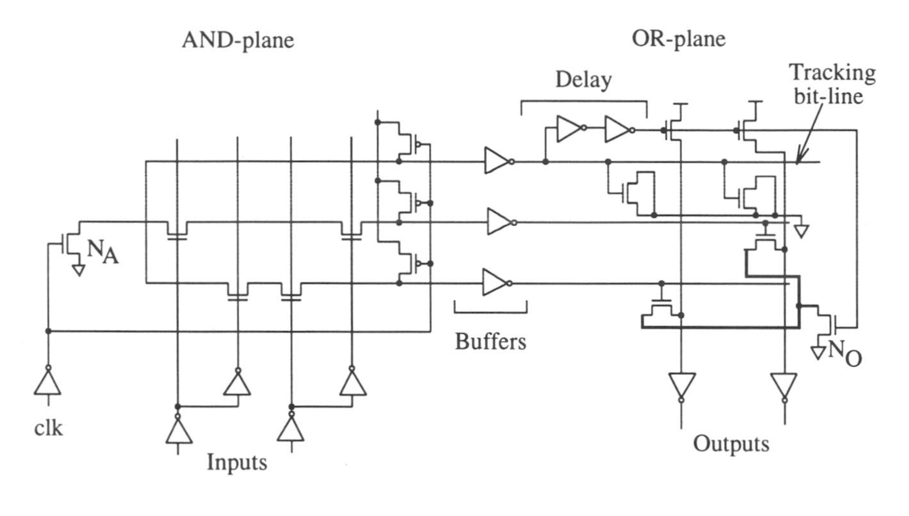

Figure 7.45 Self-timed dynamic PLA using AND-NOR style.

- The number of inverters for buffering is reduced by half.

The switching activity of the NAND implementation is also lower than that of NOR implementation, resulting in lower power in the AND plane. One problem associated with this structure is that the use of NAND may result in a large discharge time.

Another dynamic PLA combines the pseudo-NMOS and dynamic logic design styles [19]. Fig. 7.46 shows an example of such a structure. The AND plane uses a predischarged pseudo-NMOS NOR style, while the OR plane uses a conventional dynamic precharged style. During the precharge phase, the clock signal is high and the bit-lines in the AND are predischarged to ground. In the OR plane, the bit-lines are precharged to V_{DD}. The input signals to the OR plane are low. During the evaluation phase ($clk = 0$), the PMOS loads in the AND plane are ON, and the plane behaves as pseudo-NMOS logic. In this case, the PMOS device should be sized correctly to ensure safe operation when the output stays at a low level. The product terms are evaluated and then the outputs. During this evaluation phase, the PLA dissipates a static power mainly by the AND plane. Then the power is increased by this DC component.

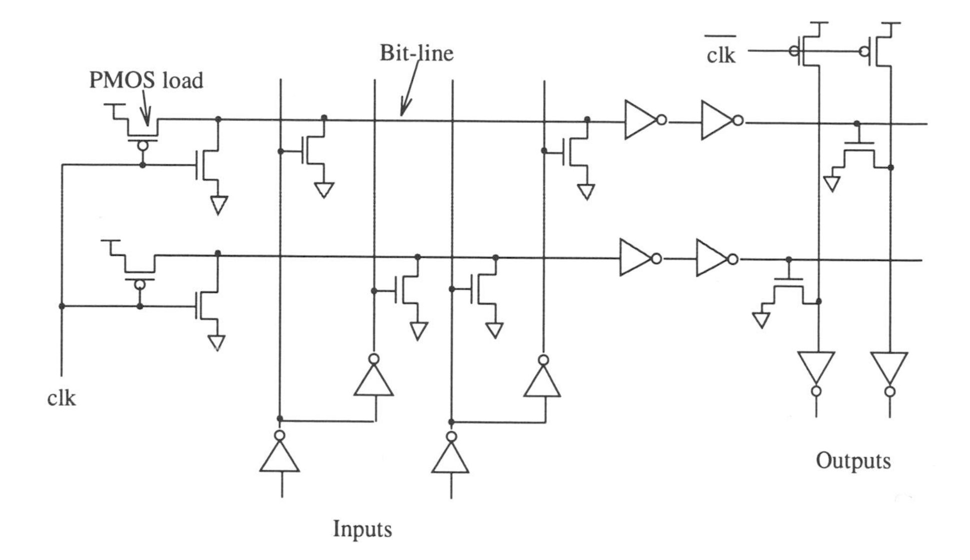

Figure 7.46 Dynamic/pseudo-NMOS PLA.

This PLA does not need the self-timing technique used previously. Also it was shown that this PLA has a fast operation [19].

When implementing smaller controllers, it is sometimes more interesting to use random logic. The implementation consists of two or more levels of logic gates using a standard cell library. It is much less regular than a PLA structure and it can have lower power dissipation.

7.4.2 Read Only Memory

Read Only Memory (ROM) is used in many applications. In DSPs, for example, it can be used as table lookup to store coefficients. Also it is often used in VLSI processors as a microcode controller. In this case, the ROM contains the microprogram instructions. Typical micro-ROM size is 2k words of 64 bits. The read-out cycle of the ROM limits the speed of the processor. Conceptually, the structure of a ROM is quite similar to that of a PLA.

Fig. 7.47 shows a simple ROM circuit architecture using NOR logic design. The state of the memory array is retained even if the ROM is not powered. The

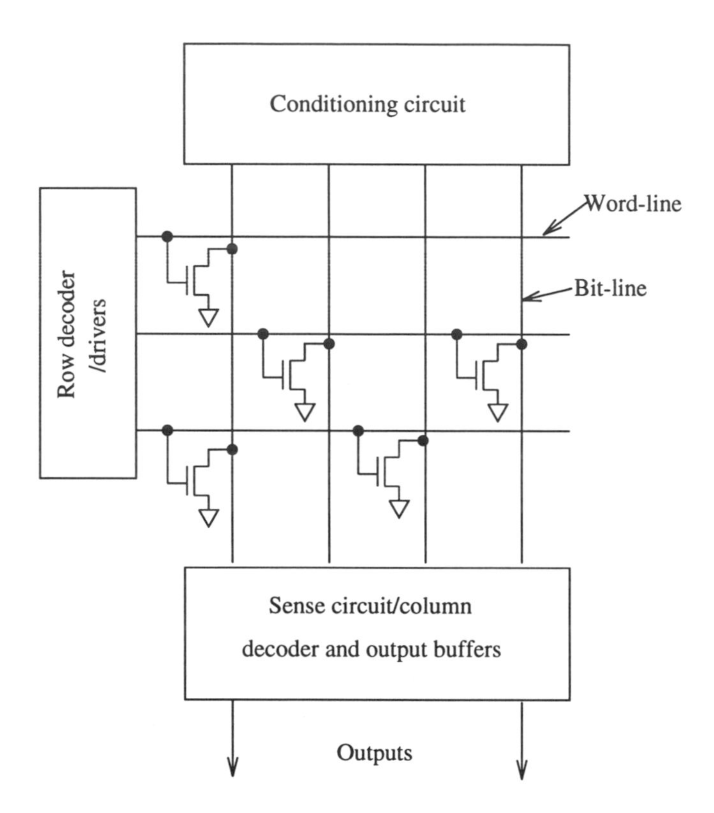

Figure 7.47 Basic ROM architecture.

decoder is necessary to select one of 2^n words by decoding the n-bit address. Decoders can be single-level circuits, but if the bit number of the address word is high, then multi-level circuits are used. This issue is discussed in Chapter 6. The conditioning circuit permits precharging of the bit-lines. The data is read from the bit-lines using sense amplifiers. Sometimes, before sensing, MUXs are added to select from the bit-lines the desired data part.

The memory cell is the same as the one used in CMOS technology. It is implemented with only one transistor which stores the state indefinitely. The W/L of the memory cell NMOS transistor is too small to obtain dense layout. Fig. 7.48 illustrates the layout of the memory cell. Metal1 and metal2 are used to reduce the resistance of lines, hence the delay caused by the word-line is reduced.

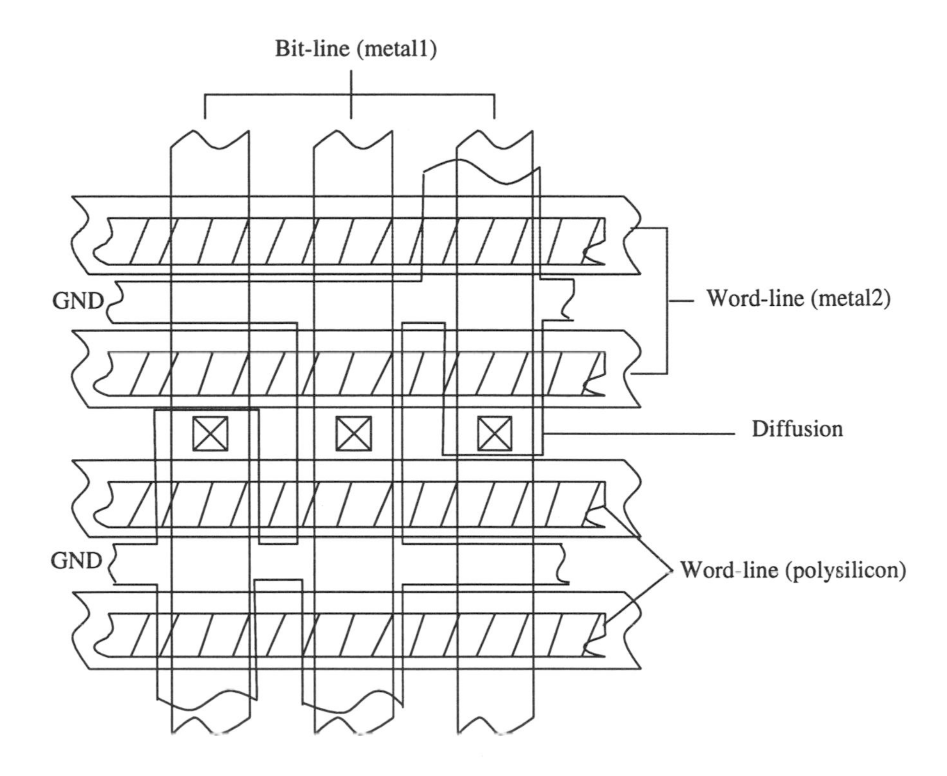

Figure 7.48 Layout of a ROM memory cell.

The ROM can be implemented in both styles: static and dynamic. In static style, the pseudo-NMOS logic, similar to that of static PLA, can be used. Fig. 7.49 shows an example of a small ROM using pseudo-NMOS circuit style. The conditioning circuits use PMOS devices, with their gates grounded, and the sense amplifier circuit is simply an inverter. The column decoder is also shown. One of the column decoders selects one of the two bit-lines. Then, node A is initially at V_{DD}. If the selected bit-line is discharged, then node A is discharged and the output is pulled up to V_{DD}. The pseudo-NMOS is easy to design and does not need a careful design, however, the power dissipation may be significant due to the DC current. For a relatively large ROM, like the one used in microcontrollers, the power dissipation can be significantly reduced using the low-power techniques of SRAMs[3]. They include pulse mode operation using address transition detection, and small swing sensing, etc.

[3] These techniques are discussed in more detail in Chapter 6.

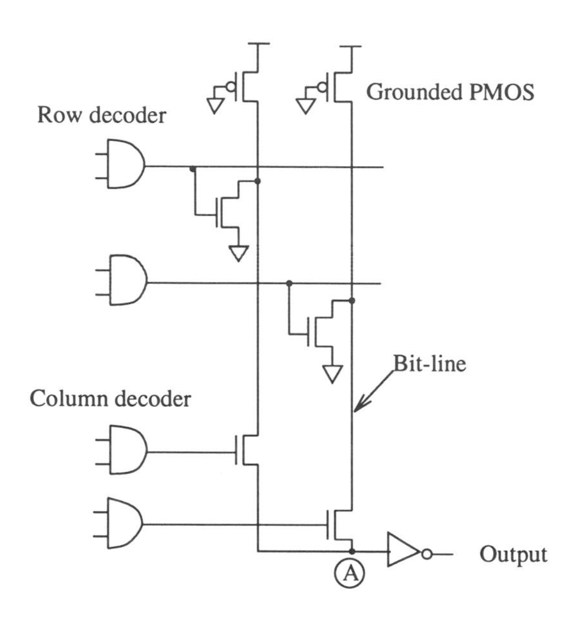

Figure 7.49 Pseudo-NMOS ROM circuitry.

A dynamic version of the ROM is shown in Fig. 7.50. During precharge phase, $clk = 1$ and the bit-lines are precharged to $V_{DD} - V_T$, where V_T is subject to the body effect. Node A is also precharged by the PMOS transistor P_p. The select lines $Sel1$ and $Sel2$ are controlled by a column decoder. All the word-lines are predischarged to ground. During evaluation, $clk = 0$ and if the bit-line is discharged to ground, node A is also discharged. Then the output of the inverter I is pulled up. If node A is not discharged, the feedback PMOS transistor P_f permits to maintain the high level at V_{DD}. Since the swing on the high-load bit-line is reduced, the power dissipation is reduced on this line by a factor $V_{DD}/(V_{DD} - V_T)$.

7.4.3 Content Addressable Memory

A Content Addressable Memory (CAM) is an important macrocell of a Translation Lookaside Buffer (TLB) [20] and cache memory [21] circuits of computer systems. The TLB permits the translation of the virtual address of a CPU to the physical address, and the cache memory from the physical address to the memory data.

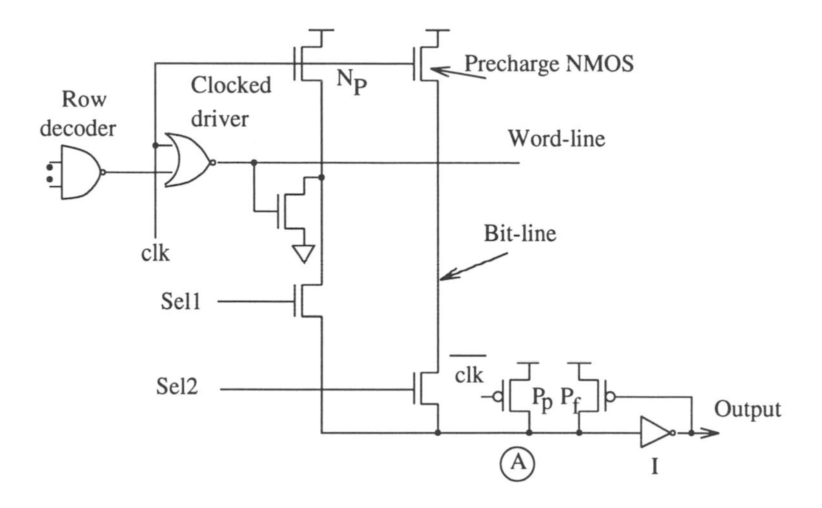

Figure 7.50 Dynamic ROM circuitry.

A CAM stores tags which can be compared against an input address word $(A_0 ... A_m)$ as shown in Fig. 7.51(a). A match detection signal is sent by the CAM if the values stored in the CAM array match with the input address word. A CMOS implementation of the CAM cell is illustrated in Fig. 7.51(b). It can be readable and writable just as an ordinary memory cell. The read/write and decoder circuits are similar to that of a RAM.

A tag word is formed by identical cells which are repeated in a horizontal array. The write lines are used to write data in the array. The comparison process is described as follows. During precharge phase, the bit-lines are predischarged low. All the write lines are low. The Match Line (ML) is precharged high. During the evaluation phase, suppose that a "1" is stored at node A. Assume that CBL line is held high and $\overline{CBL}$ line is held low. In this case, the transistors $N3$ and $N1$ are OFF, hence the ML line remains high, indicating a match at this bit location. Assume now that CBL is driven low and $\overline{CBL}$ high. The transistor $N4$ is OFF, but $N1$ and $N2$ are ON. Then the ML line is discharged, indicating a mismatch at this bit location.

For an array of n tags, there are n match lines $(ML(0)...ML(n))$. Each match line is common to m cells. If there is a mismatch in any bit of the tag word, the match line is discharged. If all the m bits match, the common match-line remains high. To detect the match signal in any of the match lines a dynamic

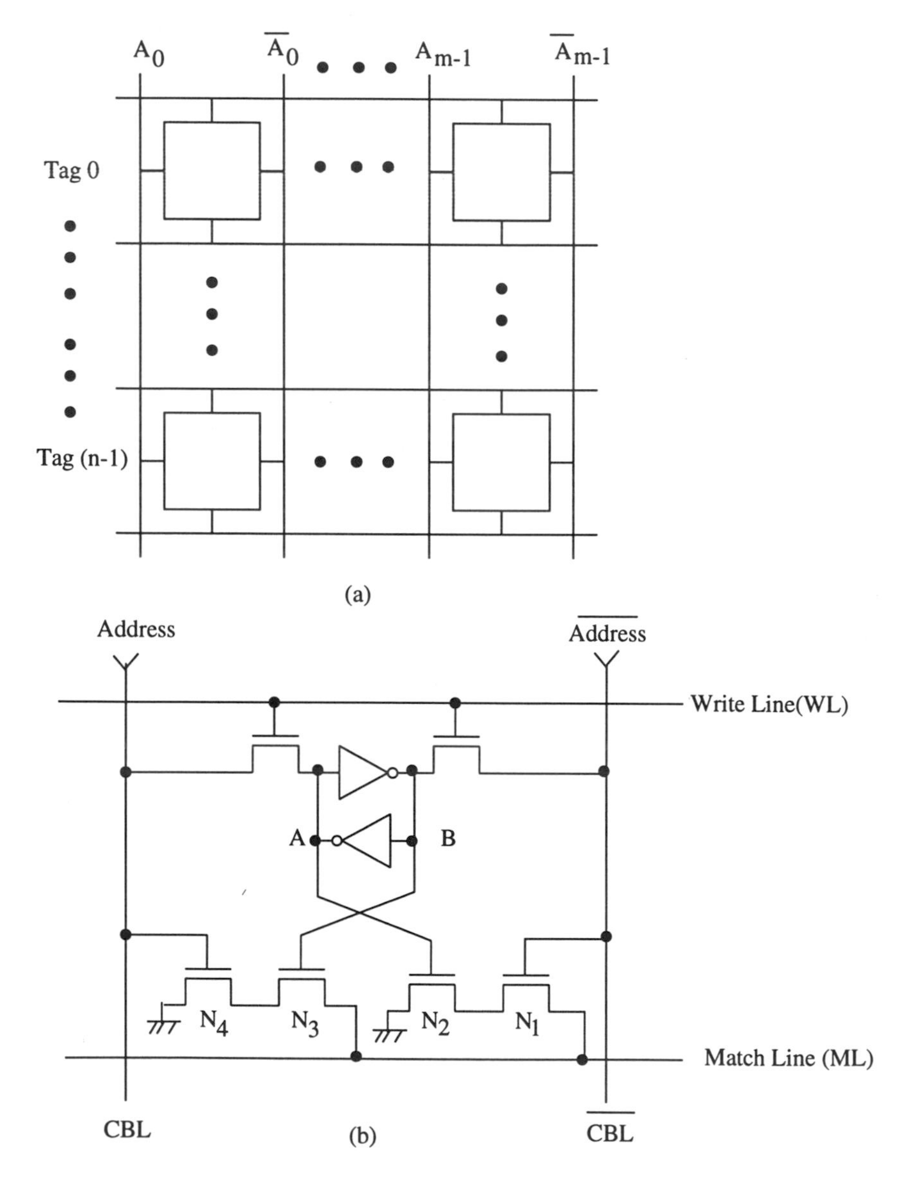

Figure 7.51 (a) CAM array; (b) CMOS CAM cell.

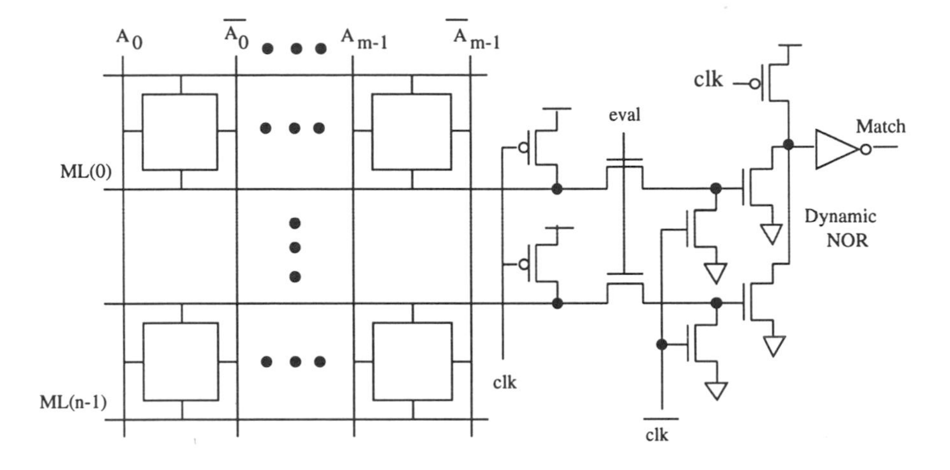

Figure 7.52 CAM array with match detection circuitry.

NOR circuit is used, as shown in Fig. 7.52. When the clock is low the NOR gate is precharged along with the match lines. The inputs to the NOR gate are predischarged to ground. When the clk signal is high (evaluation phase), one of the match lines, $ML(i)$, stays high and the others are discharged to ground. When the match lines are stable, the *eval* signal is asserted with *clk* using self-timing (similar to the PLA case). This permits keeping the dynamic NOR gate from falsely discharging. The inputs to the NOR gate must not go high until the data is stable. If one of the match lines stays high, then the NOR gate is discharged and the output match signal goes to high.

7.5 PHASE LOCKED LOOPS

Phase Locked Loops (PLLs) have many applications in digital and analog systems. In digital systems, on-chip PLLs are needed for the following reasons:

- To reduce clock skew due to clock distribution. As systems continue to demand higher clock frequencies, clock skew associated with input buffers and clock distribution becomes a significant design problem as shown in Fig. 7.53(a). The internal clock drives the output register, which in turn delivers the data to the output pad (with a buffer). The skew between the external and internal clocks is due to the clock tree. The output data is significantly delayed compared to the external clock. One main contribution is the clock skew. In Fig. 7.53(b), the internal clock is deskewed via the use of a PLL. The PLL should reduce this skew on a wide range of process, temperature and voltage variations;

- To synchronize data between chips as shown in Fig. 7.54. The PLL solves the problem of clock skew from chip to chip. An example of such an application is discussed in [22]; and
- To generate internal clocks with higher frequencies than the external clock (system clock).

There are other applications of PLL for clock recovery in serial data communications and these are not discussed in this section. Several theoretical references on PLLs can be found [23, 24, 25]. This section provides an introduction to the PLL. The CMOS circuit design of the PLL, for low-power applications, is then discussed.

7.5.1 Charge-Pumped PLL

One interesting configuration of the PLL is the charge-pumped loop shown in Fig. 7.55. It is a PLL-based frequency multiplier which consists of a Phase Frequency Detector (PFD), a Charge-Pump(CP)[4], a Loop Filter(LF), a Voltage Controlled Oscillator (VCO), and a programmable frequency divider. The feedback of the internal clock is compared to the external clock for phase and frequency error. The outputs of the phase/frequency detector are two digital signals called U (for Up) and D (for Down). The charge pump and loop filter convert these digital signals into an analog signal (control) suitable for the VCO. The VCO function of the control signal level generates a certain oscillation frequency. If the PLL generates multiples of the external clock frequency, then a frequency divider is inserted between the generated clock and the phase detector.

A simplified diagram of the charge pump and loop filter is shown in Fig. 7.56. It consists of two switchable current sources driving an impedance (LF). The pulses generated by the PFD block are used to switch the charge pump, to charge or discharge the impedance. The loop filter filters these pulses and has an analog output signal to control the VCO.

[4] The charge pump for PLL should not be confused with the one used to generate different voltages.

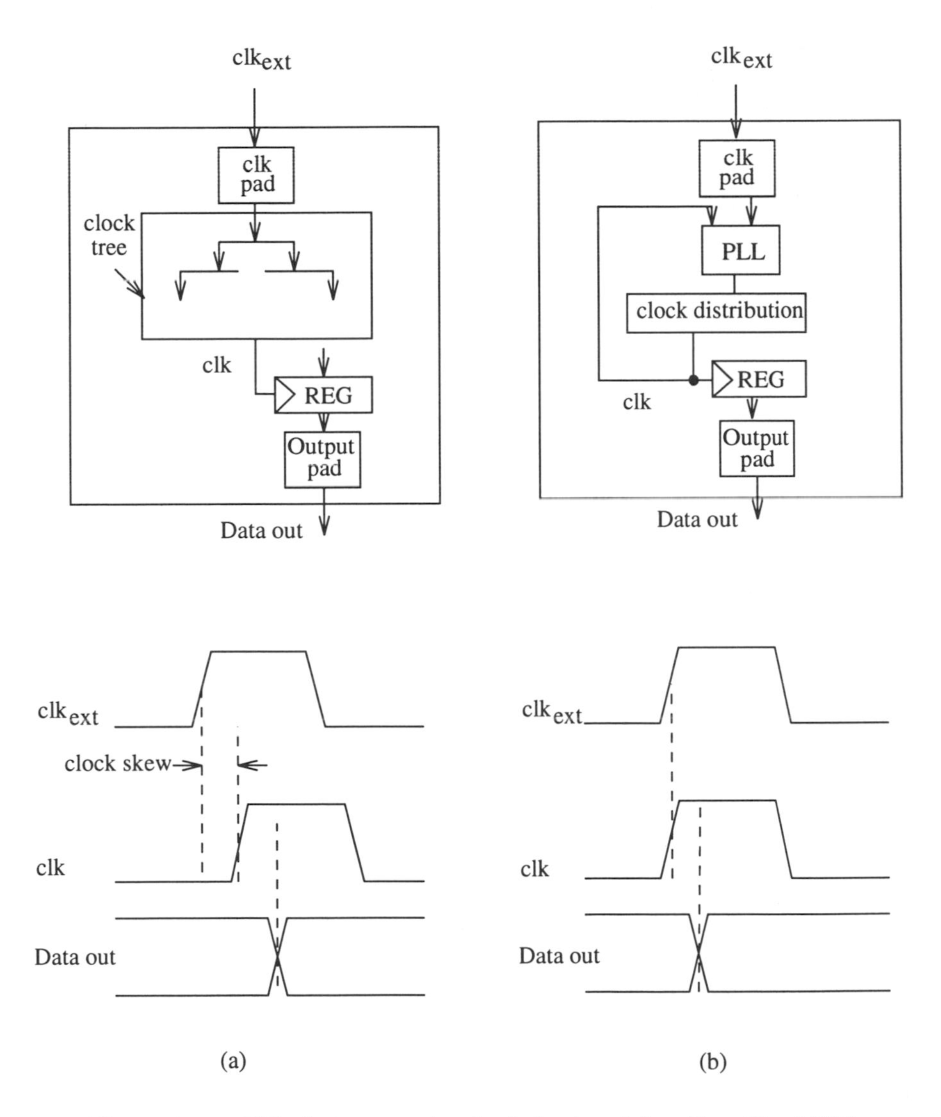

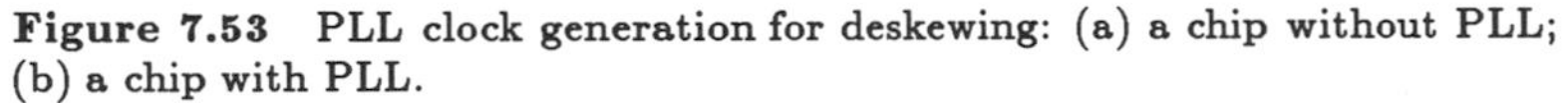
Figure 7.53 PLL clock generation for deskewing: (a) a chip without PLL; (b) a chip with PLL.

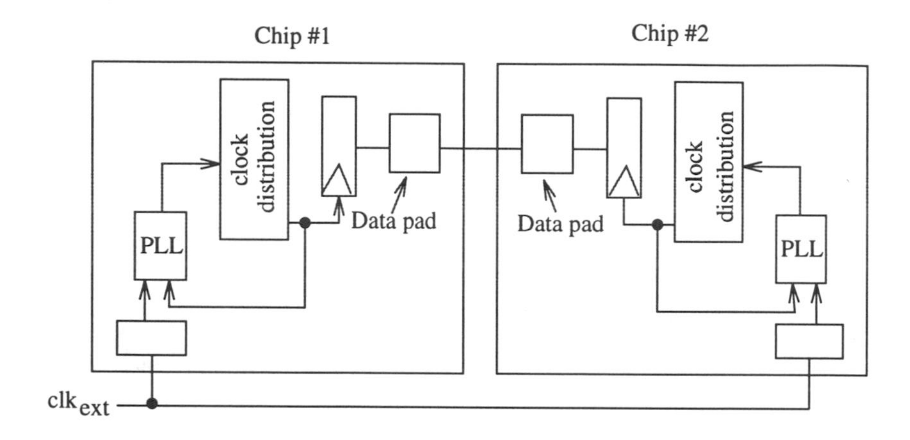

Figure 7.54 Data synchronization between chips using PLLs.

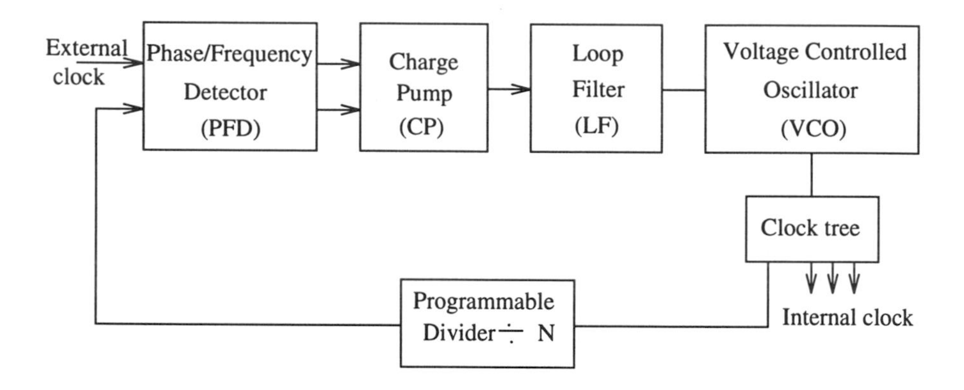

Figure 7.55 Block diagram of the PLL.

7.5.2 PLL Circuit Design

This section presents the design of the PLL components. Fig. 7.57 shows the logic diagram of the PFD circuit. It uses mainly static-CMOS NAND gates which results in good performance and low-power dissipation. The operation of this circuit using the state diagram of Fig. 7.57(c) is as follows. The circuit has three states: 1) UP, where the up signal U is asserted when the external clock clk_{ext} falls down, 2) $DOWN$, where the down signal D is asserted when the internal clock clk falls down, and 3) NOP, where the detector does not

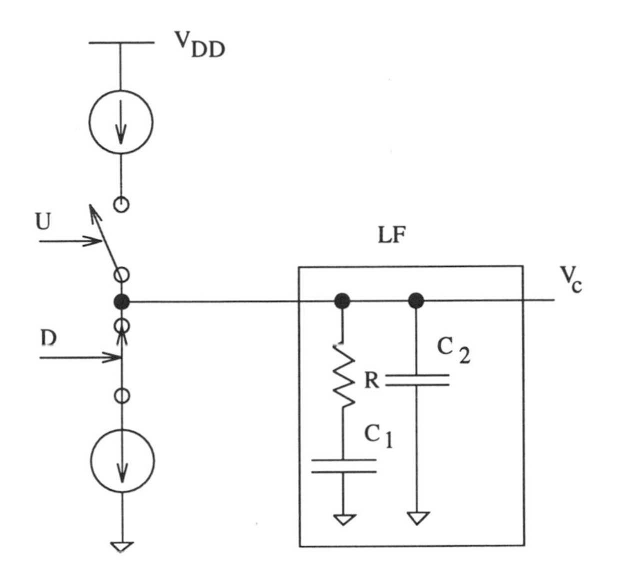

Figure 7.56 Charge pump and loop filter.

change the output control signals. In this last state both U and D signals are at zero level. The state changes whenever clk or clk_{ext} falls down. In no case U and D are both activated.

Consider that clk and clk_{ext} have the same frequency but the falling edges of clk_{ext} (clk) leads the falling edges of clk (clk_{ext}), respectively. Then, $\bar{U}$ ($\bar{D}$) is asserted with a certain duty cycle, while $\bar{D}$ ($\bar{U}$) is never asserted. In this case, the PFD is characterized as the phase detector.

Consider now the case where clk_{ext} has a higher frequency than clk. $\bar{U}$ is asserted most of the time. More falling edges of clk_{ext} signal than clk. A similar situation when clk has higher frequency than clk_{ext} and $\bar{D}$ is asserted most of the time. In this case, the PFD is characterized as frequency detector.

The $\bar{U}$ and $\bar{D}$ signals, generated by the PFD, are connected to the charge pump circuit of Fig. 7.58(a). When the signal $\bar{U}$ ($\bar{D}$) is asserted the pull-up PMOS (pull-down NMOS) transistor charges (discharges) the output, respectively. Another variation of the charge pump circuit is shown in Fig. 7.58(b). Two transistors P_{ref} and N_{ref} are added as current sources biased by a current

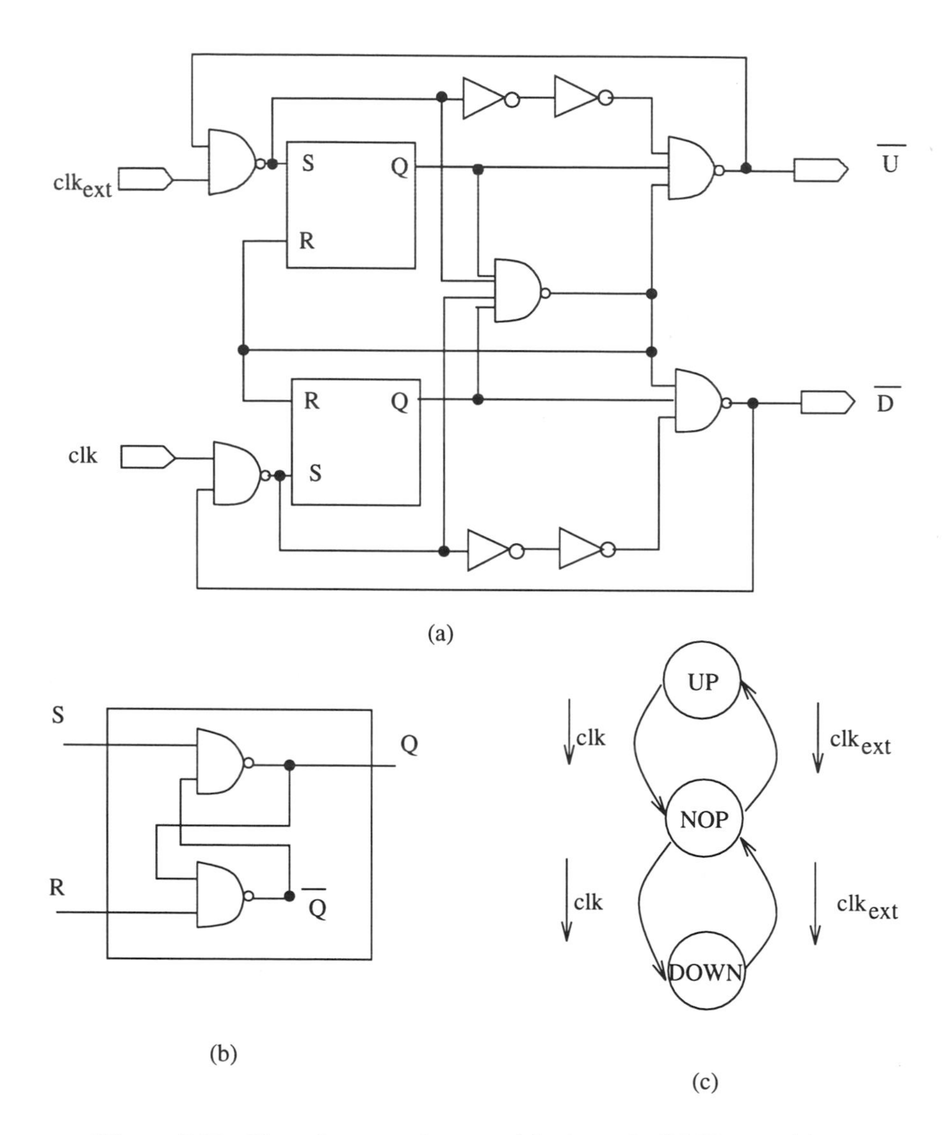

Figure 7.57 Phase-frequency detector: (a) schematic; (b) SR gate; (c) state diagram.

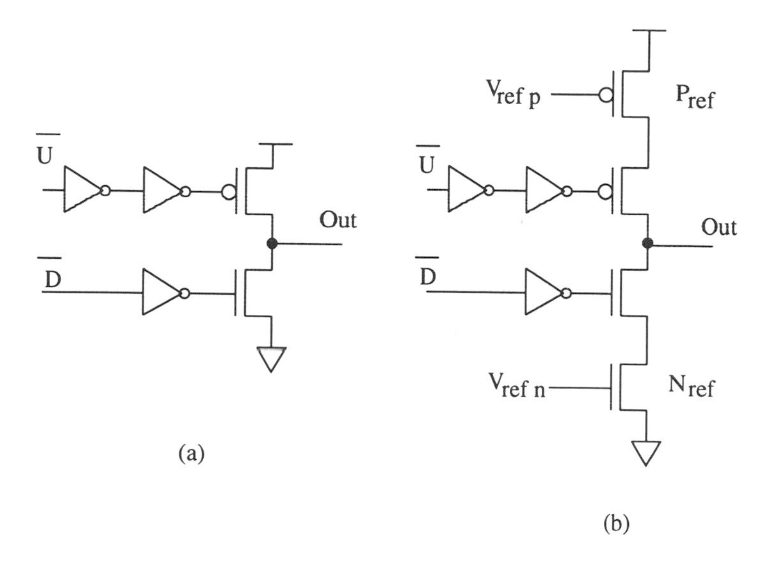

Figure 7.58 Charge pump circuits.

mirror circuit. In this situation, the output current of the charge pump can be adjusted through the control of the current mirror.

The monolithic implementation of the filter of Fig. 7.56 is shown in Fig. 7.59. The two capacitors C_1 and C_2 are in the order of tens of pF and are made with the NMOS transistors N_{C1} and N_{C2}. The resistor is made with a transmission gate in closed state. It can also be implemented with an N-well implant available in the CMOS process. The capacitor C_2 is added in parallel to the simple RC (R-C_1) low-pass filter to form a second order filter. In this case, the stability of the system is maintained even with the process variation of these on-chip components. Note that these capacitors can occupy a large portion of the PLL.

The charge pump and filter generate a control voltage for the VCO. One important parameter of the VCO is the VCO gain. When considering the characteristic frequency-control voltage, the VCO gain is the slope of this characteristic. A linear characteristic is, in general, desirable. In general the VCO is implemented using a ring oscillator as shown in Fig. 7.60. A series connection of delay inverter cells forms a tapped delay line which oscillates with a frequency determined by the delay time of the cell and the odd number

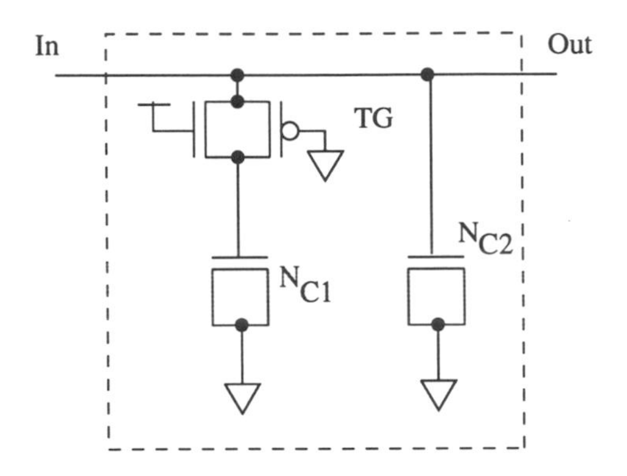

Figure 7.59 Monolithic low-pass filter.

of stages. The delay of the cell is controlled by a current which in turn is controlled by the control voltage V_c. V_c modulates the ON resistance of pull-down N_1, and through the current mirror, the pull-up P_1. All the devices of the VCO should be oriented in the same direction and have redundant contacts to reduce the jitter due to process variations. In the VCO of Fig. 7.60, maximum frequency is achieved at maximum control voltage. Typical values of the VCO gain at low power supply voltage can range from 10 MHz/V to 100 MHz/V depending on the number of stages and technology. Note that the bandwidth of the VCO presented previously is limited.

The VCO of Fig. 7.61 has an excellent bandwidth characteristic, where a wide range of frequency can be generated [26]. It is used for video signal processors and covers a wide range of applications. The frequency range can change by one order of magnitude from 50 MHz to 350 MHz. In fig. 7.61, by turning ON and OFF 8 CMOS TGs with control signals, the number of ring oscillator stages can be selected among eight values (7,9,11,15,21,29,39,51). Each stage of the ring oscillator combines an inverter in parallel with a current-controlled inverter. The inverter increases the frequency of oscillation of the VCO, whereas the current-controlled inverter permits tuning of the frequency of the VCO.

The generated clock frequency can be N times the external clock frequency (reference frequency). This clock then feeds the clock driver and tree. Since the PLL discussed here is intended to be integrated on-chip, it is then sensitive to the noise generated on the power lines (called power-supply-induced clock jitter). If the power supply changes by 100 mV the skew or phase error will

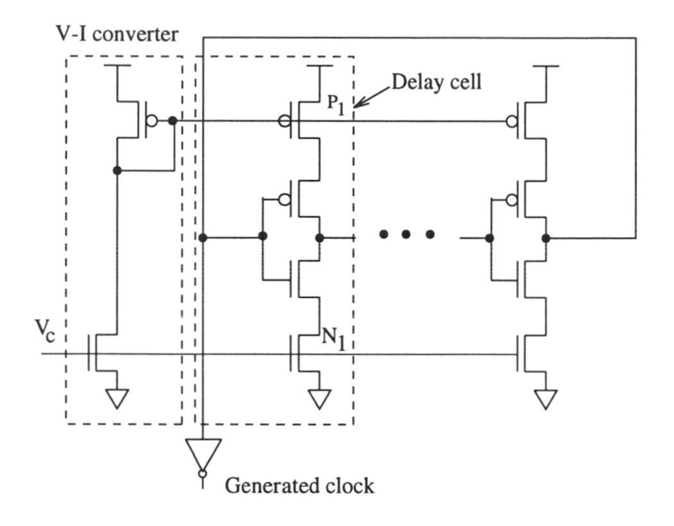

Figure 7.60 VCO using current controlled CMOS ring oscillator.

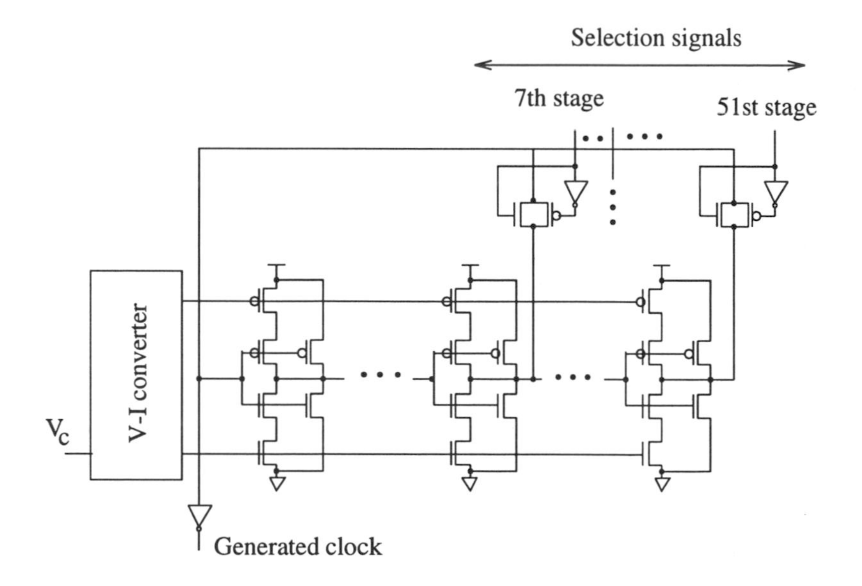

Figure 7.61 VCO with selectable characteristics.

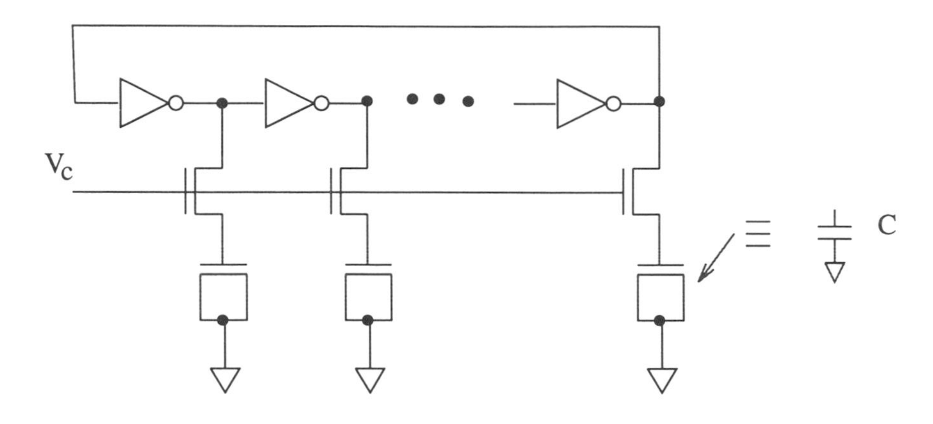

Figure 7.62 A VCO based on VCDL.

be important before the PLL has time (tens of clocks cycles) to correct this error [27]. One way to reduce the effect of this problem is to dedicate an analog power supply pin to the VCO and the charge pump. At the circuit level, a new VCO delay cell was proposed by Young [27] to reduce the phase error.

Another VCO alternative is shown in Fig. 7.62. It is similar to the Voltage-Controlled Delay Line (VCDL) [28]. The control voltage, V_c, is used to vary the amount of the effective load seen by each inverter output. The frequency-control voltage characteristic of this VCO has a negative slope. Then the minimum frequency of oscillation is limited by the maximum V_{DD}. Therefore, the minimum frequency is increased with reduced V_{DD}. A positive slope is, in general, desirable so the minimum frequency is not set by V_{DD}.

The frequency divider can be implemented using toggle flip-flops. Fig. 7.63 shows an example of a divider with division rates of 1, 1/2, 1/4, and 1/8. The PLL, so far discussed, is not completely digital. Only the PFD, charge pump and the frequency divider are digital. While, the LF and VCO are analog and operate as continuous-time systems.

7.5.3 Low-Power Design

In sleep mode, the on-chip PLL may be controlled for low-frequency operation, or it may be disabled to reduce its power dissipation to the leakage currents.

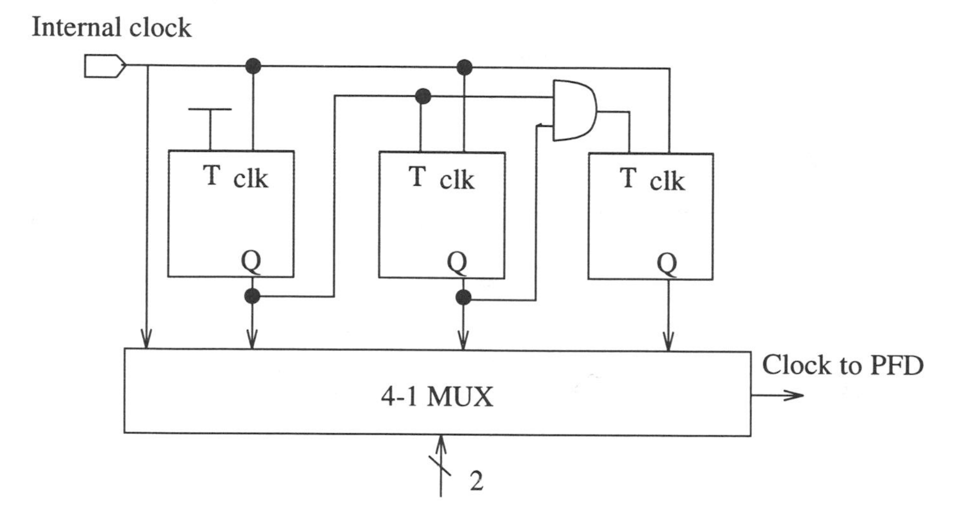

Figure 7.63 A frequency divider using T-flip flops.

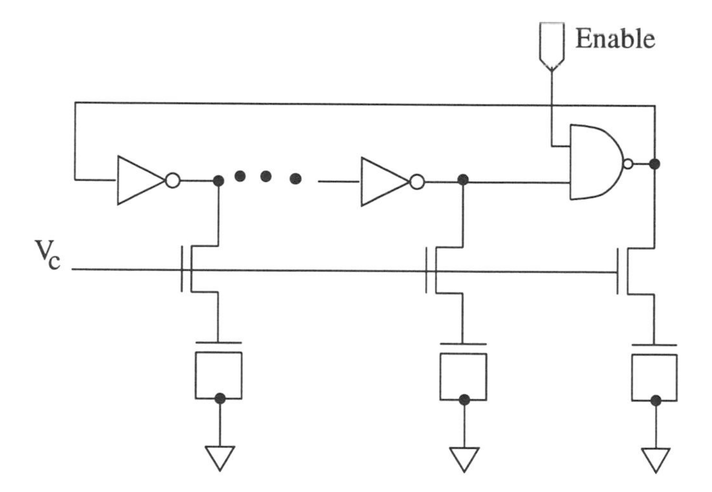

Figure 7.64 A VCO controlled by enable signal for low-power mode.

As an example, to disable the PLL, is to shut down the VCO and disable the external clock. Fig. 7.64 shows the same VCO of Fig. 7.62 but with one inverter transformed to a two-input NAND gate. One of the inputs is controlled by the *Enable* signal to shut down the PLL when it is low. The NAND gate can be used for any of the VCOs presented previously. Also the enable signal can be used to disable any current source used in the PLL to eliminate any DC current. A typical power dissipation of a PLL, at 3.3 V, is in the range of tens of mW depending on the frequency.

7.6 CHAPTER SUMMARY

This chapter has presented the design of several subsystems used in VLSI chips. Many circuit alternatives are discussed which trade area, speed and power. The reader can construct these options and compare their performance in terms of power, delay and area. The power dissipation issue is stressed more. Also several building blocks of VLSI chips using advanced circuit techniques have been investigated. These include

- High-speed addition.
- Multiplication techniques.
- PLL and clock deskewing technique.

REFERENCES

[1] J. Mori, et al., "A 10-ns 54 × 54-b Parallel Structured Full Array Multiplier with 0.5-μm CMOS Technology," IEEE Journal of Solid-State Circuits, vol. 26, no. 4, pp. 600-606, April 1991.

[2] J. Sklansky, "An Evaluation of Several Two-Summand Binary Adders," IRE Transactions on Electronic Computers, vol. EC-9, pp. 213-226, June 1960.

[3] J. Sklansky, "Conditional-Sum Addition Logic," IRE Transactions on Electronic Computers, vol. EC-9, pp. 226-231, June 1960.

[4] I. S. Abu-Khater, R. H. Yan, A. Bellaouar, and M. I. Elmasry, "A 1-V Low-Power High-Performance 32-b Conditional Sum Adder," IEEE Symposium on Low-Power Electronics, Tech. Dig., San Diego, pp. 66-67, October 1994.

[5] T. Sato, et al., "An 8.5ns 112-b Transmission Gate Adder with a Conflict-Free Bypass Circuit," IEEE Journal of Solid-State Circuits, vol. 27, no. 4, pp. 657-659, April 1992.

[6] K. Ueda, H. Suziki, K. Suda, Y. Tasujihashi, H. Shinohara, "A 64-bit Adder By Pass Transistor BiCMOS Circuit," IEEE Custom Integrated Circuit Conference, Tech. Dig., pp. 12.2.1-12.2.4, May 1993.

[7] K. Hwang, "Computer Arithmetic: Principles, Architecture, and Design," John Wiley and Sons, 1979.

[8] J. J. F. Cavanagh, "Computer Science Series: Digital Computer Arithmetic," McGraw-Hill Book Co., 1984.

[9] M. Nagamatsu, S. Tanaka, J. Mori, T. Noguchi, and K. Hatanaka, "A 15-ns 32x32-bit CMOS Multiplier with an improved Parallel Structure," IEEE Custom Integrated Circuits Conference, Tech. Dig., pp. 10.3.1- 10.3.4, May 1989.

[10] N. Ohkubo, M. Suziki, T. Shinbo, T. Yamanaka, A. Shimizu, K. Sasaki, and Y. Nakagome, "A 4.4-ns CMOS 54x54-b Multiplier using Pass-Transistor Multiplexer," IEEE Custom Integrated Circuits Conference, Tech. Dig., pp. 599-602, May 1994.

[11] R. Bechade, et al., "A 32b 66MHz Microprocessor," IEEE International Solid-State Circuits Conference, Tech. Dig., pp. 208-209, February 1994.

[12] C. A. Mead, and L. A. Conway, "Introduction to VLSI Systems," Addison-Wesley, 1980.

[13] R. W. Sherburne, et al., "Data path Design for RISC," Proc. Conf. Advanced Research in VLSI, pp. 53-62, 1982.

[14] R. W. Sherburne, et al., "A 32-bit NMOS Microprocessor with a Large Register File," IEEE Journal of Solid-State Circuits, vol. SC-19, no. 5, pp. 682-689, October 1984.

[15] K. J. O'Connor, "The Twin-Port Memory Cell," IEEE Journal of Solid-State Circuits, vol. SC-22, no. 5, pp. 712-720, October 1987.

[16] R. D. Jolly, "A 9-ns, 1.4-Gigabyte/s 17-Ported CMOS Register File," IEEE Journal of Solid-State Circuits, vol. 26, no. 10, pp. 1407-1412, October 1991.

[17] H. Shinohara, et al., "A Flexible Multiport RAM Compiler for Data Path," IEEE Journal of Solid-State Circuits, vol. 26, no. 3, pp. 343-349, March 1991.

[18] A. R. Linz, "A Low-Power PLA for a Signal Processor," IEEE Journal of Solid-State Circuits, vol. 26, no. 2, pp. 107-115, February 1991.

[19] G. M. Blair, "PLA Design for Single-Clock CMOS," IEEE Journal of Solid-State Circuits, vol. 27, no. 8, pp. 1211-12113, August 1992.

[20] H. Kadota, et al., "A 32-bit Microprocessor with On-Chip Cache and TLB," IEEE Journal of Solid-State Circuits, vol. SC-22, no. 5, pp. 800-807, October 1987.

[21] A. J. Smith, "Cache Memories," Computing Surveys, Vol. 14, pp. 473-530, September 1982.

[22] L. Ashby, "ASIC Clock Distribution using a Phase Locked Loop (PLL)," in IEEE International ASIC Conference and Exhibit, Tech. Dig., pp. P1.6.1-P1.6.3, September 1991.

[23] F. M. Gardner, "Phase Lock Techniques," John Wiley and Sons, 1979.

[24] F. M. Gardner, "Charge-Pump Phase-Locked Loops," IEEE Transactions on Communications, COM-28(11), pp. 1849-1858, November 1980.

[25] R. E. Best, "Phase-Locked Loops," McGraw Hill, 1984.

[26] J. Goto, et al., "A Programmable Clock Generation with 50 to 350 MHz Lock Range for Video Signal Processors," IEEE Custom Integrated Circuits Conference, Tech. Dig., pp. 4.4.1-4.4.4, May 1993.

[27] I. A. Young, J. K. Greason, and K. L. Wong, "A PLL Clock Generator with 5 to 110 MHz of Lock Range for Microprocessors," IEEE Journal of Solid-State Circuits, vol. 27, no. 11, pp. 1599-1607, November 1992.

[28] M. G. Johnson, and E. L. Hudson, "A Variable Delay Line PLL for CPU-Coprocessor Synchronization," IEEE Journal of Solid-State Circuits, vol. 23, no. 5, pp. 1218-1223, October 1988.

8

LOW-POWER VLSI DESIGN METHODOLOGY

This chapter presents Low-Power (LP) design methodologies at several abstraction levels such as physical, logical, architectural, and algorithmic levels. All the power reduction techniques discussed are related to the dynamic power dissipation. It is shown that LP techniques, at the high-level (algorithmic and architectural) of the design, lead to power savings of several orders of magnitude. Many examples are included to give the reader a quantitative picture of LP issues. Several LP techniques, particularly at the circuit level, have already been discussed in Chapters 4, 6, and 7 including those related to static power considerations. However, they are not reconsidered in this chapter. The power estimation techniques at the circuit, logical, architectural and behavioral levels are overviewed. Power analysis at high-level allows an early prediction and optimization of the power of a system. The LP concepts such as switching activity, glitching, etc., discussed in Chapter 4 are used throughout this chapter.

8.1 LP PHYSICAL DESIGN

There are several techniques to reduce the power at the physical design (layout) level. Some of these issues have been discussed in Chapter 4 for full-custom and semi-custom designs. In this section we present two approaches for low-power physical design.

8.1.1 Floorplanning

Floorplanning of a circuit is the first step in VLSI layout design. It permits the allocation of space on a chip for a given set of modules. A module can be *rigid*, e.g., the module is in the library and its dimension and power dissipation are known, or *flexible*, e.g., it has not been designed and has a list of parameters such as different shapes and power consumptions for feasible implementations. Floorplanner for low-power design should choose a suitable implementation for each module such as the total power/area of a chip are optimized [1].

8.1.2 Placement and Routing

The placement and routing of a VLSI circuit is performed on standard cells, gate arrays, functional blocks, etc. All the different modules are already laid out and well characterized in the library. Traditionally, placement refers to the process of placing modules to minimize area and delay. Placement for low-power uses the switching activity-capacitance products as a function to be minimized, in contrast with delay minimization, where the wire capacitance has to be minimized. After placement, routing permits connection of the modules with wires. High switching activity wires should be kept short using the lower parasitic capacitance layer. A CAD tool for placement has already been developed [2].

8.2 LP GATE-LEVEL DESIGN

The low-power design methodology should also be applied to logic design. To achieve this goal, power is traded for speed and area. In this section, we discuss a number of techniques to reduce the switching activity and internal capacitances during technology-independent and technology-dependent phases of logic design.

8.2.1 Logic Minimization and Technology Mapping

The area and power optimization of logic structures (both combinational and sequential) have matured considerably. The power optimization task benefits from these techniques. The objective of logic minimization is to reduce the boolean function. For low-power design, the signal switching activity is minimized by restructuring a logic circuit during the technology-independent phase [3]. It is assumed that at the higher-level of abstraction, decisions regarding the power supply voltage and the clock frequency have already been made. The power minimization is constrained by the delay, however, the area may increase. During this phase of logic minimization, the function to be minimized is

$$\sum_i P_i(1 - P_i)C_i \tag{8.1}$$

where P_i is the probability of the node i being a "1" $(1 - P_i)$ is the probability that node i is a "0", and C_i is the capacitance of this node. For more information on this model see Section 8.5.2.1. To minimize the above equation, one has to first evaluate the current value of P_i and then change it by making P_i close to 0 or close to 1. Also in [3], zero-delay approximation is assumed. This implies that the glitching power is neglected.

To minimize the switching activity, some techniques that can be used are:

- Use don't-cares to minimize the probability P_i of a function. Indeed, the signal probability of a gate can change by altering the ON-set or the OFF-set by adding points from the don't-care set.
- Collapse nodes that are not on the critical path. The intermediate signal lines are implemented as single node. The delay may increase, however this does not affect the overall performance of the circuit.

Power dissipation can be improved by as much as 60%, at the expense of an 8 % area increase [3] and with no delay degradation. More typical power reduction would be in the range of 10-20% [4].

The technology mapping step for low-power refers to the process of transforming a logic function into a technology-dependent (e.g., CMOS) circuit with minimized power consumed. This technology dependent step uses a target technology. The first step in technology mapping is to decompose each logic function into two-input gates. The objective of this decomposition is to minimize the total power dissipation by reducing the total switching activity. Fig. 8.1 shows an example of a four-input AND gate decomposition into two different implementations. The probabilities of inputs being at "1" logical are also shown in Fig. 8.1. Primary inputs are assumed to be uncorrelated. The switching activity at each internal node is also shown in Fig. 8.1. A two-input (i, j) AND gate is given by

$$\alpha = (1 - P_i P_j) P_i P_j \tag{8.2}$$

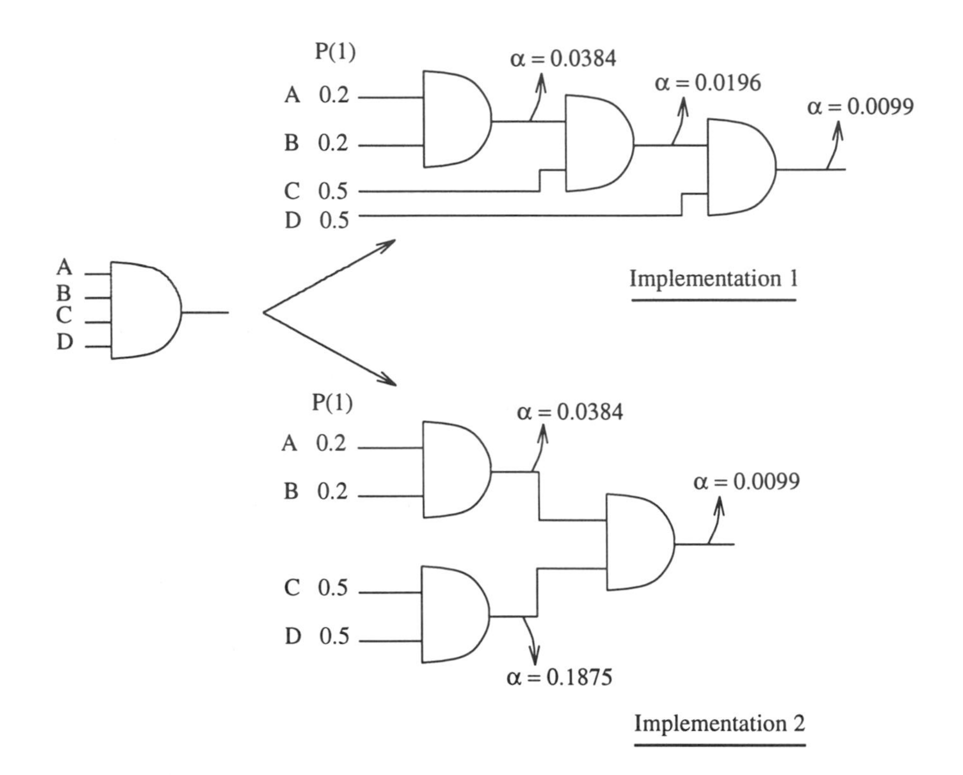

Figure 8.1 Technology decomposition of a four-input AND gate.

We assume also that the gate delays are zero to ignore the power due to the glitching phenomenon. The total switching activities for implementations 1 and 2 are 0.888 and 1.056, respectively. Therefore, implementation 1 is better than implementation 2. This problem of decomposition was addressed by [5, 6]. In [5], the power dissipation, associated to glitching, is neglected while in [6] it is not. Taking into account the power dissipation of glitches is very important as is discussed in Section 8.2.2.

The concept of technology mapping of logic optimization is an important step for standard cells and gate arrays (or sea of gates) circuit design. All the cells in the library are characterized in terms of area and speed. Another parameter to be added for low-power design is the characterization of the internal power of the gate and its output parasitic capacitance. Hence the process of technology mapping is to search, using a target library, the best possible implementation following constraints such power, area and delay.

In this section we do not consider the algorithms for technology mapping. The reader can consult references [5, 7]. We illustrate this concept of technology mapping by the following example. Fig. 8.2 shows an example for implementing the logic circuit of Fig. 8.2(a) into two implementations. The first implementation [Fig. 8.2(a)] is for minimal area design using OAI (OR-AND-INVERT) gate. The second implementation [Fig. 8.2(b)] is for minimal power design where the high switching node N of Fig. 8.2(a) is hidden using a more complex gate.

Thus the process of technology mapping is to first decompose the logic function such that the total switching activity is minimized. Then, to hide any high switching activity node within complex gates so that the capacitance of that node is minimized. However, making a gate too complex can trade the delay for low-power. Typical reduction in power dissipation is on the order of 20% without any degradation in performance but at the expense of small area penalty.

The quality of the targeted cell library can considerably impact the results of mapping [8]. For example, the availability of cells with different drive strengths and double-rail outputs (signal and its complement) gives more flexibility for logic optimization. A good library can result in 20-30% of power dissipation reduction.

8.2.2 Spurious Transitions Reduction

Due to the finite delays of logic gates, signal races in static logic designs can result in dynamic hazards. Hence, a node can have transitions in one clock cycle before stablizing to the correct logic level. These unnecessary switching transitions (glitches) can consume power dissipation in the order of 20-40% [9, 10, 11].

To reduce this power the first approach is to balance the path delays by changing the logic structure (e.g., tree) as explained in Section 4.5.5. Another technique is to balance the delay of the paths by sizing down the gates in the fast paths [12]. However, this approach can increase the delay of the circuit. Also insertion of buffers (delay elements) in the fast paths can balance the delay. However, the added buffers increase the power dissipation.

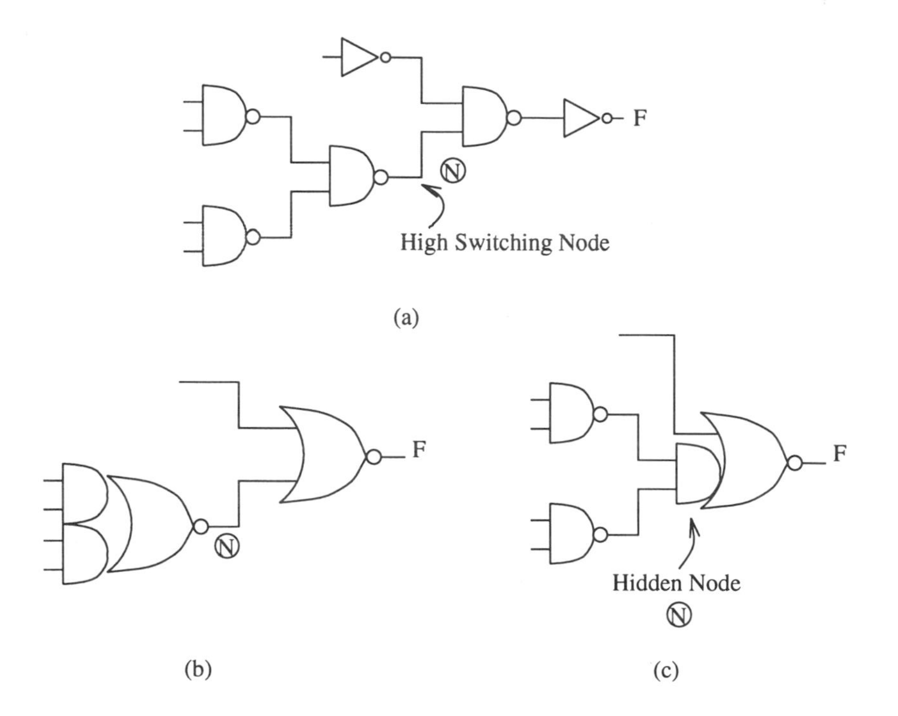

Figure 8.2 A simple example showing technology mapping for low-power: (a) original function; (b) minimal area implementation; (c) minimal power implementation.

Another technique employs self-timing techniques to reduce the logic depth and then the glitching power [9, 11]. The self-timed circuit should save more power than what it introduces. As a circuit example that exhibits spurious transitions, is an adder. The sum signals can have false transitions before they are stable. If the load capacitances on the outputs are relatively large, then the power due to the glitches can be important.

A conventional self-timed method for an adder is shown in Fig. 8.3. A Transition Detector (TD) similar to the one discussed for SRAMs in Chapter 6 is used. For each set of inputs (A_i and B_i) there is one transition detector. If A and B are both n-bit wide, then n TDs are required for the parallel adder. For any transition at the inputs, the TD generates a pulse for the self-timed function. This self-timed circuit delays the pulse by an amount equal to the critical path of the adder. The delayed pulse then feeds the clock of a D-Flip-Flop (DFF) or a gated circuit for the sum function. Consequently, the output

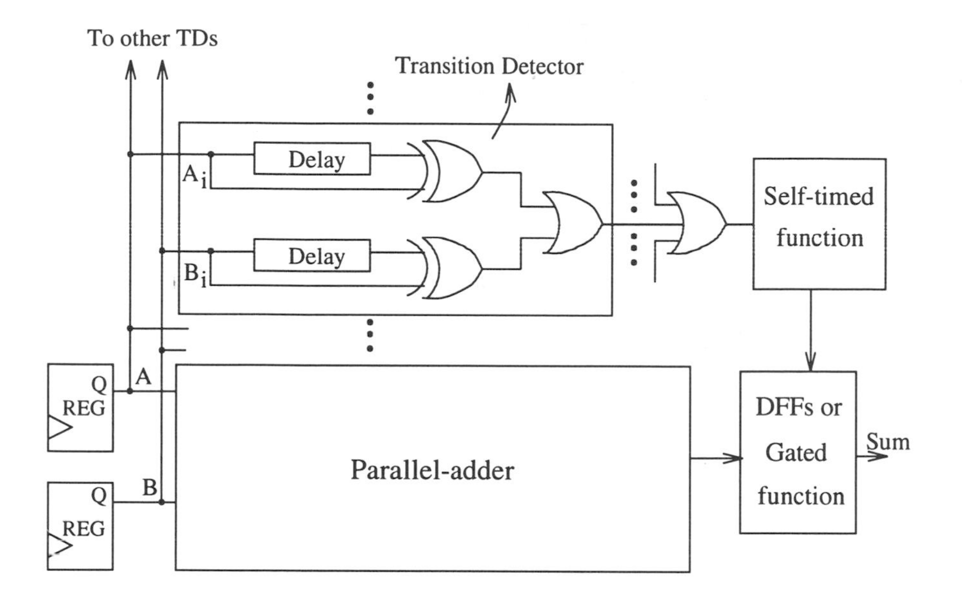

Figure 8.3 Conventional self-timed parallel adder using transition detectors.

sums are not switched until they are evaluated. The additional circuitry in the conventional approach can consume more power than it may save.

Another approach based on self-timing to reduce the spurious transition was proposed by [11]. Fig. 8.4 shows a parallel adder using simple self-timed circuitry. When input signals are written into the registers A and B, a single register bit is used to generate an "Input Valid" signal to the self-timed function. For an n-bit parallel adder, only a one-bit register is required, as shown in Fig. 8.4. The self-timed function is implemented using a series of inverters with dual-rail. Two enable signals E and $\bar{E}$ are generated by the self-timed circuit. They feed the gated sum XOR gates. Also the enable signal, E, controls the one-bit register to disable the input valid signal. This technique has resulted in 25% power reduction [11].

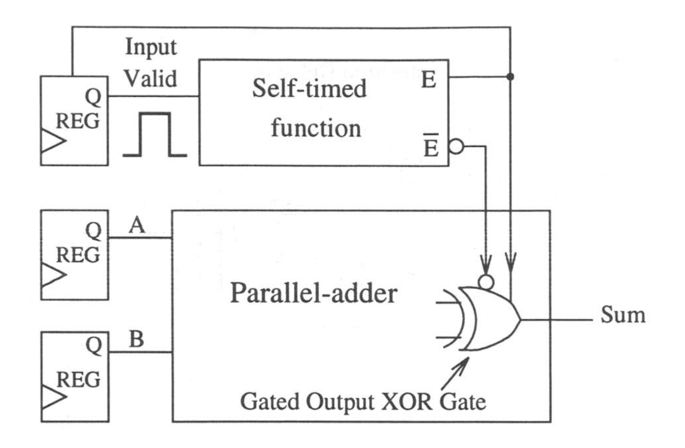

Figure 8.4 Self-timed parallel adder using a simple pulse generator.

8.2.3 Precomputation-Based Power Reduction

Consider the original circuit of Fig. 8.5(a). $R1$ and $R2$ are two registers at the input and output of a combinational logic block. The idea of precomputing is to pre-evaluate the output values of the circuit one clock cycle before they are required, to disable a part of the input register $R1$, then to reduce the internal switching activity in the succeeding clock cycle [13]. Fig. 8.5(b) illustrates a simplified architecture of the precomputing concept. This technique can be applied to several circuits such as: Finite State Machines (FSMs), pipeline circuits, etc.

To illustrate this technique, consider the example of an n-bit comparator that compares two n-bit numbers A and B and computes the function F that indicates that $A > B$. Fig. 8.6 shows the application of precomputing technique to the comparator. If the most significant bit, A_{n-1} and B_{n-1}, are different, then F can be performed from the 1-bit MSB comparator and the registers $R2$ and $R3$ are disabled. Therefore, the (n-1) comparators are shut-down. If the inputs have a uniform probability equal to 0.5, the enable signal has a probability of 0.5 to be at the logical level "1" or "0". Therefore, for a relatively large n the power saving can be quite significant even if we include the power due to the additional circuitry.

This technique of precomputation can be synthesized for logic optimization. The selection of sub-set of input signals for which the output is precomputed

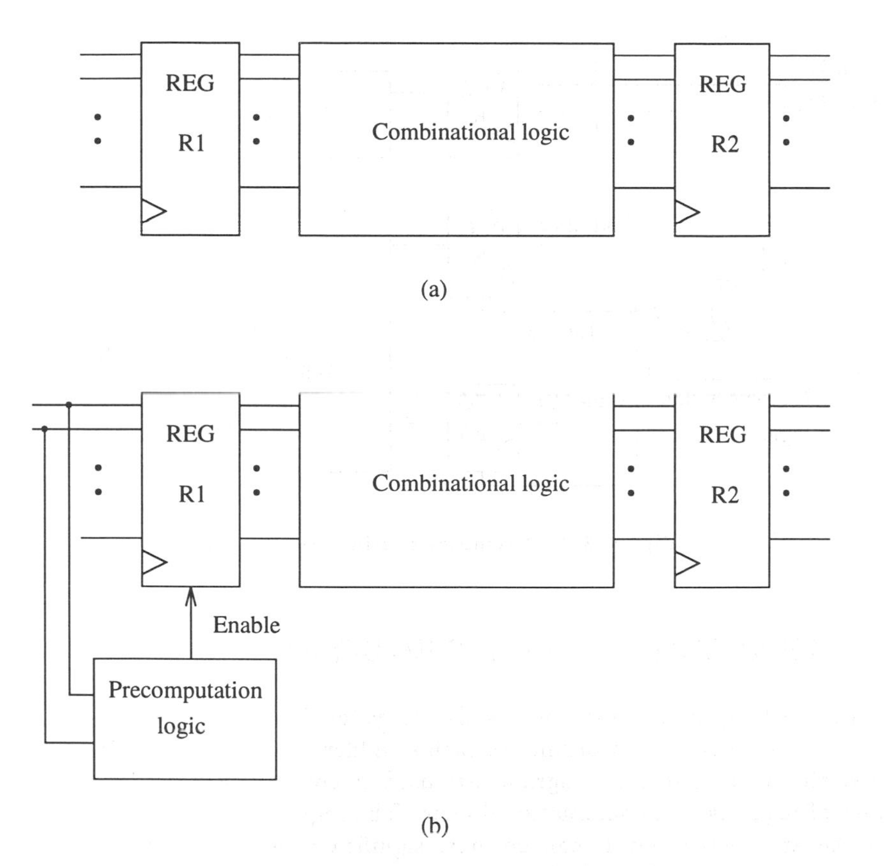

Figure 8.5 (a) Original circuit; (b) precomputation architecture.

is critical for power savings. Otherwise, the additional circuitry can dissipate a relatively important power. Note that this added logic slightly increases the area of the circuit and may also increase the clock cycle. The precomputation technique can be applied to a multiple output function. However, if the logic has a large number of outputs, then it may be worthwhile to selectively apply precomputation technique to a small number of complex outputs. This selective partitioning will add a duplication of combinational logic and registers and this may offset the power savings.

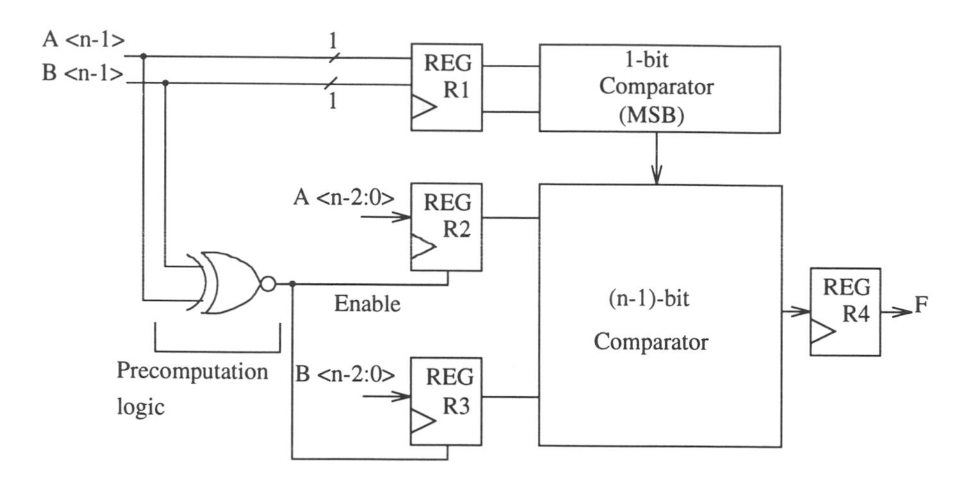

Figure 8.6 A comparator using precomputing.

8.3 LP ARCHITECTURE-LEVEL DESIGN

In this section, architecture means also Register Transfer Level (RTL). The architecture uses a set of primitives such as adders, multipliers, ROMs, register files, etc. RTL synthesis programs are used to convert an RTL description to a set of registers and combinational logic. The impact of low-power techniques on the architecture level can be more significant than the gate level as will be shown in this section. Techniques to reduce the power dissipation discussed are: parallelism, pipeline, distributed processing and power management.

8.3.1 Parallelism

Parallelism can be used to reduce the power dissipation at the expense of area while maintaining the same throughput [10]. To illustrate this, the quantitative example of Fig. 8.7 is considered. In Fig. 8.7(a), a register supplies two 16-bit operands to a 16×16 multiplier. We refer to this architecture to reference one and we use the ref notation for frequency, power supply voltage, power dissipation, etc. This register is clocked at a maximal frequency $f_{ref} = 50\ MHz$. We assume that the worse case delay of the multiplication is 20 ns at $V_{ref} = 3.3\ V$ power supply voltage. It is clear that we cannot reduce V_{ref} to reduce the

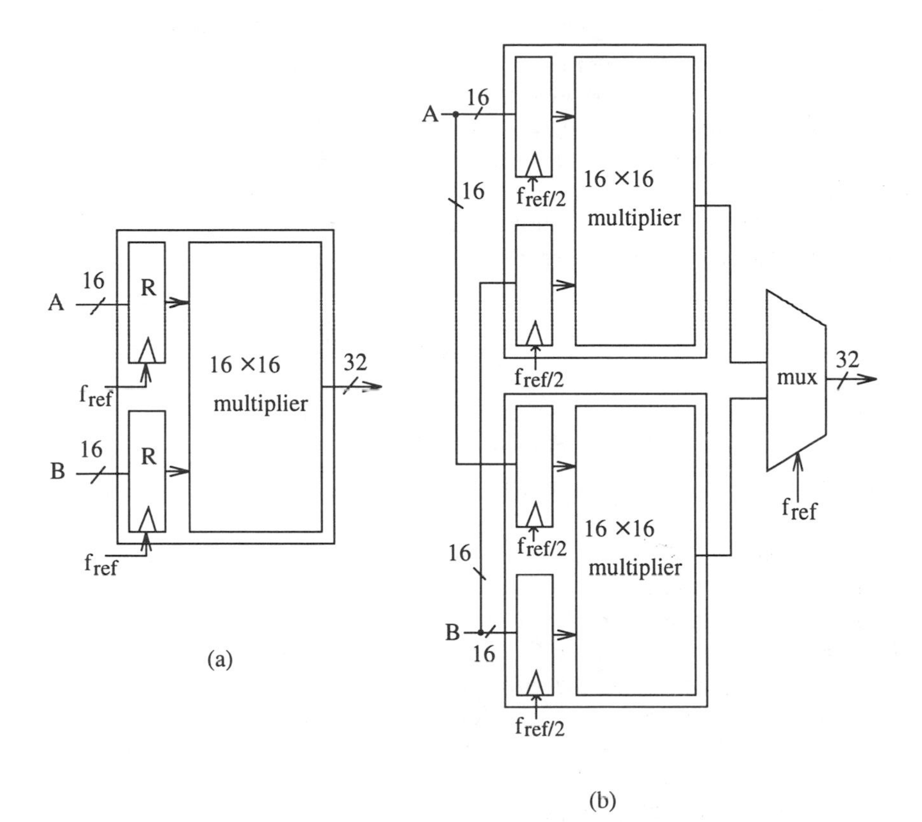

Figure 8.7 Parallelism for low-power: (a) reference architecture; (b) parallel architecture.

power dissipation. The estimated power dissipation is given by

$$P_{ref} = C_{ref} V_{ref}^2 f_{ref} \tag{8.3}$$

where C_{ref} is the total effective switching capacitance which is given by the sum of the product of switching activities by the node capacitances. So

$$C_{ref} = \sum_i C_i \alpha_i \tag{8.4}$$

Fig. 8.7(b) shows that the computation of the multiplier is parallelized by duplicating the multiplier two times. This architecture can maintain the same throughput as in the case of Fig. 8.7(a). The input registers are clocked at $f_{ref}/2 = 25\ MHz$. Therefore, the power supply can be reduced to achieve a worst case delay of 40 ns. With the same 16×16 multiplier, the power supply can be reduced from $V_{ref} = 3.3$ V to 1.8 V ($V_{ref}/1.83$). This value can be determined from the simulation of the two architectures. The effective capacitance has increased by a factor of 2 due to the duplication. However, due to the extra routing to both multipliers, this effective capacitance is around 2.2 C_{ref}. Thus, the estimated power dissipation is given by

$$P_{par} = 2.2C_{ref}\left(\frac{V_{ref}}{1.83}\right)^2 \frac{f_{ref}}{2} \tag{8.5}$$

Hence

$$P_{par} = 0.33P_{ref} \tag{8.6}$$

Thus, the power dissipation is significantly reduced.

The key to this power savings is the duplication of the hardware in parallel configuration. In general, N processors can be parallelized by duplication, with each processor running with slower clock (by a factor of N). In this case, for the same throughput, the power dissipation can be reduced with the increase of N. Therefore, the power supply voltage (V_{DD}) can be aggressively reduced to meet a worst case delay almost equal to the reference delay divided by N. To exploit this power supply reduction, the threshold voltage (V_T) should also be reduced to limit the degradation of the delay as V_{DD} approaches V_T. Keep in mind that the scaling of V_T is also limited by the static current considerations.

When the number N is relatively large, the parallelism can lead to several problems. A highly parallelized configuration can result in a drastic increase of the occupied area. In addition, there is routing overhead to distribute the input and output signals. This also increases the area and the wiring capacitance. Therefore, the power dissipation also tends to increase and then limits the utility of parallelism.

8.3.2 Pipelining

Pipelining is another architecture that can reduce the power dissipation [10]. As an example, let us consider the case of the 16×16 multiplier presented in Section 8.3.1. The 50 MHz multiplier is broken into two equal parts as shown in Fig. 8.8. A set of pipeline registers (or latches) is inserted, resulting in a 2-stage pipelined version of the multiplier. Architectures with more pipeline stages can

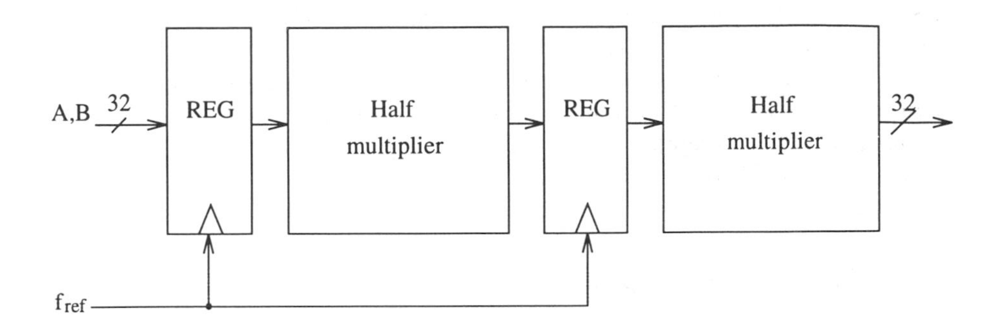

Figure 8.8 Pipelining of the multiplier for low-power.

be realized. Since the hardware between the pipeline stages is reduced then the reference voltage $V_{ref} = 3.3$ V can be reduced to 1.8 V ($V_{ref}/1.83$) to maintain a worst case delay of 20 ns (50 MHz). The estimated power dissipation is given by

$$P_{pipe} = 1.2C_{ref}\left(\frac{V_{ref}}{1.83}\right)^2 f_{ref} = 0.36P_{ref} \tag{8.7}$$

The switching capacitance has increased slightly due to the pipelining. Thus, the power dissipation is reduced by a factor of almost 2.8 which is approximately the same as the parallel case. Also the area increase is relatively low and the area penalty is due only to the additional registers (or latches). As the pipeline registers reduce the logic depth, the power dissipation, due to the glitches, is also reduced.

In general, if a processor is pipelined with N stages of registers, then the delay between the pipeline stages is reduced by almost a factor N while the clock frequency is maintained. Then, the power supply voltage can be scaled aggressively. Consequently, the power saving is large.

Note that as in the case of parallelism, an architecture with a large number of pipeline stages can result in an offset in power and area. The added registers must be clocked and hence the load on the clock network can be important, with increased pipelining. One drawback of the use of the pipeline is that more latency is added to the output signal.

The combination of pipelining and parallelism can result in further power reduction, because the power supply voltage can be reduced aggressively. Also the frequency of operation is reduced. However, the area would increase significantly. For low-voltage, the threshold voltage should also be reduced to reduce the power dissipation, otherwise the power supply voltage reduction is limited. Indeed, at low-voltage, V_{DD} approaches V_T and the delay increases drastically. To maintain the throughput with parallelism/pipelining, the threshold voltage should be reduced compared to V_{DD}.

8.3.3 Distributed Processing

To reduce the power dissipation of a centralized processor, a distributed processing technique can be utilized. This concept of distributed processing is explained by the example of the Vector Quantized (VQ) image encoder [14] presented in reference [15]. First we review the VQ algorithm for the video compression, then in the next section the power reduction at the algorithm level of the VQ is discussed.

A video image, represented by a group of pixels, is vector quantized by breaking it into blocks (vectors) of pixels that are mapped to a codebook of probable vectors using Mean Square Error (MSE) as the distortion measure. For the example given in [15], the image is segmented into 4 × 4 pixel-vector (vector size is 16). The VQ employs a codebook of 256 levels. The input data is represented on 16 × 8-bit and the output (8-bit) represents the index of the best match as shown in Fig. 8.9 [16]. Then the compression ratio is 16:1. To process 30 frames/s, a vector must be compressed every 17.3 μs (each frame is 128 × 240 pixels). The MSE (distortion metric) between a vector X of 16 pixels and a codebook vector C is given by

$$MSE = \sum_{i=0}^{15}(C_i - X_i)^2 \tag{8.8}$$

To compute this algorithm, a large number of memory access to the codebook and arithmetic operations is needed (see Section 8.4). The number of computations can be reduced by using differential search a priori combined with Tree-Search (TS) between two vectors a and b at the same level of the tree. The distortion difference between the two vectors a and b at the same level of the tree is given by

$$MSE_{ab} = MSE_a - MSE_b \tag{8.9}$$

Then,

$$MSE_{ab} = \sum_{i=0}^{15}(C_{ai} - X_i)^2 - \sum_{i=0}^{15}(C_{bi} - X_i)^2 \tag{8.10}$$

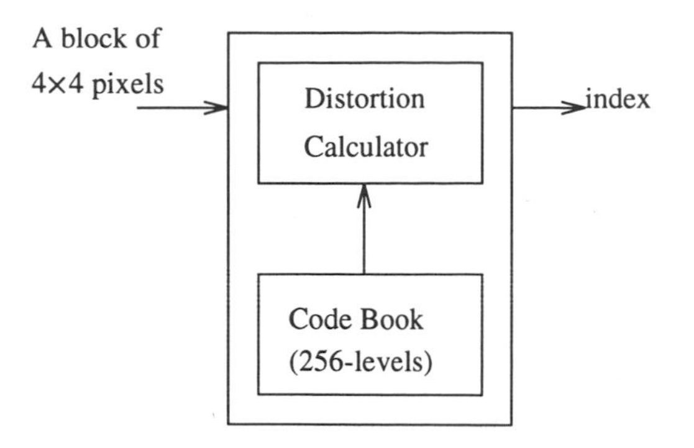

Figure 8.9 Vector quantization video encoder.

Finally,

$$MSE_{ab} = \sum_{i=0}^{15}(C_{ai}^2 - C_{bi}^2) + \sum_{i=0}^{15} 2X_i(C_{bi} - C_{ai}) \tag{8.11}$$

The two terms $(C_{ai}^2 - C_{bi}^2)$ and $2(C_{ai} - C_{bi})$ are computed a priori and stored in a memory to reduce the number of operations.

Fig. 8.10(a) shows the centralized implementation of the VQ. It has a centralized memory, processing element, and controller. This architecture is time-multiplexed, which performs operations sequentially over a large number of clock cycles. In TSVQ, each level of the tree has specific code vectors that are found only at that level. Therefore, the memory can be partitioned into separate memories for each level of the tree. Fig. 8.10(b) shows the distributed implementation of the VQ. The memory size from one module to the other increases. The architecture is pipelined allowing the clock frequency and supply voltage to be reduced.

The distributed memory architecture has lower switched capacitance when reading the code vectors than the centralized case. This distributed implementation has eight controllers and processing elements, but since they are clocked at lower frequency, with low supply voltage, the energy dissipated per vector does not change [15]. Through this partitioning, the power dissipated, of the centralized implementation, was reduced by a factor ~ 17 at the expense of an area increase by a factor of ~ 2.

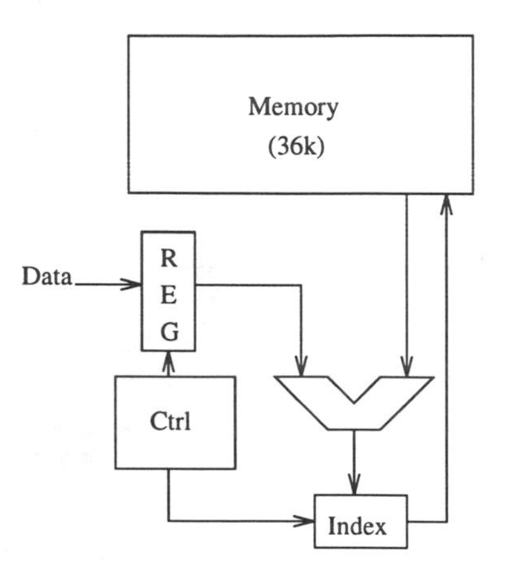

(a)

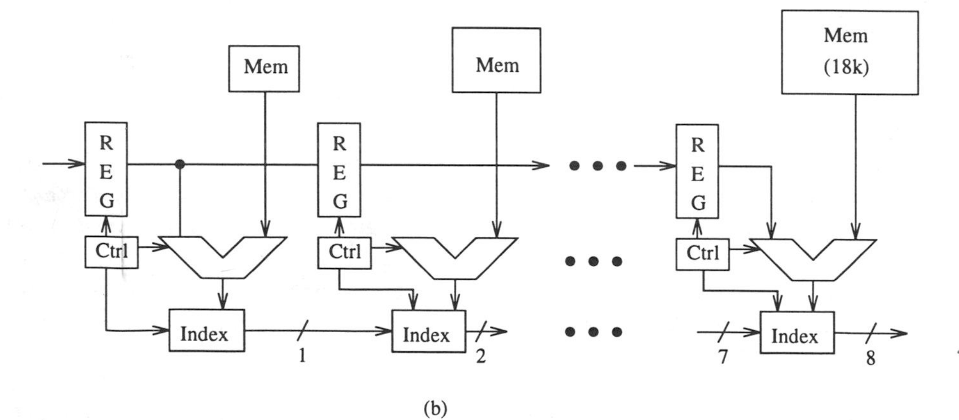

(b)

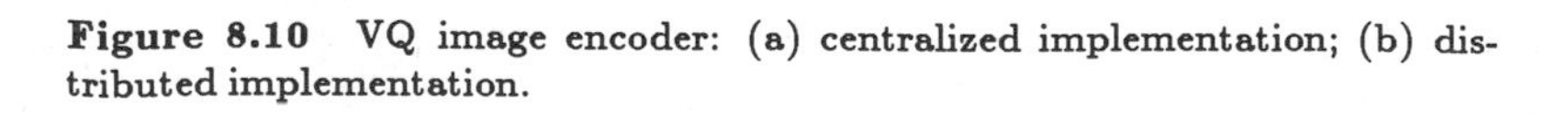

Figure 8.10 VQ image encoder: (a) centralized implementation; (b) distributed implementation.

From this example we can learn that proper design of the architecture, through distributed processing, is more power-efficient than the centralized processor. In the distributed implementation, the different local hardware resources can be optimized more efficiently than the global hardware in the centralized implementation. The application of this technique depends on whether the executed algorithm can be partitioned. Keep in mind, that the power saving trades the occupied area, while the throughput is maintained.

8.3.4 Power Management

In old designs of microprocessors, DSPs, ASICs, etc., there was wasted power due to the clocking of blocks which are idle for a significant period of time. Recently, power management methodologies are playing an important role to avoid wasted power in normal and standby modes of operation [17, 18, 19, 20]. In this section, only some of the power management techniques are discussed.

There are two types of power management: i) dynamic and ii) static. Dynamic Power Management (DPM) allows selective shut-down of different blocks of the chip based on the level of activity required to run a particular application. Different blocks of the chip may be idle for a certain period of time when running different applications. For example, the floating point unit can have 100% idle time when the processor is executing integer applications. The DPM requires additional logic on the chip. This logic is controlled by signals of idle periods.

In the **PowerPC**[1] 603 [21], the DPM mode is enabled by software. The DPM logic automatically stops the clock switching of a specific unit generated by clock regenerators. The clock regenerators produce two clocks, $C1$ and $C2$, which feed master and slave latches. Two "freeze" input signals control the clocks, $C1$ and $C2$, as shown in the timing diagrams of Fig. 8.11. The logic needed for DPM does not introduce any performance degradation and it consumes 0.3% of the total die area in the **PowerPC**. The DPM provides a power saving of 10-20% depending on the application to be executed. The DPM can be implemented at either high-level (e.g., execution unit) and low-level (e.g., a block inside a unit) of hardware.

Static Power Management (SPM) permits the saving of the power dissipation in the standby mode. In this case, the activity of the entire system is monitored rather than a specific unit (or block). When the system remains idle for a

[1] **PowerPC** 603 is from IBM Corp.

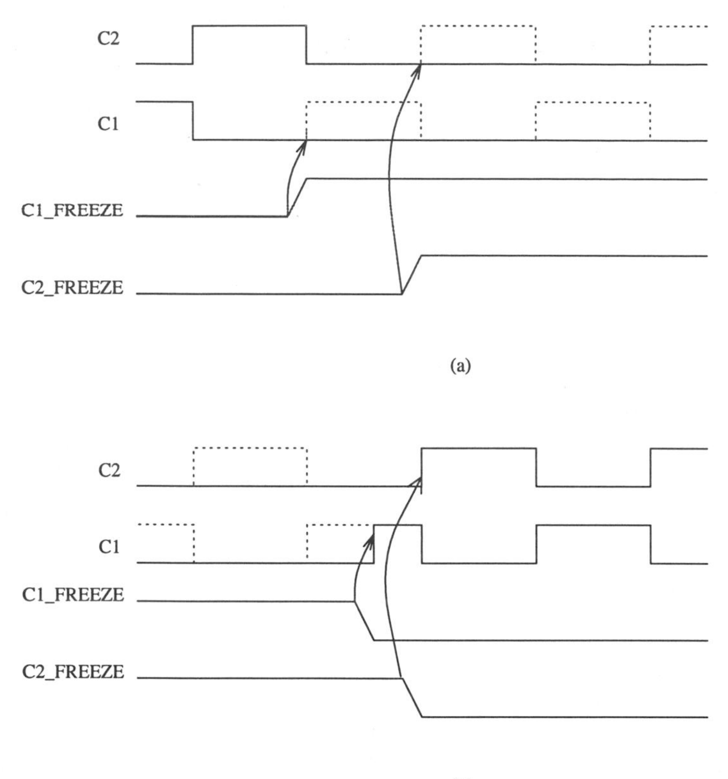

Figure 8.11 Freeze timing of $C1$ and $C2$: (a) turning the clocks OFF; (b) turning the clocks ON.

significant period of time, then the entire chip is shut-down[2]. The SPM may have several modes depending on whether the entire chip is shut-down or a part of it. For example, the **PowerPC** 603 has three modes which are programmable through a hardware but controlled by software (operating system). In this microprocessor, one mode is called *sleep mode* which allows a maximum power savings by disabling the clocks to all units. In this mode the PLL and external input clock are disabled to bring the power dissipation down to the leakage levels. The power of **PowerPC** 603, in the sleep mode, is as low as 1.8 mW [20].

[2] This mode is called sleep mode.

8.4 ALGORITHMIC-LEVEL POWER REDUCTION

Algorithm optimization can have a significant impact on the power consumption of a system. Design decisions, made at this level, combined with the architecture level, may lead to a large power saving. In this section, we discuss two approaches that reduce the power dissipation at the algorithm level. The first one is based on the reduction of the switched capacitance, by minimizing the complexity of the system. The second method exploits data coding for the purpose of low switching activity.

8.4.1 Switched Capacitance Reduction

The power dissipated by an algorithm can be measured, for example, by counting the number of operations required to execute such an algorithm. To reduce the power of an algorithm, the number of primitive operations such as: memory access, ALU operations, etc., should be minimized. The different types of operations do not consume the same amount of power. For example a multiplication operation consumes more power than an addition operation. Thus, when minimizing the number of operations of an algorithm, the type of operation should be taken into account. Keep in mind that high performance systems use complex algorithms that require a large number of operations.

To illustrate this consideration, the computation complexity for three methods of the VQ algorithm are presented. Remember that the distortion metric between the input data (vector X) and a codebook vector C is given by Equation (8.8). One method to evaluate the distortion and find the best match is to use a full search through the codebook. Thus, the distortion is computed for the 256 levels of the codebook. Each level requires 16 memory access to perform 16 subtractions, 16 multiplications, 15 additions, etc. Hence a large number of primitive operations are needed.

In the binary TSVQ already presented in Section 8.3.3, the codebook is organized into a tree structure as shown in Fig. 8.12. The input vector is compared with two code vectors at each node. Based on this comparison, one of the two branches is chosen and the codebook search space is reduced compared to the full search, since a reduced number of code vectors (16) is utilized. For each comparison, at a specific level, an index bit is generated as shown in Fig. 8.12. The process of comparison through the tree is repeated until a leaf node is reached. For a codebook of 256 levels, the tree has a depth of 8 (d=7). Compared to the full search, the number of memory access and executing operations

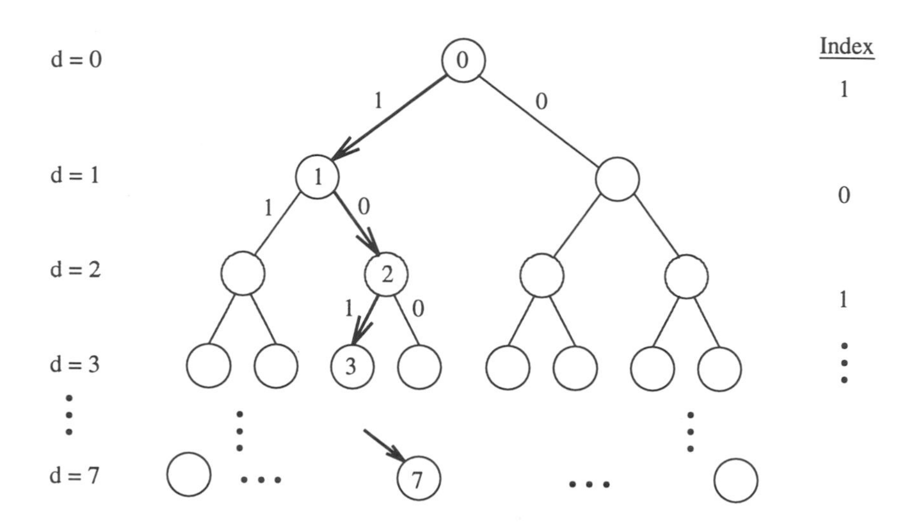

Figure 8.12 Traversal of a binary tree of depth 8.

is reduced considerably since only 16 code vectors are used in the TSVQ algorithm. One VLSI implementation of the TSVQ algorithm uses systolic arrays [22].

The number of computations can be further reduced by using the differential search of the TSVQ [see Equation (8.11)]. At each level (i) of the tree the differential distortion between the left (vector a) and right (vector b) code vectors connected to the level ($i-1$) is computed. Therefore, the number of operations is reduced. Table 8.1 [15] shows the computation complexity of the three methods of the VQ. The differential TSVQ results in a lower number of operations to be executed for each type.

8.4.2 Switching Activity Reduction

Minimizing the switching activity, at high level, is one way to reduce the power dissipation of digital processors. This can have an influence on the power reduction, especially when the switching signals have a large capacitance. One method to minimize the switching activity, at the algorithmic level, is to use an appropriate coding for the signals rather then straight binary code.

Table 8.1 Computational complexity of the VQ algorithms.

Algorithm	Memory Access	Multi-plication	Add/ Substract
Full Search	4096	4096	8448
Tree Search	256	256	520
Differential Tree Search	136	128	136

In [23], Gray-coding has been used for the address lines of a microprocessor, for both instructions and data accesses, to reduce the switching activity of the nets. The advantage of Gray code over binary code is that Gray code changes by only one bit as it sequences from one number to the next. In other words, if the memory access pattern is a sequence of consecutive addresses, then each memory access changes only one bit at its address bit. Due to instruction locality, during program execution, most of the memory accesses are sequential. Therefore the Gray code eliminates the simultaneous switches of a significant number of bits.

Table 8.2 shows a comparison of 3-bit representation of the binary and Gray codes. Note that the Gray code have only one transition for sequential change

Table 8.2 Binary and Gray-code representation.

Binary Code	Gray Code	Decimal Equivalent
000	000	0
001	001	1
010	011	2
011	010	3
100	110	4
101	111	5
110	101	6
111	100	7

while the binary code may have many transitions.

In [23], the switching property of the address coding was measured using the number of bit switches per executed instruction. For instruction accesses, both the Gray and binary coding were compared using benchmark programs. The maximum reduction in bit switches was found to be as high as 58% and the average reduction was equal to 37%. The same study was also carried out for data addresses. The average reduction of bit switches was $\sim$ 8%.

8.5 POWER ESTIMATION TECHNIQUES

Power estimation means, in general, the techniques of estimating the average power dissipation of circuits. The goal of this section is to present an overview of power analysis techniques and tools at the circuit, gate, architectural, and behavioral levels of abstraction. Measuring the power consumption is critical for low-power design as it permits the designer to optimize power, meet requirements, and know the power distribution through the chip.

8.5.1 Circuit-Level Tools

The most straight-forward method of power estimation is by circuit simulation; perform a circuit simulation of the design and measure the average current drawn from the supply. Therefore, the average power can be estimated. The disadvantage of this approach is that the results are strongly dependent on the input patterns to the circuit (*pattern-dependent technique*) also called dynamic[3] power simulation. If the circuit has a large number of inputs, then the circuit simulation would be time consuming and even impractical.

The most accurate power simulator to date is still **SPICE**. However, it can handle only very small circuits (e.g, hundreds of transistors). **SPICE** accurately takes into account non-linear capacitances (junction and gate) which cannot be captured by higher level tools. Also, it can accurately measure short-circuit and leakage currents. The latter is very important for low-V_T applications. **SPICE** cannot be used to estimate the power of large circuits or chips, due to the time consuming nature of the simulator. It is a pattern-dependent power analysis tool.

[3] Dynamically computed power should not be confused with dynamic power.

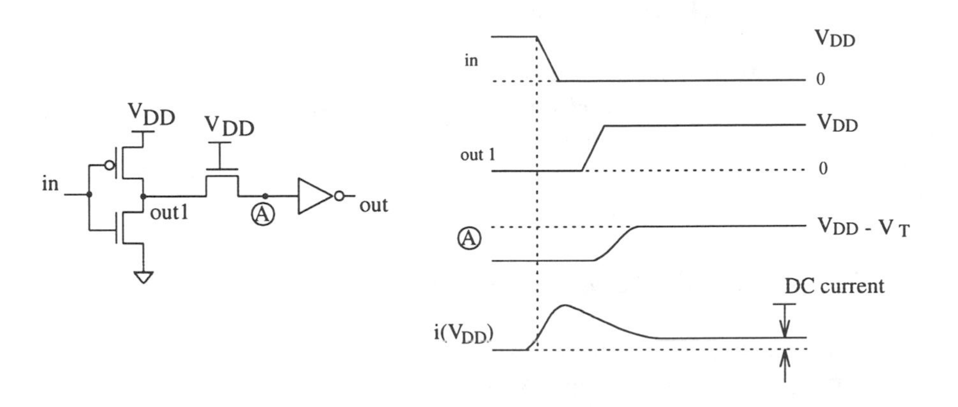

Figure 8.13 Partial-swing detection.

Another transistor-level power simulator/analyzer is **PowerMill**[4] [24]. It applies an event-driven simulation algorithm to increase the computation speed by two to three orders of magnitude over **SPICE**, with an acceptable level of accuracy (within 10%). Also, it uses table lookup to determine the terminal current of the device from the applied voltages.

PowerMill can also identify the *hot spots* (which consume more dynamic power) and *trouble spots* (which consume unexpectedly large amounts of leakage current). Moreover, elements with excessive short-circuit are detected. This allows the designer to resize the circuit to reduce the rise/fall time. Static reduced-swing nodes are detected as shown in the example of Fig. 8.13. The node A is charged to $V_{DD} - V_T$ when the input is low.

Another approach for power estimation is the use of statistical techniques. The work in [25] suggested the use of *Monte Carlo* simulation to estimate the total average power of the circuit. Basically, this statistical technique is based on applying randomly generated input patterns, at the primary inputs, and monitoring the convergence of the power dissipation. The simulation is stopped when the measured power is close enough to the true average power.

This approach, based on the Monte Carlo method, requires simulation over a large number of measurements. The advantage of the statistical techniques is that they can be built around existing simulation tools.

[4] **PowerMill** is from EPIC Design Technology.

8.5.2 Gate-Level Techniques

In order to overcome the shortcoming of power analysis tools, at the circuit level, recently several gate-level estimation tools have been proposed. In this section, we present two techniques for power estimation at the gate-level. The first approach relies on the probabilistic method, while the second one is based on event-driven simulation.

8.5.2.1 Probabilistic Power Estimation

The power dissipation can be analyzed using *pattern-independent* approach when the signals are represented with probabilities (also called *static techniques*). This approach permits to overcome the shortcomings of simulation-based techniques. The user supplies the probabilities of the primary inputs to a logic network. The average power dissipation of a logic network is estimated as

$$P = V_{DD}^2 f \sum_{i=1}^{N} \alpha_i C_i \tag{8.12}$$

where N is the number of nodes in the network. With a total physical capacitance C_i. α_i is the switching activity (or called transition probability, P_s) given by

$$\alpha_i = P_i(1 - P_i) \tag{8.13}$$

where P_i is the probability that the node i is at high level. In this expression of activity it is assumed that the circuit input and internal nodes are independent (*spatial independence*). Also the values of the same signal, in two consecutive clock cycles, are assumed independent (*temporal independence*).

If the input probabilities to a network are provided, then they are propagated through the circuit to evaluate the transition probability at each node. For example, for a 2-input AND function: $y = x_1.x_2$, the probability of the output to be at high level is given by: $P_y = P_{x_1}.P_{x_2}$. The computation of the probabilities for different gates is discussed in Chapter 4.

One tool (**LTIMES**), based on probabilities, was first proposed in [26]. In this work, the temporal and spatial independences of signals are assumed. Practically, the signals may be correlated. Also a zero-delay model was assumed, which leads to an error in estimating the power, since the glitching power is not accounted for.

Probabilistic power estimation approaches that compute the power, due to glitches, and apply a real delay model have been proposed [27, 28]. In [27], the switching activity computation is based on the *transition density*. The assumption made in [27] is the spatial independence of the signals. A power estimator tool, based on the transition density, has been called **DENSIM**. The transition density of a node is defined as the average number of nodal transitions per unit time. If y is a boolean function with inputs, x_i, then the boolean difference of y, with respect to x_i, is defined by

$$\frac{\partial y}{\partial x_i} = y(x_i = 1) \oplus y(x_i = 0) \tag{8.14}$$

It was shown in [29] that if x_i are spatially independent, then the density of the boolean function is given by

$$D(y) = \sum_i P\left(\frac{\partial y}{\partial x_i}\right) D(x_i) \tag{8.15}$$

where $P(z)$ is the equilibrium probability of the signal over time. Equations (8.14) and (8.15) are used to propagate the density through the boolean network. $\partial y/\partial x_i$ is one if a transition at x_i will cause a simultaneous transition at y. As an example, consider the case of a 2-input AND gate with $y = x_1 x_2$. In this case, $\partial y/\partial x_1 = x_2$ and $\partial y/\partial x_2 = x_1$, so that $D(y) = P(x_2)D(x_1) + P(x_1)D(x_2)$. Hence, from the probability and density values, at the primary inputs of a logic network, the density at the output can be computed. The boolean differences of a logic network are calculated using *Binary Decision Diagrams* (BDDs) [30].

Note that the average power dissipation is computed by

$$P_{avr} = \frac{1}{2} V_{DD}^2 \sum_i C_i D(i) \tag{8.16}$$

The factor 1/2 is added to account for the double transition per clock period.

This model, based on transition density, ignores the spatial correlation of the signals and computes, approximatively, the power due to glitches. The work in [28] attempts to handle both spatial and temporal correlations. One disadvantage of the approach in [28] is that the use of BDDs, for the whole circuit, tends to limit the size of the network that can be analyzed.

The probabilistic techniques have the advantage that the user does not have to supply simulation patterns and they are claimed to have fast computation time. However, they do not account for the internal power of the gates and static power dissipation. These techniques can be used, for example, as a fast power estimator for logic synthesis. They might also be suited for comparing various subsystem structures.

8.5.2.2 Event-Driven Simulation

Another gate level power analysis approach has been proposed for semi-custom design [31]. The environment of the system is shown in Fig. 8.14. The system uses a cell library that has been characterized for static and dynamic power dissipation with the **Entice**[5] (ENergy and TIming Characterization Environment) cell characterization system [32]. The dynamic power includes the power due to the short-circuit and the one due to the load capacitance. **Entice** characterizes each cell taking into account the following parameters: input signal slope, output capacitive load, operating voltage, temperature, and process parameters. **Entice** uses SPICE as a circuit simulator to model each cell for power.

A set of *power vectors* describes all possible events where power can be dissipated by the cell for dynamic and static cases. With SPICE these power events are accurately characterized. There are two types of power vectors: i) dynamic and ii) static. A dynamic power vector describes an event in which power is dissipated due to a signal switching at the cell inputs. For example, for a 2-input (A and B) AND gate, when $A = 1$ and B makes a transition from 0 to 1, an energy is dissipated. A static power vector describes the conditions of logic signals under which leakage power occurs.

The designer creates a design from the cell library at gate level then it is input to the **Aspen**[6] (A System for Power EstimatioN) system. Also the stimulus to drive the logic simulator and the interconnect loads, representing the intercell connectivity (estimates or actual values provided by back-annotation from layout) are specified. A logic simulator such as **Verilog-XL**[7] is used as an even-driven simulator. Upon invocation, **Aspen** monitors the power event occurrence (node activity) of each cell and computes the total power dissipation as the sum of the power dissipation of all the cells in the power vector paths. Multiple time windows can be specified for simulation to compute the average power over different time periods. Note that **Aspen** uses the power vectors of a cell to compute the total power.

---

[5] **Entice** is from Motorola Inc.
[6] **Aspen** is from Motorola Inc.
[7] **Verilog-XL** is from Cadence Design Systems Inc.

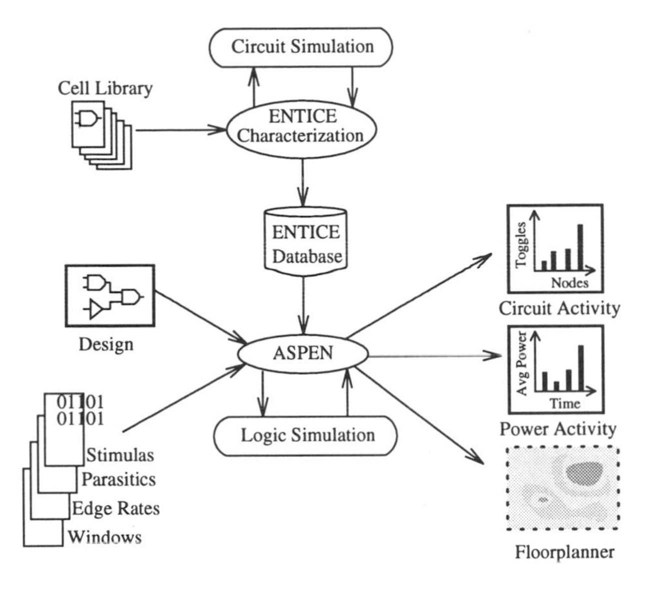

Figure 8.14 The Entice-Aspen power analysis environment.

The dynamic power of each cell is computed by multiplying the number of power events (transitions' count) by the energy dissipation per transition event of a cell. This process is applied to all dynamic power vectors for a cell to obtain the total energy dissipated. The total dynamic power of a cell, over a certain time period, is equal to the total energy divided by the time period.

The static power vector is used to compute the leakage of a cell. Note that the static power of a cell is dependent on the logic state of a cell, as shown in Fig 8.15. To compute the static power dissipation, the duration of activation time of the corresponding static power vector is measured. A transition of net signal may cause a static power vector to be activated and another vector to be deactivated. Vectors are time stamped during activation and upon deactivation. Then the total time length in which the vector is active is found. The activation time length of the static power vector is multiplied with the power dissipation value (per time unit) to obtain the static power of the vector. Again the static power dissipation for all vectors associated with a cell instance is summed to derive the total power dissipation.

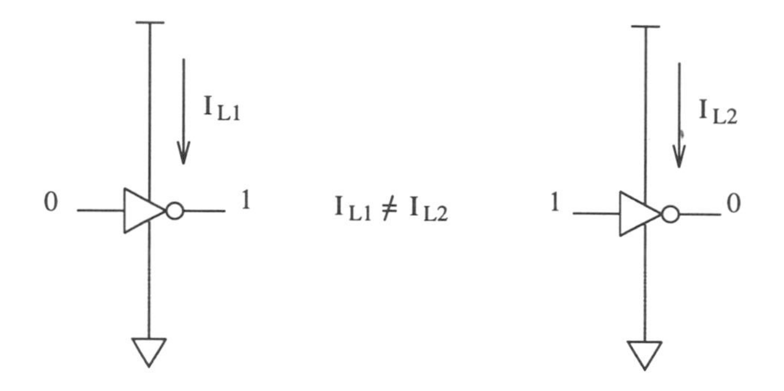

Figure 8.15 Logic state dependence of the static leakage current.

The results reported by **Aspen**, such as the switching activity of nodes, can be used to drive floorplanning, placement and routing tools. Also **Aspen** can handle chips with a complexity of several hundred thousand gates and is four orders of magnitude faster than SPICE. It produces results within 10% accuracy of SPICE results. One disadvantage of **Aspen** is that it cannot handle power due to the glitches.

8.5.3 Architecture-Level Power Estimation

The architecture of a design is represented by functional blocks and the complexity of the design at this level is relatively low compared to the circuit and gate levels. In this section, several approaches and techniques for power modeling and analysis at the architectural level are reviewed.

8.5.3.1 Gate Count Method

One tool developed for architectural power dissipation estimation is based on equivalent logic count, memory size, logic circuit styles (dynamic or static), interconnection busses, clock network and layout style (full-custom or semi-custom) [33]. The complexity of an architecture is described in terms of average number of logic gates such as a 3-inut AND (buffered NAND) gate connected to three identical AND gates at the output node (i.e, fanin=fanout=3) as shown in Fig. 8.16. The total power of the logic part is roughly equal to the number of gates multiplied by the power of a gate using a user specified switching activity. This activity factor is assumed fixed across the design.

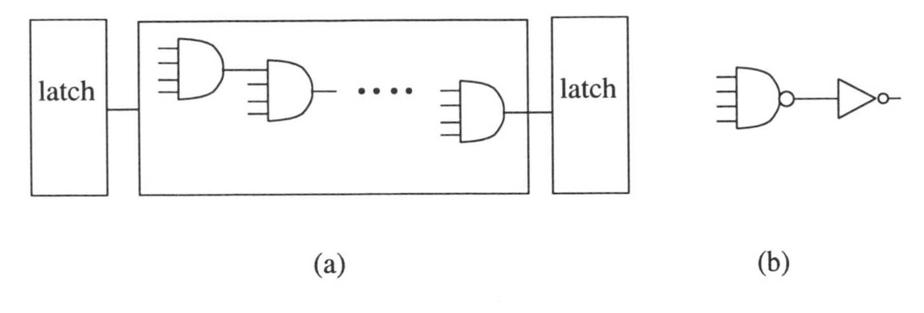

Figure 8.16 (a) A function built by 3-input AND as an average logic gate; (b) buffered logic gate.

The power of the on-chip memory is modeled for a certain memory architecture. The interconnections are defined in two categories, local and intermediate, and global busses. The local interconnection is defined as interconnections within a logic gate. The intermediate interconnections are used for connection between gates or functional blocks (sub-systems). The global bus includes data, control, and address busses. The lengths of local and intermediate interconnections are modeled by the Rent's rule [34]. Then the power can be computed from the lengths using a fixed switching activity equal to the one specified for the logic. The global interconnect is determined from the dimensions of the chip and the number of drivers/receivers connected to it.

The power model of the clock network is based on the H-tree [34] and the chip dimensions. The power of on-chip drivers are also modeled in two components. One is the power used to drive the off-chip total capacitance. The other is the power consumed by the pad driver itself. The activity factor for the pads is assumed fixed and is equal to 1 [33].

The tool developed in [33] is used as a power estimator in the early stage of the design. It requires some technological parameters (feature size, gate oxide thickness, parameters of the interconnection layers, etc.), the supply voltage, the chip area, the switching factor and the gate count. This tool can only be used as a rough estimator of the total power of the chip since the switching activity is assumed fixed through the design. Therefore the power partition between the different units can be incorrectly estimated.

8.5.3.2 The Power Factor Approximation Method

The Power Factor Approximation (PFA) technique is another method to estimate the power dissipation [35]. It has been used for DSP's architectures. The total power dissipation[8] of a functional block such as: multipliers, adders, memories, etc. can be modeled by the following approximation

$$P_{avg} = \sum_{i=1}^{G} \alpha_i \left[C_i V_{DD}^2 + V_{DD} \int_0^T i_{sc,i} \right] f \tag{8.17}$$

where G is the number of the logic gates composing the functional block, α_i is the switching activity of the *ith* gate, C_i is the load of the *ith* gate, $i_{sc,i}$ is the short circuit component, and f is the frequency. This power equation can be expressed in more compact form as

$$P_{avg} = \kappa G f \tag{8.18}$$

where κ is the PFA constant and can be related easily to Equation (8.17). G can also be looked at as the hardware complexity factor instead of a number of gates. The parameter κ has different values for different blocks. For example for an n-bit multiplier, the factor G can be approximately equal to n^2 as shown in Fig. 8.17. This is due to the number of adder cells in the multiplier. Then, we have

$$P_{mult} = \kappa_{mult} n^2 f_{mult} \tag{8.19}$$

The power supply voltage is included in the parameter κ. This parameter is extracted empirically from measured or simulated power values at a fixed power supply voltage.

For a VLSI chip, composed of several functional blocks, the total power dissipation can be determined by summing the power of all blocks. We have

$$P_{tot} = \sum_{all\ blocks} \kappa i G_i f_i \tag{8.20}$$

Thus, this PFA technique is based on modeling precharacterized functional blocks. Each block has a PFA factor independent from the other. Hence this technique provides some general methodology compared to the gate equivalent model of Svensson and Liu discussed previously. The PFA factor is extracted using *independent Uniform White Noise* (UNW) inputs (i.e, random inputs). UWN inputs mean that the input's bit are uncorrelated in space and time and

[8] Without the static power dissipation.

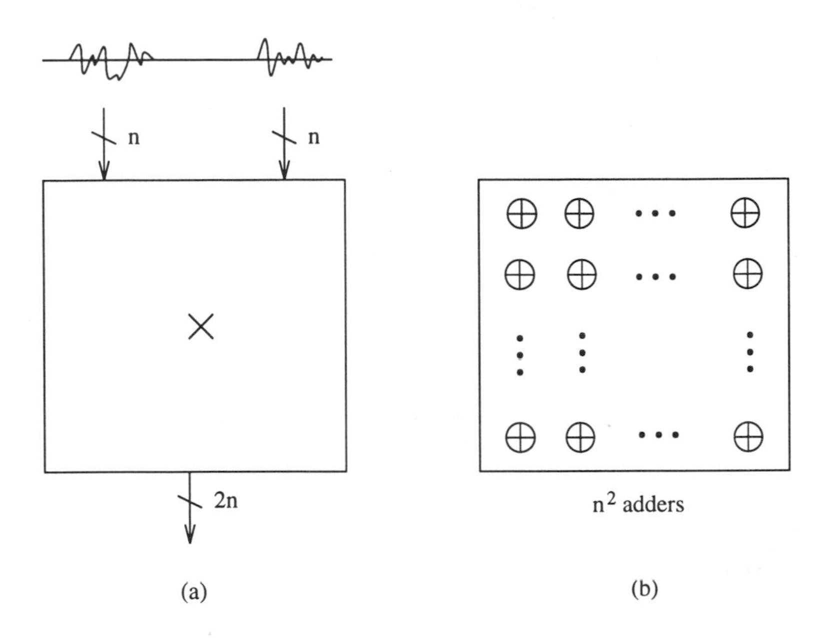

Figure 8.17 (a) An $n \times n$ multiplier with random inputs; (b) n^2 adders in the multiplier array.

independent of the data distribution. The signal and transition probabilities of each i bit of the input are given by

$$P_i(1) = 0.5 \qquad and \qquad P_i(0 \rightarrow 1) = 0.25 \tag{8.21}$$

Consequently, this technique does not account for the strong dependency of power consumption on the statistics of the input data [36]. The next section treats the case of power modeling, taking into account the correlated behavior of the bits.

8.5.3.3 Dual Bit Type Model

In digital signal processing, correlation can exist between values of a temporal sequence of data. The UWN model can lead to an error in estimating the power of a circuit even if the bit-width utilization is maximized. To take into account the data correlation, the Dual Bit Type (DBT) data model has been proposed in [36, 37]. The DBT data permits accurate estimation of the power dissipation.

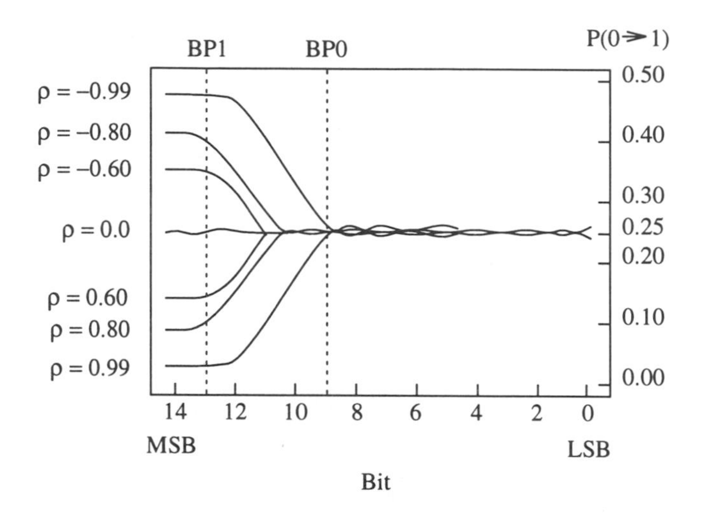

Figure 8.18 Transition activity versus bit for a data stream with varying temporal correlation.

Fig. 8.18 shows the transition activity for several different two's complement data stream versus the bit (for an n-bit word). In this figure, each curve corresponds to a different temporal correlation given by

$$\rho = \frac{cov(X_{t-1}, X_t)}{\sigma^2} \tag{8.22}$$

where X_{t-1} and X_t are successive data (in time) and σ^2 is the variance. $\rho = 0$ corresponds to the white noise case, where $P(0 \rightarrow 1) = 0.25$. From Figure 8.18 it is evident that the UWN model, while sufficient for describing activity in the Least Significant Bits (LSBs), is inadequate for the Most Significant Bit (MSB) region. The UNW model works correctly for the LSBs up to the break point $BP0$. The MSB region corresponds to the sign bits and consequently, the signal and transition probabilities of these bits are far from random. $\rho > 0$ corresponds to a lower activity for positively correlated signals, while $\rho < 0$ corresponds to a higher activity for negatively correlated signals. The MSB region starts from the break point $BP1$. The region between $BP0$ and $BP1$ can be modeled by linear interpolation. $BP0$ and $BP1$ can be determined from the word-level statistics [37].

The power estimation of the architecture modules is based on a black-box technique of the switched capacitance. Typical modules are: adders, multipliers, shifters, RAMs, ROMs, etc. The power dissipation is modeled for each module by

$$P = CV_{DD}^2 f \tag{8.23}$$

where the switched capacitance C is related to the complexity and the activity of the module. For example of an n-bit ripple-carry subtractor, the switching capacitance is modeled by

$$C = C_{eff} n \tag{8.24}$$

where C_{eff} is a capacitive coefficient (in fF/bit) determined from the DBT model. C_{eff} can be a single coefficient for the UWN case. The DBT model employs several coefficients for C_{eff}, which reflect the data representation and signal statistics. For the case of the subtractor, for example, a table of C_{eff} is generated as a function of all possible data transitions, i.e., sign bits transitions and LSB bits random transitions.

To extract the capacitive coefficients of each module, the library should be characterized. This operation is performed one-time for one library. The process of extraction consists of several steps:

- Pattern generation. Input patterns to a module are generated based on the DBT data model. Both random (UWN) and sign data streams should be used. The input patterns containing the UWN component must be simulated for several cycles. This allows convergence of the average capacitance.
- Simulation. The generated patterns are fed to a simulator (such as a circuit simulator) from which the switching capacitances are extracted.
- Capacitive coefficient's extraction. The simulation step produces the average effective switching capacitances for the entire series of applied input transitions such as: $U \rightarrow U$, $S \rightarrow S$[9] , etc. The capacitive coefficients are extracted from the effective switching capacitances and the complexity parameters.

Based on this methodology, a power analysis tool, at the architectural level, has been developed [36].

[9] U and S mean UWN and sign parts of the input bits, respectively.

8.5.4 Behavioral-Level Power Estimation

A behavioral representation describes the function of a system versus a set of inputs. The behavior can be specified, for example, by algorithms (in Verilog, VHDL, etc.) or by boolean functions. The power estimation, at the behavioral level, relates the consumed energy to the execution of an algorithm. Decisions at the system and behavioral levels can influence the final power dissipation of the circuit by several orders of magnitude.

One approach for power estimation, at the behavioral level, has been proposed in [38]. It is based on the combination of analytical and stochastic power models. In this work, a class of applications such as real time DSPs is considered for the power estimator. In the behavioral context, the power consumed by a hardware resource is given by

$$P = N_a C V^2 f \tag{8.25}$$

where N_a is the number of accesses to the resource over the period of computation. C is the average capacitance switched per access and f is the computation frequency.

In [38] the power of some hardware resources, such as execution units, registers, etc., are analytically modeled (using Equation (8.25)) from the Control/Data Flow Graph (CDFG) which is used to represent the design. The average capacitance switched, per access, for a particular hardware is estimated from the white noise data model. The power consumed by hardware resources such as controllers, interconnects, and clock network is difficult to estimate. Statistically a large number of realized chips is used to estimate the switched capacitance of these hardware resources.

8.6 CHAPTER SUMMARY

Low dynamic power techniques at several levels of abstractions have been presented. Algorithmic and architectural decisions can influence the power dissipation of a circuit by orders of magnitude. Therefore, CAD tools that help the designer to analyze the power of the circuit at these levels are needed. At lower levels of the design, the power reduction techniques offer some savings but less than the one expected at higher levels. Several power estimation tools have been discussed at the different levels of the design. Keep in mind that the circuit simulators provide a high accuracy for power analysis and take into account all power components.

REFERENCES

[1] K-Y. Chao, and D. F. Wong, "Low Power Considerations in Floorplan Design," Proc. of the International Workshop on Low Power Design, pp. 45-50, April 1994.

[2] H. Vaishnav and M. Pedram, "PCUBE: A Performance Driven Placement Algorithm for Lower Power Designs," Proc. of the EURO-DAC'93, pp.72-77, September 1993.

[3] A. Shen, A. Ghosh, S. Devadas, and K. Keutzer, "On Average Power Dissipation and Random Pattern Testability of CMOS Combinational Logic Network," Proc. of the International Conference on Computer-Aided Design, pp. 402-407, November 1992.

[4] K. Keutzer, "The Impact of CAD on the Design of Low Power Digital Circuits," IEEE Symposium on Low Power Electronics, Tech. Dig., pp. 42-45, October 1994.

[5] C-Y. Tsui, M. Pedram, and A. M. Despain, "Technology Decomposition and Mapping Targeting Low Power Dissipation," 30th ACM/IEEE Design Automation Conference, Tech. Dig., pp.68-73, June 1993.

[6] R. Murgai, R. K. Brayton, and A. Sangiovanni-Vincentelli, "Decomposition of Logic Functions for Minimum Transition Activity," Proc. of the International Workshop on Low Power Design, pp. 33-38, April 1994.

[7] V. Tiwari, P. Ashar, and S. Malik, "Technology Mapping for Low Power," 30th ACM/IEEE Design Automation Conference, Tech. Dig., pp.74-79, June 1993.

[8] K. Scott and K. Keutzer, "Improving Cell Libraries for Synthesis," IEEE Custom Integrated Circuits Conference, Tech. Dig., pp. 128-131, May 1994.

[9] C. Lemonds and S. Mahant Shetti, "A Low Power 16 by 16 Multiplier using Transition Reduction Circuitry," Proc. of the International Workshop on Low Power Design, pp. 139-142, April 1994.

[10] A. Chandrakasan, S. Sheng, and R. W. Brodersen, "Low-Power CMOS Design," IEEE Journal of Solid-State Circuits, vol. 27, no. 4, pp. 472-484, April 1992.

[11] U. Ko, P. T. Balsara, and W. Lee, "A Self-timed Method to Minimize Spurious Transitions in Low Power CMOS Circuits," IEEE Symposium on Low Power Electronics, Tech. Dig., pp. 62-63, October 1994.

[12] R. I. Bahar, H. Cho, G. D. Hachtel, E. Macii, and F. Somenzi, "An Application of ADD-Based Timing Analysis to Combinational Low Power Re-Synthesis," Proc. of the International Workshop on Low Power Design, pp. 139-142, April 1994.

[13] M. Alidina, J. Montiero, S. Devadas, A. Ghosh, and M. Papaefthymiou, "Precomputing-Based Sequential Logic Optimization for Low-Power," IEEE Transactions on Very Large Scale Integration Systems, vol. 2, no. 4, pp. 426-436, December 1994.

[14] A. Ghersho, and R. Gray, "Vector Quantization and Signal Compression," Kluwer Academic Publishers, MA, 1992.

[15] D. B. Lidsky, and J. M. Rabaey, "Low-Power Design of Memory Intensive Functions," IEEE Symposium on Low Power Electronics, Tech. Dig., pp. 16-17, October 1994.

[16] A. P. Chandrakasan, A. Burstein, and R. W. Brodersen, "A Low-Power Chipset for a Portable Multimedia I/O Terminal," IEEE Journal of Solid-State Circuits, vol. 29, no. 12, pp. 1415-1428, December 1994.

[17] J. Schutz, "A 3.3 V 0.6 μm BiCMOS Superscalar Microprocessor," IEEE International Solid-State Circuits Conf., Tech. Dig., pp. 202-203, February 1994.

[18] N. K. Yeung, Y-H. Sutu, T. Y-F. Su, E. T. Pak, C-C Chao, S. Akki, D. D. Yau, and R. Lodenquai, "The Design of a 55SPECint92 RISC Processor under 2W," IEEE International Solid-State Circuits Conference, Tech. Dig., pp. 206-207, February 1994.

[19] D. Pham, et al., "A 3.0W 75SPECint92 85SPECfp92 Superscalar RISC," IEEE International Solid-State Circuits Conference, Tech. Dig., pp. 212-213, February 1994.

[20] G. Gerosa, et al., "A 2.2 W 80 MHz Superscalar RISC Microprocessor," IEEE Journal of Solid-State Circuits, vol. 29, no. 12, pp. 1440-1454, December 1994.

[21] S. Gary, C. Dietz, J. Eno, G. Gerosa, S. Park, and H. Sanchez, "The PowerPC 603 Microprocessor: A Low-Power Design for Portable Applications," Proc. of COMPCON'94, Tech. Dig., pp. 307-315, February 1994.

[22] R. K. Kolagotla, S-S. Yu, and J. F. JaJa, "VLSI Implementation of a Tree Searched Vector Quantizer," IEEE Transactions on Signal Processing, vol. 41, no. 2, pp. 901-905, February 1993.

[23] C-L. Su, C-Y. Tsui, and A. M. Despain, "Low Power Architecture Design and Compilation Techniques for High-Performance Processors," Proceedings of COMPCON'94, Tech. Dig., pp. 489-498, February 1994.

[24] A-C Deng, "Power Analysis for CMOS/BiCMOS Circuits," Proc. of the International Workshop on Low Power Design, pp. 3-8, April 1994.

[25] C. M. Huizer, "Power Dissipation Analysis of CMOS VLSI Circuits by Means of Switch-Level Simulation," Proc. of the European Solid-State Circuits Conference, pp. 61-64, 1990.

[26] M. A. Cirit, "Estimating Dynamic Power Consumption of CMOS Circuits," IEEE International Conference on Computer Aided Design, pp. 534-537, November 1987.

[27] F. Najm, I. Hajj, and P. Yang, "An extension of Probabilistic Simulation for Reliability Analysis of CMOS VLSI Circuits," 28th ACM/IEEE Design Automation Conference, Tech. Dig., pp. 644-649, June 1991.

[28] A. Ghosh, S. Devadas, K. Keutzer, and J. White, "Estimation of Average Switching Activity in Combinational and Sequential Circuits," 29th ACM/IEEE Design Automation Conference, Tech. Dig., pp. 253-259, June 1992.

[29] F. N. Najm, "A Survey of Power Estimation Techniques in VLSI Circuits," IEEE Transactions on Very Large Scale Integration Systems, vol. 2, no. 4, pp. 446-455, December 1994.

[30] R. E. Bryant, "Graph-Based Algorithms For Boolean Function Manipulation," IEEE Transactions on Computer-Aided Design, pp. 677-691, August 1986.

[31] B. J. George, G. Yeap, M. G. Wloka, S. C. Tyler, and D. Gossain, "Power Analysis for Semi-Custom Design," IEEE Custom Integrated Circuits Conference, Tech. Dig., pp. 249-252, 1994.

[32] B. J. George, G. Yeap, M. G. Wloka, S. C. Tyler, and D. Gossain, "Power Analysis and Characterization for Semi-Custom Design," Proc. of the International Workshop on Low Power Design, pp. 215-218, April 1994.

[33] D. Lui, and C. Svensson, "Power Consumption Estimation in CMOS VLSI Chips," IEEE Journal of Solid-State Circuits, vol. 29, no. 6, pp. 663-670, June 1994.

[34] H. B. Bakoglu, "Circuits, Interconnects, and Packaging for VLSI," Addison-Wesley, Reading, MA, 1990.

[35] S. R. Powell and P. M. Chau, "Estimating Power Dissipation of VLSI Signal Processing Chips: The PFA Technique," VLSI Signal Processing IV, pp. 250-259, 1990.

[36] P. E. Landman, and J. M. Rabaey, "Power Estimation for High Level Synthesis," EDAC-EUROASIC, Paris, France, pp. 361-366, February 1993.

[37] P. E. Landman, and J. M. Rabaey, "Black-Box Capacitance Models for Architectural Power Analysis," Proceedings of the International Workshop on Low Power Design, Napa, CA, pp. 165-170, April 1994.

[38] R. Mehra, and J. Rabaey, "Behavioral Level Power Estimation and Exploration," Proceedings of the International Workshop on Low Power Design, Napa, CA, pp. 197-202, April 1994.

INDEX